on or before

Concept Research in Food Product Design and Development

Concept Research in Food Product Design and Development

Howard R. Moskowitz, Ph.D.
President
Moskowitz Jacobs Inc.
White Plains, New York, USA

Sebastiano Porretta, Ph.D.
Experimental Station for the Food Preservation Industry
Parma, Italy

Matthias Silcher, M.A.
Moskowitz Jacobs Inc.

With contributions from
Hollis Ashman—*The Understanding and Insight Group*
Jacqueline Beckley—*The Understanding and Insight Group*
Roberto Cappuccio—*Illy Café s.p.a., Italy*
Jeffrey Ewald—*The Optimization Group, Inc.*
Laurent Flores—*CRM Metrix, Paris, France*
Alex Gofman—*Moskowitz Jacobs Inc.*
Angus Hughson—*Consultant, London, UK*
Bert Krieger—*Moskowitz Jacobs Inc.*
Tracy Luckow—*University College Cork, Ireland*
Andrea Maier—*Nestle Research Center, Lausanne Switzerland*

 **Blackwell** Publishing

Blackwell Publishing Professional
2121 State Avenue, Ames, Iowa 50014, USA

Orders: 1-800-862-6657
Office: 1-515-292-0140
Fax: 1-515-292-3348
Web site: www.blackwellprofessional.com

Blackwell Publishing Ltd
9600 Garsington Road, Oxford OX4 2DQ, UK
Tel.: +44 (0)1865 776868

Blackwell Publishing Asia
550 Swanston Street, Carlton, Victoria 3053, Australia
Tel.: +61 (0)3 8359 1011

Authorization to photocopy items for internal or personal use, or the internal or personal use of specific clients, is granted by Blackwell Publishing, provided that the base fee of $.10 per copy is paid directly to the Copyright Clearance Center, 222 Rosewood Drive, Danvers, MA 01923. For those organizations that have been granted a photocopy license by CCC, a separate system of payments has been arranged. The fee code for users of the Transactional Reporting Service is ISBN 13: 978-0-8138-2424-6; ISBN 10: 0-8138-2424-9/2005 $.10.

Printed on acid-free paper in the United States of America

First edition, 2005

Library of Congress Cataloging-in-Publication Data
Moskowitz, Howard R.
   Concept research in food product design and development / Howard R. Moskowitz, Sebastiano Porretta, Matthias Silcher.—1st ed.
      p. cm.
   Includes bibliographical references and index.
   ISBN 13: 978-0-8138-2424-6
   ISBN 10: 0-8138-2424-9 (alk. paper)
  1. Food—Sensory evaluation.   2. Commercial products—Testing.   I. Porretta, Sebastiano.
II. Silcher, Matthias.   III. Title.

TX546.M66 2005
664'.072—dc22

                          2004017608

The last digit is the print number: 9 8 7 6 5 4 3 2 1

Dedications

Howard's Dedication

To the memory of my dear mother, Leah Moskowitz, 1914–2003. To you mother, I owe so much that it would take another lifetime to repay it all. Thank you.

Sebastiano's Dedication

I want to express once again my deep gratitude both to my mother, who has strongly supported my cultural development, and to my father, who passed to me his professional passion. Thank you both.

Matthias's Dedication

In memory of my dear mother, Vera Silcher, who once told me, "Son, if you see two ways leading into the forest, one path with many footsteps and one way nobody ever walked, take the second one and leave your own track." I think her words are very appropriate for this book.

Table of Contents

How the Book Was Born—Sebastiano's Tale

Even if I am quite sure he feels the same of me, I always realized that Howard is a little bit *"unique"*. There are many anecdotes supporting this fact. I never understood how Howard could ever reply in a minute to my e-mails sent in the middle of the night/morning/afternoon/summer/winter. . . .

Since we published together different papers in scientific journals and participated in symposia and conferences we decided in February 2003 that the time was right for a book that summarized our professional passions: early-stage product development.

In an Italian coffee shop in Parma, over wonderfully strong lattes and espressos, we decided the chapters and the deadline for publishing. By the following morning the first pages were ready. The picture below was taken to testify to our excitement that day.

Preface

Concepts are critical for the development and marketing of products and services. Concepts constitute the blueprint for these products and services, albeit at the level of consumers rather than at the technical level. Market-research practitioners and their corporate clients are well aware of the need to create winning concepts. To that end a very large practice has emerged in the business world, designed to understand the current competitive environment, identify opportunities, and convert these opportunities into concepts. Look at any prospectus by a market-research company and there immediately emerges a self-proclaimed expertise on concept research.

In actuality, however, the scientific underpinnings of concept research are quite weak. There is a dearth of both practical and scientific information about how to create and evaluate concepts. Although practitioners provide services for evaluation and optimization, the publications offered by the scientific community and the academic business researchers do not go into depth about how to create and measure concepts, leaving corporations and their developers/researchers in the lurch. This gap between practical needs and absence of solid scientific information can be traced to the fact that, until recently, both concepts and concept research have been relegated to the realm of application. There has been little focus on establishing relevant, scientifically accessible knowledge bases for concepts, although the scientific literature will attest to keen interest in the general topic of food product development. Concept development

in the new product development literature is conventionally relegated to the so-called fuzzy front end, yet it is the concept that often directs product development and marketing. A good product concept can help make the product a success by guiding developers and advertising in the right direction.

The lack of information about creating good concepts is most pronounced at the corporate research and development (R&D) function. Most R&D professionals either work with the concepts that marketers provide or are instructed to come up with a winning new concept and then, in turn, create a product to match the concept. For the most part people assigned to this job are poorly prepared and have nowhere to turn. The aforementioned books on product development give short shrift and superficial treatment to concept work.

This book remedies that problem by providing a unique treatment of concepts for business professionals as well as for research scientists. The book begins with simple principles of concepts, moves forward to methods for testing concepts, and then moves onto more substantive areas such as establishing validity, testing internationally and with children, creating databases, and selling new methods for concept testing. The book combines a how-to business book with a detailed treatment of the different facets of concept research. As such, the book represents a unique contribution to business applications in food and consumer-research methods. The book is positioned specifically for foods to maintain a focus on a coherent set of topics.

Target Audience

This book should appeal to a variety of audiences. The lack of any book devoted to concepts, the increasing recognition that innovation/concepts are important to business success, and the increasing focus of R&D on the business of product development all bode well for wide acceptance. The book should be promoted to R&D, marketing, and universities alike. R&D will be most interested in the background to create strong concepts. Marketing will be interested in the topic of concepts because that is their business. Students at universities are becoming more interested in the consumer connection, and concept research is certainly part of that connection. Those interested in sensory analysis, for example, will find this book a natural extension of their interest in product features. The fact that there are no other books like this one needs to be stressed to the audience.

Concept Research in Food Product Design and Development

Chapter 1

The Business Environment and the Role of Concept Research in that Environment

The Business Environment

Concepts comprise written statements about products or services. At first glance it seems intuitively easy to deal with concepts because they are similar to the stimuli that one has encountered during one's life: the written word applied to the description of a product or a service. To create the concept one might need to do some quick writing, jot down the features of a food or beverage, talk perhaps about the way the product is packaged and consumed, perhaps move on to a statement about health benefits and price, and somewhere within the description bring in emotional elements or a tie-in with a famous person (Graf and Saguy 1991; Brody and Lord 2000). Furthermore, there does not seem to be the technical barrier to research on concepts that pervade research on actual products. The investigator need not be a scientist; there are no esoteric formulas to master and no complex equipment or formidable chemical or physical principles to understand. Certainly, concepts should be easy to understand, and concept development and research should be straightforward. In light of the importance of concepts in the business world there should be no impediments to becoming a master of concept research. Or are there? Do impediments lurk, ready to trip us up?

Despite the apparent simplicity of concepts, there are no general principles about what makes a good concept or indeed even what a concept happens to be (Fuller 1994). Is a concept a flowery presentation of a product, couched in fanciful language designed to sell to readers? Some researchers aver, often ve-

hemently, that only "fully fleshed out" vignettes of a product (or a service) deserve the name *concept*. These concepts are almost print advertisements or really new products (Urban et al. 1996). Other researchers believe just the opposite and may be, in contrast, virtual minimalists. This other camp of researchers avers, equally vociferously, that the essence of a concept is the communication points. In their mind the adornment of language and graphics only distract from the key messages that describe the product or service.

Both of these viewpoints are correct. Each has merit, each has its own adherents, and each is practiced in the consumer research community. Indeed, there is no standard way to deal with concepts, thus allowing, if not openly encouraging, such leeway. Despite the importance of concepts as blueprints to products and services there is a dearth of literature on concept development in academia, either of a product or a positioning nature. Academic literature generally is limited to data about the market structure and the competitive nature of the different brands, as well as the reasons why consumers choose one brand over another. Product features and positioning features may be discussed in published articles. Concepts, however, are more limited to momentary, tactical combinations of product features, benefits, and so on. They are not of much interest to the academic world, because they represent a focused attempt at a single time period for a product or service. They incorporate the aspects of benefits and features, but otherwise represent localized, limited efforts that are not the focus of academics. It is for that reason that there is no large-scale archival literature dealing with concepts, per se.

The Importance of Concepts in the Business World

Concepts are the building blocks of products. A consumer-driven food product development process is likely to produce more successful products (Stewart-Knox and Mitchell 2003). A careful perusal of the food science or food-marketing literature quickly reveals a noticeable absence of books and articles dealing with food and beverage concepts in particular and concepts in general. According to Stewart-Knox and Mitchell (2003), only five studies appear so far to have addressed issues relating to product development in the food sector: one is a qualitative study (Parr et al. 2001), two are surveys (Hoban 1998; Iiori et al. 2001) and two are very recent, predictive models of food product development (Kristensen et al. 1998; Stewart-Knox et al. 2001). Whereas there are dozens of books on the physical properties of foods, and an increasing number of books and articles dealing with the testing of food products by using physical and subjective methods, very few articles discuss concepts. Even the business literature talks about methods for measuring concepts but says little about creating good concepts and the science behind understanding concepts (Crawford 1987; Shocker and Srinivasan 1979; Bhattacharya et al. 1998; Dahan and Srinivasan 2000; Ulrich and Eppinger 2000).

One consequence of the missing science of concept research is that concepts to be tested come in many forms, without rules, without norms, and without rationales. Some concepts are simple combinations in phrases that scarcely look like anything but skeletons of ideas. Other concepts may resemble complete advertisements. Some concepts comprise text alone. Still others contain a great deal of graphics. More recently, concepts have in some cases become simple combinations of graphics. No matter what the form of the concept may be, however, the concept fundamentally represents an idea about a product or service.

Why Do Research on Concepts? If We're in Business We Should Know the Answers

A lot of business runs on intuition and "gut feeling." Years ago, and not many at that, the role of research was considered to be a luxury, often to be tolerated or perhaps to be paraded to the investors as an example of one doing the appropriate job, but rarely to be considered as a strategic business weapon. This assertion is, of course, hard to quantify in published research reports, since such statements are made at conferences, by executives, but not saved for the archival literature. Concept information was *nice to know*, but rarely delivered in such a way that it could effectively impact the decision process.

Research in those years was an appendage, often funded so poorly that the data emerging would be at best qualitative and generally useless. The senior author has been to a variety of such treatments of research by companies, beginning in the late 1960s and continuing even today. At an anecdotal level, it appears in retrospect that most of the individuals averse to research were not averse to the idea of research as much as opposed to the idea of taking valuable time and resources to investigate a problem systematically. For them, one or two focus groups, a quick consult with their industrial colleagues, and decisions made on experience and intuition were the norm. That "cowboy" approach was quite prevalent, far more in fact than one might realize.

Today's environment, in the beginning of the 21st century, differs considerably from even a decade ago and certainly differs dramatically from two and three decades ago. Calatone and Cooper proved during the late

1970s that product success depends on the product being unique and superior; good understanding of consumer wants, needs, and preferences; and effective product marketing and launch (Calatone and Cooper 1979). From the results of Cooper's and Kleinschmidt's work, one can clearly see that two aspects have had a significant positive influence on the success of new products. These are (a) the proficiency of activities carried out in the individual phases of NPD (new product development) especially in development, test marketing, and market introduction, and (b) the use of market information along the entire NPD process (market orientation) (Cooper and Kleinschmidt 1995b, 1996). This difference affects how business views research in general and concept research in particular.

In today's business environment, knowledge is king. This *knowledge*, occasionally elevated to *insights*, refers to the quality of market research with reference to the understanding and evaluation of customer needs (Griffin and Hauser 1993), the accurate prognosis of the market potential (Balbontin et al. 1999), the observation of the competition (Mishra et al. 1996), new ways to develop optimal product profiles in the context of competition (Green and Krieger 1989, 1991), and the execution of test markets (e.g., see Dwyer and Mellor 1991a, 1991b), and so on. Ideally, this information should be updated over the entire NPD process (Rothwell et al. 1974). Some of the new reliance on research is simply the shift from one superstitious behavior to another. A diffident manager may refuse to make a decision that could impact a career and employment, unless that decision can be supported by a mountain of numbers. If the decision is incorrect, then this mountain transforms itself to refuge, because it can be claimed that the data clearly pointed in a certain direction. However, the real reason for the explosion of research is the demand that the decision be correct, be made swiftly, and be sufficiently impactful to add revenues to the

corporation (see Noble and Mokwa 1999, and Varadarajan and Jayachandran 1999).

The quality of planning before entry into the development phase has become increasingly important for the success of new products (Ernst 2002). Concepts are part of this planning. The necessary preparations for the project include, in particular, the first broad evaluation of ideas, the execution of technical and market-directed feasibility studies. Beyond this, the product concept, the target market, and the relative utility gain for the customer by using the new product as opposed to the competing product all need to be clearly described (Dwyer and Mellor 1991a, 1991b; Mishra et al. 1996; Calatone et al. 1997). All but the most resistant managers in the corporate environment respect the value of research as providing the true knowledge needed to make decisions. "We're seeing more integration of marketing, market research and research and development in the form of product-development teams. But the key is understanding what the consumer wants, and describing the product's advantages to the consumer," concludes Nancy Smith, Ph.D., vice president and food industry consultant at Arthur D. Little (Morris 1993).

The Good, the Bad, the Disappointing

Technology is today's engine of research, just as the competitive business situation is today's motivator. Technology enables researchers to do powerful data collection and follow that with insightful, meaningful, and action-oriented data analysis. The research can be done quickly, powerfully, and effectively. We can thank the happy combination of the intensely competitive environment and the maturation of research-savvy businesspeople for that growth in the reliance on research.

Research in the business environment is not a panacea, however, even in times where *knowledge is king*. Other aspects, such as integrating the research function into the corporate culture, are important. As the environment was heating up rapidly several years ago, and as management was beginning to recognize the true value of research, researchers found themselves in an awkward position. Managers who had not accepted the value of research suddenly found themselves under attack by competition and willy-nilly handed over a lot of their decision making to ill-prepared or even incompetent researchers. Occasionally, this change in the environment and the panic response of management resulted in research debacles. Researchers in such an environment could do more of the job than they had originally done, and test concepts and products in the slow paced way. The additional request by management to produce more and better data simply sped up their slow pace in order to accommodate management directives. The incorrect decisions were still being made, only more quickly.

Despite the large amount of research, the vast majority of new food products (72%–88%) continue to fail (Buisson 1995; Lord 1999). Nancy Smith, the food industry consultant at Arthur D. Little, estimates cost to the industry of failed products at more than $20 billion per year (Morris 1993). In a 1991 survey of 27 Chicago-based companies conducted by Chicago management consulting firm Kuczmarski and Associates, the "lack of market research" per se ranked third, right after a "demonstrated market need" and "no market need or changing needs," as the major reason for new product failure (Kuczmarski 1992)

In the literature, one finds a steadily accumulating amount of evidence that product failure is most closely linked to inadequacies within predevelopment activities (Cooper and Kleinschmidt 1987; Cooper 1993; Davis 1993; Dyer et al. 1999). This low rate of innovation, coupled with the high failure rate

of food products following market launch, implies that the methodology for new food product development urgently needs improvement (Stewart-Knox and Mitchell 2003). It is reasonable to inquire as to whether time to make the decision was the only cause of research failure, or whether there were other, perhaps more fundamental, causes. From the authors' viewpoint, the cause for the research failures and occasional debacles stemmed from a structural problem. Researchers could not provide the powerful early-stage research needed to direct the creative process of concept creation, beyond, of course, the ever-popular focus group. It took a number of years before the misalignment of demand on researchers for powerful developmental guidance and their ability to deliver that guidance was straightened out, and researchers found themselves able to deliver what management demanded from them. Concept research played and continues to play a major part for researchers as they attempt to answer management's demands for better guidance.

About Concepts

Product Concepts versus Positioning Concepts

There are two fundamental types of concepts, although in practice the components of these concepts may be mixed in the specific stimulus that a respondent evaluates. The first is a *product concept*, which describes the product or service. In its pure form the product concept comprises simple declarative statements. There is no adornment, no attempt to *sell* readers on the benefits to be obtained by purchasing the product or service. The product concept is simply designed to test whether the idea is acceptable. To the degree that a researcher uses flowery language, selling the idea to a respondent with communication beyond the simple idea itself, the product con-

Table 1.1. A product concept compared with a positioning concept for yogurt

Product concept
   Introducing Health Yogurt
   All the nutritional features of regular yogurt
   With active cultures
   A low-fat yogurt
   Available in plain, fruit, and tropical flavors

Positioning concept
   Yogurt designed for good nutrition
   Its active cultures improve digestion
   Specially formulated to be low fat so you don't need to feel guilty
   The array of flavors, plain, fruit and tropical, specially selected for your sensory delight

cept will depart from its objective. Table 1.1 presents an example of a product concept.

At the other end of the *concept continuum* is the positioning or selling concept. The positioning concept tries to sell a respondent on the benefits of the product or service. The emphasis tends to be on what the respondent will feel and receive, etc., from the product or service, rather than on the product itself. Table 1.1 also provides an example of a positioning concept.

In the pragmatic world of business, few concepts are purely product, and few are purely positioning. Most concepts comprise some of each. The reason for this mixture is quite simple. People do not respond meaningfully to either pure product description or pure selling in such a clear way that one could easily tell the difference. In the *real world* of one's daily environment, concepts or advertisements comprise descriptions of the product or service, along with benefits to be gained by using it. Most people do not react well to simple descriptions, because there is no call to action. Most people, in turn, do not react well to pure selling concepts because nothing in the concepts backs up the selling message. People often reject pure appeals to emotion. This recognition that concepts comprise both information and motivation stands behind some of the methods that advertising agencies use, such as *problem solution*. The problem provides the motivating statement.

The solution provides the specific product or service feature that solves the problem.

Job Roles: Who Creates the Concepts?

Since concepts constitute a blueprint for a product or service, or a reason to buy, it is relevant to ask about *who* in the corporation creates the concept. Unlike product research where research and development (R&D) is clearly separated from marketing, almost anyone can create a concept; that is, concept creation does not entail specialized technical knowledge, as do, for example, products or packages. There are no rules that limit concepts, although there are rules, or at least guidelines, about technical feasibility for product concepts or business realities for positioning concepts.

As a consequence of many people legitimately being able to create concepts we see concept development occurring in the corporation at many levels and occasionally outside the corporation, as well. Concept development is not a strategic capability in the hands of only one group. Quite often in the food industry, marketers are the main group who create concepts because marketers have a good sense of the corporate need for new ideas and new sources of revenues. The marketer must identify the new opportunities in the business and then turn those opportunities into products. Occasionally, the marketer relies on the advertising agency as well. The advertising agency typically works outside the corporation, on a long-term contract or on a short-term project basis. Often the account people who manage the agency-client relation come up with the concepts. The agency may even dedicate a team of individuals to creating the concepts. Occasionally, companies hire outside specialists to create their concepts. These specialists typically position themselves as creativity experts, or trend watchers, and by so doing claim the ability to produce better concepts. The extensive consultation by food companies with agencies and the involvement of expertise beyond the company have had a wide, positive impact on the success of food products (Balbontin et al. 1999; Stewart-Knox et al. 2001)

Quite often the R&D product developer is charged with creating the concepts. R&D professionals usually have a good idea about the consumer's needs and in turn can create the product concept. In recent years R&D professionals have participated increasingly frequently in the creation of product concepts, driven by management's need to maintain a competitive position in the marketplace and by the recognition that R&D often has its pulse on the competitive environment. R&D does not, however, often create positioning requirements. The positioning concepts are usually softer in nature, do not deal with the product per se, or deal with the product but only in passing. The R&D professionals often judiciously *pass* on the opportunity to create these positioning concepts, leaving that task either to the marketers or more frequently to the advertising agencies.

Traditional roles in concept creation have been changing, however, in an effort to improve the chances of success. There has been a recent shift in the organization of product development in practice, and many firms have adopted a team structure in which the

traditional functional divisions are less pronounced (Ettlie 1997). More recent research has indicated that cross-functional teams are even more effective than coordinator-led or matrix approaches (Cooper and Kleinschmidt 1996; Jenkins et al. 1997; Karlsson and Ahlstrom 1997). This implies that companies that bring together individuals from different departments and from beyond the company to work cohesively together are more likely to be more successful. (Stewart-Knox and Mitchell 2003). As a result, organizational process tools such as cross-function teams and colocation (Allen 1986) have been developed. A number of works verify that the project team should comprise members from several areas of expertise who can make substantial contributions to the development of a new product (Pinto and Pinto 1990). This team includes, above all, members from R&D, marketing, and production (Song et al. 1997). The impact on concept performance of this broad change in responsibilities and ways of working remains to be quantified, however. To date there has been little in the way of measuring the strength of the concept as a function of the way it was created within the different types of corporate structures and responsibilities.

Testing Concepts: Why, When, How?

Concept creation constitutes a limited effort, in a specific time period, focused on developing a new idea having commercial potential. Since concepts represent efforts to create the blueprints for a product or service, much of the literature and expertise surrounding concepts focus on measuring the reactions to these concepts. As a consequence much of the expertise surrounding concepts deals with measurement tasks and issues rather than on deep understanding of the nature of concepts and how they should be written and interpreted. The measurement task provides a metric for how well the concept will do in the marketplace. Measurement, a systematic task that can be specified, regulated, and analyzed, generally takes over and becomes the central focus. Much of what we know about concepts, therefore, is what we know about concept testing.

The key issue about concept testing has been elegantly and simply stated by Hyman (2002). Specifically, "how can an organization translate consumers' needs and preferences into a product [i.e., goods-service-idea mix] that consumers acquire willingly, use beneficially, and dispose of with minimal environmental stress?" The "hows" of testing concepts are well established. Even the inquiry into best practices has been dealt with by practitioners and academics alike. Over more than two decades Cooper and Kleinschmidt (1986, 1987) have encouraged product developers to consider not only what they do but also how to go about it.

By now, the creation and measurement of concepts appear to have landed squarely in the domain of marketing research, with other groups (marketers, product developers, and academics) involved tangentially. Marketing researchers are experts in the measurement of concepts. Their measurement may include grading the concept on its interest, on its communication, and/or on its persuasiveness. Researchers often do far more with concepts than measure overall reactions. Some researchers who specialize in measuring the deeper emotional reactions to concepts are interested in the creation of concepts and the ability of a concept to satisfy deeper, unexpressed, unmet needs. They may approach concept work from general psychoanalytic principles. Other researchers, who specialize in the measurement of concepts to generate sales and market share, express the needs of the marketers and are not as much interested in the subjective aspects of a concept as they are in its economic impact. Interestingly, few researchers can span the range between measuring a concept as a reflection of personal need and measuring

a concept as a generator of revenue. The range of ability required to perform the two types of measurement—individual emotional response versus objective estimates of future market performance—is simply too great.

The specific methods to measure responses to concepts are usually left up to researchers. These methods are lumped together in the category of methods called *concept tests*. Many such test methods have been offered over the past decades, generally by practitioners in the business of measuring the viability of concepts for their business clients. Based in part on the question to be answered; in part on the sheer inventiveness of the researcher in coming up with new, proprietary methods; and in part on theory-based assumptions, researchers have devised innumerable methods to measure the appeal of concepts. Considerations include the following.

The Number of Concepts That a Single Person Should Evaluate

Some practitioners permit a respondent to assess only one concept, feeling that otherwise the respondent will be biased and the data will be invalid. To these people, only one exposure generates valid data. They also feel that the respondent may become bored if presented with too many concepts, so one concept is ideal because it avoids boredom. Other practitioners feel that a respondent should test many concepts. To this second group the justification is that the single concept exposure produces noisy data because in the end people vary, perhaps more than do responses to concepts. These practitioners, who want a person to test many concepts, feel that the pervasive interindividual variation will generally mask the differences among the concepts. Having the same respondent evaluate all or most of the concepts will reduce this interpersonal variability by averaging out the variability. The former approach, wherein a respondent evaluates only one concept, requires many more respondents, and that limited exposure is called the *monadic* approach. The latter approach is called the *within subjects* design and enables a researcher to work with fewer respondents.

Types of Concepts That a Person Should Evaluate

Battles royal break out again and again over this issue. The advertising agency opts for concepts that appear close to finished advertisements and frequently and heatedly disputes any attempts to test concepts in a simpler, "white card" format, which lacks all the "bells and whistles." Other professionals hold the opposite viewpoint, stating just as adamantly that the appropriate test for a concept is the message, not the execution. In the opinion of this second group, the white-card, or purely descriptive, nonselling version of the concept, is the more appropriate test stimulus. In actuality both viewpoints have merit and can be correct, depending on the situation. The white concept board, bereft as it is of execution, allows one to test the attractiveness of an idea. The fully executed, almost-print-ad format allows one to test the attractiveness of an idea and the execution. Of course, the difference in ratings between the two formats measures indirectly the addition of the execution.

Types of Ratings

The most important rating is, of course, the measure of acceptance, whether that measure is couched in scales of liking, purchase intent, or something else. The other types of rating include profiles of communication and perhaps even projective image attributes (viz., assigning a rating about the type of person that might be interested in this concept). By and large most decisions are made on some sort of acceptance measure, such as purchase intent. Many professionals, how-

ever, insist at the same time that they need "diagnostic" attributes, which are nothing more than profiles of other aspects beyond interest in purchasing.

When Should the Evaluative Question Be Asked?

This may sound like a trivial question, and to many people it makes no logical difference whether the overall evaluative question (e.g., purchase intent) is asked at the start of the concept ratings, in the middle, or at the end, when all the other attributes have been rated for that concept. As is so often the case in research, the smallest issues engender the largest fights, perhaps because most people want to argue about concept research but feel intimidated until they identify a *stock question* that is easy to formulate and to understand. Certainly, position of the evaluative question belongs in this latter category. There is no correct answer, however. Some professionals feel that the initial attribute rating should be overall evaluation, when the concept is fresh in the respondent's mind. Other professionals, arguing their point with just as much vehemence and with the same sense of research rightness, feel that the evaluative rating should best be asked after the respondent has profiled the concept on the other rating scales. This rating in the latter position ensures that the respondent's answers are mutually consistent for a single concept; that is, by placing the evaluative rating as the last rating scale for the concept the researcher ensures that the thoughtful respondent will try to make everything mutually consistent. From the authors' viewpoints it really makes little or no difference to a respondent. First, respondents do not typically try to be consistent. The thought that a respondent sits and ponders the answer is a researcher's fantasy. Most respondents are not that involved. Second, even if a respondent attempts to be consistent, it is hard to be consistent when the respondent rates the separate concepts on attributes. If a respondent were choosing between two concepts, then perhaps the notion of consistency might be important. If there are only two concepts, and a respondent always prefers concept A to concept B on attributes, then the consistent respondent would be expected to prefer concept A to concept B on the overall measure. This expectation is irrelevant when there are multiple concepts and the rating scales are amount of (i.e., magnitude) rather than expression of choice.

Base Size or Number of Respondents

Mythology or its dual, *best practice* in consumer research, somehow focuses on the number 100 as an appropriate minimal number of respondents for a study. The data for concept evaluations may stabilize at a smaller number of respondents, but the number 100 is psychologically comforting (Moskowitz et al. 2000). For some claims tests, however, a base size of 300 respondents is necessary. One must consider the issue of base size within the context of the issue involved. Network ratings use the number 300 to validate the representativeness of a claim (Council of Better Business Bureaus 1990). The number of concept ratings must ensure that there is a sufficient number of respondents to represent the ultimate population and that the data from the respondents are sufficiently robust and stable so that the conclusion is not affected by random noise should the study be repeated. This issue is dealt with later in this book, under the topic of *reliability and validity*. For the purposes of this introductory discussion, however, it suffices to keep in mind that the issue of base size is always raised and has many different facets attached to it. The issue is not simple, even if the question is straightforward.

Analysis of Ratings

Practitioners often berate academicians for being "increasingly out of touch with what practitioners actually do" (O'Driscoll and Murray 1998), being excessively focused on basic research, new research methods, and articulating concepts, but insufficiently focused on problem-oriented research (Razzaque 1998). From this perspective, academicians are "overly focused on theory (dis)confirmation rather than theory creation" (O'Driscoll and Murray 1998) and "too subjective and nonpragmatic" (Razzaque 1998). There is a big difference between scientists and business-minded market researchers. Although it may not seem important, the intellectual heritage of scientists compared with that of market researchers predisposes one to different types of analyses and encourages ongoing arguments. Scientists are trained in statistics and use the average or mean to describe a lot of their data, as well as the standard deviation or variance to describe the scatter of the data around that mean. Consequently, an R&D scientist intuitively would look at the average rating that a concept achieves, just as the scientist looks at the average rating of product acceptance on a 9-point hedonic scale (Peryam and Pilgrim 1957). Indeed, the norms for performance on acceptance scales are couched in mean ratings, and the statistics are *inferential* ones appropriate for these metric, mean ratings (e.g., the standard deviation and the significance value of the mean). In contrast, market researchers with a heritage in sociology and public opinion polling deal with *incidence* statistics. Their focus in concept testing is not so much on the average rating achieved by a concept as on the distribution of the ratings and the proportion of respondents falling into the acceptor class versus the rejector class. A key statistic for market researchers is the proportion of respondents who say that they will definitely or probably purchase the product or, in more stringent cases, who say they will definitely purchase the product. Market researchers also consider as important the proportion of respondents defined as *rejectors*; that is, who say they will definitely or probably not purchase the product. Market researchers thus deal with percentages of respondents falling into a class, rather than with mean values, because they look at mass behavior rather than at individual behaviors.

What Should the Concept Comprise: Pricing, Brand Name, or Something Else?

The rules for creating concepts are not fixed. There is no legislation about the proper contents of a concept, leaving open opportunities for concepts to contain all sorts of information. For example, some individuals responsible for concepts aver, quite strongly, that unless the concept has a price and a name, it is not worthy of the name *concept*. These individuals do not go so far as to say that testing is inappropriate. They do, however, say that the data obtained from consumers for concepts without price, and in some cases without brand, are meaningless. In their opinion consumers can only decide about the purchase intent in a concept if the concept has a price, for there is always a trade-off between price and purchase intent. Other professionals, with just as strongly held opinions, feel that a concept need not have a price in order to be validly rated on purchase intent. Field research as discussed by this latter group shows that consumer respondents appear to have no trouble rating the concepts on purchase intent, even if the concepts lack a price point. There is no real answer to this question. Observation by the senior author (H.R.M.) suggests that respondents in fact appear to have little or no problem reliably rating purchase intent with concepts that have no price and, in the same study, rating purchase intent of concepts having a price. The issue of brand name is easier to deal with. Respondents appear to have absolutely no difficulty rating a concept that lacks a brand name, and indeed most consumer respondents appear to have little problem at all. Thus, the price issue may be valid, but the brand-name issue in the concept is simply a matter of opinion, and the different positions cannot be supported either by data or by common sense.

Agonizing Over Poor Performance, and Self-defense Tactics

When researchers, creatives, and marketers discuss concepts, different viewpoints emerge regarding how much analysis is appropriate. Researchers pride themselves on understanding the different nuances and, if truth were to be known, many researchers would opt to explore reactions to concepts in depth. Often, in focus groups the reactions to concepts are probed in agonizing detail, forcing out viewpoints and reactions that are merely incidental to the concept itself. Some of this depth analysis leads to new insights, and in the end the insight improves the concept. Some of the depth analysis is simply self-indulgence and in the end is an exercise without other purpose. In contrast, most marketers are less interested in the nuances of concepts and instead focus on performance. If left to their own devices, then, these marketers would throw away all of the diagnostic information in the concept and simply deal with its performance on the one key evaluative dimension, such as percent top-2 box purchase intent (i.e., percent of the respondents who say that they would probably or definitely purchase this product). Casual readers might dismiss this issue as a tempest in a teapot and simply relegate the whole thing as a turf battle. The issue becomes more important, however, when the ratings for the concept are not as high as management had hoped for. The marketer might simply opt to try it again—to create the new concept and return to test the revised concept. The typical researcher, in contrast, will return to the same data again and again, looking for the reason why the concept performs weakly. The researcher's goal is consistency, not necessarily improvement. In the worst of cases the researcher will flail around, analyzing the data in many ways, weighting the respondents, and exhausting himself or herself, the client, and everyone else's patience in an attempt to achieve cognitive relief from the poor performance. Ultimately, however, it is this author's feeling (H.R.M.) that, when the concept scores poorly, one should make some effort to understand why, but then move on quickly.

Moving On to Better Concepts

From a business perspective, the timely and consequent termination of unprofitable NPD projects is an important success factor (Cooper and Kleinschmidt 1995a). At the Food Plants '91 conference, executives from Kraft USA, Campbell Soup, and J.M. Smucker described time compression as the main factor in speeding up product development (Morris 1993). Recognition and action in the face of poor concept performance are critical here. It does little or no good to stick with the poor performance and explain it by whatever means possible in this effort to achieve consistency and perhaps validate one's own research efforts. If the necessary outcome does not appear quickly, with so-called ocular trauma (hitting one squarely between the eyes), then the more prudent action is to proceed with the next iteration. Understanding the problem doesn't cure it. Only better concepts cure the problem. According to the "Product Development Funnel" of Dahan and Hauser, the key management ideas at this stage are (a) that it is much less expensive to screen products in the early stages than in the later stages, and (b) that each stage can improve the product and its positioning so that the likelihood of success increases (Dahan and Hauser 2002).

Strategies in Concept Development

Lots of Ideas Up Front

Organization models of product development, whether applied to food or not, consistently link product success to *up-front* activities

such as consumer testing and the subsequent feeding through of consumer need into technical development (Rudder et al. 2001; Dahan and Hauser 2002). It has been shown again and again that the preparatory work for the project in the early phases of the NPD process (*initial screening* and *preliminary market and technical assessment*) is decisive for the success of new products (Ernst 2002). A lot of concept testing occurs in the so-called *fuzzy front end*, where the goal is to identify new products (Khurana and Rosenthal 1997; Doll and Zhang 2001). The organizational goal of the *fuzzy-front team* in product development is to reduce uncertainty during the design team's search for winning product concepts. The uncertainty can be reduced by accurately capturing customers' viewpoints and communicating customer preferences to the design team (Dahan and Hauser 2002). Many of the more progressive, result-oriented companies test lots of ideas early and often, looking for the ideas that have "sticking power" and promise. To these companies the notion of concept testing is a rapid-screening device. They are not generally worried about the representativeness of the sample or the degree to which the concept is polished and precise but, rather, focus on whether the idea has any promise. Other companies, with just as much professionalism, feel that the concepts should be better structured, and that concept tests should occur only when the concepts are further along. These two strategies differ radically and represent different worldviews. The first strategy assessing many concepts uses concept evaluation as a screening tool. The second strategy assessing only finished and promising concepts uses concept testing for a go/no go decision. In the past decade or two, however, a great deal of the rigidity of concept testing, embodied by the second approach, has disappeared as marketing and development professionals continue to face competition. In a fast-moving environment, it is no longer possible to conduct rigid concept tests at a stately pace and with the statistical rigor once so proudly proclaimed. As with market testing of new products, one no longer enjoys the luxury of controlled test markets, evaluations of products over months, and then heralded rollouts from one market to many markets in a phased, gradual, systematic manner. The environment is so demanding that the screening approach to concept work, like the screening approach to product testing, is flourishing all over because it is pragmatic and is appropriate for the times. The work of Srinivasan and Lovejoy (1997) argues that, with the new economics of product development (e.g., declining costs of prototyping, and more powerful computer-based tools), it may be optimal to pursue multiple concepts and select the best design later in the process. In later work at the project level the contents of the NPD process are subdivided into more detailed phases (Ernst 2002). The results show that the existence of a formal NPD process, which is comprehensive and characterized by professionalism throughout, especially in terms of evaluation and selection of new ideas, has a positive effect on the success of new products (Kotzbauer 1992). Bhattacharya and colleagues (1998) also find that finalizing specifications later may be desirable in dynamic environments.

The Role of Experimental Design in Concept Development

For many years, researchers avoided the issue of systematically varied concepts. Indeed, in food product design it took many years for the professionals to accept the reality that one could profitably use the method of experimental design to systematically vary the physical features of a food, test the physical combinations among consumers, acquire consumer ratings, and then identify optimal combinations of food features. When this battle was won in the 1970s, the issue of experimentally designed combinations in concept research was still to be fought. Certainly, those involved in concept development often were

poised on one of two camps. One camp of professionals held quite strongly that it was simply impossible to vary the components of the concept systematically. These professionals, aided by agency and marketing professionals, felt very strongly that the creation of concepts was an artistic endeavor, so the use of systematic variation was akin to creating Shakespeare's works by means of some numerical system. The thought was simply repulsive to them. On the other hand were arrayed some professionals who had been schooled in scientific research (Green and Wind 1973). They could not *prove* that the systematic approach was better and, for many years, could adduce only a few arguments to support their position. Over time, however, and as the research community recognized the usefulness of systematic exploration in concept research, the arguments died down. A better way to describe what happened is that the fear died away as the *creatives* recognized the systematic approach simply as a technique by which they could identify what aspects of the concept were more likely to generate consumer acceptance. Systematic exploration no longer comprised "painting the concept by the numbers," which in some circles had become the battle cry for resistance to these new ideas. Experimental design simply became a tool, one of many in the arsenal of the "creatives" faced with the real-world problem of producing a strong concept or advertising copy.

Can Concept Development Be Taught or Is It an Art?

Closely allied with the previous topic of experimentally designed concepts is the issue of whether concept creation is an art form or a scientific discipline, or perhaps both. If one fancies oneself an artist when it comes to concepts, then of course there should be no issues. This extreme viewpoint is held by many individuals working in advertising agencies, who categorically state that the creation of concepts is best left to those creatives who have been publicly recognized and accepted as the arbiters of good ideas. To their chagrin, however, ongoing practice in advertising and marketing, as well as in the technical world, belies this self-proclamation. In company after company the concepts to be tested are, in actuality and all too often, created by advertising accounts executives, by marketers (often the junior-role marketer), and in some rare instances by research professionals. So much, therefore, for the oft-stated importance of the *creative* in the process. The creative person at the agency may not participate as much in concepts as in the final execution. The concept is merely the blueprint and may not be sufficiently important to require a creative's involvement.

Interpreting Concepts: What Do the Scores Really Mean, and Does Anyone Know What to Do Next?

Previously we dealt with the different types of measures: percentages versus means. By interpretation of these measures, however, we don't mean the actual, literal interpretation of the proportion of the respondent population who would be interested in the concept. That is a rather self-evident conclusion, straight from the data. Rather, we mean *how the information will be used*. Sometimes the information is used to guide further development. The concept is a representation of the product idea. Responses to that idea are then used to determine whether the product idea is relevant and whether it is productive for the corporation to invest in further development. For instance, if the idea of food sterilized by electrical current is positive, then the researcher and marketer, as well as product developer, might feel this to be a profitable area for development. The art of interpreting concept data may be as important, in fact, as the art of creating the concepts in the first place. According to Kotler (2003), at every stage the executives have to make one of four decisions: "Go," "Kill," "Hold," or "Recycle." Kotler's dictum applies just as strongly to concepts, in and of themselves, as it applies to the entire development and marketing cycle.

References

Allen, T.J. (1986). Managing the Flow of Technology. Cambridge: MIT Press.

Balbontin, A., Yazdani, B., Cooper, R., and Souder, W.E. (1999). New product development success factors in American and British firms. International Journal of Technology Management, 17: 259–280.

Bhattacharya, S., Krishnan, V., and Mahajan, V. (1998). Managing new product definition in highly dynamic environments. Management Science, 44: 50–64.

Brody, A.L., and Lord, J.B. (2000). Developing New Food Products for a Changing Marketplace. Lancaster, PA: Technomic.

Buisson, P.D. (1995). Developing new products for the consumer. In: Marshall D., editor. Food Choice and the Consumer. London: Blackie Academic and Professional.

Calatone, R.J., and Cooper, R.G. (1979). A discriminant model for identifying scenarios of industrial new product failure. Journal of Academy of Marketing Science, 7: 163–183.

Calatone, R.J., Schmidt, J.B., and Di Benedetto, C.A. (1997). New product activities and performance: the moderating role of environmental hostility. Journal of Product Innovation Management, 14: 179–189.

Cooper, R.G. (1993). Winning at New Products: Accelerating the Process from Idea to Launch, 2nd edition. Boston: Addison Wesley.

Cooper, R.G., and Kleinschmidt, E.J. (1986). An investigation into the new product process: steps, deficiencies, and impact. Journal of Product Innovation Management, 3: 71–85.

Cooper, R.G., and Kleinschmidt, E.J. (1987). New products: what separates winners from losers? Journal of Product Innovation Management, 4: 169–184.

Cooper, R.G., and Kleinschmidt, E.J. (1995a). Benchmarking the firm's critical success factors in new product development. Journal of Product Innovation Management, 12: 374–391.

Cooper, R.G., and Kleinschmidt, E.J. (1995b). New product performance: keys to success, profitability and cycle time reduction. Journal of Marketing Management, 24: 315–337.

Cooper, R.G., and Kleinschmidt, E.J. (1996). Winning businesses in product development: the critical success factors. Research Technology Management, 39: 18–29.

Crawford, M. (1987). New Products Management. Homewood, IL: Irwin.

Dahan, E., and Hauser, J.R. (2002). Product development: managing a dispersed process. In: Weitz, B., and Wensley, R., editors. Handbook of Marketing. London: Sage.

Dahan, E., and Srinivasan, V. (2000). The predictive power of Internet-based product concept testing using visual depiction and animation. Product Innovation Management, 17(2): 99–109.

Davis, R.E. (1993). From experience: the role of market research in the development of new consumer products. Journal of Product Innovation Management, 10: 309–317.

Doll, W.J., and Zhang, Q. (2001). Clarifying the Fuzziness in the Concept of Front End Fuzziness: A Dual Theoretical Rationale. Toledo, OH: University of Toledo.

Dwyer, L., and Mellor, R. (1991a). New product process activities and project outcomes. R&D Management, 21: 31–42.

Dwyer, L., and Mellor, R. (1991b). Organizational environment, new product process activities, and project outcomes. Journal of Product Innovation Management, 8: 39–48.

Dyer, B., Gupta, A.K., and Wilemon, D. (1999). What first to market companies do differently. Research Technology Management, 42: 15–21.

Ernst, H. (2002). Success factors of new product development: a review of the empirical literature. International Journal of Management Reviews, 4: 1–40.

Ettlie, J.E. (1997). Integrated design and new product success. Management, 15: 33–55.

Fuller, G.W. (1994). New Food Product Development: From Concept to Marketplace. Boca Raton, FL: CRC.

Graf, E., and Saguy, S. (1991). Food Product Development: From Concept to the Marketplace. New York: Van Nostrand Reinhold.

Green, P.E., and Krieger, A.M. (1989). Recent contributions to optimal product positioning and buyer segmentation. European Journal of Operational Research, 41: 127–141.

Green, P.E., and Krieger, A.M. (1991). Product design strategies for target-market positioning. Journal of Product Innovation Management, 8: 189–202.

Green, P.E., and Wind, J. (1973). Multiattribute Decisions In: Marketing: A Measurement Approach. Hinsdale, IL: Dryden.

Griffin, A., and Hauser, J.R. (1993). The voice of the customer. Marketing Science, 12: 1–27.

Hoban, T.J. (1998). Improving the success of new product development. Food Technology, 52: 46–49.

Hyman, M.R. (2002). Revising the structural framework for marketing management. Department of Marketing, New Mexico State University; http://cbae.nmsu.edu/~mktgwww/hyman/8Ds_FullPaper_10212002.PDF.

Iiori, M.O., Oke, J.S., and Sanni, S.A. (2001). Management of new product development in selected food companies in Nigeria. Technovation, 20: 333–342.

Jenkins, S., Forbes, S., Durrani, T.S., and Banerjee, S.K. (1997). Managing the product development process. Part II: Case studies. International Journal of Technology Management, 13: 379–394.

Karlsson, C., and Ahlstrom, P. (1997). Perspective: changing product development strategy—a managerial strategy. Journal of Product Innovation Management, 14: 473–484.

Khurana, A., and Rosenthal, S.R. (1997). Integrating the fuzzy front end of new product development. Management Review, 38: 103–120.

Kotler, Ph. (2003). Marketing Management, 11th edition. Englewood Cliffs, NJ: Prentice Hall.

Kotzbauer, N. (1992). Erfolgsfaktoren neuer Produkte: Der Einfluss der Innovationshöhe auf den technischer Produkte. Frankfurt a. Main: Lang.

Kristensen, K., Ostergaard, P., and Juhl, H.J. (1998). Success and failure of product development in the Danish food sector. Food Quality and Preference, 9: 333–342.

Kuczmarski, Th.D. (1992). Managing New Products. Englewood Cliffs, NJ: Prentice Hall.

Lord, J.B. (1999). New product failure and success. In: Brody, A.L., and Lord, J.B., editors. Developing New Food Products for a Changing Marketplace. Lancaster, PA: Technomic.

Mishra, S., Kim, D., and Lee, D.H. (1996). Factors affecting new product success: cross-country comparisons. Journal of Product Innovation Management, 13: 530–550.

Morris, C.E. (1993). Why new products fail. Food Engineering, 65(6): 129.

Moskowitz, H.R., Gofman, A., Tungaturthy, P., Manchaiah, M., and Cohen, D. (2000). Research, politics and the Internet can mix: considerations, experiences, trials, tribulations in adapting conjoint measurement to optimizing a political platform as if it were a consumer product. In: Proceedings of ESOMAR Conference: Net Effects3, Dublin. Amsterdam: ESOMAR [World Society of Market Research], pp. 109–130.

Noble, C.H. and Mokwa, M.P. (1999). Implementing marketing strategies: developing and testing a managerial theory. Journal of Marketing, 63: 57–73.

O'Driscoll, A., and Murray, J.A. (1998). The academy-marketplace interface: who is leading whom & does it really matter? Irish Marketing Review 11: 5–18.

Parr, H., Knox, B., and Hamilton, J. (2001). Problems and pitfalls in the food product development process. Food Industry Journal, 4: 50–60.

Peryam, D.R., and Pilgrim, F.J. (1957). Hedonic scale method of measuring food preferences. Food Technology, 11: 9–14.

Pinto, M.B., and Pinto, J.K. (1990). Project team communication and cross-functional cooperation in new program development. Journal of Product Innovation Management, 7: 200–212.

Razzaque, M.A. (1998). Scientific method, marketing theory development and academic vs. practitioner orientation. Journal of Marketing: Theory Practice, 6: 1–15.

Rothwell, R., Freeman, C., Horlsey, A., Jervis, V.T.P., Roberston, A.B., and Townsend, J. (1974). SAPPHO updated: project SAPPHO phase II. Research Policy, 3: 258–291.

Rudder, A., Ainsworth, P., and Holgate, D. (2001). New food product development: strategies for success. British Food Journal, 103: 657–671.

Shocker, A.D., and Srinivasan, V. (1979). Multiattribute approaches for product concept evaluation and generation. Journal of Marketing Research, 16: 159–180.

Song, X.M., Montoya-Weiss, M.M., and Schmidt, J.B. (1997). Antecedents and consequences of crossfunctional cooperation: a comparison of R&D, manufacturing, and marketing perspectives. Journal of Product Innovation Management, 14: 35–47.

Srinivasan, V., and Lovejoy, D. (1997). Integrated product design for marketability and manufacturing. Journal of Marketing Research, 34: 154–163.

Stewart-Knox, B., and Mitchell, P. (2003). Trends in food. Science and Technology, 14: 58–64.

Stewart-Knox, B., Parr, H., and Bunting, B. (2001). Model of 'best practice' for the food industry [Abstract]. Proceedings of the British Nutrition Society, 60: 169a.

Ulrich, K.T, and Eppinger, S.D. (2000). Product Design and Development, 2nd edition. New York: McGraw-Hill.

Urban, G.L., Weinberg, B., and Hauser, J.R. (1996). Premarket forecasting of really-new products. Journal of Marketing, 60: 47–60.

Varadarajan, P.R., and Jayachandran, S. (1999). Marketing strategy: an assessment of the state of the field and outlook. Journal of the Academy of Marketing Science, 27(2): 120–143.

Part I

Nuts and Bolts, Raw Materials, and Ratings

Chapter 2

Single Benefits Screening (Promise Testing) and More Complex Concept Testing

Introduction

The 1992 Innovation Survey, conducted by Group EFO Limited of Weston, Connecticut, with responses from 166 managers of new products who represented 112 companies of food or beverage giants such as Coca-Cola, Campbell Soup, and Kraft General Foods revealed that just 8% of new-product projects at major companies survived to reach the marketplace—an "internal mortality rate" of 92% (Morris 1993). Yet, most of the marketing and product development professionals in these companies would have averred, quite strongly, that they took the necessary steps to understand consumer needs prior to launch, and that the failure must be traced to factors outside the corporate marketing process.

One of the more challenging decisions faced by a new-product development team is concept selection, or the narrowing of multiple product concepts to the best single design (Dahan and Srinivasan 2000). These considerations will be important for benefits screening, on the one hand, and for full concept testing, on the other. If the concept does poorly, then the blueprint for the product is flawed, and the chance for success is even more dismal than the aforementioned 8%.

Benefits Screening (Promise Testing)

When a manufacturer wants to test ideas, quite often the first approach is to test the ideas one at a time. This is called *benefits screening* or in some cases *promise testing,*

At the early stage of development the first goal often is to generate as many new ideas as possible. Typically, 50–100 or more new-product ideas and concepts result from early-stage brainstorming. The next crucial step is to evaluate these embryonic ideas and decide which are sufficiently promising to warrant further development.

The basic, organizing principle behind promise testing is that the researcher should test each element in isolation. The researcher presents the consumer with the different ideas, each idea as a simple statement. The consumer rates the different ideas, one at a time, on a set of scales. One of the scales is typically *acceptance,* whether the scale be purchase intent or general acceptance. Researchers don't stop at these simple ratings of acceptance, however. Often a researcher wants to measure other things about the specific idea, such as uniqueness, and communication of specific characteristics. For instance, Hershey Foods uses new-product testing to examine a myriad of concept elements such as flavors, names, and benefits positioning. According to David Hoover, Hershey staff research analyst, "You need to test maybe 100 concepts to get one good product that might make it to market" (Hoover 2002). Andy Gibbs, the president and CEO of PatentCafe.com, answers the question about how to convert a concept into a marketable commodity with the special note that one should break such a questions down into a few pieces: concept, marketability, and presentation (presumably, with licensing in mind). In his opinion, concepts or ideas are basically not marketable: "First, that's because they're little more than an imagination. Second, because, as the common wisdom states, 'Ideas are a dime a dozen'" (Gibbs 2000). Both Hoover and Gibbs recognize the importance of elements as precursors to ideas that will be viable in business.

Depending on the predilections of the researcher the benefits test can be as Spartan as a simple rating of the concept element or as complex as a full-blown profile of the same single element. For example, at the more complex end, Urban and his colleagues (1996) describe a powerful method of forecasting new-product success called *information acceleration,* in which virtual product representations are combined with a complete virtual shopping experience.

It is worth noting here that, as benefits screening has become more accepted, the tendency has been to pile question on question into the promise test, so that the researcher obtains volumes of information from a single test. It is tempting to feel smug and secure with all that data, but the reality is that one or two attribute ratings suffice for making decisions about most concept elements. Indeed, for the most part, a researcher really concentrates on acceptance, with the other rating attributes simply used and reported, but often ignored.

A good sense of the type of data that benefit screening generates can be found in the results from a study shown in Table 2.1. The panelists rated each of nine different statements on acceptance and two additional rating scales. From the table, one can see what types of elements do well and what types of elements do poorly. Note that the data are presented as means rather than as percentage of respondents who would score the benefit very high (i.e., 7–9 on a 9-point scale).

What Do Researchers Look for When Analyzing Benefit Screen Data?

When faced with the type of summary data presented in Table 2.1, what do typical marketers, product developers, or researchers look for? How are these data approached? Is there a specific order of questions that one might ask? Is there a recommended way to approach the data, using them to answer those questions? Certainly, when one observes an experienced professional dig through the data, one sees some evidence of a plan, although the questions asked may seem to jump around a bit. A novice who is shown the same type of information will be quite stumped or, more than likely, will ask some of the correct questions. One has the nagging feeling, though, that the novice is blindly feeling around this mass of data and asking questions that intuitively should be asked,

Table 2.1. Results of testing nine benefits on three rating scales, across 200 respondents: data are from a sauce to be added to meat

		1 = 9 =	Low interest High interest	Daily use Special use	Modern Traditional
1	The same piece of meat, in four different ways		7.1	5.4	4.2
2	Ideal for those who like a change		6.8	5.6	4.7
3	This turns your gravy as if by magic into a light and creamy sauce		6.6	5.6	6.4
4	With a subtle herb/spice package		6.4	6.1	5.0
5	For something special on ordinary days		5.9	5.5	5.7
6	The ordinary piece of meat with just a touch of difference		5.3	5.8	5.0
7	With bits of onion		5.1	5.5	5.7
8	For an exotic piece of meat		4.8	7.0	3.8
9	Ideal for modern people		4.5	4.6	4.8

but not asking these questions in a fashion that makes one feel that there is knowledge, experience, and expectation behind it.

Moving on with this scenario, our expert might ask some questions, in approximately this order:

How do the different benefits score? Are they separated by a wide range or by a narrow range? This pair of questions really deals with the signal-to-noise ratio. Table 2.1 shows that the respondents differentiated among the various benefits (i.e., the potential concept elements discussed in later chapters). However, differentiation does not occur all the time. Sometimes many of the benefits show the same response profile. If the benefits score similarly, then this result should set off a warning signal. There is less of a signal in the respondent's mind, meaning that the respondent sees fewer differences among the benefits than the researcher would have liked. Testing nine benefits rather than one or two decreases the odds that all the benefits will score similarly. If the benefits do score similarly, even across nine different options, then this is an important result. The respondent simply does not see differences. One has the same type of expectation in music. A theme might be heard once or twice, to fix the theme. When one hears the same theme four times, one gets bored. The same theme four times begs the question of whether there are subtle, relevant differences that listeners should identify and to which they should attend. The same expectation applies to concept research. Two stimuli scoring the same do not disappoint. They constitute a finding. Four stimuli scoring the same do disappoint. They constitute a failure to change the stimulus so that the change is perceived by the respondent and reflected in the ratings.

Is there a pattern in the data that might be expected on the basis of characteristics inherent in the benefits? Experts look for patterns. For example, if the benefits present varying prices, then one might expect that a rating scale such as "value for the money" to covary inversely with stated price. Of course, there are other statements or features in the benefit statements that could modify this pattern. Nonetheless, an expert in concept testing would look for evidence of this pattern. Certainly, if the highest-priced statement scored very high in value for the money, or in purchase intent, then this finding would give pause. The pattern of ratings does not make sense and, if they are correct, then something else is going on worth exploring. Researchers are accustomed to expensive items scoring lower on purchase intent or perceived value for money, unless the item is a luxury one wherein price denotes quality. An unexpected finding does not, however, mean that the data are incorrect. The finding simply gives one pause and demands some time to reflect on the data.

Do the subgroup data make intuitive sense? In concept research the researcher can obtain information about the respondent beyond the ratings themselves. This additional or exogenous information comes either from having the respondent profile himself or herself on a set of attributes in the classification questionnaire or from knowing something about the respondent because one has additional information about the particular respondent from another, third-party source, such as an external database. That source could be the company that provided the respondent's name. The company could know a lot about the respondent, such as purchase history. In any event, the researcher computes the conventional statistics about the concept promise on both the total panel and on key subgroups that are deemed relevant. Such breakouts of the population may be gender (males vs females), age (defined in groups, such as 10-year intervals, or older vs younger), market, usage pattern, or brand used, etc. When the data are presented for the total panel and all of these groups, the researcher looks for consistency across subgroups, ever vigilant to the possibility that some unexpected variation may be generated by a particular subgroup. The researcher looks for patterns that may signal something important, such as lower-income respondents really disliking the promise elements with the higher price. This capacity to scan the subgroups is more intuitive, more qualitative in nature. The expert looks at the subgroup data to get a sense that there are not wildly unexpected patterns that could hint at a major finding that was unexpected. One's expertise here is important, because quite often the expert cannot even articulate what catches his eye and what is simply the normal variation that is passed over and disregarded.

When Do Companies Use Benefits Screening, and Who in the Company Uses It?

Many companies use benefits screening because it is easy and makes intuitive sense. Typically, researchers who use benefits screening do so because there are a number of different statements to reduce to a limited set. The limited set of elements will then be inserted into concepts. The nature of these single elements makes the screening particularly attractive at the very early stage of development.

Benefits screening becomes especially compelling at the very early stages in research, where the participating corporate parties are reluctant to write concepts and simply want to know what individual ideas are most promising. For example, companies may hire a firm to facilitate brainstorming of new ideas. Dozens, if not hundreds, of ideas may emerge that can be reconstructed into small promises and individually rated for relevance, interest, uniqueness, and so on. The test execution is so simple, the analysis so straightforward, and the selection criteria so clear that many companies favor this type of idea screening as the very first research phase in concept development.

Most researchers are not ready to work with this many elements. Researchers are happy to have a method that allows them to eliminate the less promising methods. Hence the term *benefits screening*; it is not a test of benefits, per se, but the elimination of unpromising elements that give the method its name.

Most benefits screens are run by researchers in research and development (R&D) and marketing research—the same individuals who run full-scale concept evaluation tests. This is no surprise, because benefits screening is really a preliminary version of concept testing. The screening is run because researchers feel either that the stimuli are not appropriate for full-concept research or because they, all too often, feel uncomfortable with more complex and yet appropriate approaches, such as conjoint analysis (experimentally designed combinations). One-element-at-a-time research, with single elements, feels more comfortable to these individuals. If we look at the type of individual who runs the benefits screen, we find that it is not necessarily the age or the experience of the person that dictates the test. Rather, the personality type who finds the benefits screen attractive tends to be one who needs to move on with the data and who must make a decision. Because the screening approach is so simple, inexpensive, and easy to interpret, the individual most attracted to it tends to be a person who just simply wants an answer, such as the rank of the different elements. That person will then disregard the poor-performing elements and work only with the strong-performing ones.

There are both positives and negatives about benefits screening:

1. *Positive: Simplicity.* Benefits screening is very simple. There is no arcane, difficult, and possibly incorrect mathematics. Respondents simply rate the elements on a scale or a series of scales.

2. *Positive: Easy to analyze.* The key goal is to put the differences into a rank order of elements, from the top that are worth further attention to the bottom that are worth throwing back.

3. *Negative: Too simplistic.* Benefits screening stops researchers from thinking about new ways to understand the issues involved. Benefits screening is primarily clerical. Once the researcher has done one screening study, the template is set. There is very little thought. The research process all too often then becomes automatic, despite the loud protestations of the people behind the testing, who swear that they think, whereas they really shut down intellectually. Benefits screening is so simplistic that it often stops one from doing better research on full concepts. The thinking that it stops is not the thinking about the problem. If anything, benefits screening pushes that problem forward so it must be addressed later, preferably by other professionals involved in the business issue.

The Mechanics: Reporting the Results

Benefits screening results are typically reported by means of a cross tabulation. The columns represent the different groups of consumers who participate, the rows represent the different phrases, and the numbers inside the table are either the percentage of respondents who feel that the phrase is acceptable or the mean of the phrase on the rating attribute. These two measures typically are correlated very highly, so decisions made using the percentage acceptors versus decisions made using the mean will usually be quite similar to each other.

As already noted, the type of numbers reported comes from a researcher's intellectual history. Marketing researchers trace their intellectual history to a different sociology and generally focus on the percentage of individuals who express interest in an idea. Sociologists, who focus on the proportion of respondents answering in a certain way, are not interested in the depth of feeling for an idea as felt by a single person but rather the proportion of individuals in a group who feel that way. In contrast, many R&D-oriented researchers report the average or mean rating in the body of the table. These researchers trace their heritage to the sciences, especially physics and chemistry, or even to psychophysics, that branch of experimental psychology dealing with the relation between perceptual magnitude and physical stimulus level. They are interested in the intensity of feeling rather than the percentage of people who exhibit a feeling.

It is important to remember that the majority of the statistical analysis is done on the key evaluative measure of interest/acceptance. In general commercial practice, the greatest interest focuses on the analysis of interest, and less interest focuses on the other attributes/scales. Those attributes (e.g., uniqueness) are less important to businesspeople, because the primary goal of the research is to identify the winners. The other information is of secondary interest only. Usually, this other information is put into an appendix. The information is either used to guarantee that the particular phrase communicates what it should communicate or, in some cases, used to choose between two equally performing phrases. Occasionally, researchers defend the use of other scales beyond evaluative ones as help in tie-breakers, but the specific way that these scales help to break ties is rarely defined and virtually never standardized from one study to another.

Analyzing the Data and Understanding the Meaning of the Results

Various statistical tests are appropriate for the benefits data. Consider the data for yogurt that are provided in Table 2.2. They are from 246 respondents who rated the benefits on interest (1 = definitely not interested to 9 = definitely interested).

Question 1: Statistical Summaries That Prepare for Decision Making

What is the average of each of these four different promises? The simplest answer comes from the averages, as presented in Table 2.2.

Table 2.2. Means and standard deviations for four yogurt benefits, rated singly in a benefit screen*

Benefits	Text	Average rating (total panel)	Standard deviation of the ratings	Top 3 boxes on the 9-point scale
Ben1	100% organic	5.12	2.33	30
Ben2	A quick and easy snack or meal	5.41	2.19	35
Ben3	Thick with lots of real fruit on the bottom	5.68	2.33	43
Ben4	The delicious, classic fruit flavors like raspberry, strawberry banana, and blueberry	6.08	2.05	47

*The data come from the responses of 246 individuals who rated each of the benefits on a 9-point scale for acceptance. The ratings were converted to top-3 box percentages.

From the average a researcher can very easily rank order from the most promising to the least promising statement if only one rating scale is involved. With more than one scale, however, each scale would generate its own rank order, based on the average on that scale. There is absolutely no reason to believe that the order (best to worst) on one scale will be the same as the order on another scale.

At a more serious level, however, one should ask the question and then perform the relevant statistical analysis. For example, we have two particular benefits or features that could be appropriate for the product. One benefit is "100% organic," which scores an average of 5.12. The other benefit is "A quick and easy snack or meal," which scores a 5.41. The first question that arises is whether these two numbers really differ from each other. If we were to repeat the study many times, would we see a difference between these two benefits of 0.29 (5.41–5.12)? We know that all data are subject to variability. We can run a simulation to answer this question. Let us assume that we can sample small groups of 50 respondents from these data. Let us perform this sampling 2000 times, which is easily done with a computer. Each sampling represents a *simulated study* using the data we just obtained. We will find that the difference is not always 0.29. Sometimes the difference is larger and sometimes smaller. On average, however, if we perform this study again and again, under comparable conditions, benefit 2 will score 0.29 points higher than benefit 1. Figure 2.1 shows

this expected distribution from the 2000 simulated studies, based on the data. We see from Figure 2.1 that repeating the study by computer simulation generates a distribution of differences. Looking at one simulation study at a time, we would say that, in about 75% of the cases, benefit 2 would score higher than benefit 1.

Researchers typically use *t* tests to answer the question as to whether two benefits differ. The *t* test enables researchers to look at the ratio of the difference between two benefits compared with the variability of the difference of those benefits (specifically, the standard error of the mean of the differences). This ratio distributes itself according to a specific pattern given by the *t* distribution. The specific ratio or *t* value computed for the dataset can be used along with an already-computed and available table of the *t* distribution to estimate the likelihood of observing this specific magnitude of difference or a higher magnitude, if in actuality the two benefits were to score similarly. Table 2.3 shows the analysis of these two benefits using the *t* test and confirms the fact that *by sheer chance alone* we would not see this difference or a larger difference. The probability of seeing a difference of −0.29 or a more negative difference (benefit 1 minus benefit 3) is virtually 0. If we switch gears for a moment and look at the percentage of respondents who rate each benefit as 7–9 and belong in the top-3 box group, we see the probability as 0.03 or lower if the two benefits had originally come from

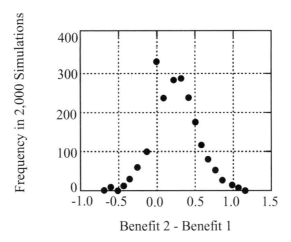

Figure 2.1. Distribution of difference in interest for two benefits statements about yogurt. The distribution is taken from a bootstrap analysis of the original data, with a sample size of 50.

the same distribution of concept acceptors, defined as member or nonmember. We conclude, therefore, that benefits 1 and 2 do differ from each other. There is no reason to assume that the original means are the same and that the 0.29-point difference arose from chance alone for the 9-point rating scale or similarly that the 4.87% difference in top-3 box arose from chance alone.

When dealing with many benefits, concepts, or products, researchers often use more comprehensive statistics, such as the analysis of variance (ANOVA), which handles a larger number of test items. ANOVA approaches the problem differently, but the objective is the same. The t test looked at the odds of seeing a specific difference between two benefits by using the difference in the means and the variability of that difference to estimate the likelihood. ANOVA examines the ratios of variabilities. It allocates the variability in the data to as many sources as it can, but always allocates some of the variability to error and some of the variability to "signal" or factors that are under the researcher's control, or at least of interest. Once the total variability in the dataset is allocated, the ANOVA looks at the ratios of

Table 2.3. Analysis of the acceptance of benefits 1 and 2 by using the t test and either the actual 9-point ratings or ratings converted to top 3 boxes

Difference between benefits 1 and 2 on the 9-point scale			
Mean ben1	=	5.12	
Mean ben2	=	5.41	
Mean difference	=	−0.29	
SD* difference	=	1.43	$t = -3.16$
Degrees of freedom	= 246	Change probability	= 0.00
Difference between benefits 1 and 2 on the top-3 box scale			
Mean topben1	=	30.15	
Mean topben2	=	35.01	
Mean difference	=	−4.87	
SD* difference	=	34.58	$t = -2.21$
Degrees of freedom	= 246	Change probability	= 0.03

*SD, standard deviation.

two variabilities. One of the variabilities—the denominator—is the error or unexplained variability corresponding to the variability that cannot be allocated to the test stimuli or to whatever other factors the researcher wishes to study.

The ratio of the two measures of variability (benefits and error) provides a statistic, the *F* ratio, whose properties and, especially, whose distribution are known mathematically. Specifically, one can estimate the odds of observing a specific *F* ratio or higher if the two variabilities are equal (namely, if the error variability is as big as the variability ascribable to the four benefits).

ANOVA enjoys a great deal of popularity in research because it indicates whether a significant variation exists among treatments under an experimenter's control. In this case the treatments are the four benefit statements for yogurt. A sense of the relative magnitude of variabilities is apparent in Table 2.4, which is a conventional ANOVA table. What is important, however, is the ability to identify specific pairs of benefits that differ from each other.

The key results from Table 2.4 can be summarized as follows:

1. The ratio of variability due to benefit versus error variability is very high for both the 9-point scale and the membership in the acceptor group. We conclude, therefore, that there is probably a significant difference between some pairs of benefits, although we do not yet know which pair of benefits differs. We must still use the t test to determine which pairs differ.

2. The *F* ratio for the original 9-point data is higher than the *F* ratio for the recoded data (8.01 vs 6.15). Recoding may come up with the same order of performance, but loses some information in the data. This information loss makes intuitive sense because the recoding converts everything to a binary scale and hides some of the finer-grained differences.

3. Benefit pairs 1–2 and 3–4 are most similar in their scores for both types of meas-

ures: the 9-point scale and the binary recode, respectively. We see similarity from the table of probability values. These numbers are probabilities of obtaining this specific difference or greater if the means were actually the same. When the probability value is low, the conclusion is that the odds of seeing this difference would be 0 if the original means of the benefits were the same. When the probability value is low, we conclude that this difference has a low chance of appearing by chance alone. Thus, based on our current observation, we *reject* the hypothesis that the original means of the benefits are the same.

Question 2: Next Steps

Now that we have the data, what should we do? What should we talk about? What decision should the researcher make, based on these results? Research is done for a reason. Most often the decision is whether to go forward with a specific message; for instance, to create a product or to put the message into advertising for an existing product. When doing research it is important to avoid simply churning out lots of statistics. In all too many instances the detailed analysis of data and the conclusions from that detailed analysis are either ignored completely in favor of rapid decisions or simply added to other information without the research user paying much attention. There is a reason for this cavalier attitude toward a lot of the research results. A great deal of the statistical analysis is done as a self-indulgence to protect oneself against punishment for being incorrect. Thus, much of the research simply adds a seal of approval through external statistical testing.

A researcher can go further by looking at other pieces of information beyond the rating assigned by the total panel. For example,

Subgroups. If there are subgroups, such as users of different yogurts, then the researcher might want to know whether these subgroups rate the benefits in the same way. As Table 2.5

Table 2.4. Analysis of variance (ANOVA) and post hoc tests of difference applied to the four benefits and to their recoded values

ANOVA: scores of four benefits on the 9-point acceptance scale

	ANOVA				
Source	Sum of squares	df*	Mean square	F ratio	P
Benefit	119.4096	3	39.8032	8.0159	0.0000
Error	4866.2480	980	4.9656		

Fisher's least-significant-difference test
Matrix of pairwise comparison probabilities that two benefits score the same on the 9-point scale:

	1	2	3	4
1	1.0000			
2	0.1512	1.0000		
3	0.0064	0.1957	1.0000	
4	0.0000	0.0011	0.0477	1.0000

ANOVA: scores of four benefits on membership in acceptors (obtained from recoding the 9-point scale into 0 or 100, respectively)

	ANOVA				
Source	Sum of squares	df*	Mean square	F ratio	P
Benefit	43201.2195	3	14400.4065	6.1584	0.0004
Error	2.29159E + 06	980	2338.3524		

Fisher's least-significant-difference test
Matrix of pairwise comparison probabilities that two benefits score the same in terms of proportion in the top-3 box acceptor group:

	1	2	3	4
1	1.0000			
2	0.2635	1.0000		
3	0.0039	0.0768	1.0000	
4	0.0001	0.0053	0.3.54	1.0000

*df, degrees of freedom.

shows, these subgroups rate the benefits differently. The pattern that appears to emerge is that users of brand A like the statements about fruit ("Thick with lots of real fruit on the bottom" and "The delicious, classic fruit flavors like raspberry, strawberry, banana, and blueberry"). Users of brand B like these benefits but substantially less. Such differences among brand users occasionally emerge, although later on in this book we will see more powerful ways to divide consumers (called concept response segmentation; see Chapter 7)

R&D input. Quite often developer scientists and consumer researchers deal only with the consumers. A great of deal of attention

and analysis are paid to the consumer data, but far less attention is paid to the response of others involved in the development, manufacturing, and trade aspects of the problem. In the case of yogurt it may be possible to create a very highly acceptable product at the bench, but the scale-up to manufacturing may be difficult and occasionally impossible. The product may be unstable, hard to process, and have a short shelf life. It may be difficult to source ingredients that deliver the desired sensory and formula quality. All of these considerations require that the experts play a role in rating these promises and do so with respect to their expertise. The experts may not need to participate in the up-front screening, where the consumer evaluates dozens, if not hundreds, of promises. At some point, however, the expert must get involved, if only to ratify the consumer's response by agreeing that, indeed, the products are feasible to make, ship, and sell. By bringing the information from R&D, and even from packaging and the trade, the researcher can get a deeper feeling about the performance of the benefits on other criteria besides consumer acceptance. If R&D can assess the benefits on aspects such as cost impact (i.e., product cost) or time to produce a yogurt with those features, then the researcher has created a dataset with substantially greater value. The data can be used to determine which benefits combine the features of consumer acceptance, low expected cost of goods, and high R&D feasibility to deliver within the development and marketing schedule.

Testing Complete Concepts

The shift from benefits screening to tests of complete concepts is both a small jump and a large leap. Full concepts comprise a variety of elements in tandem, including promises, graphics, pricing, and the like. Full concepts comprise multiple ideas, not just single benefits that might be encountered in the conventional benefits screen. Full concepts come in a variety of formats. The test concept can comprise simple white-card concepts with no embellishment, all the way to more elabo-

rate, almost finished, print ads that have excruciating detailed embellishments.

At the very simplest level are the *white-card concepts*. These concepts typically present new product or service ideas, such as those shown in Figure 2.2A. The name *white card* or *white board* concepts is simply a descriptor term, coming from the fact that the concepts can be typed up on a white piece of paper and given to the respondent, who then rates the concept. Of course, the paper need not be white, and the respondent can get the concept from the interviewer in person or by mail or through the Internet, and can rate the concept on one or several attributes. The essence of the white-card concept is that it presents the idea simply, most typically in relatively unembellished form. Sometimes the white-card concept has a selling slant to it—there is really no single way that the concept is written.

A more embellished format comprises a concept with pictures, possibly with specific typeface, with pricing, and so on (see Figure 2.2B). This more embellished format more closely approximates a print ad. Some researchers, especially those in the advertising agency, feel that only the more embellished, more finished concept is worthy of the name concept testing, and that the white-card concept can be relegated to the area of very preliminary research.

A study using the more embellished, finished concept is often conducted quite differently from a study using the white-card con-

Table 2.5. Performance of the four yogurt benefits from the perspective of consumers and subgroups, and from that of research and development (R&D)

| Benefits | Creative text | Consumer | | | R&D | |
		Total	User A	User B	Cost	Time
Ben1	100% organic	5.13	4.83	5.43	High	Long
Ben2	A quick and easy snack or meal	5.42	5.75	5.09	Low	Short
Ben3	Thick with lots of real fruit on the bottom	5.68	6.02	5.34	Medium	Medium
Ben4	The delicious, classic fruit flavors like raspberry, strawberry banana, and blueberry	6.08	6.32	5.84	Low	Medium

Now you can savor the delicious taste of homemade
treats anytime, with new Orchard Desserts.

Orchard Desserts combines real fruit with a crumb topping
for a homemade taste.

Simply pop one in the microwave for a tasty, warm, treat anytime.

Choose from three deliciously distinctive varieties:
Cranberry, Apple, or Blueberry.

Comes in a package of three convenient 4 oz. portions for $2.39

A

Now you can savor the delicious taste of homemade
treats anytime, with new Orchard Desserts.

Orchard Desserts combines real fruit with a crumb topping
for a homemade taste.

Simply pop one in the microwave for a tasty, warm, treat
anytime.

Choose from three deliciously distinctive varieties:
Cranberry, Apple, or Blueberry.

Comes in a package of three convenient 4 oz. portions for $2.39

B

Figure 2.2. Examples of concepts. **A:** A simple, white-board concept. **B:** A simple, white-board concept with a picture.

cept, perhaps due to the more finished nature of the latter and the significantly greater cost. The concept format and test execution are not really part of the research per se, but they are worth noting if only for the difference in research styles. The finished concept is produced on a high-quality matte or gloss paper (see Figure 2.2C). Very often the finished concept is printed in color, and the artwork more nearly approximates what one might call final artwork. The size of the finished concept is also larger—often larger than an 8 × 10 sheet. That finished concept can be blown up to fit a 16 × 20 sheet. Finally, and more for executional issues in the interview, the finished concept is mounted on Styrofoam so that it becomes, in colloquial research terms, a *concept board* (see Figure 2.2D). The notion of a concept

Introducing **Orchard Desserts...**

Now you can savor the taste of homemade treats anytime!

There's nothing quite like the satisfying taste of homemade snacks and desserts. But sometimes it's hard to find the time to make them. Now you can savor the delicious taste of homemade treats anytime, with new Orchard Desserts.

Orchard Desserts combines real fruit with a crumb topping for a homemade taste. Just sprinkle the crumb topping over the fruit and enjoy. Simply pop one in the microwave for a tasty, warm treat anytime. And because they don't need to be refrigerated they are great for on-the-go.

Choose from three deliciously distinctive varieties: Cranberry, Apple, or Blueberry.

Comes in a package of three convenient 4 oz. portions for $2.39

C

Introducing **Orchard Desserts...**

Now you can savor the taste of homemade treats anytime!

There's nothing quite like the satisfying taste of homemade snacks and desserts. But sometimes it's hard to find the time to make them. Now you can savor the delicious taste of homemade treats anytime, with new Orchard Desserts.

Orchard Desserts combines real fruit with a crumb topping for a homemade taste. Just sprinkle the crumb topping over the fruit and enjoy. Simply pop one in the microwave for a tasty, warm treat anytime. And because they don't need to be refrigerated they are great for on-the-go.

Choose from three deliciously distinctive varieties: Cranberry, Apple, or Blueberry.

Comes in a package of three convenient 4 oz. portions for $2.39

D

Figure 2.2. Examples of concepts. **C:** A concept board without a picture. **D:** A fully elaborated, finished test concept.

board in contrast to white-card concepts has nothing to do with the importance of the research. Rather, the finished concept is produced in a limited quantity instead of being run off in batches of dozens on a copying machine. The finished concept board is relatively expensive to produce, so fewer can be created within the research budget. Furthermore, the final, finished concept board is not left to the respondent to handle, but rather shown by an interviewer in either a one-on-one personal interview or in a hall setting with a number of respondents. The same concept board is shown to a number of people in sequence. Therefore, the concept board itself must be created to be physically robust, which is why the concept is pasted on a Styrofoam board and why the stimulus is called a concept board.

More Complex Concepts Can Generate Richer Response Data

In contrast to promise testing, concepts enable researchers to ask more detailed questions. Since a concept comprises different features, including description of the service, benefit, price, graphics, and the like, and even layout, the concept paints a more detailed word-and-graphics picture of the new product or service.

One major goal of concept testing is to measure the acceptance of that word-and-graphics picture. Another goal, just as important, is to create a profile of the concept on other, evaluative as well as nonevaluative attributes, to quantify what the concept communicates and how strongly. The other attributes may deal with value for the money, an attribute that combines different types of impressions, such as acceptance (Would I buy this product?) versus economics (Is this an appropriate price for the product?).

A concept and the typical array of questions and ratings are presented in Table 2.6. The array shows the richness of the information that one may glean from respondents. Each researcher has a favorite set of questions. Often in companies a limited

Table 2.6. The yogurt concept, the anchored rating scales, the mean rating across 200 respondents, and the percentage top 3 boxes on anchored, 9-point rating scales

Concept
Introducing Taster's Heaven Yogurt
100% all natural—nothing added
Naturally low in calories
With a delicious creamy texture, and in decadent flavors such as honey orange, fudge chocolate, and bourbon
    sweet vanilla
Priced at an affordable $3.29 for a 16-oz. container

	Rating scale	Mean	Top 3 boxes
Evaluative scales			
How interested are you in purchasing			
this yogurt?	1 = Definitely not purchase		
	9 = Definitely interested	5.16	33
How much do you like this yogurt?	1 = Hate		
	9 = Love	6.12	51
How unique is this yogurt?	1 = Not at all unique		
	9 = Very unique	3.89	11
How do you rate the "value for the			
money" for this yogurt?	1 = Far too cheap		
	5 = Just right	5.48	66
	9 = Far too expensive		
Communication scales			
Nutritional value	1 = Low nutritional value		
	9 = High nutritional value	4.88	34
Good taste	1 = Poor taste		
	9 = Great taste	6.93	27
Storage	1 = Hard to store		
	9 = Easy to store	5.36	60
For adults versus children	1 = For children		
	9 = For adults	7.34	76
For daily versus special occasions	1 = For daily use		
	9 = Special occasions	7.22	74

number of such questions carry over from one study to another, with other questions added on an ad hoc basis to cover specific issues pertaining only to the particular product being studied.

One very interesting behavior in concept research at the corporate level might best be labeled attribute creep or laundry-list creep. In years gone by, researchers in a corporation may have been satisfied with a limited number of concept attributes. A lot of the work was done by hand, so the issue of creating a massive database for one or two concepts was not a particularly attractive option. Judicious researchers limited themselves to the questions that would be deemed appropriate for the particular study and especially those that could answer the question in an expeditious, straightforward fashion. This limitation, similar to the limitation imposed by the memory and processing power of early personal computers in the 1970s, led to thoughtful questionnaires. Over time, however, with experience and with the passing of the guard over to new and perhaps less experienced researchers, the nature of the concept questionnaire changed, as it changed for products as well. The change, subtle at each stage, resulted in the slow, unheralded, but unstoppable addition of questions, but rarely the matching deletion of questions. Eventually, the concept questionnaires become longer and longer because no one wanted to delete a question. From instructing respondents to rate the concept on a few attributes, including purchase intent, researchers evolved into asking respondents many dozens of questions, ranging from interest in the concept, to profiles of expectations about the product, to even profiles of the concept/product on a variety of personal scales. The widespread use of computers, the relegation of the fieldwork to outside agencies, and the ability to do graphic presentations of the concept results to summarize the findings all conspired to generate massive, often unwieldy, concept studies.

Concept Testing Answers Some Questions that Benefits Screening Cannot

The ability of researchers to test many benefits or simple statements in a very easy, cost-efficient fashion leads often to the logical question about what full-concept testing accomplishes that benefits screening does not. A further question regards the types of analytic tools available in concept testing that are not appropriate for benefits screening.

If we distinguish between simple benefits and concepts, we come up with the following two differences that can answer the foregoing questions.

Concepts Present More Complete Descriptions of a Product or Service

Those who work with benefits typically get involved at the early stage of the development cycle. Benefits screening is used as a screening device to identify what looks promising and what does not. With benefits screening it is not reasonable to estimate share, volume, and the ability of the particular benefit to steal share from a competitor. As much as researchers try to expand the scope of benefits screening, the stimuli remain simply statements. The test stimulus does not and cannot present a complete picture of a new product. Perhaps that fuller picture might be presented in a setup page that introduces the benefit. In such a case, however, one should consider the study to be a concept test, with some elements in the concepts substituted, but with the main portion of the concept remaining the same.

Benefits Screening Cannot Deal with the Prediction of Simple Sales and Share of a New Product

It is important to remember that at the early development stage it is very difficult to span the range between considerations of development (early-stage promise and effort) and subsequent market share. These disciplines are not typically merged with concept research, per se. Market share of new products traditionally has focused on one idea, with many different aspects of marketing execution. The more complex, subsequent market behavior has more to do with a fuller selling proposition rather than with a single benefit. Responses to systematically varied white-board concepts probably would have a difficult time predicting sales and share. Benefits screening, coming as it does in the early development stages, cannot easily and simultaneously sort through many ideas and for each one provide a measure of potential share. To perform such a pair of tasks simultaneously is to invite disaster for one of the two, and perhaps for both. There are entirely different disciplines devoted to this important topic of share.

References

Dahan, E., and Srinivasan V. (2000). The predictive power of internet-based product concept testing using visual depiction and animation. Journal of Product Innovation Management, 17: 99–109.

Gibbs, A. (2000). Developing profitable concepts: our invention marketing expert shows you how to make your bright idea a reality; http://www.entrepreneur.com/Your_Business/YB_SegArticle/0,4621,278253,00.html.

Hoover, D. (2002). A bold commitment pays off. Marketing News, by Catherine Arnold; http://www.marketingpower.com/live/content.php?Item_ID=16174&Start=1&Category_ID=5115.

Morris, C.E. (1993). Why new products fail. Food Engineering 65: 129.

Urban, G., Weinberg, B.D., and Hauser, J.R. (1996). Premarket forecasting of really-new products. Journal of Marketing, 60: 47–60.

Chapter 3

Ideation Strategies and Their Deployment in Concept Development

Introduction: How Really Fuzzy Is the Fuzzy Front End?

Early-stage decisions on product characteristics, design, and markets have strategic importance (Clark and Fujimoto 1991). The real key to product development success lies in the performance of the front-end activities (Khurana and Rosenthal 1998). Creating new product visions with high value to customers and speeding this vision through concept development, product design, process engineering, and market introduction with focus and discipline has become the hallmark of competitive advantage (Doll and Zhang 2001).

In recent years it has become de rigueur for business writers to talk about the *fuzzy front end* (e.g., conferences like "Bringing the Fuzzy Front End into Focus" at the Institute of International Research, 2001), defined as that beginning in the development process when the little universe of the concept is filled with possibilities, and where the ideas have not taken shape. The fuzzy front end, like the Wild West of American folklore, is an invention of business writers—a dream where possibilities have not yet given way to the constrictions of reality. The product development literature associates front-end fuzziness almost exclusively with uncertainty in the external environment (Bacon et al. 1994; Mullins and Sutherland 1998; Gardner and Buzacott 1999). It is often associated with ad hoc decisions and ill-defined processes (Montoya-Weiss and O'Driscoll 2000) or the need to reduce environmental uncertainty (Moenaert et al. 1995). Daft and Lengel (1986) argue that the greater the un-

certainty of the task environment is, the greater is the information-processing burden.

Certainly, there is that wonderful moment at the starting gate of concept and product development where the possibilities are great, where the directions are many, and where the potentials are unmeasured. All that wonderful opportunity is meaningless, however, unless a method for systematic creation can be established. For sure, one can take advantage of serendipity in the creation of a new product, but realistically this fuzzy front end needs to be tamed to be productive. Recently, there has been a growing realization that the causes of many product failures can be traced back to this fuzzy front end (Khurana and Rosenthal 1997). Smith and Reinertsen (1991) first noted that over half of the time from idea generation to market introduction is idle front-end time where the product idea floats around, but there is no organized effort to develop it.

This chapter discusses *ideation*—the creation of new ideas that are precursors to the concepts. Unlike most business processes, which thrive on standardization and the elimination of idiosyncrasy, however, the fuzzy front end thrives on relaxed discipline and revels in idiosyncrasy. It is in the unusual that many new products find their niche and thrive. The business goal is how to nurture that region and time, giving them just enough discipline to have the fruits, but not so much business discipline as to kill it off.

A whole cadre of professionals now realizes the exceptional opportunity at this early stage. Business literature dealing with this fuzzy front end underscores its importance for success in efforts to develop new products. The sources of front-end *fuzziness* have been identified as uncertainties in the external environment [i.e., uncertainty about customers' needs, markets, competitors, and technology (Doll and Zhang 2001)]. The business of learning what to do from the consumer point of view as a whole goes under the name of *ideation*—the production of ideas that can serve as precursors. This chapter highlights some of the methods, the issues, and the outputs. It moves the creative task, generating the ideas in almost a factory-like mode, to measuring these ideas, with the goal of culling out the poorer ones and identifying the most promising ones.

How Do Ideas Announce Themselves for New Product Concepts?

Many times when the senior author (H.R.M.) leads a discussion with college students or visitors the question arises about the origin of ideas to put into concepts. To most consumers the notion that there could be a system underlying the development of new food concepts is startling, yet systems are in place and practitioners have formalized these systems and made businesses out of them. For example, Kuczmarski (2001/2002) talks about a "guaranteed innovation system" with seven integral, and intricately linked, parts (e.g., problem orientation, because breakthrough innovations must squarely address real problems; and creating a strategy for the way innovation programs will work or forming innovation teams). To be sure, none of the more aware individuals thinks for a moment that somehow these ideas really appear out of the ether, take root in the corporate mind, and are seamlessly transformed into concepts and then into products. It is just that in daily lives of consumers people don't stop to think about the act of creating products,

unless it happens to be one of those situations where they need something new for a particularly pressing, momentary need.

Occasionally, novices believe that being in a company endows them with the ability to come up with ideas, because the company is about new ideas, new products, and service to customers. Contrary to what many young college students think, one's first job in a company does not confer the ability to understand consumer needs. Being in a corporation, being surrounded by fellow workers in the deep throes of business issues, and being challenged to produce do not automatically make the novice marketers, product developers, or sensory scientists any better at coming up with ideas. One could certainly bluff one's way through a meeting with ideas picked out of the blue and, for a short time, make these off-the-cuff ideas sound impressive. At the end of the day, though, these new ideas may not test very well, and if, heaven forbid, one of them is launched by force of personality and fails miserably, well . . . so goes the nascent career of the developer.

Companies are risk averse. They like having systems for everything. That *everything* can range from the invention process, to the testing process, to the launching process. Note the word *process*. By process the corporation can control the flow—from thinking to creating, fabricating, launching, and so forth. Unfortunately and probably distressingly so, the front end, fuzzy or otherwise, typically has none of the traditional "handles" that managers use to control a process (Smith and Reinertsen 1991). It is not surprising, therefore, that over the past 50 years the corporate style for concept development has evolved from what may have been maverick insight to a streamlined, occasionally effective, always fun process. Indeed, the word *fun* and the term *fuzzy front end* seem to be inextricably linked, as if the development of new ideas is equivalent to releasing the child in the adult, through play.

So where do ideas come from? One source is the *idea generation* process. Idea generation processes vary across the board. Unlike consumer research procedures with concepts, wherein there are certain standardized methods for setting up the stimulus and testing, measuring, and analyzing it, idea generation is somewhat free form. This difference between the ideation process that leads to the concept and the testing process that measures the concept should not surprise. Two different needs are being met, and two different constituencies are involved.

Copying with Slight Modification

No one likes to admit being a copycat. Few business leaders boast about their ability to imitate. Most like to be thought original. Yet, in the end, many products resemble one another. But fear of the failure of new products has resulted in low rates of innovation in the food industry, with many companies preferring to redevelop old products to create new products in the attempt to increase success rates (Van Trijp and Meulinberg 1996). Kuczmarski cites three major internal barriers to new product development: risk aversion and short-term orientation; lack of necessary resources, such as people and funding; and poor market understanding (Morris 1993). If a *new* food product is defined as "one that is new to the consumer," only 7%–25% of food products launched can be considered truly novel (Lord 1999).

The ideas for these products probably start out as being very similar, with perhaps a caveat or two (e.g., change the cap; change the color; modify the flavor). Copying old ideas with such small modification is a perfectly respectable task. According to the Georgia Manufacturing Survey of 2002, fewer than 15% of Georgia manufacturers were involved in copying and modifying existing competitor products (MVS 2003). Despite the goal of ideation to produce new products, a company may want to limit the risk of newness. Line extensions and copying competitors with minor changes both represent low-risk actions in the ideation business. With respect to the food business, a growing body of evidence indicates also that original concepts are more successful than copycat or "me too" products (Hoban 1998; Knox et al. 2001). This is further corroborated by a recent survey of food company practices in the USA, which has indicated that the failure rate for truly new food products is only 25% (Hoban 1998). According to Stewart-Knox and Mitchell (2003), new innovative products are more likely to succeed because food product markets can become rapidly overcrowded. New and improved technologies are increasingly being used in food innovation to differentiate products successfully (Katz 1998; Stewart-Knox and Mitchell 2003).

Ideation

The ideation process is more free form than simply copying with modification. The corporation sets up ideation processes to be structured and systematic. By its very nature, that structure fights against the notion of the free construction of ideas. People who are so good in process, moving the corporation forward and making sure all the parts fit together and work, are probably not particularly good in letting themselves go and coming up with unusual, so-called out-of-the-box ideas. It simply is not part of their personality. It takes a special people, with special talents. Mike Vance, the former Dean of Disney University, is a good example. Vance has a vast wealth of knowledge on building a creative environment in companies and has firsthand knowledge of how the Disney Company builds and fosters creative culture, yet he is not the standard manager that one typically finds in a company. That type of person would not be able to do the job (Vance and Deacon 1995).

Deconstruction of the Competitive Environment

Deconstruction refers to the systematic assessment of the different products and messages that currently exist. Chapter 17 discusses concept deconstruction in detail, so this section is just a précis. In product research, deconstruction is known as *category appraisal*. The objective of this appraisal is to understand the features of products and how they drive acceptance by the structured assessment of what exists in the marketplace today. The parallel approach in concept research is simply deconstruction. Through systematic analysis of the current messaging, the marketer and product developer begin to understand what is being talked about. This understanding is a tremendous aid to the ideation process. Quite often, and prior to an ideation session, the participants will be given the assignment to go out into the marketplace and identify the different products and messaging currently available. Somehow, in the back of the consumer's mind, this information is churned and novel ideas come out.

Trend Studies: Reading Published Material in Different Areas

Professionals in the business of developing concepts are often intellectually omnivorous. They do not limit themselves to the task of coming up with ideas in a formalized way; nor in fact do they necessarily depend on serendipity alone. Many of these professionals are trend watchers, subscribing to newspapers, magazines, newsletters, etc., that look for trends. Through monitoring the environment they recognize new opportunities. They search for stimulation and ideas in many different areas, not only in their own specialty. This omnivorous appetite and viewpoint should come as no surprise. As our society becomes increasingly complex, the nature of food may become more complex. Besides good taste, foods may be designed for occasions, for specific nutrition, for certain kinds of personalities, for special needs, etc. Only a well-read, observing individual will recognize some of these behaviors as fertile areas for new developments.

Creative Consumers

During the past several decades a number of researchers, such as Foy Conway of Conway Milliken (www.conwaycreative.com), have been promulgating the use of specially selected consumers for idea generation. Thus, consumer involvement helps the product development team clarify the product definition and project targets as well as develop a sense of shared team purpose (Doll and Zhang 2001) Consumer involvement can help ensure that the product's design remains consistent with customer needs. Not content with the usual run-of-the-mill consumers, Conway and similar-minded professionals have suggested that they can screen the population to discover consumers who score high on intuition, problem solving, and the like. Whether the screening uses conventional test instruments from psychology, ad hoc tests from the researcher, or even interviews with the consumer is not important for this discussion. The key thing to keep in mind is that in the mind of these professionals an important aspect of creativity is to select the correct individual. It is the individual who is important. The professionals do not purport to understand the mechanism of creativity, but they do maintain that having the correct participant in a creative session enhances session performance and productivity.

Lead Users

Lead users, according to Von Hippel of MIT, are individuals who use products in a new way, perhaps to solve problems, perhaps in a new application. Von Hippel and others describe how to identify lead users and then how to incorporate their insights into the product design process in a five-step process (Von Hippel 1986; Dahan and Hauser 2002):

1. Identify a new market trend or product opportunity (e.g., greater computer portability).
2. Define measures of potential benefit as they relate to customer needs.
3. Select lead users who are ahead of their time and who will benefit the most from a good solution (e.g., power users).
4. Extract information from the lead users about their needs and potential solutions and generate product concepts that embed these solutions.
5. Test the concepts with the broader market to forecast the implications of lead-user needs as they apply to the market in general.

By working with these lead users in a category, the marketer and the developer might well be able to identify opportunities for new products because they see how these individuals use products in new and novel ways.

The Contribution of Human Resources to New Product Ideation

The combination of an innovation-friendly climate in the organization with risk-taking behavior may enhance the chances for the development of new products (Voss 1985). It is at this level that corporate functions such as human resources (HR) make their greatest contribution to the development of new products. HR managers keep their eye on the internal workings of the corporation. From the HR group comes the recognition that employees probably know the corporation product and issues better than anyone else does. To this end comes the very popular suggestion box, ever-present in many corporate headquarters and branch offices. To what degree

this suggestion box works with real ideas, and to what degree it represents another version of the corporate principles tacked up on the wall, remains for analysis on a case-by-case basis. It is important to note that the suggestion box usually gets suggestions that pertain to better performance of corporate processes rather than new products, although some companies like 3M feel that the suggestion box and the employee suggestions generate a great number of new product ideas as well. Cooper and Kleinschmidt (1995, 1996) suggest that HR legislates *free time*, where technical employees are offered free time or *scouting time* (up to 10%–20% of their workweek) to do creative things. A policy such as this, sponsored by HR, can help the ideation because it frees the employees to roam around mentally rather than shackling them to their job every minute.

Brainstorming for Concept Ideas versus for Full Concepts

Brainstorming and all of its ancillary methods is a time-honored way for researchers to obtain ideas. Some brainstorming approaches require the participants to create ideas, but not necessarily complete concepts. Other brainstorming approaches require complete concepts. Still other methods begin with an idea and let other people in the session add to the idea until it is perfected. One could write a very large chapter on brainstorming methods alone, but in the end the objective is to create ideas. Most of the variation among brainstorming methods deals with the different ways of preparing, motivating, and cheerleading the participants rather than around the output of the sessions themselves. Indeed, the financial issues involved in brainstorming as a commercial, marketing-services enterprise dictate that the various up-front preparations be differentiated to attract clients and establish market position through one's uniqueness in the practice.

Ethnographic Observation

Observing everyday behavior is a standard tool in an anthropologist's repertoire. Recently, however, observation of behavior at home, in a restaurant, at a store, etc., has become a popular research tool (Schensul and LeCompte 1999). Ethnographic research records behavior of people in the context of their everyday lives: where they work, live, shop, or play. Although the information is expensive to acquire, the results allow researchers to better understand how the people interact with the food. As a consequence, researchers who do the observation can identify newly revealed unmet needs. The observation itself can reveal how the observed respondent interacts, and that can immediately generate a new product idea. Inspiration is another way. Researchers can identify the person–product interaction in a situation and come away inspired with a product idea. Ethnographic observation is the latest approach to spawn different business services, as practitioners begin to recognize the value of the observational methods.

Collaborative Filtering

In recent years, and with the power of the Internet, a new class of approaches has been developed that share the rubric *collaborative filtering*. Quite simply, the brandDelphi system as embodied by one leading practitioner, CRM Metrix in Paris, is set up in a very simple, but powerful sequence to maximize the flow of ideas (Flores et al. 2003).

1. The participant reads a question. The question is open ended.

2. The respondent looks at a selection of answers provided by previous participants and selects a requested number of these ideas (at most 4–6) that seem relevant. The selection procedure is simply checking. The reason for the selection is that the brandDelphi can begin to cascade, generating literally hundreds of elements

in a few hours, as hundreds or thousands of respondents participate.

3. The respondent is then asked to offer two of his own elements.

4. Finally the respondent is provided with another set of elements, previously offered by other participants, and instructed to rate them (e.g., on interest, but the rating could be on applicability to a given end use).

5. The collaborative filtering approach generates ideas that are somewhat more consumer-oriented than the other methods, because the ideas that emerge from the approach have been "vetted" by other consumers. Only the ideas that continue to meet the standard of interest by other consumers are maintained.

A good example of collaborative filtering comes from a study run by A. Maier, L. Flores, and the senior author (Maier et al. 2003). The objective was to collect the thoughts of 300+ consumers to create new ideas for bread and bread packaging. Table 3.1 shows results of the first set of 40 from the 500+ elements that were generated by the brand-Delphi ideation or collaborative filtering system. The open-end question was to provide ideas for a new variation of healthy bread with better packaging. The specific question generating these elements was the packaging question. It is clear that some of the respondents stayed with the specific question, whereas others responded with different ideas—some new, but not appropriate to the question that was asked. This departure from the task is not unusual in ideation. The same forces that generate new ideas also promote lateral thinking, so respondents often offer these nonrelevant ideas in the computer or interpersonal ideation session.

The table has four columns

1. The element as provided by the respondent.

2. The total number of times it was selected by other respondents as being important

(Sel). This is part of the collaborative filtering.

3. The proportion of times that the element appeared and was selected (Selp). This is a relative measure that shows the relative frequency of being selected. The proportion corrects for the fact that the element was not shown to everyone.

4. The importance of the element, rated by the respondents who saw it. The respondent used a 1–9 importance scale (Rate).

We can deduce a few principles from looking at the table:

1. It is clear that ideas about packaging are important, but so are ideas about health. Ideas about new flavors do not appear.

2. Depending on the specific instructions to the respondent, the researcher can turn the flow of new ideas toward a variety of topics. In *this* particular question on bread, the objective was to get ideas about a health product, not a flavored product. Had the instructions focused more on flavors and *inclusions* such as nuts or cranberries, the respondents would have provided this type of element more frequently.

3. The same idea may appear in different forms from different respondents. There may be fewer ideas than one might expect, based on the large volume of elements that this method and other ideation methods generates. Other methods, including simple visual inspection, must be used to edit the elements to a more limited number. The editing is generally subjective and comes from reading through the list, creating categories, and then identifying both membership of an element in a category and redundant elements.

Table 3.1. First 40 of 500+ ideas generated from Internet-based collaborative filtering*

Elements	Sel	Selp	Rate
The bread stays fresher	81	64	8.5
Stay fresh zip lock	76	75	8.5
Sealable with something that collapses as the bread is used and a closure other than a twist tie that is secure and you can't lose.	62	61	8.4
Be nice to have bread that doesn't mold in like 3 days.	61	67	8.5
A bag that would be easier to reseal and keep bread tasting fresher for a longer period. It is too easy to misplace the little wire tie and plastic tie-tops sometimes break or are hard to replace.	60	74	8.6
A see thru label so you can tell if the product is drying out, mouldy, etc	58	55	8.3
Easy reading for calories and nutrition	57	56	8.0
It should have good stay fresh package.	56	73	8.6
Added calcium	50	45	8.0
Health and taste	48	48	8.6
I would like different loaf sizes . . .	47	56	8.2
A bread that stays fresh because of quality packaging	46	60	8.5
Closes easy	46	42	8.2
Packaging I can see through to make sure the bread is OK. And nutritional information in larger, easier to read print for the whole thing, not just the good stuff.	45	51	8.1
I want to know what is the benefit of any added nutrient. Why is it important to have fiber in the diet, etc.?	45	34	7.8

(continued)

Table 3.1. First 40 of 500+ ideas generated from Internet-based collaborative filtering* *(cont.)*

Elements	Sel	Selp	Rate
A clear expiration date in a clear package so you can see the bread.	44	60	8.2
I would prefer a clear and easy to understand labeling of genetically engineered (GM) ingredients (e.g., corn) and potential health concerns.	44	40	7.9
Clear package on top where you can see the whole loaf with just the name across part of it then on the bottom in bigger writing ingredients	43	39	8.2
Better taste	43	37	7.8
Resealable packaging that will maintain freshness	43	67	8.7
A Ziploc bag would be nice for freshness and taste.	42	55	8.5
I really don't have any ideas on the information labeling, but a resealable bag would be nice. I don't care too much about the tie wrap.	42	44	8.0
Flavor & Vitamins	42	38	8.3
WE DO NEED A ZIPLOC BAG INSTEAD OF THE TWISTIES	41	62	8.5
Just plain and simple like my grandma used to make. With a list of all the ingredients on the bottom.	41	42	8.4
I would like to see the diet exchanges given by the American Heart Association listed on the packaging.	41	38	8.1
I would like to see bread packaged in a wrapper that could be resealed (zipper type) so it would stay fresher.	40	78	8.6
To be lower in calories, & higher in vitamins.	40	57	8.3
I wish bread would stay moist longer	40	53	8.3
Visible and understandable freshness labels	40	51	8.1
Easily resealable packaging that helps bread stay fresh longer	40	70	8.8
I would love to have bread recipes included in the package.	39	37	7.3
Bread that would not cause a high caloric intake.	38	48	8.2
On the front would like the bread to have a longer shelf life	38	35	8.0
Moist	37	47	8.2
I would like to see the ingredients on the top of the bread as well and a bread that when you store it in the refrigerator it doesn't get hard and stale tasting	37	43	7.8
Clear packaging so the break can be seen from outside. Large printing so nutrients and calories can be seen	37	34	7.7
Low carb breads	37	32	8.2
Language everyone understands, so you do not have to use your college education to read the label	37	28	7.9

*The elements are presented in the exact format they were typed. Sel, the total number of times an element was selected by other respondents as being important; Selp, the proportion of times that an element appeared and was selected; and Rate, the importance of the element, rated by the respondents who saw it (the respondent used a 1–9 importance scale). *Source:* brandDelphi approach; courtesy of CRM Metrix, Paris.

The Nature of the Ideas

Are All Idea Generations Really the Combination of Smaller, Preexisting "Idealets"?

A recurring theme in the history of psychology is the generation of ideas. We are not much removed from that issue. How do ideas emerge? One could go back as far as Plato and assume that the ideas preexist in the mind of a person. The Greek philosophers assumed that people who "saw" the correctness of a mathematical theorem were actually tapping into preexisting knowledge (e.g., dialogues such as Plato's *Theatetus* and Plato's *Sophist*). More recently, however, psychologists and philosophers alike, such as John Locke, have assumed that knowledge and ideas are combinations of smaller units. Thus, idea generation is merely the recombination of old ideas into new groups. If this is so, then ideation and concept development should focus on identifying these pieces of ideas and facilitating their combination. If, however, the ideas are not mere recombinations of components, but rather, like Venus, emergent creatures that spring fully formed from the head of Zeus, then perhaps the better strategy would be to encourage the creative thought itself, the generation of the fully formed idea. One should then spend much less time and expend less effort on the combination of *idealets*. This dichotomy in viewpoints plays out day after day in industry, with some practitioners favoring the creation of fully formed ideas and others favoring the creation of parts of ideas and their subsequent combination by other means, such as computer-based interviewing. The approaches and their respective proponents merely carry forth ideas and controversies that are thousands of years old.

How Good Are the Ideas?

An ongoing issue in the fuzzy front end and the business of ideation is to identify how well these new ideas actually perform. It is one thing to boast about the ability to extract hundreds of ideas from one's ideation session. Certainly, that capability is no mean feat. What is more critical, however, is to identify with a little more certainty which of these work products from ideation represent the kernel of a good idea and which are simply poor ideas. Furthermore, which of the following two alternatives is better?

1. A system that produces a few good ideas,
2. A system that produces many ideas, most of which are not particularly good, but some of which are good and others of which suffice to trigger yet new ideas? This is an unanswered question, but it is certainly worth reviewing.

A recent client-sponsored study by Jeffrey Ewald at the Optimization Group showed that many of the new ideas that ideation teams developed did not do particularly well when these ideas were converted into elements and tested in a conjoint analysis task among consumers. Figure 3.1 shows the utility values of 1800 of these ideas.

The Nature of the Participants

Who Can Provide Ideas?

Although we talked before about creative consumers, in actuality ideas come from everyone, not only from these experts. Anyone who has had experience with the product or with the situation can offer an idea. Whether the idea is particularly good, relevant, or even feasible is another issue and remains for professionals and concept tests to decide.

Does the Person Make the Product or Does the Situation Make the Product?

The notions of creative consumers and lead users were raised previously in this chapter. One of the key issues in ideation is the nature of the participants versus the nature of the situation. What leads to creativity: having a

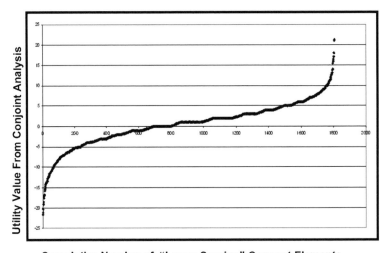

Cumulative Number of "Lower-Scoring" Concept Elements

Figure 3.1. Cumulative distribution of 1800 ideas or concept elements, from 50 different IdeaMap.net studies. Utility values of 10 or higher are typically considered to be strong performing elements. Only a very few elements do that well, at least on a total panel basis.

need or having a bright person? This question is not new and is not limited to new products. The English playwright James M. Barrie in his play *The Admirable Crichton* underscored the idea that leadership is not socially determined, but rather internal. A similar situation occurs for creativity and new ideas. Is it nature (i.e., the person) or is it nurture (i.e., the situation) that stimulates creativity? The oft-heard truism "necessity is the mother of invention" speaks to that question, as well. This truism would have us believe that the creative consumer and the lead user are probably happenstance creations, and that it is the situations in which people find themselves that generate the creativity. The answer to the question—person or situation, nature or nurture—has two very practical business ramifications for the development of concepts:

1. If nature—the person—determines creativity, then it behooves product developers to cultivate a cadre of consumers with defined creativity and communication skills. This cadre of creative consumers provides a unique reservoir for the development of new ideas. To some extent Delphi groups with experts appeal to the unique creative guru, and other people-oriented solutions are part of this answer.

2. If nurture—the person—determines the creativity, then it behooves product developers to develop methods to monitor the environment and identify problems and then swoop in with research tools to identify solutions. To some extent the ethnographic approach is part of this answer. It is ordinary lives, ordinarily lived, with conventional problems that provide the ever-fresh source of new ideas. The job is simply to monitor this life with the best tools available.

Should Technical People Get Involved in the Moderation of the Session and in the Analysis of the Creative Ideas and, If So, When?

A great deal of the expertise in creativity comes from the ability to analyze the data within a framework that promotes new ideas. In most ideation work, researchers have relatively little technical expertise, and without experts participating, many of the ideas that emerge are naïve and ultimately impractical. They may sound good when presented and may add to the bounty generated by the ideation session, but they may have no value at all.

More and more, however, technically trained people are entering into the ideation business, especially in the food and beverage industry. This trend means that in many of the ideation sessions the research professional conducting the sessions brings to bear other insights beyond those associated with the ideation itself. Words, notions, and even doodles brought up in the ideation session can trigger insights into the product. A technical person who either acts as a moderator or who gets involved in editing the rough cut of first ideas may well be able to bring more value to the session. The value comes from the perspective and from the ability to marry the ideas offered in the ideation session to existing technology.

The Role of "Homework" and Pasteur's Dictum: Chance Flavors the Prepared Mind

A great deal of the mystique in ideation comes from the feeling that the participants are tapping into some great unknown, all-knowing mind. There is a sense of the mystical in ideation. The social interaction of people coming up with new ideas, the adrenalin generated, and the excitement that is so palpable in the first few minutes lead one to feel that these off-the-cuff ideas could be breakthroughs. Unfortunately, this is simply not the case, except in the most unusual, fortuitous circumstances.

Many professionals in the ideation business feel that some level of preparation is best, so that the ideas that emerge can be better and more readily recognized. The unprepared mind, the excited novice, or the energetic type who participates in the ideation session without thinking through the problem may come up with ideas. However, the prepared mind may better create connections because of some familiarity with the project. The

traditional Delphi method, using experts, instinctively uses the mind of the person familiar with the topic, because that mind is more likely to have snippets of ideas that can be recombined (Brown 1968). The novice, unprepared, unpracticed, with little information, may come up with ideas, but in some sense may be on the path to rediscovering the commonplace in the ideation session. The expert, with the greater depth, may discard these even before stating them, because the expert, familiar with the situation, knows that these ideas are already commonplace and represent nothing new. *Homework* in this respect prepares the mind and gives the participant a base of raw material from which the psychological processes then take over to help creativity. The novice is less of a novice after doing the homework and becomes somewhat more prepared.

The Nature of the Activity

Do Fun Exercises Increase Creativity?

The business world has an entire industry devoted to practitioners of methods that increase creativity. Some individuals insist that the participants in a creativity session travel from restaurant to restaurant, location to location, to see how people eat and what they eat. This is great fun and actually opens up one's eyes to new ideas. Other professionals, with clients having an even greater budget, stage affairs to show their clients what people eat. Still others, approaching the limit of credibility and good taste, have their clients go to sandboxes and recreate the freedom of youth, when imagination was unfettered, vision unchallenged. Whether these minor dramas and stage sets work is another issue altogether. Practitioners need to distinguish themselves. In some ways these exercises become the signature—the leitmotif—of the practitioner. Whether they work becomes irrelevant. Like the placebo effect in drug research they reassure the patient, or in this case the client, that indeed the client is in a regimen deigned to promote creativity. Sometimes that is all the push one needs to become truly creative.

How Long Is an Ideation Session, and How Can Its Productivity Be Measured?

To what degree can a session's creativity be measured? Are there indices, such as ideas? Do shorter sessions do better than longer sessions because the respondents are fresh? Or, conversely, do longer sessions produce better information because the participants push through the "low-hanging fruit," the easy ideas, onto better, but more subtle, ideas that would have been previously unavailable due to the short session length? There is no clear answer. From the authors' experience, sessions lasting under an hour are not particularly productive. There is just not enough time for ideas to germinate, for people to be stimulated by the ideas of others, and for they themselves to contribute modifications of those offered ideas. On the other hand, after 3 or 4 hours, many of the participants seem tired and complain that they simply cannot provide more elements. In some of the senior author's ideation sessions for IdeaMap, the goal is to provide single ideas rather than full concepts. The typical time pattern followed by the ideation session instructs. The session begins with the participants providing categories or buckets. Then, the respondents go around the room in a fixed order, each person identifying a bucket and providing one idea and one idea only. This is usually a phrase. Within 10 minutes this nice order breaks down. Within 1 hour the group is quite lively. By 2 hours, however, the contributions are beginning to slow down. The session participants are still able to provide many ideas, but with less demonstrable vigor. By 2½ hours the rate of contribution has slowed dramatically. Even if a break is called for the session and resumption is scheduled 2 hours later after lunch, the second session never has the

energy and productivity of the first. Thus, it appears that a concentrated 2- to 3-hour session probably produces the most energy and the fastest rate of element production. Whether it produces the best ideas, however, remains for empirical analysis to determine.

The Measurement Task

Informal Methods

A well-run ideation group can generate hundreds of ideas. Indeed, many practitioners of ideation pride themselves on the enormous productivity of their development sessions. Some may, from time to time, place advertisements guaranteeing a certain minimum number of ideas generated. Whether this approach of volume works to produce better ideas is open for debate. Occasionally, folk wisdom in marketing states (without much proof) that it takes 100 ideas to generate 10 interesting ones, and 10 interesting ideas to generate one winner in the marketplace. Clearly this ratio of 100/10/1 is for illustrative purposes, but it does show the developer's recognition that there will be more success with more concepts up front.

More concepts up front mean more screening. Which ideas are good? Which are bad? The simplest method—informal screening—comprises the act of sitting at a table, going through the volumes of elements and ideas, keeping the ones that are interesting, and discarding those that are uninteresting. If truth be known, this approach is perfectly reasonable. And why not? What an individual finds disinteresting often mirrors what the entire population finds uninteresting. These are the key times when the individual does not mirror the group:

1. *Radically different criteria.* Sometimes the situations in which developers find themselves differ dramatically from those of consumers and vice versa. Situations that are quite different generate different judgment schemes and rules.

2. *Profound mind-set segmentation across people.* Segmentation by pattern of interest can make a concept or element very interesting to one person but boring to another. In foods this segmentation is especially prevalent, meaning that the researcher and developer must be aware that the segment into which they fall predisposes them to be interested in a concept element that people in other segments would find unappealing or just plain uninteresting.

3. *Changing focus during the task.* Anyone who has ever gone through hundreds of ideas in an effort to cull them down can sympathize with the problem of changing focus. When the marketer or developer begins, all of the ideas are interesting. As the marketer goes through the elements, some common themes emerge again and again. At one level, all of these elements can be put into one group, and all but one discarded. Yet, at another level, some of these elements that have been discarded really have interesting nuances that make them a different idea. An overwhelmed, challenged professional dealing with the ideas may adopt a lenient criterion, letting many of these ideas in and, in an instant, switch from a lenient criterion to a stringent one, eliminating the next batch of similar ideas.

4. *Active screening versus passive screening.* Although we often think that the ideation process obtains new ideas from the participants, the ideation can actually spur on new ideas from the professionals, as well. There is no reason to assume that the only good ideas are those coming from the respondents when a statement by the respondent can spur on thoughts. Some good ideas may slip through the cracks if the researcher adopts a purely limited rule of processing what the respondent said and culling respondent ideas. The researcher simply looks at what has been said. Active screening requires that, on a post hoc basis, the researcher (or whoever does the sorting of the elements and their culling) insert himself or herself psychologically into the situation and provide even new ideas stimulated by the ideas previously suggested.

Formalized Concept Testing: Qualitative Methods (Focus Group, Depth Interview)

The testing process requires a structure in which to operate. There are two major forms of testing. One is qualitative testing, such as focus groups or depth interviews; the other is quantitative testing.

Qualitative Methods for Idea Generation

The focus-group interview has been adopted as a major data-gathering technique by market researchers who are interested in the appeal of advertising strategies or in consumer product preferences (Axelrod 1975). In many focus groups the experts sit behind one-way mirrors and watch respondents. Sometimes, experts sit with the consumers. Occasionally, the process is inverted, and the consumers sit with the experts. For example, Doll and Zhang (2001) suggest that customers can also participate as members of the design team.

Focus groups, comprising small groups of individuals, or even depth individuals with one person, facilitate the free exchange of feelings and reactions to concepts. In a discussion of the use of group interviews in marketing research, Calder (1977) asserted that this technique is an excellent vehicle to establish what Alfred Schutz (1967) called *intersubjectivity*, or ordinary descriptions of reality shared by actors. The interviewer, trained in eliciting information from respondents, can probe a respondent about reactions to concepts and ask other members of the group to comment. The interaction can be very fruitful. Those involved with the concept can sit with the group members or, more frequently, as already noted, sit in a more comfortable viewing room, away from the respondents and shielded by a one-way mirror. The concept writer, the marketers, and the researchers listen to the dialogue between the focus-group moderator and the respondent, noting key ideas where they emerge. At the end of the focus-group session, or usually at the end of a series of group sessions, conducted in different markets, with different respondents, the moderator writes records observations of reactions to the concepts and other relevant information. Often the reactions provide a wealth of information about what consumers like and don't like about the concept. These reactions are often accompanied by suggestions about what specifically to change. Occasionally, and more frequently today than before, the moderator may use the focus group to obtain consumer ratings of the concepts. There are too few respondents, at least in the mind of a quantitative researcher, but the practice continues to occur.

Formalized, Quantitative Concept Evaluation

There is a very large and growing business in the formalized evaluation of concepts, beyond the focus-group stage. For years, researchers have had the task of measuring the appeal of new ideas, either at the very early stages or in terms of potential market share or volume. Depending on the business issues, concept evaluation can range from the remarkably simple to the esoterically complex.

At the very simplest level, one needs to determine whether the concept is appealing. Do the consumers who look at the concept like what they see or not? Is the concept unique, different from everything current in the market, or is it similar and just another rehash of the same product themes that are so popular. What about the price of the product—is it too high or too low? If there is no price, then what is the respondent's guess about what price is fair? If there is a product that can be shown or eaten, then do the concept and the product fit together? All of these

questions bear on the different aspects of the concept. Acceptance is not enough: It is important to understand the concept in greater depth.

Formalized Concept Testing: Volumetric Projections

Marketers are not only interested in the performance of concepts, per se, but in the sales expected to be generated by a product. It is this type of performance, in market, that constitutes the success or failure of a product. Concept testing, per se, does not provide this, except at the most rudimentary level. In concept research, one can ask the respondent to estimate such behaviors as frequency of use, degree to which the product would replace or augment currently used products, and so on. In this way the researcher attempts to fit the new product into the respondent's life and estimate total sales dollars. Usually, the results are fairly qualitative; one can get an idea about the new product in terms of whether it would be frequently or infrequently used. This type of information helps to locate the product against other products, but doesn't necessarily predict volume. In most cases, estimated volume is a guess, but with norms and with competitor benchmarks for other products the researcher might be better able to estimate the approximate rate of consumption.

Other methods, such as bases, for predicting volume are currently available from commercial vendors. These methods use the results from concept research and, in some cases, actual product tests. The methods work with norms obtained over years of research and locate the product concept within that framework. The output is an estimate of the trial (year 1) and the sustained repeat rate (years 2 and perhaps 3). The commercial vendors of these predictions claim high accuracy in their methods and, as a result, many companies purchase the services for their concept testing. It is worth noting that these types of predictions are best made with fully formed concepts rather than with the bare-boned concepts created in early-stage research.

Overview

The fuzzy front end of concept development is an emerging area of interest in business. As the competitive environment grows increasingly frenzied, and as products have shorter life cycles, it is incumbent on prudent marketers to develop more ideas, develop them faster, and make them better. The ideation methods described in this chapter help marketers in that task. They do not guarantee success, but do provide the mechanism and machinery by which the ideas can be generated. Beyond idea generation, however, is the testing of these ideas. This book deals with the analysis of such ideas through consumer research tests. It is well worth keeping in mind that the testing of ideas in a valid way is just as important as the creation of the ideas. People can spew forth ideas, sometimes at blinding speed, often in a jumbled array. It takes discipline and insight to measure the potential of these ideas, separate the gold from the dross, the wheat from the chaff, and the less promising from the truly great.

References

Axelrod, M.D. (1975). Marketers get an eyeful when focus groups expose products, ideas, images, ad copy, etc. to consumers. Marketing News, 8: 10–11.

Bacon, G., Beckman, S., Mowery, D., and Wilson, E. (1994). Managing product definition in high-technology industries: A pilot study. California Management Review, 36: 32–56.

Brown, B.B. (1968). Delphi Process: A Methodology Used for the Elicitation of Opinions of Experts. Santa Monica, CA: Rand.

Calder, B.J. (1977). Focus groups and the nature of qualitative research. Journal of Marketing Research, 14: 353–364.

Clark, K., and Fujimoto, T. (1991). Product Development Performance. Boston: Harvard Business School Press.

Cooper, R.G., and Kleinschmidt, E.J. (1995). Benchmarking the firm's critical success factors in new product development. Journal of Product Innovation Management, 12: 374–391.

Cooper, R.G., and Kleinschmidt, E.J. (1996). Winning businesses in product development: The critical success factors. Research Technology Management, 39: 18–29.

Daft, R.L., and Lengel, R.H. (1986). Organizational information requirements, media richness and structural design. Management Science, 32(5): 554–571.

Dahan, E., and Hauser, J.R. (2002). Product development: managing a dispersed process. In: Weitz, B., and Wensley, R., editors. Handbook of Marketing. London: Sage.

Doll, W.J., and Zhang, Q. (2001). Clarifying the Fuzziness in the Concept of Front End Fuzziness: A Dual Theoretical Rationale. Toledo, OH: University of Toledo.

Flores, L., Moskowitz, H.R., and Maier, A. (2003). From weak signals to successful product development: using advanced research technology for consumer driven innovation. In: Proceedings of the Second ESOMAR Conference. Cannes: Technovate.

Gardner, D.T., and Buzacott, J.A. (1999). Hedging against uncertainty in new technology development: the case of direct steelmaking. IEEE Transaction on Engineering Management, 46: 177–189.

Hoban, T.J. (1998). Improving the success of new product development. Food Technology, 52: 46–49.

Katz, F. (1998). How major core competencies affect development of hot new products. Food Technology, 52: 46–50, 52.

Khurana, A., and Rosenthal, S.R. (1997). Integrating the fuzzy front end of "new product" development. Sloan Management Review, 38: 103–120.

Khurana, A., and Rosenthal, S.R. (1998). Towards holistic "front end" in new product development. Journal of Product Innovation Management, 15: 57–74.

Knox, B., Parr, H., and Bunting, B. (2001). Model of best practice for the food industry [Abstract]. Proceedings of the British Nutrition Society, 60: 169a.

Kuczmarski, T. (2001/2002). Innovation unmasked. Context, December 2001/January 2002.

Lord, J.B. (1999). New product failure and success. In: Brody, A.L, and Lord, J.B., editors. Developing New Food Products for a Changing Marketplace. Lancaster, PA: Technomic.

Maier, A.S., Flores, L., and Moskowitz, H.R. (2003). Accelerating development by understanding weak signals and strong communications: the case of a wholesome bread that delights consumers. (Presented at the Sensory Evaluation Symposium, IFT.) Food Quality and Preference (in review).

Moenaert, R.K., De Meyer, A., Souder, W.E., and Deschoolmeester, D. (1995). R&D/marketing communication during the fuzzy front end. IEEE Transactions on Engineering Management, 42: 243–258.

Montoya-Weiss, M.M., and O'Driscoll, T.M. (2000). From experience: applying performance support technology in the fuzzy front end. Journal of Product Innovation Management, 17: 143–161.

Morris, C.E. (1993). Why new products fail. Food Engineering 65: 129.

Mullins, J.W., and Sutherland, D. (1998). New product development in rapidly changing markets: an exploratory study. Journal of Product Innovation Management, 15: 224–236.

MVS (2003). Georgia manufacturing vital signs. http://www.ceds.gatech.edu/newsletters/Focuson-Manuf/MVSAugust2003.pdf.

Schensul, J.J., and LeCompte, M.D. (1999). Designing and Conducting Ethnographic Research: Ethnographer's Toolkit 1. Lanham, MD: Rowman and Littlefield (non NBN).

Schutz, A. (1967). The Phenomenology of the Social World. Evanston, IL: Northwestern University Press, xv–xxix.

Smith, P.G., and Reinertsen, D.G. (1991). Developing products in half the time. New York: Van Nostrand Reinhold.

Stewart-Knox, B., and Mitchell, P. (2003). What separates the winners from the losers in new food product development? Trends in Food Science and Technology, 14: 58–64.

Van Trijp, H.C.M., and Meulinberg, M.T.G. (1996). Marketing and consumer behaviour with respect to foods. In: Meiselman, H.L., and McFie, H.J.H., editors. Food Choice, Acceptance and Consumption. London: Blackie Academic and Professional.

Vance, M., and Deacon, D. (1995). Think Out of the Box. Franklin Lakes, NJ: Career.

Von Hippel, E. (1986). Lead users: source of novel product concepts. Management Science, 32: 791–805.

Voss, C.A. (1985). Determinants of success in the development of application software. Journal of Product Innovation Management, 2: 122–129.

Chapter 4

From Questions and Scales to Respondents and Field Execution

Introduction

There are always issues involved in research. A lot of the issues are matters of opinion, so the answers given are as much conjecture and viewpoint as they are based on experience. Thus, this chapter provides a mix of fact and opinion about what is appropriate, what is not appropriate, and how researchers should set up, execute, and analyze a concept study. We begin with a short refresher discussion on concept format/information and move on to the meat of the matter: field execution and basic analysis.

Up-front Issues

How Much Information Does a Concept Need to Contain to Be Valid?

Depending on who one talks to, the advertising agency or the researcher, the concept to be tested *must* take on different forms. To an advertising agency schooled in concepts, the more complete the concept is, the better the stimulus is. In many cases the belief is that the closer the concept is to a finished print advertisement, the better the concept will be. This strongly held belief does not mean to the advertising agency professional that simple, *white card* concepts are invalid. Rather, it means that, in the agency's opinion, the closer one moves to a real advertisement, the more realistic is the data that emerge.

In contrast to this viewpoint is that of researchers, who can live either with fully executed print advertisements as concepts or with white-card concepts that list the idea in an un-adorned form. If truth be known, both formats are valid. The white-card concept presents the unadorned idea. When a researcher measures reactions to the white-card concept, the outcome is a measure of the validity of the idea. Even brand name is often eliminated from the white-card concept in order to measure the strength of the idea alone. A brand name can severely affect the concept ratings

Emotional Language versus Descriptive Language

How simple versus how evocative should a concept be? That is, should the concept be expressed in flowery language similar to what one might read in an advertisement or should the concept be a simple communication? This question is slightly different from the question regarding descriptive versus selling concepts. The question here is about the nature of the concept itself. For the most part, the safest route to follow is to express the concepts in simple English or whatever other language. Concepts are not usually developed to test the execution of an idea (i.e., the way an idea is expressed). Rather, concepts are usually developed to test the viability of an idea. To the degree that a researcher can use simple language, the results of a concept test will be crisper and cleaner.

Text versus Pictorial Elements in Concepts

The role of graphics in concepts, like the very nature of concepts themselves, constitutes a battleground among researchers. There is not necessarily as much light as there is heat in the argument, again because most likely the role of graphics in concepts is more opinion and less fact. Such opinion-driven issues generate lots of discussion and serve as a safe region for battling viewpoints. Those who feel that the concept has to have a picture aver strongly that the graphics communicate aspects of the product that cannot be easily communicated by words. This may be true. However, that communication does not correlate with strong performance of the graphics as driving interest in the concept. Figure 4.1 shows the distribution of utility values for elements of a coffee study discussed at length in Chapter 8. The data are taken from a large-scale study with 273 elements, of which 38 were graphics. Across eight countries this provides a very large array of elements and utility values. By breaking out the utilities according to text phrases and pictures, one can see at a gross level whether the pictures drive interest. They don't do any differently in their performance than text elements do. This means that, when the criterion is *interest* or *utility*, the picture brings little in the way of additional "convincing power." The picture may contribute, but it is not necessary for persuasion.

What Are the Criteria for an Appropriate Rating Scale?

In concept work as in all other research a limited number of factors go into a good scale:

1. *Respondent friendly.* The scale should be easy for the respondents to understand. If the scale is complicated, then it will be hard to execute in the field and lead to errors. Typically, unipolar scales are very easy to understand (e.g., "Does not talk about health" vs "Talks a great deal about health"). Bipolar scales are also easy to understand if the two ends of the scale are similar but opposite in meaning (e.g., hate vs love; definitely not purchase vs definitely purchase). Bipolar scales that do not feature opposites are more difficult to understand, but respondents can still rate concepts on them (e.g., more for good health vs more for good taste).

2. *Intuitively obvious.* The scale should be easy for researchers to interpret. As management becomes increasingly pressed for time, attention span decreases. The stress reduces the focus to what can be communicated easily. Easy scales are more powerful than potentially better but harder-to-explain scales. One cannot minimize the importance and power

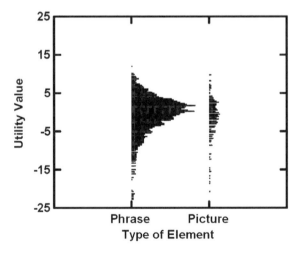

Figure 4.1. Utility values for coffee elements. The data come from eight countries, each of which generated utilities for 235 phrase/text elements and utilities for 38 picture elements. The elements were the same across all eight countries.

of simplicity in business. Easily understood scales typically use a limited number of scale points, with meaningful and intuitively simple anchoring phrases at the end.

3. *Statistics friendly.* The rating scale should be statistically robust and amenable to standard statistical analysis. Nothing so irritates a businessperson as having to sit through a presentation whose statistics are esoteric. All of the value disappears once the statistic becomes a "black box." The more open the statistic is, the more standard the analysis is, the better is the place of the particular statistic in the researcher's repertoire. With the plethora of statistical packages available today, having an esoteric statistic is not a strong positive. Usually, the simpler scales are the more robust scales.

What Is the Appropriate Numerical Rating Scale?

Every researcher has a point of view about scales. Perhaps the widespread difference of opinion on scales is a safe arena for professionals to express themselves. It is difficult to talk about philosophical issues in concept research, especially for practitioners and young professionals. On the other hand, it is not particularly difficult to talk about scales or other mechanical and executional issues. There is no loss of face in an argument about scales and execution, because there is no body of knowledge about scaling whose contents are accepted by all professionals. The lack of a coherent, accepted body of knowledge makes everyone's opinion as valid as a colleague's opinion.

Most researchers grow up in concept research using a 5-point purchase-intent scale, ranging from 1 = definitely not buy through 3 = might/might not buy to 5 = definitely would buy. The purchase-intent scale is only one of many evaluative scales that measure the interest of respondents to a concept. Another very popular scale is the 9-point hedonic scale, ranging from 1 = dislike extremely to 9 = like extremely. A number of these scales chosen from different researchers are listed in Table 4.1.

When choosing a rating scale, or even when thinking about the nature of a scale and its history, one should keep in mind the different issues that the researchers have faced

Table 4.1. Verbal descriptors for hedonic scales

Scale points	Descriptors
2	Dislike, unfamiliar
3	Acceptable, dislike, (not tried)
3	Like a lot, dislike, do not know
3	Well liked, indifferent, disliked (seldom if ever used)
5	Like very, like moderately, neutral, dislike moderately, dislike very
5	Very good, good, moderate, tolerate, dislike (never tried)
5	Very good, good, moderate, dislike, tolerate
5	Definitely buy, probably buy, might/might not buy, probably not buy, definitely not buy
9	Like extremely, like very much, like moderately, like slightly, neither like nor dislike, dislike slightly, dislike moderately, dislike very much, dislike extremely
9	FACT scale (Schutz 1964): Eat every opportunity, eat very often, frequently eat, eat now and then, eat if available, don't like—eat on occasion, hardly ever eat, eat if no other choice, eat if forced

Adapted from Meiselman (1978).

and the reasons underlying the use of each scale. Some of these issues and rationales may seem very inconsequential today, but they may have been quite relevant years ago when they were offered. Some of these evaluative scales talk about purchase, others talk about interest, others talk about frequency, and so on. There are differences in the language and in the meaning. Some scales comprise an even number of scale points, whereas some comprise an odd number. The scales may differ in the balance of positive and negative scale points.

Purchase Intent versus Liking

Both of these scales are evaluative. *Evaluative* means the scale requires a respondent to assign an overall rating to the concept, denoting acceptance or rejection. The covert assumption is that a concept that achieves a higher evaluative rating is somehow better or has more promise in the marketplace. *Purchase intent* scales instruct respondents to put themselves into the position of someone who will buy the product. Ratings of *liking*, in contrast, ask the respondent to scale how much he likes the product, without any hint of purchase.

Quite often the purchase and the liking scales correlate highly, especially when there is no price information attached to the concept. Showing a respondent a concept without price and instructing the respondent to rate either liking or purchase intent yield the same pattern across many concepts. Concepts that score high on liking will score high on purchase intent. When one adds a price, either to the concept (e.g., offered at x) or attached to the rating scale (purchase intent at x), the differences between the liking and the purchase-intent ratings emerge. If the price is too high, a respondent may like the product very much but would never think of buying it. If the price is exceptionally low, the respondent may buy the product even without liking it very much.

Anchored versus Unanchored Scales

The assignment of anchors or verbal labels to scales is a favorite pastime among researchers. The in-going thought is that the respondent will find the scale easier when words are associated with it. Once the decision has been made to incorporate words and numbers, or even words alone, the search proceeds rapidly to the specific scale words. It is assumed that numbers are equally spaced, so that the psychological distance between 1 and 2, 2 and 3, and so on, is the same. This may or may not be true, but it typically doesn't lead to problems. What does lead to problems and fierce discussions is the *choice* of the particular words to accompany the scale points. Table 4.1 provides just a simple taste of the array of terms that researchers consider.

The scale terms are easy to disagree with. There is relatively little information on the psychological distance between scale words, leading any disgruntled researcher to focus his irritation and dismay on the scale itself. Consequently, it is not unusual to see violent disagreements among the parties regarding the words, even if, in the end, they all use the same scales. They do so because it is easier to use a common scale, but it is just as ego gratifying to put one's opinion into the fray.

Short versus Long Scales

Whether the scale should be long, comprising many points, or be short, comprising few points, never ceases to fascinate researchers. Many feel that they should modify the scale to fit a respondent's scaling behavior. Often the researchers find that the respondents modify their use of the scale as a function of the number of points. For instance, with most scales, respondents avoid using the end points, usually for fear of running out of numbers. This is the so-called *end effect*. Observing this end effect again and again inspires researchers to suggest reducing the number of available categories by truncating the scale at both the bottom and the top ends. Sometimes an enterprising researcher offers the converse suggestion: lengthen the scale by adding an extra point. This elongation will ensure that the more internal scale points will be used. As for truncating the low or high part of the scale to drive discrimination, there is simply no data regarding the wisdom or efficacy of this strategy.

Balanced versus Unbalanced Evaluative Scales

The 5-point purchase-intent scale and the 9-point hedonic scale are balanced. They both comprise an equal number of positive and negative scale values. Most concepts that people test tend to be positive or are at least thought so by the agencies and companies that proffer them for testing. Thus, the ratings for these concepts cluster in the positive area. Occasionally, a respondent feels that a concept is unacceptable. The converse situation also occasionally arises, but more rarely, wherein most of the concepts are simply poor and down-rated. An enterprising researcher, observing this state of affairs, feels that it would be better to allow the respondents more scale points for the positive side and correspondingly fewer scale points for the negative side. The reasoning makes sense from an intuitive level. Typically, there are more positive-scoring concepts than negative-scoring concepts. If a researcher wants to discriminate between the concepts, then shouldn't the scale comprise more positive categories than negative categories? In that way it is possible for the researcher to identify small differences between concepts, since there are many more positive categories to use, allowing for better discrimination.

Unfortunately, this accommodating approach, although well meaning and designed to increase discrimination, ultimately fails. Once the scale is unbalanced, each researcher feels free to create an individualized form of an unbalanced scale. Eventually, the scales are jury rigged to accommodate the particularities of each experiment. Rather than using a fixed scale, comparable from one study to another, allowing a normative database to be developed, the ever-vigilant and motivated researcher will change the scale, depending on the demands of the marketing client and the business situation. Such relativity of scale value, with size of scale modified to meet current scientific and political needs, reduces the normative value and interpretive value of the scale. The actual scale changes from end use to end use, preventing anyone from becoming a true expert in the use of the scale.

Direction of Scale Points

In which direction should the scale go? That is, if the scale comprises 5 points, should the most acceptable scale point be *5* or *1*? This is really a minor issue, but it does continue to fascinate researchers, attracting them and their arguments like moths to a flame. There is no really persuasive recommendation for direction of the scale. Some researchers feel that the value 1 should be reserved for the best concept, because 1 occupies a special place, at the start of the rating system. To be number 1 is to be special. The senior author prefers the opposite: to use the highest number on the scale for the most acceptable concept, and the lowest number for the least acceptable concept. Other options include a bipolar scale, in which negative numbers denote dislike (e.g., -2, -1), 0 denotes neutral, and $+1$ and $+2$ denote liking.

Descriptive Scales

One uses other scales beside evaluative ones. Many of these scales are of secondary import; that is, they are not really used for making a marketing decision about whether to proceed with or to improve a product. Rather, these descriptive scales are often used as a check on the communication of the concept. The scales are not typically analyzed particularly deeply, because rarely is a management decision predicated on the scores of these descriptive scales.

Can Consumers Really Profile Their Expectations of a Concept as If It Were a Product?

Although much concept research focuses on performance (e.g., purchase intent), often the researcher and the product developer try to determine what sensory aspects a concept promises. For instance, if the ultimate marketing goal is to provide a product that is sophisticated, then does the concept promise a sophisticated taste? More to the point, if the marketing goal is to provide a spicy flavor, then is the concept consistent with that spicy flavor? Based on the concept description, do consumers have an expectation of the sensory characteristics of a product? This agreement between concept and product becomes important for product development for two reasons:

1. *Concept-strategy fit.* If there is an ultimate marketing strategy, then does the concept generate a sensory expectation that is consistent with the strategy?

2. *Development guidance.* After a concept is selected for development work and prototypes are created to match the concept, can the researcher validly ask the respondent to rate whether the prototype delivers more of what the consumer expects based on the concept, delivers what the concept promises, or delivers less than what the concept promises. This type of directional testing is done very often in development laboratories. It assumes that the respondent has a sensory expectation of the product, based on reading the concept.

Respondents can profile a concept, just like a product, on the just about right (JAR) scale. This scale requires respondents to rate a stimulus as having too much versus too little of an attribute. The JAR scale is com-

monly used for food products, but not commonly used for concepts; that is, one might ask the respondent to "rate the degree to which this concept communicates flavor: 1 = far too weak a flavor . . . 5 = just right . . . 9 = far too strong a flavor." If the researcher is successful in using a directional scale in concept work, then it becomes possible to diagnose problems with the way the concept communicates its messages.

One adaptation of the JAR scale presents respondents with systematically varied products and asks respondents to rate directionality against the concept, based on reading the concept. Rather than asking respondents to rate the amount of a sensory characteristic in an absolute sensory way, researchers can instruct the respondents to rate the delivery of the product on the characteristic, with respect to the concept. If a respondent has a sensory expectation, then the data should make sense. As the physical intensity increases, the product should go from *underdelivery* (at low sensory intensities) to *just about right* to *high sensory level*.

Although it seems a reasonable task to treat concepts as products and ask respondents to profile the concept as a product, there is little information in the literature on what to expect. This paucity of information may result from preconceptions about research methods that hinder exploration of new approaches. Following are two actions that a researcher might take to integrate products and concepts:

1. The researcher should test multiple concepts on these directional scales to ensure that respondents really can discriminate concepts by using the scale. If the ratings for all of the concepts look similar on the scales, this would suggest that the respondent does not understand how to use the directional scale to evaluate concepts. This failure to understand suggests that the directional scale is probably not meaningful in the particular context.

2. The researcher might wish to test actual products and concepts together by using

the scale, so he or she can compare concepts and products. Do the concepts and products show similar types of directional results, or do the concepts always score *just right*, whereas the products overdeliver or underdeliver? This approach appears not to have been done, but may provide a fertile area for research in both methodology and in actual products and concepts.

Open Ends

Concept research goes beyond measurement scales. Frequently, and most especially in both early-stage focus groups and in written questionnaires, researchers ask respondents to explain verbally what the good points are about the concept and what the bad points are. These are *open ends*. Respondents have no problem identifying problems that they feel inhere in a concept and also have no problem identifying what they like about a concept. Just as the case with actual foods, however, the open ends must be taken with a grain of salt. Consumers do know what they like, but may not be sensitive to nuances in the concept beyond the specific message; that is, respondents may be subtly influenced by factors extraneous to the message, such as the type of language, graphics, and brand. Some of the more astute respondents, more verbally articulate, might report that they are swayed positively or negatively by subtle concept elements. For the most part, however, respondents pay little conscious attention to many aspects of a concept and instead describe only the most immediate concept features; namely, what is directly communicated. Fortunately, in-depth probing by a skilled researcher can force a respondent to peel away the different layers of impressions, so that the researcher can report, in the respondent's own words, the real driving force. This ability to reveal more than the superficial is a talent that must be cultivated by a researcher. An untrained researcher usually cannot elicit this type of information, except perhaps in a moment of inspired, untrained, serendipity.

Testing One Concept versus Many Concepts: Which Strategy Is Better?

In research there are two viewpoints regarding evaluating concepts (or products or packages). One viewpoint holds that it is vital that a single respondent evaluate a single concept, because that situation is better or more validly represents what happens in nature. According to this opinion a respondent simply would not be exposed to many executions of the same idea in the ordinary environment. This opinion holds that it is better to replicate the ordinary environment. This viewpoint also applies to product work, where the measurement takes place only on one product. If another product enters into the test, then the researcher needs to have another group of respondents.

Some researchers with a different worldview aver that the best data are obtained when a respondent rates many different concepts (or products), each separately, but in the same session. This test design is called the *sequential monadic method*, which tends to produce more discriminating data. Indeed, the first sample or concept shows the least discrimination; that is, if we look only at the concept tried first and compare the ratings of different concepts when each was in the *tried first* position, we discover relatively few differences across different concepts. On the other hand, when we look at the same set of concepts in the *tried second* position and in subsequent test positions, we discover more concept-to-concept differences.

The question of one concept versus many is not purely academic. It affects the cost and the sensitivity of the research. By stringently requiring a researcher to test each concept alone, the researcher requires more respondents for two reasons:

1. To achieve the same base size of N respondents, with M concepts, requires $M \times N$

respondents. In contrast, if each of the N respondents tests all of the M concepts, then the research requires only N respondents, not $M \times N$ respondents.

2. If, in fact, all concepts are in the tried-first position, and if this position shows the least sensitivity across concepts, then the researcher will have to use even more respondents. The additional number of respondents will compensate for the low discrimination of the tried-first position.

Can Children Evaluate Concepts?

The majority of concept research is done with adults, some less with teens, and a great deal less with children. The covert assumption in marketing research circles is that the appropriate audience for concept research is a woman who is the head of household. Other groups, such as men, may be relevant for specific male-relevant products. Almost never, at least historically, have researchers done extensive concept testing with children, except for products specifically designed for them, such as toys. Children are occasionally respondents in concept tests, especially among the more advanced manufacturers.

Many companies that commission concept, product, or package research studies among children try to create test situations that they believe will be appropriate for children. These adjustments to the regular procedure may include nonverbal rating scales and watered-down concepts. In actuality, daily life belies this concern with a child's intellectual capacity. Children who can read are able to function in society. Certainly, older children, ages 9 and above, can function quite well in school, where they are responsible for completing tasks that are far more demanding than simple concept research. That children can validly evaluate concepts appears, therefore, reasonable. Perhaps children are not as articulate as adults, but we know from observing children that they can read, and they do like or dislike

products and things that they read. Therefore, it should be fairly straightforward to work with them in concept evaluation tests. The big problem is not the children, but rather the researchers, who approach the issue of child research with outdated biases. Chapter 11 deals with concept research and children.

Field (Test Execution) Issues

The heart of any research project lies in its field execution. The soul of the project is the design. Design and execution work together; neither can stand separately. Although many feel that almost anyone can design and execute a concept study, this belief is simply not true. Certainly, it is possible to field a concept study, and many non-researchers field perfectly fine concept studies at the very simplest level. However, good fielding and study execution is an art that should be mastered by anyone going into the profession and who wants to obtain data that can be relied on for knowledge and informed decisions.

Poor fielding of a study may not invalidate the results, but it can cast suspicion and cause damage in several ways. Good fieldwork can, in contrast, increase the credibility of the research.

Specific Observations and Issues
Good Fieldwork Promotes Face Validity

Poor fielding can destroy a client's belief in the research, which means that the data will not necessarily be acted on, no matter how valid and powerful it may be. Since concept research is generally done to address a business problem, and since there is no conventional archival, scientific literature backing up concept findings, the research must be executed without generating suspicion. No research is ever flawless, so problems that arise must be addressed in a professional manner for the data to be believed. There is no body of basic scientific evidence about concept research to which researchers can turn to demonstrate validity. Perception is reality, because concept research is an art rather than a science. It is also worth noting the inverse. If the fieldwork is executed perfectly, and the study appears to run without a hitch, then this high-quality execution lends face validity to the study, even if the results are, in actuality, not valid. Again, perception is reality, or at least perception dictates the reality.

Choreograph the Interview to Be Feasible and Not to Be Onerous to Respondents

Over the past decades, researchers have been pressed to learn more and more from consumers, using one interview to generate this learning. Time and cost pressures militate against the relaxed approach that was common 30–40 years ago. Researchers are asked, cajoled, begged, and occasionally ordered to incorporate all sorts of additional, "necessary" questions in the interview. For concept research, the interview must be focused to achieve the best data. The concepts cannot simply be thrown in as part of a much larger interview.

Define a Specific Time or Task for the Concept Evaluation

If a respondent knows that the interview will last 30 minutes, then the respondent will begin to pace nervously as the time approaches. The data at the end of the session will not be as good as the data at the start. It is better, in this case, to tell the respondents that the interview will be longer, so the respondent expects to test more concepts than are actually presented. Ending before the respondent finds the task too long is always to be preferred to the opposite, which is ending the interview afterward with an unhappy, irritated respondent.

Limit the Number of Concepts Tested to Fit the Interview Timing and Structure

The more concepts a respondent evaluates, within reason, the better the data will be. This is especially the case for conjoint measurement, where the objective is to create an individual utility model based on responses to a large number of concepts. Yet, if the respondents must evaluate many concepts, make sure the concepts differ from one another so that the respondents do not feel that they are rating the same concept again, and again, with only small changes. Minimize the number of attributes on which the concept is rated, especially if the respondents rate many concepts. Try to balance the better data that will be generated from more concepts from the respondent with the loss of motivation resulting from an overly long, potentially boring interview.

Do Not Lie to Respondents About the Length of the Task

Many interviewers feel that, in order to fill their quota of interviews, they must represent to the respondents that the interview time will be relatively short. Clients also like to hear that they will get everything they want in this short interview. Nothing so irks a respondent as being told that the interview will last a few minutes, only to have it drag on as the interviewer presents another concept. For longer interviews, prerecruit respondents to participate. This will cost most up front because of the recruiting costs and require that the respondents be paid. In the end, however, it will pay out in better data.

Limit the Number of Attribute Scales

Along with the desire to extract the information necessary for modeling, there is the perennial tendency for researchers to burden respondents with an onerous number of rating scales. It is tempting to have respondents rate the same concept on many scales, often dealing with the same evaluative issue, such as interest, purchase probability, and other ways of saying the same thing—"I like" or "I dislike." From the senior author's viewpoint (H.R.M.), by the time a respondent has profiled the concept on three or more scales, the respondent has started to pay less attention. This situation is not particularly damaging if the respondent evaluates one or two concepts. However, when the respondent participates in a conjoint study with dozens of concepts, it is prudent to limit the respondent's scale to one, two, or three attributes only. Beyond the first few attributes the respondent's attention wanders and the data are not particularly good. The respondent stops paying attention to the rating scales.

Make the Rating Scales Meaningful

Some scales, such as purchase intent, are intuitively meaningful. Other scales may be less meaningful, such as the number of times that a person feels he or she will purchase the product or use it within the next year or month. Quite often a researcher uses these frequency scales as input for an estimate of total volume. Perhaps the smart respondent knows the frequency of purchase. Yet, for the most part, people can only guess. That estimate is at best really a guess. It might be more productive for the researcher to treat the frequency question as a rough indication rather than as an accurate measure. It is tempting, however, for the researcher to obtain the frequency estimate and plug that estimate into a model to estimate shipments or share.

Vary the Stimuli in a Single Interview or Accept That the Ratings for Similar Stimuli May Be Biased Because of Boredom or Overattention

In concept research, whether with full concepts or promises, there are many conflicting objectives. Sometimes the research deals with the evaluation of a wide range of ideas. This range does not bore respondents. Rather, it is interesting. Other times, and especially in categories where there are legal issues, the promises and concepts are more "close in" and resemble each other. To those who create the concept and who focus on its nuances, these concepts are as different as night and day. To respondents, however, the concepts are the same. Testing these similar concepts one after another is a guarantee to bore respondents. The same problem occurs when product developers modify a product only slightly in a few different ways. They may feel that the product has been significantly changed, but respondents often complain that there is some trick going on because the products are so similar. Whether with similar concepts or similar products, respondents often try to please the interviewer. They may deliberately accentuate differences that they perceive, magnify the differences in ratings among the concepts, and by so doing feel that they have justified their participation in the study. What happens, in turn, is that the stimuli are rated as very different by respondents who are trying to evidence some discrimination. The ratings belie that the concepts are really far more similar.

Ensure Respondent Comprehension Through a Short, Warm-up Exercise

Good fieldwork requires that the respondents understand what is expected of them. How do researchers ensure in concept work that the respondents actually understand the concept and understand the rating question and the scale? The issue of comprehension is less of a problem with food itself. An interviewer can take a respondent through a practice task in which the respondent eats the food and rates the orientation food on a variety of characteristics. A sensory or other product researcher expects that the respondent may not understand how to open the food container, how to use the food, what attributes to attend to, what the scale means, and so forth. This care and concern about respondent comprehension is, unfortunately, missing in concept research. The absence of this practice exercise can be easily remedied by having the respondent rate a practice concept ahead of the other concepts. The explanation of attributes, and the like, is more problematic. Researchers are, in fact, afraid, and perhaps rightly so, of biasing the respondent by pointing out the specific attributes and their meaning on the questionnaire. They fear that they will sensitize the respondent to those characteristics in the concept itself. Up to now (2004) this issue of practice concepts in a concept test has not been addressed or, if it has, the decision has not been to go with a training concept for the reasons of bias.

Ensure Valid Interviews and Avoid the So-called Kitchen-table Research Syndrome

Most researchers go into the business with honest intentions. Often, however, and for reasons beyond the control of the field, things go wrong in execution. Occasionally, less than honest behavior occurs. The recruiting of respondents may become more lax. This is not particularly bad if the client approves the reduction in recruiting stringency. What is really, problematic, though, is the rare but troubling instance of field research being entirely made up. This is called colloquially *kitchen-table research*, because the interviews are fabricated. In Germany, Elizabeth Noelle-Neuman mentions one famous case. Almost 20 years ago the unemployed Heiner Forroch and his family survived because he worked for seven institutes and faked interviews systematically (Noelle-Neuman and Petersen 1998). Researchers must be ever on the guard. The owner and manager of the field service may be absolutely honest, try his or her best, and deliver superb results. The actual interviews, though, are in the hands of

people paid minimum wage who know that they will not get paid unless they fill the quota of interviews. To this end they may occasionally fabricate their interviews in order to make their quota. Market researchers have long recognized this unethical behavior. Companies have been created based on the researcher's need to check whether the respondents have actually done what they say they have. Since the interviewer has to provide the respondent's telephone number, a third party can *validate* the interview by calling the respondent to determine whether the respondent actually participated in the interview at the time and place stated. Other organizations can track participation in multiple interviews by the respondent's telephone number. When respondents are recruited, the instructions are to obtain respondents who have not participated for some previous period, such as 3 months. Often, however, the same respondents participate in study after study, and focus group after focus group. Some people are even proud of making their living this way. The validation services often ferret out these repeat offenders.

Conduct Regular Exit Interviews to Uncover Any Problems

Sometimes, researchers are loath to discover what is wrong with their interviews. Many researchers have a blind spot to their own research. Exit interviews with respondents are standard operating practice among many large-scale research buyers such as Kraft Foods. These interviews help the researcher to improve the interview experience. They do not guarantee perfectly valid data, nor do they ensure better-quality results if the study design is poor. The interviews identify areas to improve in the interview process itself. That improvement should lead to happier, more involved respondents, which in turn might well lead to better-quality data. Be careful, however, not to rely on a respondent's emotional reaction to the interview.

Often respondents will say that they became confused during the interview and could not remember their answers. Respondents will volunteer that they feel their interviews are invalid. This mea culpa syndrome itself needs to be ignored. Some of the senior author's best data has come from respondents who, in a conjoint task, did very well but, in their personal opinion, thought they were answering randomly. The respondents should not be the sole judges of their own data quality. They should be listened to only with respect to execution problems in the interview, such as that the interview seemed to last twice as long as they were promised it would last.

Criteria for Selecting Respondents

Concept research is conventionally planned and executed by the marketing research staff rather than by research and development (R&D). Marketing researchers have been educated in the sociological tradition or, if not formally educated, trained on the job. Early market researchers laid down some of the *rules* or at least mores of concept testing in the 1930s and 1940s. These rules required that the sample represent the population to whom the product is targeted, that there be sufficient respondents for statistical tests, and that the respondents be chosen in different markets to represent the geographic spread (at least in the United States).

Representative Sample

The sample of respondents must match the ultimate consumer group, at least to a reasonable degree. This sounds perfectly reasonable, and it is. The requirement is drilled into every novice's head. Market researchers would not think of working with a sample that does not reasonably represent the target population, except in those instances where, for convenience, cost, or a dozen other rationales, they select a *general population*. Cer-

tainly, however, market researchers typically avoid using a nonrepresentative sample. Attempts to create a representative sample can range from simply screening the relevant group of product users to screening with specific quotas and nesting. *Nesting* refers to one quota within another (e.g., 60% users of product A, 40% users of product B as the major quota and, within those two groups, 66% males, 34% female as the nested quotas).

Appropriate Base Size

Market researchers deal with issues of testing proportions rather than testing means. Their conclusions deal more with the number or proportion of respondents who state a certain attitude than with dealing with the depth of feeling about that attitude. Market researchers use incidence statistics. Any person is simply a *0* if the person does not exhibit the behavior or a *1* if the person exhibits the behavior. There is relatively little information in the data from any single individual. Therefore, to understand the population means not to understand a single person but rather to understand the distribution of these simple responses. *A great deal of quantitative research in concept testing focuses on the number of respondents who feel positively about the concept, not on the nature of the feelings of an individual.* Quantitative research requires base sizes that will generate stable results. Such bases sizes begin at 100 individuals (a magic number in research if there ever were one) and increase to many hundreds. Having thousands of respondents in research is also unusual, owing to the researcher's perception that the appropriate number of respondents should be affordable. Research does not observe responses from large numbers of people, like sociology does, but rather experiments with hundreds of people. For the nonquantitative, focus group or in-depth group, however, smaller base sizes of 6–12 are acceptable, because the investigation looks for richer information from the person rather than just a vote.

Appropriate Distribution of the Respondents in the Markets and by Gender, Age, and So Forth

In the United States there is the ever-present pressure to test across the country, or at least in a number of markets. This pattern of testing in multiple markets occurs in the research conducted worldwide, especially in the large countries (large in population; large in area). The reason given is that the distributed evaluations across the different markets ensure that no regional biases occur. In reality the biases due to region, gender, and age have, with the exception of voter alignments and cleavage structures, not really been well established (Lipset and Rokkan 1967); that is, most research with which the senior author is familiar fails to show consistent market bias, except perhaps in those foods that are native to a region's cuisine (e.g., Cajun food in New Orleans, or hot foods in the South). For the most part, when a researcher pulls out the results from the different markets there are few clear differences. If a difference occurs, it is usually hard to understand the reason under-lying the unusual performance of that concept in that market. A good example of the lack of differences appears in Figures 4.2 and 4.3, respectively. The data in both figures come from the Crave It! study, discussed in Chapter 20. Each point corresponds to 1 of 20 foods and 1 of 9 elements dealing with the description of the product. There are thus 180 points in Figure 4.2, each one being the utility of a single concept element. The results suggest no difference in utility values between men and women. Figure 4.3 shows the same analysis, this time by age. The respondents were divided into five age groups, with the oldest age of that group shown on the graph (e.g., the group called Topage60 comprises respondents ages 51–60). There is no clear difference in the utility value by age either. What one age likes, the other age tends to like. In general, if there are differences in groups, then the differences are probably more profound than effects that can be traced to the simpler geodemographics such as market, age, or gender. The differences tend to be in terms of concept-response segments.

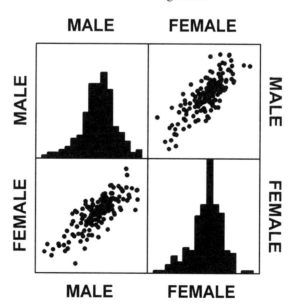

Figure 4.2. Utility values for nine elements for each of 20 foods, by male and female respondents. Data courtesy of It! Ventures, Inc.

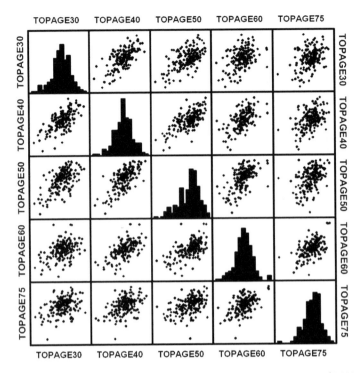

Figure 4.3. Utility values for the 20 foods, nine elements each, by age. Data courtesy of It! Ventures, Inc.

Can There Ever Be Random Samples?

In consumer research the rare-achieved gold standard is a fully random probability sample, where the sample matches the target population. For concept work, and even more for product work, this gold standard is simply not possible to achieve, for a variety of reasons. One reason is that the concept research is done very early in the process. In early-stage research the objective is to get a sample that can provide the necessary guidance, not to measure the response of the final universe of consumers. Second, the cost is very high for a fully probability sample. The research is just not that important. Third, the effort is too great.

Execution Venues and Their Issues

In the execution of any research there are always problems to be reckoned with that en-dure because of the nature of the research medium. Each of these media—whether mail, telephone, person to person/door to door, mall intercept (central location), computer-aided personal interview (CAPI), or Inter-net—brings with it a host of problems or, stated less alarmingly, a host of issues to be recognized, addressed, and hopefully solved.

Mail

Mail interviewing is a standard approach for concept research. The respondent receives a booklet of concepts, or even one concept, in the mail, evaluates the concepts in the order presented by the booklet, and returns the an-swers, either by mail or through a follow-up telephone interview. With an incentive to complete the interview often enclosed in the package (it's cheaper that way), the mail in-terview is cost effective and study is scala-ble. For example, it is easy to do 5000 interviews because the researcher need only send out the material to 5000 different people who qualify based on the criteria. There are list brokers who, for a small fee, can provide mailing addresses and even labels for the target respondents.

The key issues for mail interviewing are that the client must produce many copies of the test book, and it may take the researcher a month or so to do the study. The respondent data can be tabulated and presented only at the end of the fieldwork, after the results are keypunched. Furthermore, there is no one to guarantee that all of the respondents are whom the they say they are, nor is there any way to monitor that all actually evaluate the concept rather than just randomly recording ratings on the answer form. Often the researcher sends out a full book of concepts. Sometimes the book is simply a foil to disguise the real single concept inside. Sometimes the research company creates efficiencies by having multiple clients test their concepts within the same book so that the cost of doing the study is reduced for any single client (so-called *omnibus testing*). Most of the research cost is in the production and mailing, with great economies of scale to be realized with a dozen or more concepts in single, mailed booklet instead of one concept.

Telephone

Telephone interviewing for concepts is not particularly appropriate when the concept comprises visuals or when the concept is long. Benefits screening testing can be conducted by telephone because the benefits comprise simple phrases. In general, the telephone is not used for concept research because the venue does not allow for visual stimuli and for a relaxed setting. Telephone interviews are interruptive; most people today find telemarketers and interviewers annoying, and there is a growing refusal rate. This refusal rate—the proportion of contacts declining to participate in the interview—

will become even more critical because of the developing privacy laws and the *do-not-call legislation*. People don't want to be bothered during their free hours with an interview and so, quite often, they simply hang up. Occasionally, they do so rudely. Today, many consumers are either unreachable or unwilling to participate in a survey, with nearly 30% of all US households maintaining unlisted telephone numbers (www.world opinion.com/the_frame/2001/nov_2.html). Telephone interviewing using concepts can be effective when the response to a short, spoken concept is used as a *screening device*, but in this case we are not dealing with concept research, per se, but rather with the concept as a mechanism to identify appropriate respondents.

Door to Door

In the United States, door-to-door interviewing was once the dominant form of research. This was especially true for those markets where the interviewers could proceed safely in a neighborhood. Many sampling specifications for national probability samples were written with door-to-door interviewing in mind. The interviewer brings the concepts, shows them to the respondent, and gets the answers. The interview format permits more detailed discussion of the concept and a richer interchange with the respondent. In countries outside the United States, such as in Latin America, door-to-door interviewing is still widely done and may constitute the dominant form of research. In the United States and in Europe, door-to-door, personal interviewing is relatively quite expensive. Door-to-door interviewing is not easily scalable because of the cost of data production. For every respondent there must be a corresponding interviewer. Consequently, door-to-door research is less frequent where cost is an issue. Door-to-door research remains popular in countries and cultures where face-to-face interviewing is socially desirable.

Mall Intercept (Central Location)

Central location interviewing, typically inside a mall or in a trailer parked outside the mall, continues to be popular. Beginning in the 1960s with the spread of shopping malls comprising stores, cinemas, and other entertainment venues, as malls turned into a destination location for shoppers, central location research has evolved into a staple in the market researcher's arsenal of research venues. An interviewer is trained regarding how to conduct the interview, stands at the mall in a prominent location, and invites passersby to participate. To ensure that the respondents at least appear to satisfy the criterion, the interviewer can visually screen them before offering the invitation. Like the door-to-door approach, the central location interview cannot be scaled. It requires that a single interviewer deal with a single respondent. Production efficiencies are obtained by distributing the interview task to many interviewers at one mall and across many malls. The problems endemic to mall intercept interviewing are the possible covert signals passed by the interviewer to the respondent, the low response rate (people are too busy), and the lack of respondent motivation since the respondent is at the mall to accomplish a specific objective. The interview is a distraction for the respondent. Advantage? The respondent at the point of sale is definitely a target group!

Computer-assisted Personal Interviews

A growing number of researchers are migrating toward computer-assisted personal interviews (CAPIs). The technology is available for better CAPI, and the software ensures both proper presentation of stimuli and automatic data collection in an errorless format. There are really very few issues with CAPI, other than the tendency for researchers to pack more questions into the CAPI format than would be prudent or would be the case for personal interviews. In a personal interview, the researcher has to state the question. This requirement cuts down on the amount of information that can be obtained in the interview because there is the inevitable back-and-forth time to present the question and get the answer. In CAPI the computer flashes the question on the screen, and the respondent quickly replies. The absolute simplicity of this stimulus-response format suffices to encourage the researcher to obtain more information, since there is no time wasted on the recitation of the question.

Internet Interview

Internet-based interviewing is becoming increasingly popular for concept work. Concept research is a natural fit with the Internet because the Internet provides both text and graphics. Furthermore, the ability to have a self-administered test executed at very low cost among a large number of respondents makes concept tests of all types very attractive as a business proposition. Internet-based conjoint analysis and new adaptive methods have the potential to increase the effectiveness and reduce the cost of the marketing input to concept selection and refinement. Dahan and Hauser (2002) have written that they expect the costs to drop by a factor of 10 and the time to completion of the market input to drop from 6 weeks to a few days. These developments will further enhance the ability of the new products team to design and engineer concepts. To this end, many companies such as Procter and Gamble and General Mills are migrating toward the Internet. There are, of course, still reservations about using the Internet, because the respondent population may be different, the respondent must have an Internet account with a service provider, the respondent must feel comfortable working with the Internet, and so on. To a great degree the Internet issues are the same that are faced by CAPIs (computer-aided personal interviews). With the introduction of any new survey mode, initial suspicions regarding the efficacy of novel and groundbreaking techniques are fairly commonplace. During the late 1960s, the same held true for telephone surveys as they gradually supplanted traditional face-to-face interviewing. Since then, however, research on telephone surveying has supported the claim that this mode is an effective means to collect accurate and meaningful data (Dillman 1978, 1999). Fortunately, over time there is an increasing acceptance of the Internet research as a valid venue for interviewing. However, some recurring problems with the Internet are worth mentioning:

Interview Length

The first problem is the issue of length. As MacElroy (2000) has reported, longer interviews on the Internet lead to more respondents dropping out. This should come as no surprise. In any lengthy interview, more people will become disinterested as the task progresses, unless the topic is so riveting as to be almost entertainment. The problem with the Internet, however, is greater because the Internet interview is conducted in the privacy of one's own home. Thus, a respondent who gets bored can drop out. A sense of the proportion of the respondents that drop out can be seen from Table 4.2. The data in Table 4.2 were provided by Open Venue for general interviews conducted on the Internet. These studies represent the typical type of survey, whether concept, attitude and usage, etc. The results reveal that one must send out a large number of e-mail invitations in order to have a reasonable sample. The response rate is less than 10% and often less than 5%.

Where Are the Men?

Gender representation is a recurrent problem in Internet interviews, with gender emerging as a cause célèbre. Based on the senior author's observations in several hundred concept studies, usually more women than men partic-

ipate (Beckley and Moskowitz 2002). Just because women participate far more frequently than do men does not mean that the data are invalid. When the topic is very interesting to a man, such as sports, he will participate.

The clearest evidence from food and beverage studies regarding interest compared with participation comes from the Drink It! mega-study (Moskowitz unpublished data). When it comes to red wine, men comprise 32% of the sample. When it comes to flavored alcohol products, men are hardly in sight, comprising 8% of the respondents for coolers and 7% of the respondents for flavored, low-alcohol drinks. It is clear from these frequencies of participation that men choose the study in which they want to participate and that, although they may constitute a minority, they can comprise up to 30% or more of the respondent population. The trick is to discover what men find interesting and oversample them for those studies rather than expecting a 50% completion by men in a study that they find boring.

A sense of the differences in representation between men and women can be obtained from the data in Table 4.3 showing the log-ins, completes, and the proportion of men for the 2002 Crave It! study, and in Table 4.4 showing the same statistics for the 2002 Healthy You! study. Both studies used Internet-based interviewing. The respondent was sent an e-mail invitation from an e-mail house. The respondent who answered was taken to a wall where he could choose a study in which to participate. The interview was also administered on the Internet. As Table 4.3 shows, the proportion of men who participate in a study varies by the nature of the study. Food topics of interest to men, such as BBQ ribs, steak, and hamburger, attract a higher proportion of men than do food topics such as cinnamon rolls, fresh fruit, salad, and chocolate candy. Table 4.4 shows the same pattern, with far fewer males participating for health-oriented products. The underlying reason for the low showing

Table 4.2. Invitations versus completes for a representative array of studies

Study topic	Length of survey, minutes	Gender weighting of invites	Gender weighting of completes	No. of invites	No. of completes	Notes
Baked goods	10	50% men 50% women	26.9% men 73.1% women	50,000	3250	18–55
Personal care	10	90% men 10% women	51% men 49% women	27,025	500	Specific age and gender quotas including teenagers
Soft drinks	5	80% men 20% women	60% men 40% women	25,000	1285	50% under 18
Computer software	25	NA	NA	15,000	172	Extremely low incidence rate
Radio-station listeners	5	NA	NA	4000	150	Specific counties in New York
Shoes	10	50% men 50% women	29% men 71% women	7500	400	Wanted 70% women, 30% men Sample income > $75k
Banking	15–20	NA	NA	20,000	895	No gender quotas, income > $25k
Homeowners	30	NA	NA	14,000	200	Specific states, income > $65k
Postal services	20	NA	NA	12,000	1248	No selects

Source: Open Venue Ltd. Courtesy of Chris Keeling.

Table 4.3. Number of log-ins, percentage of completes, and percentage of men completing each of the Internet-based Crave It! studies

Database = Crave It! 2002	Log-ins	% Completes	% Men
BBQ ribs	409	60	41
Steak	386	62	40
Hamburger	358	68	36
Hot dogs	452	53	36
Nuts	366	66	34
Meat loaf	450	53	32
Ice cream	489	66	31
Pizza	494	65	31
Bacon	387	62	30
Chicken	370	65	30
Cola	460	59	30
Shellfish	365	68	28
Coffee	397	69	27
French fries	398	60	27
Gravy	417	57	26
Donuts	442	54	25
Snack mix	404	59	24
Tacos	412	58	24
Tortilla chips	409	58	24
Cheese	436	55	22
Iced tea	400	61	21
Potato chips	438	55	21
Chocolate chips and cookies	501	50	20
Bread	393	61	19
Popcorn	366	66	19
Pretzels	376	64	19
Cheesecake	384	66	17
Mashed potatoes	365	66	17
Cinnamon rolls	428	56	16
Fresh fruit	426	60	16
Chocolate candy	516	64	14
Salad	362	66	13

Adapted from Beckley and Moskowitz (2002). Data courtesy of It! Ventures, Inc.

of males is probably the basic interest in the food products. The Healthy You! products are not meat proteins.

The Role of Incentives

Marketing research and, for that matter, sensory evaluation have generally tried to run themselves as a business by minimizing variable costs. One of these variable costs is the *co-op* payment to respondents that remunerates them for their participation. In the past, respondents were happy to, or at least effectively cajoled to, donate their time for a

study, especially when they found the subject interesting. Today, however, paying respondents rather than getting free data has become the norm. The key issue is how the respondents are paid (cash, sweepstakes, points in loyalty programs) and how much.

No one questions the need to pay respondents a reasonable amount of money to participate in focus groups or in other supervised central location tests. Respondents are paid relatively handsomely to attend because researchers know that otherwise the respondents simply will not participate. The inconvenience is too great. The issue becomes

Table 4.4. Number of log-ins, of percentage of completes, and percentage men completing each of the Internet-based Healthy You! studies

Database = Healthy You!	Log-ins	% Completes	% Men
Nuts	250	56	29
Peanut butter	246	56	27
Frozen fish	242	55	27
Water	246	50	25
Milk	249	52	25
Crackers	249	51	25
Coffee	250	50	25
Cheese	244	48	25
Salsa	244	54	24
Energy or grain bars	243	57	23
Soup	242	51	22
Juice (citrus)	250	55	22
Canned beans	243	52	22
Milk-based shakes	240	48	21
Bread	251	47	20
Pretzels	241	52	19
Pasta	244	42	19
Cold breakfast cereal	247	47	19
Tea	250	55	18
Frozen meals	242	48	17
Soft nutrient chews	242	54	17
Margarine or spreads	248	56	16
Juice (noncitrus)	246	52	16
Vegetarian burgers	241	49	16
Pasta sauce	242	49	15
Chocolate	250	47	15
Yogurt	246	53	14
Flavored rice mixes	243	60	14
Canned/jarred fruit	241	56	13
Salad dressing	245	52	12

Adapted from Moskowitz (unpublished). Data courtesy of It! Ventures, Inc.

more heated and problematic for shorter interviews, such as those conducted by mail or at a mall. Some practitioners feel that paying respondents will destroy the financial base of the industry. Other practitioners feel that the increasing refusal rate in the field must be alleviated by the one incentive (i.e., money) that they know will work. Whether the incentive is a small, token amount such as a one-dollar bill inserted into a mail survey or a five-dollar bill or some other gift at the central location intercept is an empirical issue that can be answered by experimentation. It is clear, however, that the days of free research are nearing an end. Researchers must recognize that the information they obtain

from respondents cannot be obtained with an outlay of zero dollars.

Strategies to Incentivize Respondents in Internet Interviews

One of the key ongoing issues in Internet-based research is that there is no control over the interview situation. Thus, disinterested respondents may drop out. There is no interviewer to cajole the respondents to participate. This is one reason why there have been concerns about the length of the interview. As already noted, longer interviews tend to yield more dropouts (MacElroy 2000). One logical consequence of this behavior is that researchers are driven to create a shorter interview in order to minimize the dropouts and thus increase interview *yield*. Another logical consequence is that researchers must provide incentives to the respondents in the form of direct payment, opportunity to win a sweepstakes, or some type of information. The senior author has found that a sweepstakes for Internet research produces a relatively high response rate of 10%–20%. This is a low response rate for critical, hard-to-reach respondents, but for concept research with consumers the sweepstakes approach appears to work, at least today (2004). Whether sweepstakes eventually fail, or whether the completion rate continues to drop even in light of the sweepstakes, remains an empirical question that can be answered in the years to come.

Overview

Although it might seem daunting at first to determine the right questions, find the right respondents, and execute the study correctly, most research works out just fine in the end. Today's booming research industry worldwide, numbering almost $17 billion dollars according to the newsletter *Inside Research*, suggests that most practitioners are doing the right things and doing things right. The issues raised in this chapter are not meant to frighten the novice, nor to chide the practitioner, but rather to provide some hints and warnings about issues that inevitably plague all research efforts at one time or another.

References

Beckley, J., and Moskowitz, H.R. (2002). Databasing the Consumer Mind: The Crave It!, Drink It!, Buy It! & Healthy You! Databases. Anaheim, CA: Institute Of Food Technologists.

Dahan, E., and Hauser, J.R. (2002). Product development: managing a dispersed process. In: Weitz, B., and Wensley, R., editors. Handbook of Marketing. London: Sage.

Dillman, D.A. (1978). Mail and Telephone Surveys. New York: John Wiley and Sons.

Dillman, D.A. (1999). Mail and Internet Surveys: The Tailored Design Method, 2nd edition. New York: John Wiley and Sons.

Lipset, S.M., and Rokkan, S. (1967). Cleavage structures, party systems, and voter alignments: an introduction. In: Lipset, S.M., and Rokkan, S., editors. Party Systems and Voter Alignments. New York: Free Press.

MacElroy, B. (2000). Variables influencing dropout rates in Web-based surveys. Quirks Marketing Resarch Review; www.quirks.com, paper 0605.

Meiselman, H.L. (1978). Scales for measuring food preference. In: Petersen, M.S., and Johnson, A.H., editors. Encyclopedia of Food Science. Westport, CT: Avi, pp. 675–678.

Noelle-Neuman, E., and Petersen, T. (1998). Alle, nicht jeder: Einführung in die Methoden der Demoskopie, 2nd edition. Munich: Deutscher Taschenbuch.

Schutz, H.G. (1964). A food action rating scale for measuring food acceptance. Journal of Food Science, 30: 202–213.

Part II

Experimental Designs, Graphics, Segments, and Markets

Chapter 5

Systematic Variation of Concept Elements and the Conjoint-analysis Approach

Introduction

One of the increasingly popular methods for understanding respondent reactions to complex stimuli such as foods is known as *conjoint analysis* (Wittink and Cattin 1989). Conjoint analysis has become a popular tool in the marketing literature. A search on Google in late June 2002 revealed approximately 27,000 hits containing the words *conjoint analysis*. The number of hits increases monthly as researchers, marketers, and developers discover its power. Conjoint analysis has traditionally been used to identify the driving elements for rational concepts, such as product descriptions, where interest focuses on the rational features versus price. More recently, however, interest has focused on the use of conjoint analysis to understand emotional drivers, as well as rational drivers. Within the framework of a concept the respondents do not necessarily differentiate on an intuitive level among the features offered, the price required, and the emotional or end benefits of a more intangible, yet real, nature. It is the marketers, and to some extent the product developers, who make this differentiation and who have typically limited conjoint measurement to the arena of rational features.

History

Conjoint analysis refers to a class of research procedures that measure the contributory value of components to a mixture, based on measures of the mixture rather than on measures of the specific components. The mathe-matical foundations of conjoint measurement are presented in an historic article by Luce and Tukey (1964). For the purposes of this book, we need not delve into either the mathematical underpinnings of conjoint measurement or into the different variations of the approach, ranging from pairwise trade-off methods to full profile, and from nonmetric analysis to metric analysis. These deeper treatments of conjoint analysis appear in mathematical psychology texts and in books devoted to the topic (e.g., see Gustafsson et al. 2000). The user manual for Systat (1997), a well-known statistical analysis system deployed on personal computers, has a good introduction to the different methods and some demonstration exercises that one might perform with the Systat data system.

In the 1960s, the notion of conjoint analysis as a powerful, productive, and popular research procedure would have been hard to envision. If we retroject ourselves into the 1960s, we will see an intellectual ferment taking place, with mathematical methods becoming increasingly respected in psychology. Psychology had always used statistical procedures for data analysis, but it was only in the decade from the late 1950s to the late 1960s that researchers became interested in the foundations of measurement to represent subjective processes. Conjoint analysis, the study of components from responses to their mixtures, was only one of a variety of approaches of interest to psychologists. The focus of these psychologists was on grounding the psychological methods in truly valid, mathematically grounded procedures for measurement. The early papers on conjoint measurement were, therefore, highly mathematical, seeking to ground the approach in rigorous, axiom-based methods. Certainly, in their seminal article on conjoint analysis, neither Luce nor Tukey could have imagined the degree to which the approach would find its warm, welcoming home in applied marketing research some forty years later.

Conjoint analysis received its initial thrust into application through the efforts of the very well-known research group at Bell Laboratories in Murray Hill, New Jersey. Eminent psychometricians such as Doug Carroll, Joseph Kruskal, Roger Shepard, and their associates explored the use of these newly emerging psychometric methods, including conjoint analysis. At first, some of the applications were purely demonstrative, for example, to show that the analysis by new nonmetric methods reproduced data structures that were already known. The strategy of using easily understood and often old, well-known datasets, reanalyzing them by new techniques and showing new results, worked well with the conjoint-measurement procedure. The demonstrations revealed that researchers could easily recover the known, underlying additive structure in the data through the use of the analytic approaches. However, the demonstrations did not generate the demand for the technique, perhaps because in the end the user must be shown applications that directly affect his or her research.

To create demand for ideas, one often needs an encouraging push from those who live outside the theoretician's world and who are not necessarily as technically proficient as they are visionary. Practical applications of the method, especially those accompanied by financial reward, are generally far more effective in helping to spread a new technology. Thus, the history of conjoint measurement took a radical, far more impactful and productive turn in the early and the middle 1970s. The impetus for this development can be traced to Professor Paul Green and Professor Yoram (Jerry) Wind, both at the Wharton School of Business, University of Pennsylvania (Green and Wind 1973). Wharton had attracted a variety of scholars and research practitioners, many of whom maintained ties with Bell Laboratories and enjoyed the best of the business and the psychometrics worlds. These scholars, especially Green and Wind, quickly recognized that the simple notion of identifying the part-worth utilities of a mixture from responses to that mixture had implications far beyond simple theory. They could apply the approach to business issues. To the trained business mind sensitive to psychometrics, conjoint analysis represents a way to understand the algebra of a concept or, more specifically, how the components of a concept interacted to drive a rating such as purchase intent (Johnson 1974).

The notion of relating components to response is straightforward and immediately obvious to us today. For a better understanding we must put ourselves back in the late 1960s. The business world of marketing and the academic worlds of psychometrics were just starting to converse with each other. In the 1960s, to a lot of researchers and practitioners in business, quantitative methods meant understanding how to conduct tests of significant difference or, for the most advanced professionals, knowing how to use the newly popular methods of factor analysis to understand the respondent mind. The contributions of Green and Wind in popularizing conjoint measurement as a truly new tool with substantial business implications must be viewed against that background of excitement about the discovery of a new world of analytic power.

Trade-off Analysis

The initial applications of conjoint analyses used trade-off procedures. Respondents would be presented with two options and instructed to select which of the two they preferred. From the array of trade-offs made by the respondents, the researcher would then create a scale comprising a utility value for each stimulus such that the pattern of choices would be recaptured if the respondents were to have used those utility values to make their pairwise choices.

Trade-offs make historical sense when viewed against the background of psychometrics and marketing research, even if to us today they seem tortuous and agonizingly indirect. Experimental psychologists in the latter part of the 19th century assumed that human judges were incapable of acting as measuring instruments. With this presupposition, therefore, it was left to the ingenuity of the newly emerging psychological scientist to devise a method for measurement that did not use human judgment as a measurement of magnitude. Gustav Theodor Fechner (1860), the founder of modern psychophysics, averred that a better, and probably a more valid, way to measure sensory magnitude was by analyzing the behavior of the judge who tried to compare two stimuli, either to make a direct match or to say which of the two was more intense. One act of matching consisted of adjusting two stimuli so that their perceived magnitudes were equal. Fechner believed that one could measure the error of adjustment and subsequently, through analysis, one could then convert the observed error to a measure of underlying subjective magnitude. Fechner's strategy—measurement error in the matching of two stimuli—would need an algorithm that converted measurement error into underlying psychological units. Another method was to have each panelist identify which of two stimuli appeared more intense, was more acceptable, or in general had *more* of a specific aspect. The degree to which people differed in their choice (viz., a measure of error across people or a measure of an individual's error across stimuli or replications) was assumed to relate to subjective magnitude. Two stimuli that were very close together in subjective magnitude were thought to be confused more frequently; hence, the converse—the degree

of confusability was assumed to parallel the similarity in subjective magnitude.

This paired comparison approach is provided in Figure 5.1, which presents a screen showing two concepts for a soft drink. The concepts comprise both picture and descriptor terms. Each respondent selects one of the two and, in some cases, rates the magnitude of preference. From the selection patterns made to a set of such pairs, the researcher can erect a model showing the part-worth contribution of each of the elements. The methods for creating this type of utility value vary from processing the rank orders of the choice to regression modeling and are described in books devoted to conjoint analysis (e.g., see Gustafsson et al. 2000).

When dealing with trade-offs, it is important to keep in mind that researchers assume that the respondents are unable to act as a measuring instrument, per se, and thus cannot directly rate element utility. Rather, the respondents are assumed to be able to judge pairs of products and to make a selection. In the history of experimental psychology this approach is called the *indirect method of subjective measurement*. The *indirect* refers to the fact that the respondents never really assess utility, per se. Rather, the utility is deduced from the choice patterns. The choice patterns need not be perfect. The indirect method recognizes that the choices of people are error prone. To the degree that the choice behavior is *noisy* and full of inconsistencies, the utility values of the different elements will be close together. To the degree that the choice behavior is *not noisy* and the patterns are cleaner, the utility values will be farther apart and show a clear hierarchy.

It is worth noting that the indirect methods, exemplified by paired comparison and subsequent processing, continue to be well accepted by market researchers. Often, when one hears the term *conjoint analysis*, one also hears the accompanying clarification that the method is a *trade-off*. A lot of people accept the trade-off approach. Indirect methods are

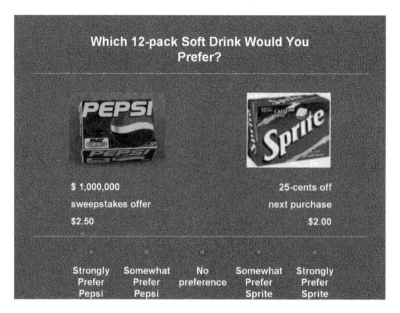

Figure 5.1. Example of a choice study in conjoint measurement where the respondents choose between two stimuli. Utilities are constructed from the pattern of choices.

often justified by the researcher's conviction that respondents really don't know the utility values of the components in concepts. Thus, in light of this argument, the researcher moves inexorably toward the use of trade-off methods, no matter how tortuous and cumbersome they may be, and no matter how involved the subsequent analytics may become.

The history of product testing followed the same path. Over the years, researchers have used the method of paired comparisons for product testing, even though the data require extensive post hoc statistical manipulation to reduce the array of paired comparisons to a single scale. The concentration of efforts on the statistical machinery to process paired comparisons left little time for researchers to do more with the utilities than estimate them and perhaps put them in a model.

Full-profile Conjoin

An alternative way to deal with the problem of trading off requires that respondents be presented with a set of combinations already created by experimental design (Box et al. 1978). The experimental design combines these components to generate complete concepts, but these complete concepts, in turn, comprise elements that appear independently of one another; that is, although the respondents may perceive the concepts to be complete or whole ideas, similar to the way complete ideas appear in nature, the experimental design ensures that the reaction to the whole can be deconstructed into the part-worth reactions. Elements can be combined into small, easily read combinations in many different ways. Whichever way is chosen, however, the approach develops the combinations in a structured fashion amenable to subsequent regression modeling.

Figure 5.2 presents an example of the stimulus for this approach. Respondents need only look at one concept at a time, which simplifies the effort. They can rate this concept on a scale or simply choose to accept or reject the concept. When the researcher combines the elements by experimental design, it becomes straightforward through subsequent

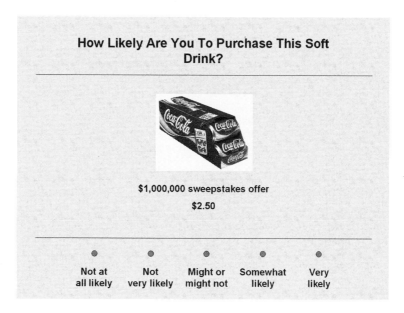

Figure 5.2. Example of a single concept as a combination of features, and the rating question. In conjoint measurement the respondents would evaluate a variety of combinations.

analysis to estimate the contribution of each concept element to acceptance/rejection rating or to the decision to select or reject. The conventional tool of multiple linear regression works fine for this analysis. The conjoint-analysis approaches dealt with in this book use regression modeling to show the application of the method in different situations involving concepts.

Analyzing Data from a Conjoint Study

A worked example shows the straightforward analytic approach to conjoint analysis. Let us consider the case of a new concept comprising a can with one of three symbols (colored blue, green, or red) or with no symbol, and one of three statements (new taste, great taste, and low calories) or no taste/calorie mention at all. Combinations can be created in many ways. Consider one particular set of 15 combinations listed in Table 5.1. The researcher creates the 15 combinations and gives them to the respondents, who evaluate

either all of them or a subset, in some random order. The table shows the levels or options for the two variables.

The last two columns in Table 5.1 list the ratings. The column labeled "9-Point scale rating" shows the rating on the 9-point scale. Finally the ratings were transformed to a binary scale, with ratings of 1–6 transformed to 0 and those of 7–9 transformed to 100. The results of the transformation are listed in the last column. We saw this approach in Chapter 1, dealing with concepts. This binary transformation follows the market-research protocol, which transforms ratings into class membership—either disinterested in the concept or interested. Market researchers like to deal with the proportion of respondents who find a concept interesting rather than dealing with the magnitude of interest as shown by a single respondent or by the average across the respondents.

By itself the experimental design in Table 5.1 does not allow for analysis. One can create a database for easy analysis by expanding the experimental design. The expansion of

Table 5.1. The 15 systematically combined concepts, comprising a colored symbol and some short text information*

Concept	Symbol	Text	9-Point scale rating	Binary rating
1	Blue	New	7	100
2	Blue	Great	5	0
3	Blue	Low	7	100
4	Green	New	3	0
5	Green	Great	7	100
6	Green	Low	4	0
7	Red	New	2	0
8	Red	Great	8	100
9	Red	Low	5	0
10	Blue	None	3	0
11	Green	None	6	0
12	Red	None	3	0
13	None	New	5	0
14	None	Great	7	100
15	None	Low	7	100

*The binary rating is a recode: 1–6 are recoded to 0, and 7–9 are recoded to 100.

the two independent variables generates six new variables (Dblue, Dgreen, Dred; and Dgreat, Dlow, Dnew). These are so-called dummy variables, which take on the value *1* if the element is present and the value *0* if the element is absent. A concept has either one or two of these elements. Table 5.2 shows the expanded design. The format of Table 5.2 becomes immediately more conducive to statistical analysis, especially of a type called dummy-variable analysis. The term *dummy variable* refers to the fact that the key independent variables (corresponding to the concept features) are either present (value 1) or absent (value 0). This convention of a binary representation typifies that representation known as dummy variable.

Creating a Simple Model: How the Presence of Concept Elements Drives Ratings

The recoding in Table 5.2 creates a data matrix that can be easily analyzed by conventional regression modeling, as shown in Table 5.3. The independent variables are the six dummy variables corresponding to the three symbols and the three text elements of the concept. The dependent variable is either the rating itself or the binary recode of the rating. We will save that analysis for later.

The additive model comprises the following eight aspects:

1. *The multiple R, which shows the degree to which the equation fits the data.* The multiple *R* is the Pearson correlation, which varies from 0 to 1.0. The multiple *R* statistic of 1.0 corresponds to a perfect linear relation. The value for multiple *R* is 0.67.

2. *The squared multiple R, which shows the proportion of variability that the model accounts for.* The squared multiple *R* is 0.45. In this dataset the model accounts for 45% of the variability.

3. *The adjusted squared multiple R, which corrects for the fact that there are many predictors.* With many predictors, one can obtain a perfect relation between the data and the model, even if the data are random. As the number of predictors approaches the number of cases or observations, the multiple *R* statistic will approach 1.0, because the predictors can account for noise or random variability. The adjustment corrects for this misrepresentation of the goodness of fit caused

Table 5.2. The 15 concepts, expanded into binary form

Concept	Symbol	Text	Rating	Binary	Blue	Green	Red	Great	Low	New
1	Blue	New	7	100	1	0	0	0	0	1
2	Blue	Great	5	0	1	0	0	1	0	0
3	Blue	Low	7	100	1	0	0	0	1	0
4	Green	New	3	0	0	1	0	0	0	1
5	Green	Great	7	100	0	1	0	1	0	0
6	Green	Low	4	0	0	1	0	0	1	0
7	Red	New	2	0	0	0	1	0	0	1
8	Red	Great	8	100	0	0	1	1	0	0
9	Red	Low	5	0	0	0	1	0	1	0
10	Blue	None	3	0	0	0	0	0	0	0
11	Green	None	6	0	0	1	0	0	0	0
12	Red	None	3	0	0	0	1	0	0	0
13	None	New	5	0	0	0	0	0	0	1
14	None	Great	7	100	0	0	0	1	0	0
15	None	Low	7	100	0	0	0	0	1	0

by an overly large number of independent variables (predictors) versus the number of cases (test concepts).

4. *The variable, which is the specific variable whose contribution is being estimated.* The variables include the additive constant) and the six variables. These are so-called dummy variables because they take on the value of 1 if present in the concept or 0 if absent.

5. *The coefficient, which shows the contribution of the variable to the rating.* The coefficient is 4.50 for the additive constant, meaning that, if there were no elements in the concept (a purely hypothetical situation), then the rating would be estimated to be 4.50. This is clearly a computed value because all of the concepts comprised at least one of the six elements. The effect for blue is 0.50, meaning that if the blue symbol is put into the concept, then we expect the rating to increase by 0.50 rating units. We add that to the additive constant to obtain a value of 5.0. In contrast, the effect for the term "low" in the concept is 1.50, meaning that if the term "low" is put into the concept, then we can expect an increase of 1.50 points on the 9-point scale, or a total of 6 (4.5 + 1.5 = 6). The regression attempts to partial out the individual effects from the response to the combination.

6. *The standard error of the regression coefficient (SE) is the estimated variability of the coefficient if the study were to be run again.* Given the data, one can estimate the variability around the different coefficients. Large values for the variability mean that the estimate of the coefficient is less stable. Small values for the standard error mean the estimated value of the coefficient is more stable.

7. *The t value or t statistic is the ratio of the coefficient to the standard error.* This *t* statistic is treated like the *t* statistic in ordinary inferential statistics. Researchers would like to obtain a significant value of *t* in order to conclude that the coefficient differs significantly from chance (coefficient = 0). Typically, researchers look for *t* values greater than at least 1.0 but sometimes greater than 1.5 or 1.96.

8. *The P value is the probability that the coefficient is actually 0.* Researchers look for *P* values approaching 0, meaning that, given the data, there is almost a 0 probability by chance of observing this particular coefficient if, in fact, the coefficient were really 0. In many cases the *P* values for a few of the variables are low, indicating significance, whereas the *P* values for many of the variables are not significant.

Table 5.3. Parameters of the regression equation relating the presence/absence of the six elements in the 15 concepts to the rating

Model rating = constant + Dblue + Dgreen + Dred + Dgreat + Dlow + Dnew				
Dependent variable: Rating $n = 15$				
Multiple $R = 0.67$				
Squared multiple $R = 0.45$				
Adjusted squared multiple $R = 0.04$				
Standard error of estimate = 1.87				
Variable	Coefficient	SE*	t	P (2-tailed)
Constant	*4.50*	*1.32*	*3.40*	*0.01*
Blue (color)	0.50	1.46	0.34	0.74
Green (color)	−0.50	1.32	−0.38	0.72
Red (color)	−1.00	1.32	−0.76	0.47
Great (text)	2.50	1.46	1.71	0.13
Low (text)	1.50	1.46	1.03	0.34
New (text)	0.00	1.46	0.00	1.00

*SE, standard error.

It is important to keep in mind that the foregoing regression approach—ordinary least squares—is a very powerful, widely used, analytic tool, with the following three features that are relevant to conjoint analysis:

1. *Portability.* Regression can be done with any computer system that has standard regression packages.

2. *Applicability.* Regression can be done with the data of the total panel or done at the individual respondent level, if the researcher has ensured that each respondent evaluates concepts that have been arrayed according to an experimental design.

3. *Flexibility.* Regression can be done with either rating data or with percentage data. If the ratings were to be replaced by percentage values (e.g., percentage of respondents interested in the concept), the analysis would have been identical. Only the interpretation of the coefficients would have changed. Rather than the coefficient showing the number of rating points on a 9-point rating scale contributed by each concept element, the coefficient would show the percentage of respondents who changed their opinion from disinterested (1–6) to interested (7–9) when the concept element was introduced into the concept.

Using Recoded Data (Accept/Reject Response) Rather Than Degree of Interest

Let us look at the recoded data. The recoding to generate a binary scale of acceptance/reject produces different values for the coefficients, as it should. After recoding, much of the information about the magnitude of interest is lost. For example, a rating of 1 and a rating of 6, very different from each other, are recoded to 0. However, the results show something different about the impact of the specific concept elements: namely, the conditional probability that a person will find the concept interesting if the element is added into the concept. Table 5.4 presents this analysis. The coefficients are estimated by ordinary least squares, although one could use methods such as probit analysis (Hosmer and Lemeshow 1989). The interpretation is quite simple:

1. The additive constant (16.67) is the conditional probability that the concept would be acceptable (i.e., rated 7–9) if no concept elements were present. This probability is 16.67%. Clearly, this is a purely estimated parameter since, as was already noted, all of the concepts comprised at least one element.

2. The coefficient for each of the six elements shows the additive or incremental *conditional probability* that the concept would be acceptable (i.e., rating 7–9) if the element were to be added to the concept. Clearly, based on Table 5.4 the elements differ from one another, as they did prior to the recoding (Table 5.3). However, the same types of decision would be made. An element that adds to the magnitude of interest also adds to the number of interested respondents. An element that subtracts from interest also subtracts from the number of interested respondents. In fact, the decisions reached using the rating scale would be pretty much confirmed by the decisions reached using the recoded binary scale. Figure 5.3 shows a comparison of 144 utilities, one per element, across four studies. In those studies the respondents used a 9-point scale. The ratings for the scale were transformed to an 11- to 99-point scale by multiplying the ratings by 11. This generated a scale more similar to a 100-point scale. (Note that the transformation was made for convenience only and simply multiplies the coefficients by the value 11, but does not affect relations among the different coefficients, or utilities). The abscissa shows the utilities from the rating scale. The ordinate shows the utility values obtained from processing the binary recoded data. The results across the 144 concept elements from four studies show a high correlation between the average utilities derived from ratings scales and the average utilities derived from the binary transformed data.

3. In this dataset the utilities for the colors are all negative, whereas the utilities for the statements are all positive. This need not be the case in ordinary datasets and generally is not. Often, the same category of element, such as text, generates both positive and negative utility values, depending on the specific set of elements.

Probit Analysis: Another Way to Model the Concept Data

Occasionally, a purist researcher might wish to opt for other methods to analyze the data matrix, arguing that the response data are no longer ostensibly continuous (i.e., the 9-point liking scale, which is a category scale) but rather constitute a binary scale. Other types of regression analysis are often more appropriate for binary data, although they are harder to interpret. We deal briefly with the

Table 5.4. Parameters of the regression equation relating the presence/absence of the six elements in the 15 concepts to the recoded rating

Model intbinary = constant + Dblue + Dgreen + Dred + Dgreat + Dlow + Dnew
Dependent variable: Tbinary $n = 15$
Multiple $R = 0.73$
Squared multiple $R = 0.53$
Adjusted squared multiple $R = 0.18$
Standard error of estimate = 44.10

Effect	Coefficient	SE*	t	P (2-tailed)
Constant	16.67	31.18	0.53	0.61
Blue	−27.78	34.47	−0.81	0.44
Green	−25.00	31.18	−0.80	0.45
Red	−25.00	31.18	−0.80	0.45
Great	77.78	34.47	2.26	0.05
Low	52.78	34.47	1.53	0.16
New	2.78	34.47	0.08	0.94

*SE, standard error. 1–6 recoded as 0, and 7–9 recoded as 100.

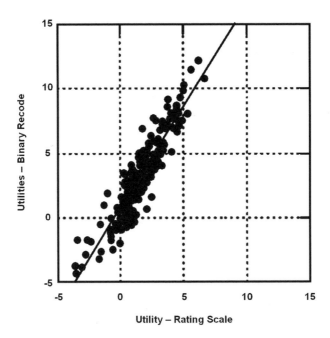

Figure 5.3. Comparison of the utility values for 144 elements from four conjoint studies before recoding to a binary scale and after recoding. Each *circle* corresponds to an element.

most appropriate of these methods—the *probit* (probability unit) method—in which the dependent variable is binary (0,1), whereas the independent variables can be either binary or continuous. In the current study on combinations, both the independent and the dependent variables are binary.

The probit analysis is run exactly like the regression. The results are interpretable as measures of statistical significance; specifically, as measures of which one of the six independent variables is a good predictor of the dependent variable. Through a different type of curve fitting, appropriate for an exponential equation, the probit analysis identifies those variables that are significant predictors. It is intuitively hard to interpret the meaning of the coefficients from the output of a probit analysis, except by reference to the equation, and the calculation of probabilities for a specific set of coefficients. As a consequence, most researchers simply use probit or its more advanced version, *logistic regression*, to identify the key predictor variables that drive the response.

Table 5.5 lists the results of the probit analysis (Systat 1997). We see in its text that "great" and "low" are both significance drivers of the binary rating (interested). The ordinary least-squares estimation shown in Table 5.4 suggested the same thing. The probit analysis is simply more appropriate statistically for the binary data, but it is far harder to interpret. It is easy to interpret the results from the ordinary least squares as the *conditional probability* that a person goes from disinterested to interested if the element is inserted into the concept. No such simple explanation can be given with probit analysis.

Working with Many Elements: The IdeaMap Approach

A great deal of current conjoint measurement works with a range of artificially simplified stimuli. The artificial limitations come from the nature of conjoint measurement, which typically requires many combinations of elements into test concepts. The rule of thumb is to create 2x to 3x test con-

Table 5.5. Results from logistic regression*

| Number with binary = 0 | 10.00 | | | |
| Number with binary = 1 | 5.00 | | | |

Results of estimation
Log likelihood = −3.82

Parameter	Estimate	SE	t ratio	P
1 Constant	−5.19	360.55	−0.01	0.99
2 Blue	−5.43	225.29	−0.02	0.98
3 Green	−5.43	225.29	−0.02	0.98
4 Red	−5.43	225.29	−0.02	0.98
5 Great	11.06	425.15	0.03	0.98
6 Low	10.20	425.14	0.02	0.98
7 New	0.07	473.44	0.00	1.00

*The key to the analysis is to look for those independent variables that are statistical significant (low P value); −2 log-likelihood ratio = 11.46 with 6 degrees of freedom. Chi-squared $P = 0.08$.

cepts for x number of concept elements. This rule of thumb ensures that the statistical analysis will be sufficiently rigorous. At the same time, however, the use of a single design encompassing all of the different elements can run into problems and requires some modifications:

1. For a few elements (e.g., 10–20), there is no problem. A researcher can create 20–40 concepts fairly easily, and a single respondent can evaluate the set of combinations.

2. For more than a few elements (e.g., 100–200), a researcher can create many more concepts (e.g., 200–400). There is more effort involved.

3. With a few concept elements, each respondent can assess all of the combinations, enabling the researcher to estimate either a model for each person or a model for the entire panel.

4. With many concept elements, a researcher cannot estimate a full model for any respondent, because each respondent evaluates only a portion of the combinations. It is very hard for a respondent to assess 200–400 combinations, except by breaking up the test session.

5. Nature is *rich*. There are many ways to state the same element. Language makes a difference. There are also many things to talk about in a concept. The richness leads, in turn, to the need to use many elements in order to capture this richness.

6. The result is that most research done at the development stage is limited to a small set of stimuli, arbitrarily selected by the researcher ahead of time, to produce what is hoped will be the best results. To work within this complex world, most researchers force themselves to follow an artificial structure. They first work with bare-boned statements, which they label *strategic*. After analyzing these simplistic strategic statements, researchers then work with the more meaty, rich, and meaningful stimuli, which are labeled *executional*.

7. What is needed is a method (or even a worldview) that does not have to be divided into strategic versus executional. The method must deal with the complexities of language, picture, and sound (music) in concept development, without having to first simplify, analyze, decide, and then resurrect and rephrase in respondent language.

8. Previous methods to extend conjoint analysis to deal with many elements have been suggested, but none of them have the ambition of dealing with hundreds of concept elements in an essentially *scalable fashion* (Green 1984; Green and Helsen 1989).

The IdeaMap Approach to Dealing with the Issue of Complexity in Nature

The IdeaMap approach was developed to extend conjoint analysis to many elements. The in-going objective was to allow conjoint analysis to be scalable. This means allowing conjoint analysis to do the following three things:

1. Deal easily with many hundreds of elements.

2. Ensure by up-front restrictions between elements that meaningless combinations will never occur in test concepts.

3. Create individual-level models for all of the dozens or hundreds of concept elements even though any single respondent only evaluates a limited number of elements in a limited number of test concepts.

The IdeaMap approach divides into 11 steps:

1. Create elements (the raw material for conjoint analysis).

2. Classify elements into categories or related groupings.

3. Define the order of appearance of the categories in a concept.

4. Dimensionalize elements by means of nonevaluative semantic differential scales.

5. Identify pairwise restrictions (elements that cannot go together or elements that must go together).

6. Use an experimental design to combine the elements efficiently at the individual respondent level so that there are relatively many combinations versus number of elements.

7. Create the combinations, which are the test concepts

8. Have respondents rate short, easily read combinations of elements (miniconcepts) by a scale (e.g., a 1–9 scale).

9. Develop individual models showing the part-worth contribution of each element to respondent acceptance and other respondent-rated attributes.

10. Estimate the utility of untested concept elements by means of an interpolation method.

11. Aggregate individual models to group models based on total panel or key subgroups.

A Case History: Popcorn

The best way to understand the IdeaMap approach is by means of a case history. This case history deals with popcorn. The company wanted to begin with the basics, which included identifying features and benefits of the product that would make it attractive. Furthermore, the features of the popcorn product had to be consistent with the company's image of a high-quality product. The issue was thus to analyze the contribution of all features (product, sensory benefit, pricing, and brand) that could influence the positioning and provide a competitive advantage in an extremely competitive category.

Create Elements

In all concept work the objective is to identify those communication elements (words, pictures, and even music) that best communicate the product or the service. It is best if the researcher creates hundreds of elements. The elements may be different names, heritages, physical characteristics, pictures of the item, etc. Some elements may differ dramatically (representing different strategies), whereas other elements may differ slightly (representing a similar strategy, but slightly different execution).

The elements are developed through ideation sessions, assessment of competitor copy, research and development (R&D) inputs, and marketing wish lists (see Chapter 3). Ideally, the researcher should investigate several hundred elements. Unlike conventional conjoint measurement, IdeaMap enables researchers to assess many different strategic as well as tactical variations in one study.

Table 5.6 lists the elements that were identified in the ideation session, after being edited to remove redundancy. Some were currently impossible to execute but were of interest to marketing in their consideration of future plans. The elements include pictures as well as words. The elements also include the messages used by competitors, as well. With competition included in the study, the analysis can uncover the expected performance of both potential elements for the new product and performance of competitor elements. The competitor's elements may be current elements and/or new elements that competitors might introduce as a response (i.e., defensive counterstrike) to the new popcorn entry.

Classify Elements into Categories or Groupings

When developing concepts it is important that there be a flow and logic to what the respondents see and read. To ensure that logic, we classify each element as belonging to a category or group. The test concepts comprise a sequence of groups, one element per group. For the most part the classification procedure is more of a bookkeeping or accounting method to ensure that the test concepts make sense and do not have five names or five benefits strung together in a sequence. Table 5.6 shows the categories to which the test elements belong. An IdeaMap study may have as few as 4–5 categories or as many as 15–20. They need not be of equal size.

When working with many elements, researchers have the luxury of investigating both qualitative statements and quantitative statements in the same study. Most traditional conjoint analyses were reserved for quantitative variation (e.g., changing prices). As Table 5.6 shows, there are different qualitative variations, as well as quantitative vari-

ations. With the large number of elements, researchers build a comprehensive database, showing the impact of changes in continuous variables such as nutrition issues (calories, fat, and price) and changes in discrete variables such as brand names, endorsements, and emotional benefits.

Determine the Order of Appearance of Categories in the Concept

In addition to classifying the elements into categories, researchers have to determine the format of the concept or, in this specific case, the order in which the categories are presented. It makes no sense to present the respondents with a randomized set of elements, without a logical order. It is critical that the researchers identify the most appropriate order of attributes (or at least an order that makes intuitive sense to the respondents, who will be reading and rating the concept).

Dimensionalize the Elements on Semantic Differential Scales

The approach presented here works on an individual-by-individual basis. In a limited interview a single respondent cannot test all elements in the set, especially when the study comprises more than 100 elements. There must be a way to estimate how a respondent would have rated elements that he or she did not test, given ratings of elements that the respondent actually evaluated in the concepts. Unlike physical products where the variables are often continuous, there are no continuous variables in language and therefore no way either to interpolate or to estimate responses to untested levels. Furthermore, it does not make sense to force all of the elements in a concept to be linked to one another numerically by selecting variables that are intrinsically numerical.

Since there is no naturally underlying metric for language, we have to create one. We locate all of the elements on a series of

Table 5.6. Concept elements for popcorn

		Utility
	Additive constant	*47*
	Visual element	
VS1	One pouch of liquid butter sauce with one popped bag of popcorn—w/zip-strip feature	4
VS2	One cup of liquid butter sauce with one popped bag of popcorn—w/zip-strip feature	4
VS3	One packet of grated parmesan cheese (unbranded) with one popped bag of popcorn—w/zip-strip feature	2
VS4	One packet of Kraft grated parmesan cheese with one popped bag of popcorn—w/zip strip feature	3
VS5	One packet of spicy seasonings (unbranded) with one popped bag of popcorn—w/zip-strip feature	0
VS6	One packet of McIlhenny spicy seasoning with one popped bag of popcorn—w/zip-strip feature	−1
VS7	Zip-strip feature converting bag to bowl—w/zip-strip feature	9
VS8	One pouch of liquid butter sauce with one popped bag of popcorn—no zip-strip feature	4
VS9	One cup of liquid butter sauce with one popped bag of popcorn—no zip-strip feature	2
VS10	One packet of grated parmesan cheese (unbranded) with one popped bag of popcorn—no zip-strip feature	6
VS11	One packet of Kraft grated parmesan cheese with one popped bag of popcorn—no zip-strip feature	6
VS12	One packet of spicy seasonings (unbranded) with one popped bag of popcorn—no zip-strip feature	4
VS13	One packet of McIlhenny spicy seasoning with one popped bag of popcorn—no zip-strip feature	−10
	Name	
NA1	American Popcorn presents new Pop N' Budder's Plus, a microwave popcorn Plus more of what you love in popcorn	2
NA2	American Popcorn presents new Pop Fun's Plus, a microwave popcorn Plus more of what you love in popcorn	6
NA3	American Popcorn presents new Pop Fun's Butter Plus, a microwave popcorn Plus more of what you love in popcorn	−1
NA4	American Popcorn presents new Pop N' Budder's Butter Plus, a microwave popcorn Plus more of what you love in popcorn.	5
NA5	American Popcorn presents new & improved Pop Fun's Butter	5
NA6	American Popcorn presents new & improved Pop N' Budder's Movie Theater Butter	7
NA7	American Popcorn presents new Pop Fun's Bigger Butter	4
NA8	American Popcorn presents new Pop N' Budder's Bigger Butter	0
NA9	American Popcorn presents new Pop Fun's Double Drizzle	21
NA10	American Popcorn presents new Pop Fun's Double Butter	1
NA11	American Popcorn presents new Pop N' Budder's Double Drizzle	1
NA12	American Popcorn presents new Pop N' Budder's Double Butter	0
NA13	American Popcorn presents new Pop Fun's Bigger Better Butter	2
NA14	American Popcorn presents new Pop N' Budder's Bigger Better Butter	21
NA15	American Popcorn presents new Pop Fun's Biggie Pop	5
NA16	American Popcorn presents new Pop N' Budder's Biggie Pop	3
NA17	American Popcorn presents new Pop Fun's Popcorn Bowl	2
NA18	American Popcorn presents new Pop Fun's Buttery Bowl	10

(continued)

Table 5.6. Concept elements for popcorn *(cont.)*

		Utility
NA19	American Popcorn presents new Pop Fun's Bigger Better Butter Bowl	−4
NA20	American Popcorn presents new Pop N' Budder's Bigger Better Butter Bowl	1
NA21	American Popcorn presents new Pop Fun's Blissfully Better Butter	4
NA22	American Popcorn presents new Pop N' Budder's Blissfully Better Butter	4
NA23	American Popcorn presents new & improved Pop Fun's Light Butter	1
	Benefit	
BE1	With kernels that pop up 10% larger than any other popcorn, American's best just got better! Bigger kernels hold more flavor for a bigger taste	4
BE2	With kernels that pop up 20% larger than any other popcorn, American's best just got better! Bigger kernels hold more flavor for a bigger taste	4
BE3	With kernels that pop up 30% larger than any other popcorn, American's best just got better! Bigger kernels hold more flavor for a bigger taste	6
BE4	With kernels that pop up 40% larger than any other popcorn, American's best just got better! Bigger kernels hold more flavor for a bigger taste	2
BE5	With kernels that pop up 50% larger than any other popcorn, American's best just got better! Bigger kernels hold more flavor for a bigger taste	8
BE6	With kernels that pop up larger than any other popcorn, American's best just got better! Bigger kernels hold more flavor for a bigger taste	5
BE7	American's best just got better! Ten years of painstaking research have provided a new popcorn hybrid that pops up larger, lighter, and fluffier than anything else available	4
BE8	American's best just got better! Our new popcorn hybrid pops up so much bigger, lighter, and fluffier that it hardly fits in your mouth	2
BE9	With 10% fewer unpopped kernels than any other popcorn, American's best just got better	3
BE10	With 20% fewer unpopped kernels than any other popcorn, American's best just got better	6
BE11	With 30% fewer unpopped kernels than any other popcorn, American's best just got better	1
BE12	With 40% fewer unpopped kernels than any other popcorn, American's best just got better	1
BE13	With 50% fewer unpopped kernels than any other popcorn, American's best just got better	4
BE14	American's best just got better! Ten years of painstaking research have provided a new popcorn hybrid where every kernel is guaranteed to pop	6
BE15	If there's one thing we can't stand, it's unpopped kernels. So we've developed a new, even better popcorn guaranteed not to leave any unpopped kernels sitting at the bottom of the bag	10
BE16	Each bag has more popped corn per bag than any other microwave popcorn	5
BE17	Each bag has 15 cups of popped corn per bag	0
BE18	Each bag has 2 cups more popped corn per bag than any other microwave popcorn	4
BE19	Each bag has 4 cups more popped corn per bag than any other microwave popcorn	7
BE20	Each bag has 6 cups more popped corn per bag than any other microwave popcorn	5
BE21	Each bag has so much popped corn that it's bursting at the seams	3
BE22	You get a lot more for your money, because there's more popcorn in each bag than any other popcorn	4

(continued)

Table 5.6. Concept elements for popcorn *(cont.)*

		Utility
	Flavor pouch	
PL1	All this Plus a separate flavored sauce packet containing only real butter and natural ingredients. You can add as little or as much flavoring as you like	9
PL2	All this Plus a separate flavored sauce packet containing only real butter and natural ingredients	4
PL3	All this Plus a separate flavored sauce packet containing real butter and artificial ingredients	−5
PL4	All this Plus a separate flavored sauce cup containing only real butter and natural ingredients	4
PL5	All this Plus a separate packet containing a spicy blend of seasonings for a little extra "kick"	2
PL6	All this Plus a separate packet containing a spicy blend of seasonings made from real McIlhenny Tabasco sauce	−7
PL7	All this Plus a separate packet of Grated Parmesan cheese	−4
PL8	All this Plus a separate packet of real Kraft 100% Grated parmesan cheese	5
PL9	All this Plus a special "zip-strip" that lets you turn the bag into a bowl for easy, no-mess eating	4
PL10	All this Plus it's made with Real butter	2
PL11	All this Plus it's made with Real Land O' Lakes butter	3
PL12	All this Plus it's made with Real USDA Grade A butter	4
PL13	All this Plus it's made with Real Farm butter	0
PL14	All this Plus it's made with Real Wisconsin butter	−1
PL15	All this Plus it's made with Real Creamery butter	3
PL16	All this Plus it's made with Real Sweet Cream butter	1
PL17	All this Plus it's made with Real Kraft butter	2
PL18	All this Plus it's made only with Real Butter and Natural Ingredients— no artificial flavors	9
PL19	All this Plus a special "zip-strip" to turn the bag into a bowl—no more messy hands from reaching down into a bag	10
PL20	All this Plus a special "zip-strip" to turn the bag into a bowl—you can eat right out of the bag	11
PL21	All this Plus a special "zip-strip" to turn the bag into a bowl—no need to wash a bowl	5
	Instruction	
IN1	Pop it in the microwave for 3–5 minutes and it's ready to eat	1
IN2	Pop the popcorn in the microwave for 3–5 minutes, drizzle the butter-flavored sauce on the warm popcorn, and it's ready to eat	5
IN3	Pop the popcorn in the microwave for 3–5 minutes, sprinkle on the butter seasoning, and it's ready to eat	6
IN4	Pop the popcorn in the microwave for 3–5 minutes. Then heat the packet of butter-flavored sauce in the microwave for 10 seconds, drizzle on the warm popcorn, and it's ready to eat	−4
IN5	Pop it in the microwave for 3–5 minutes. While the popcorn is popping, warm the packet of butter-flavored sauce in hot tap water. Once the popcorn is done popping, drizzle the sauce on the warm popcorn, and it's ready to eat	−3
IN6	Pop it in the microwave for 3–5 minutes, sprinkle on the dry seasoning, and it's ready to eat	0
IN7	Pop it in the microwave for 3–5 minutes, use the zip-strip to open the bag, and it's ready to eat	5

(continued)

Table 5.6. Concept elements for popcorn *(cont.)*

		Utility
IN8	Pop the popcorn in the microwave for 3-5 minutes, use the zip-strip to open the bag, drizzle the butter-flavored sauce on the warm popcorn, and it's ready to eat	9
IN9	Pop the popcorn in the microwave for 3–5 minutes, use the zip-strip to open the bag, sprinkle on the butter seasoning, and it's ready to eat	7
IN10	Pop the popcorn in the microwave for 3–5 minutes, then heat the packet of butter-flavored sauce in the microwave for 10 seconds. Meanwhile, use the zip-strip to open the bag. Drizzle the sauce over the warm popcorn, and it's ready to eat	6
IN11	Pop it in the microwave for 3–5 minutes. Meanwhile, warm the packet of butter-flavored sauce in hot tap water. Once the popcorn has popped, use the zip-strip to open the bag. Drizzle the sauce over the warm popcorn, and it's ready to eat	0
IN12	Pop it in the microwave for 3–5 minutes, use the zip-strip to open the bag, sprinkle on the dry seasoning, and it's ready to eat	3
	Health	
HE1	Each serving has 10 grams of fat (about 3 servings per bag)	5
HE2	Each serving has 12 grams of fat (about 3 servings per bag)	0
HE3	Each serving has 14 grams of fat (about 3 servings per bag)	2
HE4	Each serving has 16 grams of fat (about 3 servings per bag)	−1
HE5	Each serving has 18 grams of fat (about 3 servings per bag)	−2
HE6	Each serving has 20 grams of fat (about 3 servings per bag)	−2
HE7	A 5-cup serving has 12 grams of fat	3
HE8	A 5-cup serving has 15 grams of fat	1
HE9	A 5-cup serving has 18 grams of fat	3
HE10	A 5-cup serving has 21 grams of fat	−3
HE11	A 5-cup serving has 24 grams of fat	−3
HE12	A 5-cup serving has 27 grams of fat	−2
HE13	40 grams of fat per bag (15 cups total, about 3 servings)	−4
HE14	45 grams of fat per bag (15 cups total, about 3 servings)	−6
HE15	50 grams of fat per bag (15 cups total, about 3 servings)	−7
HE16	55 grams of fat per bag (15 cups total, about 3 servings)	−9
HE17	60 grams of fat per bag (15 cups total, about 3 servings)	−4
HE18	65 grams of fat per bag (15 cups total, about 3 servings)	−6
HE19	70 grams of fat per bag (15 cups total, about 3 servings)	−9
HE20	75 grams of fat per bag (15 cups total, about 3 servings)	−9
HE21	One 5-cup serving has the same fat as 12 regular tortilla chips	−3
HE22	One 5-cup serving has the same fat as 15 regular tortilla chips	1
HE23	One 5-cup serving has the same fat as 18 regular tortilla chips	−3
HE24	One 5-cup serving has the same fat as 21 regular tortilla chips	0
HE25	One 5-cup serving has the same fat as 24 regular tortilla chips	−10
HE26	One 5-cup serving has the same fat as 27 regular tortilla chips	−7
	Price	
CO1	Each box contains three bags and costs $1.99	2
CO2	Each box contains three bags and costs $2.09	3
CO3	Each box contains three bags and costs $2.19	2
CO4	Each box contains three bags and costs $2.29	−2
CO5	Each box contains three bags and costs $2.39	−1

(continued)

Table 5.6. Concept elements for popcorn *(cont.)*

		Utility
CO6	Each box contains three bags and costs $2.49	−1
CO7	Each box contains three bags and costs $2.59	−5
CO8	Each box contains three bags and costs $2.69	−3
CO9	Each box contains two bags of popcorn and two seasoning packets and costs $1.99	1
CO10	Each box contains two bags of popcorn and two seasoning packets and costs $2.09	1
CO11	Each box contains two bags of popcorn and two seasoning packets and costs $2.19	1
CO12	Each box contains two bags of popcorn and two seasoning packets and costs $2.29	1
CO13	Each box contains two bags of popcorn and two seasoning packets and costs $2.39	−2
CO14	Each box contains two bags of popcorn and two seasoning packets and costs $2.49	−2
CO15	Each box contains two bags of popcorn and two seasoning packets and costs $2.59	−2
CO16	Each box contains two bags of popcorn and two seasoning packets and costs $2.69	−6
CO17	Each box contains three bags of popcorn and three seasoning packets and costs $2.49	4
CO18	Each box contains three bags of popcorn and three seasoning packets and costs $2.59	−1
CO19	Each box contains three bags of popcorn and three seasoning packets and costs $2.69	−1
CO20	Each box contains three bags of popcorn and three seasoning packets and costs $2.79	−4
CO21	Each box contains three bags of popcorn and three seasoning packets and costs $2.89	−1
CO22	Each box contains three bags of popcorn and three seasoning packets and costs $2.99	−2
CO23	Each box contains three bags of popcorn and three seasoning packets and costs $3.09	−3

nonevaluative semantic differential scales (see the columns in Table 5.7). These semantic scales are nonevaluative because they represent the *meaning* of the element rather than a judgment of how good or bad the element may be. Elements close together on the semantic differential scale share similar meanings (albeit not identical, because we work with only a limited number of such scales).

The semantic differential scales are relevant to the category being tested. A small group of 5–10 respondents locates each of the elements on the semantic differential scales. The semantic differential *profile* accompanies the elements and becomes an intrinsic part of the database.

Identify Pairwise Restrictions

Not all elements in a concept are a good fit when placed together, either because the two elements do not make sense logically to the respondent or because the they may be incompatible from a technical aspect (even though respondents would have no problem were they to read a concept comprising these pairwise incompatible elements). There may be a few, a dozen, or several hundred. By knowing what pairs of elements cannot appear together, the IdeaMap algorithm can create combinations that never feature these incompatible pairs.

With many elements selected for evaluation it is important that the marketing (and,

Table 5.7. Partial element list for popcorn, including two elements per category, semantic profile on seven scales, and utility value

		Uniqueness	Type	Buy for	Eating occasion	Value for money	Image	Preparation
	Semantic value = 1	Similar	Home-made	Taste	Daily	Value	Traditional	Easy
	Semantic value = 0	Unique	Store bought	Health	Special	Premium	Innovative	Hard
	Visual element							
VS1	One pouch of liquid butter sauce with one popped bag of popcorn—w/zip-strip feature	6.1	5.4	2.6	4.9	4.0	4.3	4.8
VS2	One cup of liquid butter sauce with one popped bag of popcorn—w/zip-strip feature	6.9	5.1	2.7	4.6	4.1	5.1	5
	Name							
NA1	American Popcorn presents new Pop N' Budder's Plus, a microwave popcorn Plus more of what you love in popcorn	5.1	5.8	3.6	4.1	4.0	4.6	4.2
NA2	American Popcorn presents new Pop Fun's Plus, a microwave popcorn Plus more of what you love in popcorn	4.8	5.6	3.4	4.1	4.0	5	4.7

(continued)

Table 5.7. Partial element list for popcorn, including two elements per category, semantic profile on seven scales, and utility value (*cont.*)

	Uniqueness	Type	Buy for	Eating occasion	Value for money	Image	Preparation
Semantic value = 1	Similar	Home-made	Taste	Daily	Value	Traditional	Easy
Semantic value = 0	Unique	Store bought	Health	Special	Premium	Innovative	Hard
Benefit							
BE1 With kernels that pop up 10% larger than any other popcorn, American's best just got better! Bigger kernels hold more flavor for a bigger taste.	6	5.1	3.8	4.3	4.6	5.3	5.3
BE2 With kernels that pop up 20% larger than any other popcorn, American's best just got better! Bigger kernels hold more flavor for a bigger taste.	5.4	5.2	3.6	4.9	4.7	5.2	5.2
Flavor							
PL1 All this Plus a separate flavored sauce packet containing only real butter and natural ingredients. You can add as little or as much flavoring as you like.	7.9	4.8	4.1	5.4	5.7	6.2	6
PL2 All this Plus a separate flavored sauce packet containing only real butter and natural ingredients.	6.7	4.1	4	5	5.9	6.3	6.2

(*continued*)

Table 5.7. Partial element list for popcorn, including two elements per category, semantic profile on seven scales, and utility value *(cont.)*

	Uniqueness	Type	Buy for	Eating occasion	Value for money	Image	Preparation
Semantic value = 1	Similar	Home-made	Taste	Daily	Value	Traditional	Easy
Semantic value = 0	Unique	Store bought	Health	Special	Premium	Innovative	Hard
Instruction							
IN1 Pop it in the microwave for 3–5 minutes and it's ready to eat	4.1	5.3	4.7	4.2	4.6	5	4.8
IN2 Pop the popcorn in the microwave for 3–5 minutes, drizzle the butter-flavored sauce on the warm popcorn, and it's ready to eat	5.8	4.2	3.3	4.8	5.2	5.1	4.6
Health							
HE2 Each serving has 12 grams of fat (about 3 servings per bag)	3.6	5.1	5.7	4.4	5.4	4.8	4.2
HE3 Each serving has 14 grams of fat (about 3 servings per bag)	3.8	6.1	5.1	4.8	4.2	4.1	4
Price							
CO1 Each box contains three bags and costs $1.99	6.1	4.9	5.8	4.9	3.9	5.6	4.2
CO2 Each box contains three bags and costs $2.09	4.7	5.1	6	5.6	4.7	6	4.1

when relevant, the R&D) groups identify the different potential pairs of elements that might be incompatible. With 10–20 elements this identification task poses no problem. One could do the identification by brute-force methods, if necessary. With procedures that have up to 300 elements the marketers have to spend more time with the elements, identifying the pairs up front. All the incompatible pairs may not be identified at the first stage. After the first round, where many incompatible pairs are identified, the researcher can create small test combinations of allowable elements and simply inspect random combinations to determine whether any additional incompatible pairs emerge. (This task could take several hours for a base of 300 elements.)

Restrictions will emerge more strongly in some studies than others. Typically, when a study deals with quantitative aspects of product design, such as popcorn, relatively few pairs of elements are mutually contradictory. In contrast, when a study deals with different emotional statements, benefits, and the like, that are not connected quantitatively, but represent different directions, then many more combinations will be contradictory because they promise different things that just don't go together. The restrictions are written simply in the following format. Note that the examples below are just for illustration:

If X1 appears, then Y6 cannot appear.
If A3 appears, then D7 D8 E1 E2 E3 E4 U1 P3 cannot appear.
If A4 appears, then P1 P6 cannot appear.

Select the Appropriate Experimental Design to Underlie the Concepts

Since the study objective is to identify how every concept element promotes acceptance and communication, it is vital to use an experimental design or statistical plan that lays out the combinations of the concept elements.

These combinations (i.e., the test concepts) are designed so that the elements within the concept appear independently of one another in a statistical sense. There are a variety of experimental designs from which to choose, depending on the number of categories that one wishes to present and the number of elements that one wishes to test in a category.

One of the key issues in IdeaMap is the need to cover many elements with relatively few combinations. There are so many elements to deal with in IdeaMap—sometimes many hundreds—that a single respondent must be exposed to a lot of these elements, yet not be forced to evaluate too many. The two factors—many elements and ease of the respondent task—drive the experimental design toward the type that is *efficient* in terms of presenting many elements in the fewest number of combinations. The sacrifice is the relatively large number of replicates of an element. Typically, one wishes to have $2x$ or $3x$ the number of combinations for x elements. For IdeaMap, that laudable statistical goal cannot be easily attained unless the researcher wants to have the respondent participate in very long test sessions. Instead, there are other, more efficient experimental designs with more combinations than predictors, making the regression work at the individual level, but without so many combinations, making the regression a little weaker.

Table 5.8 shows one particular design: the Plackett-Burman five-level screening design. The experimental design enables researchers to investigate up to five categories in a concept and up to four elements per category. The experimental design allows for five elements per category, but the fifth element is reserved for *null* or "no element present." By allowing for a true null *zero condition*, researchers can use the regression analysis to better estimate the contribution of every element to respondent reactions. The design is efficient because it can deal with many elements with relatively few combinations and

Table 5.8. Experimental design for 25 concepts, based on the Plackett–Burman screening design*

Concept	Category 1	2	3	4	5
1	4	1	3	1	1
2	0	4	1	3	1
3	3	0	4	1	3
4	3	3	0	4	1
5	2	3	3	0	4
6	3	2	3	3	0
7	4	3	2	3	3
8	1	4	3	2	3
9	2	1	4	3	2
10	2	2	1	4	3
11	0	2	2	1	4
12	2	0	2	2	1
13	4	2	0	2	2
14	3	4	2	0	2
15	0	3	4	2	0
16	0	0	3	4	2
17	1	0	0	3	4
18	0	1	0	0	3
19	4	0	1	0	0
20	2	4	0	1	0
21	1	2	4	0	1
22	1	1	2	4	0
23	3	1	1	2	4
24	1	3	1	1	2
25	4	4	4	4	4

*Numbers in table body show which of the four elements selected from that category appears in the specific concept. The zero means that the element from that category does not appear in the concept.

still allow for a valid regression model at the individual respondent level.

Create the Concepts According to the Experimental Design

By itself the experimental design provides only a layout for the elements. The researcher must select the categories (five from the full set) and the elements within the categories (four elements per category). The order of the categories is predetermined. Let us take a simple example. Let us assume that the study has 150 elements, but each respondent will evaluate only 100 elements. This calls for the following strategy in concept creation:

1. *Respondent task.* Each respondent will be presented with five designs, comprising 20 elements each in 25 concepts. This means that each respondent will evaluate 125 concepts, comprising 100 elements. This is an efficient approach because it covers many elements with a reasonably efficient number of concepts. The 125 concepts may be either just right for a session where the respondent is prerecruited to participate and paid, or may be too long if the respondent is intercepted at the mall and invited to participate for a "short" interview. This is not a short interview.

2. *Element selection.* For each respondent, the IdeaMap algorithm creates five designs so that any single element appears only once in the five designs. Furthermore, the algorithm ensures that there are no violations of the pairwise restrictions.

3. *One set of concepts per respondent.* The algorithm in step 2 is repeated as many times as there are respondents in the study (e.g., let us assume 250 respondents in the study).

4. *Total number of designs.* The outcome of the algorithm is 250 sets of designs, one set of designs for each respondent. A total of 100 elements appear in each set of 125 concepts. The entire *master* set of designs (five designs per respondent; 250 respondents and 1250 designs in total) is managed so that the elements all appear approximately equally frequently across all the respondents, but of course each respondent sees only a partial set of elements.

5. *Next steps.* Once the designs are set up, the 125 concepts for a single respondent are thoroughly randomized and stored for the actual fieldwork.

Evaluate Orientation Concepts, Competitor Controls (Benchmarks), and Then Test Concepts

The typical IdeaMap interview lasts 30–45 minutes, during which the respondent first rates *orientation concepts*, which comprise fixed sets of elements. These orientation concepts help the respondent to feel comfortable with the task. Then, the respondent rates the benchmark concepts that, in turn, represent the full array of competitor concepts. Occasionally, the benchmark concepts are presented after the test concepts. Unlike the experimentally varied concepts, the orientation concepts and the benchmark concepts are fixed. They look like the experimentally varied concepts in structure and appearance, but do not have the systematic variation that defines the test concepts.

The concepts are presented on a computer screen to the respondents, with the computer program creating the test concepts in *real time* during the interview. Each respondent is presented with 100 different experimentally designed concepts during the 45-minute interview. Since a single study may involve 200 or more respondents, the computer actually creates all combinations in real time, minimizing researcher effort and errors. With many elements from which to choose it is unlikely that two respondents will ever evaluate the exact same set of concepts, because the categories and elements are chosen anew for each respondent for each round.

Respondents rate interest in the concept that they read (taking into account any picture in the concept, as well as any music or voice-over, should the concept comprise those types of elements). Respondents also can rate other attributes as well, such as communication and uniqueness. There is a limit, however, to a respondent's ability to rate the elements. The more scales along which a respondent rates the concepts, the fewer are the concepts that the respondent can and should rate.

Although in the past, respondents were often nervous about computers, the increased frequency with which computers are encountered and used in everyday life has reduced much of this anxiety. The orientation phase, and a gentle introduction by an interviewer

trained in the procedure, reduces interview anxiety considerably.

Create Individual Equations Relating the Elements Seen by the Respondent to the Ratings Assigned

Keep in mind that each respondent will have evaluated concepts created from several experimental designs. The property of the Plackett-Burman screening design is that the data for the respondents lend themselves to dummy-variable regression. For our popcorn example, we have 100 elements or independent variables, and 125 combinations, for a particular individual. That individual's rating can be transformed to a binary scale. By convention in IdeaMap, this is a 9-point scale. Ratings of 1–6 are transformed to *0*, and ratings of 7–9 are transformed to *100*. Afterward, the data are analyzed by dummy-variable, ordinary least-squares regression. The output of the regression is a simple equation for the particular respondent:

$$\text{rating} = k_0 = k_1 \text{ (element 1)} = k_2 \text{ (element 2)} \ldots k_{100} \text{ (element 100)}$$

Estimate the Utility of Untested Elements at the Individual Respondent Level

No one respondent can rate all of the elements in combination, especially when the number of elements is 200 or more. The interview would take too long. On an individual respondent basis, it is desirable to estimate how that respondent would have responded to elements that were present in the study but not directly tested by that individual respondent.

There is no underlying continuum of language, pictures, and sound as there is for ingredients. With ingredients, if one knows the reaction to 10 grams of sweetener, and if one knows the relation between liking and grams of sweetener, then it is straightforward to interpolate and to estimate the reaction to 12 grams. One cannot easily interpolate in language to estimate the reaction to words that have not been tested.

To interpolate in language means that the researcher has to first develop an underlying *metric* or structure, which is created by using the semantic differential scale, as described above in step 4 (*dimensionalization*). Elements that are close together in the semantic differential space should have similar values on the scale of part-worth contributions. By identifying "neighbors" in the semantic space, one can interpolate to estimate how untested elements would have scored, given the contributions of neighbors in the space that have been tested and whose contributions therefore are empirically known. Clearly, the more elements that are tested, the better will be the interpolation. Furthermore, each respondent will have a different set of elements that will be empirically measured, and therefore a different set of elements that will require interpolation. Table 5.9 summarizes the algorithm and the underlying rationale for the different steps. The algorithm generates a full utility model for each respondent. The algorithm never changes the utility values of the elements that were directly evaluated by a respondent, because those are empirically determined. On an interactive basis, though, it estimates and reestimates the utility values for the untested elements, continuing the iteration until the estimates become *tight* and do not change with subsequent iterations of the algorithm.

Aggregate the Data to Group Models

The data from many respondents can be analyzed in this respondent-by-respondent fashion. After the individual-level analyses are complete (including the estimating utilities of untested elements), the researcher can aggregate the coefficients from the total panel or from specific subgroups of respondents. These specific subgroups may be defined by external criteria (e.g., age, income, gender, market, and usage pattern).

The subgroups may be defined by the pattern relating part-worth contribution to semantic differential scales [so-called *concept-response segmentation* (Chapter 7)]. Concept-response segmentation often reveals dramatically different groups of respondents with different patterns of the utility values (Green and Krieger 1991).

Overview

Experimental design or conjoint analysis of concepts enables researchers to work more systematically with concepts. By varying the concepts in known ways, presenting the combinations to respondents and obtaining the ratings, researchers rapidly understand which elements work in a concept and which do not.

There are three key benefits to designed experiments:

1. *Bottom-up development.* The researcher works from the bottom up, not from the top down. This bottom-up approach forces the researcher to better understand the nature of the stimulus, since one has to begin with components and end up with combinations.

2. *Easy interpretation, using conventional statistics.* The output of statistical analysis from ordinary least squares is a set of coefficients, which show either the number of points added to the expected rating scale value by the individual concept element or the incremental conditional probability of a respondent falling into the acceptor class. In either case, the statistical analysis is performed by conventional methods that are easy to implement and easy to answer.

3. *IdeaMap increases the scope of conjoint to hundreds of elements.* The IdeaMap approach provides a system for conjoint analysis that is easily scalable to many hundreds of concept elements. The task of individual respondents remains unchanged: namely, to evaluate a limited set of concepts whose elements are varied by experimental design. Dimensionalization and interpolation allow the researcher to estimate the utility value at the

Table 5.9. An algorithm to estimate utilities of untested concept elements by using the utilities of the closest concept elements in a semantic space

1.	With 80–100 elements or fewer, each respondent can test a limited number of concepts that in the end cover all elements, and the data can generate an individual utility model.
2.	With more than 100 elements, each person can, in general, evaluate concepts comprising only a fraction of the elements. No one can test combinations of all elements.
3.	In many cases, we want to create individual-level models.
4.	If we want individual models with more than 100 elements, then we need a *hybrid estimator*.
5.	The hybrid estimator fills in untested utilities of untested elements. It is similar to interpolation.
6.	Using an algorithm, we estimate coefficients or utilities for elements not tested
7.	The method is taken from numerical analysis applied to engineering problems. It is called *finite element analysis*.
8.	As an example, consider the body/wing of an airplane. The goal is to estimate the stress at points on an airplane wing where there *are no direct measurement data* (blank ellipse on the left).

 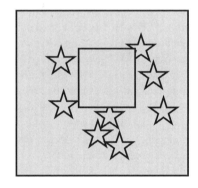

9.	Applying the same thinking to concepts enables us to use numerical analysis.
10.	Semantic scaling locates points in space; each element is rated on 5–10 scales.
11.	Interpolation estimates utilities for untested elements. Interpolation uses finite element analysis.
12.	Distances in semantic space guide interpolation (interpolated utility value = average of utilities of the eight closest points).

individual respondent level for elements not directly rated.

Note

The section on IdeaMap is based on the paper by H. Moskowitz and D. Martin: "How computer aided design and presentation of concepts speeds up the product development process." In: Proceedings Of the ESOMAR Congress, Copenhagen, September, 1993 [Amsterdam: ESOMAR (European Society of Marketing Research)].

References

Box, G.E.P., Hunter, J., and Hunter, S. (1978). Statistics for Experimenters. New York: John Wiley and Sons.

Fechner, G.T. (1860). Elemente der Psychophysik. Leipzig: Breitkopf und Hartel.

Green, P.E. (1984). Hybrid models for conjoint analysis: an expository review. Journal of Marketing Research, 21: 155–169.

Green, P.E., and Helsen, K. (1989). Cross validation assessment of alternatives to individual-level conjoint analysis: a case study. Journal of Marketing Research, 26: 346–350.

Green, P.E., and Krieger, A.M. (1991). Segmenting markets with conjoint analysis. Journal of Marketing, 55: 20–31.

Green, P.E., and Wind, J. (1973). Multiattribute Decisions In Marketing: A Measurement Approach. Hinsdale, IL: Dryden.

Gustafsson, A., Herrmann, A., and Huber, F. (2000). Conjoint Measurement: Methods and Applications. Berlin: Springer.

Hosmer, D.W., and Lemeshow, S. (1989). Applied Logistic Regression. New York: John Wiley and Sons.

Johnson, R.M. (1974). Trade-off analysis of respondent values. Journal of Marketing Research, 11: 121–127.

Luce, R.D., and Tukey, J.W. (1964). Conjoint analysis: a new form of fundamental measurement. Journal of Mathematical Psychology, 1: 1–36.

Systat (1997). Systat, the system for statistics. Evanston, IL: Systat Division of SPSS.

Wittink, D.R., and Cattin, P. (1989). Commercial use of conjoint analysis: an update. Journal of Marketing, 53: 91–96.

Chapter 6

Concepts as Combinations of Graphics

Introduction: What Packaging Is About

Packaging is exceptionally important and is omnipresent in our world. "Virtually everything grown or manufactured is packaged in some fashion. It is necessary for our modern lifestyles. It is decorative. It is a salesman. It is a protective screen" (www.virtual.clemson. edu/groups/pkgsci/) (Meyers and Lubliner 1998). Packaging plays a variety of roles. Some package design elements are optional and used for their aesthetic value whereas other package elements are required for functionality (Lingle 2003). Beyond their roles in aesthetics, manufacturability and functional tasks, product protection and rational packing, package features are, above all, powerful marketing instruments for products and brands. But, what is a good packaging? The Internet site www.baddesigns.com features dozens of poorly or badly designed packages, allowing the visitor to understand what poor design looks like. A package must meet three criteria:

1. It must protect the product from physical, chemical, and microbiological invasion.

2. It must provide a medium for presenting advertising messages and other important information to consumers. This is where concept research can help identify the specific information that most strongly influences consumers.

3. It is one of the greatest influences on a consumer's decision to try a product.

For many fast-moving consumer goods the package is the first experience a consumer has with the product. For this reason, at the point of purchase the package must present the product to clients in an attractive and desirable form, that is, reflect the quality of the product. A successful package communicates the product's approximate value. One "rule" in the consumer mind is that lower-priced products in any category have lower-quality packaging, whereas high-priced products have high-quality packaging (Debelak 1998).

Visual impact is important at the point of purchase in order to drive selection. Products need visual impact to succeed on retailers' shelves. According to Brad Young (2000), "No matter how practical your customers might be, there's no doubt visual appeal is a major selling point to retailers, who want a dynamic-looking product, and to customers, who want to look cool." For Dave Dettman (2001), the founder of Mr. Product LLC, the ultimate package design for a product should look like "A brilliant array of colors, clearly typed instructions, detailed descriptions and high-resolution photography. [. . .] A perfect package design is one that says it all and keeps the consumer's attention on all four sides of the box. Ask yourself: If I were a consumer, what would I want to see? What would make me take it off the shelf and buy it?"

For researchers, one key question is "What is a good design?" In practice, *design* as a term has many uses. To the car companies, design can mean what the styling department does with the car chassis. To a container company, design can mean what their customer's packaging people do. According to Meyers and Lubliner some key factors underlying design are establishing the image or *personality* and determining the most important features of the product to the audience. It is important that the label and package send the same intended message to the consumers. The ultimate goal is to produce a label that is educational and user friendly and that adequately markets the product within legal specifications (Meyers and Lubliner 1998). In any case, design should not be considered simply as an afterthought where industrial designers are asked to "pretty up" a product that is about ready to be manufactured (Evamy 1994). After recognizing the high failure rates of products and the consequential financial losses for their newly introduced products, many CEOs have come to recognize that product design will rise to the forefront of competitive priorities in the coming years (Dickson et al. 1995).

Packaging Research

This is the neglected child in consumer testing of price, positioning, product, and package. Until recently, management paid little attention to a product's package. Companies that launched new products might have involved their in-house package design professionals, as well as outside design houses. However, much of the research was either qualitative, such as focus groups or in-depth studies. Quite often no research was done at all. Part of this lack of research, and thus paucity of information, stems from the perception in the corporation by many people, especially marketers, that the package is not particularly important; it is just a vehicle by which to enclose the product, especially in fast-moving consumer goods.

Part of the reason for the absence of a corpus of research on packaging can be traced to the design houses, which perceive their work to be artistic and thus are incapable of judging its merit. No design house wants consumers to design a package for them. This is threatening, as is the consumer-based design of concepts. Whether or not it is a well-founded fear, the trepidation felt by design

houses has communicated itself to the marketing community. As a consequence, there is a paucity of research literature on package design, although there are scattered articles here and there, and a number of well-respected companies specializing in package research. We can compare this dearth of literature to the extremely high volume of literature on advertising testing, perhaps because many more dollars are spent on advertising, so being "right" in one's advertising is perceived to be important, at least economically. The lack of information about packaging has been noted. Hutton (2003) remarked, "Marketing literature includes very little on packaging and design. Most relevant information can be found in practitioners' publications in the fields of corporate identity and graphic design."

When it comes to package development and research, the notion of packages as concepts is relevant but not at the forefront of the packager's mind. Most package designers feel that the package is a work of industrial art, incorporating the messages for the product in graphic form and, where relevant, providing the necessary function (e.g., specific shape, handle, and ergonomics). The designers simply ignore that the package can be systematically varied in terms of its features and these variations explored for consumer acceptance. The notion of research and consumer-guided design may be tolerated, but certainly it is not to be embraced.

Designers come from a different world than researchers, which explains the designer's worldview and why up to now there has been so little consumer research on design. The designer's education is different from that of marketers and market researchers. Designers undergo rigorous training that teaches them how to design products that function well mechanically, that are durable, that are easy and safe to use, that can be made from easily available materials, and that look appealing. Clearly, many of these requirements will be in conflict, and it is up to the skillful designer to achieve all of them simultaneously. The focus is on the technical solution of packaging problems, not on the consumer response to the packages. New York-based cosmetics giant Estée Lauder recently launched a training initiative aimed at strengthening the core capabilities of 40 package developers who work in the firm's packaging group. Roger Caracappa, who heads up packaging, requires that anyone in his group who does not have a technical degree in packaging, industrial engineering, chemical engineering, or the like, complete the Certified Packaging Professional (CPP) training program organized and offered by the Institute of Packaging Professionals (www.packworld.com). This training program focuses on the technology of packaging issues, not on the research into consumer perceptions.

There is no doubt that top management recognizes the importance of design and can use it as a tool to boost its competitive advantages. However, again, recognition of the need for consumer input in package design is emerging only slowly. In half of the companies surveyed in a recent study, the CEO had primary responsibility for design decisions! (Dickson et al. 1995). Another trend was recently uncovered: ". . . large multinational companies have begun to 'unchain' product designers capable of bridging and building upon the expertise of both marketing and engineering" (Lorenz 1994). It is important to be able to give the designer specific direction. Designers, unlike top management, are not closely involved with consumers and do not live or die by consumer acceptance of their product, as does management and marketing. It is easier for a designer to create what consumers look for when management and marketing provide the designer as much information as possible on the target market, package structure, and desired image. Once the required specifications of the packaging have been determined, the design of the packaging can be created so that it performs

within these boundaries (Meyers and Lubliner 1998).

Although many package designers are not privy to business issues, they are keen trend watchers. There is a continuing intimate relationship between packaging and the user, and that relationship infuses the package designer with ideas. For example, in the last year or two there seems to have been a boom in single-portion packaging and, as a result, shelves are home to packages containing very small volumes. Manufacturers will say this is a reaction to demographic changes indicating that more people are living alone (www.packaging-technology.com).

Optimizing Package Design: Can Art and Science Mix?

The increasing popularity of fact-based decision making in business has created an opportunity for package research to grow beyond simply testing whether consumers prefer one package or the other. One of the most intriguing developments is the advance of conjoint analysis (systematic stimulus variation) into package design. As discussed in Chapter 5, conjoint measurement comprises the experimental variation of components in a concept in order to understand the underlying dynamics of how the components perform. The respondent evaluates full combinations of these components (e.g., benefits and prices), and, from the ratings, researchers estimate the part-worth contribution of each component. The same research paradigm, viz., systematic variation of components, has begun to enter package research. This time, however, the components are features of the package, for example, different names, colors, and graphics. The concepts are full packages comprising these systematically varied features. From the responses of consumers to test packages, now often created on the computer screen, the researcher and package designer quickly discover what every design feature contributes to consumer interest, communication, etc.

The Design Process and Conventional Package Testing

No single ideation is likely to generate the final design concept that will move on to prototype development. The best parts of each ideation exercise are combined into a single design in a step called *design consolidation*. The designer fleshes out as much detail as possible at this time, including decorative graphics and brand name and logo (if known). It is necessary to flesh out the detail as much as possible at the stage of design consolidation because this is typically one of the last evaluation points before the company commits a lot of financial and human resources. Generally, at this consolidation phase realistic computer-generated renderings are preferred.

Industrial designers consider several factors when deciding on the appropriateness of a design. These may include quality of user interface, emotional appeal, maintenance and repair, appropriate use of resources, and product differentiation (Ulrich and Eppinger 2000). All of these considerations are then incorporated into the package, which may either be launched or, more recently, undergo consumer testing.

After all the foregoing factors have been considered and the package has been designed at the conceptual level, the designer is ready for consumer input. This is typically the early-stage package testing. Package testing traditionally has been an evaluative discipline that follows the design consolidation phase and works with realistic test stimuli. The stimuli may be mock-ups but should approximate the final package, if not in graphics, then at least in shape and contour. Sometimes the package research is done through qualitative research. The package designer general brings the prototype packages to a focus-group facility where the respondents handle them, look at them, and then discuss their reactions. The qualitative results from this early-stage exposure are then used to modify the package, change the graphics, and otherwise incorporate the consumer feedback. Other research procedures use conventional quantitative approaches, much like the evaluation of fully formed concepts. The package is presented to consumers, who profile their reactions to the package on a variety of attributes. The respondents may use the product or simply inspect it. From the ratings the package designer is assumed to know what to do. Where relevant, the researcher might also ask the respondent to indicate what other features should be present in the package.

The foregoing approach has been traditionally called the *voice of the customer* for obvious reasons. For example, the voice of the customer was extensively used in the design of the Infiniti QX4 sport utility vehicle. Infiniti drivers and nondrivers within the target market (35-64 years old, over $125,000 household income, and willing to purchase a luxury car) were presented with five different designs. The best of these was molded into clay and fiberglass models with the additional input of dealers. As a result, sales far exceeded the expectations (Gustke 1997).

It becomes clear from the foregoing description that package testing serves a number of purposes. The most important one is to confirm or disconfirm the objectives of the package designer. Typically, package designers create packages (either graphics and/or structures) with some objective in mind, such as reinforcing the brand or communicating new benefits, thus enhancing the chances that the product will be selected. Package testing must provide some idea as to whether a new package is successful. A modicum of sensitivity to the creative process is also in order, since package testing often reflects on the creative talent and sensitivity/intuitive capabilities of the package designer.

Conventional research develops a profile of the package and, occasionally, of expectations about the product in the package, respectively. The researcher shows the package to the respondent and obtains a profile of ratings, similar to the way that the researcher obtains product ratings. The attributes can vary substantially from one product category to another. Some of the ratings may deal with interest in the package (or expected interest in the product, based solely on exposure to the package). If the respondent actually uses the product in the package, then this experience enables the researcher to obtain ratings of person–package interaction, including ease of carrying or gripping the product, ease of opening, ease of removing the product, ease of storing the package, and so on. This type of information is extremely valuable to the package designer, who needs to know whether or not the package is on target.

Eye-tracking Research for Package Designs

The goal of *eye tracking* is to determine how the consumer responds when presented with either a shelf set or a single package. Eye-tracking technology traces the location of the line of sight. The tracking method records the path of the eye wanderings and what the eye is looking at. The recording and analysis of eye movements are fundamental to a diverse set of research applications, including studies of package features on the shelf (Gitelman 2002). The typical shelf comprises a complex array of package designs over which the consumer's eye wanders. The objective of package designers is to guide consumers to a specific package. This action then constitutes the first step in selecting the product. Eye tracking enables a researcher to identify the pattern to answer questions such as the following:

1. Is the eye drawn to the client's particular package?

2. Does the eye wander away from the package (thus diminishing the chances that the customer will select the product)?

When done for a single stimulus (e.g., an over-the-counter medicine), eye-tracking technology can show if and when, and for how long, key messages are looked at, but it cannot determine whether these messages drive interest in purchasing a product. When done for the entire shelf, eye tracking can identify whether a package is even looked at and for how long. In particular, analysis of users' eye movements in a virtual reality setting can potentially lead to further insights into the underlying cognitive processes of respondents (Duchowski et al. 2002). On the other hand, though, eye tracking has several drawbacks: accurate eye-tracking equipment is expensive, often awkward for participants, and requires frequent recalibration, and the data can be difficult to interpret (Jansen et al. 2003).

T-scope Research

A very popular method to understand responses to packages looks at the speed at which the respondent correctly recognizes a package. The method has its origins in the experimental psychology of more than a century ago. Psychologists hypothesized that underlying processes are involved in people's decisions. Although it may be impossible to know what is actually going on, a sense of the complexity of the underlying processes can be obtained by measuring the speed of reaction to the stimulus. High reaction speeds (i.e., short latencies for the response) suggest that the underlying processes are fairly simple. Low reaction speeds (i.e., long latencies) suggest that the underlying processes are more complex.

The rationale behind this type of testing, called *tachistoscope (T-scope) testing*, is that typical shoppers spend relatively little time inspecting a store shelf. Measures derived from eye-movement data reveal that, during brand choice, consumers adapt to time pressure by accelerating the sequence of activities involved in visual scanning by filtering information and by changing their scanning strategy. In addition, consumers who are highly motivated to perform the task filter brand information less and pictorial information more. Consumers under time pressure filter text information more and pictorial information less (Pieters and Warlop 1999).

For a package to make its impact, it must visually "jump off" the shelf. The T-scope research deals with the *recognizability* of the package when it is presented alone, as well as with the its recognizability when it is placed within the competitive array of other products. Researchers assume that those packages that are perceived in the short interval permitted by the T-scope have *shelf presence*. If the research interest focuses on the recognizability of a single package, then one can present single packages and determine the fastest shutter speed (i.e., the shortest time needed) for the package to be correctly identified. If the research interest focuses on the *findability* of the package on the shelf, then the researcher places the test package at different locations on the shelf and systematically as-

sesses the contribution of both package design and package location on the shelf as joint contributors to findability.

Package Testing as Concept Testing

With a little change in focus we can approach the design and testing of packages as a variation of systematic concept design and bring to bear the power of concept research. Conventional concept design comprises words and pictures, systematically varied in combinations according to a statistically balanced layout. The same can be done for package features, as well, although the freewheeling approach of concept research has to be reigned in a bit. When working with experimentally designed concepts, researchers do not have to worry about the features fitting together into a coherent whole. Researchers can combine phrases in a set of bullet points.

Figure 6.1. Features of a package, deconstructed to their visual components.

Respondents provide the connections between the different phrases. The fact that the concept comprises stand-alone phrases and does not look like a paragraph means very little because respondents fill in the missing material and connectives.

For systematic variation of features in a package such as graphics in a package design, researchers have to be a little bit more careful. Researchers cannot put together features on the graphics design in an array without having a *template*, which organizes the features and ensures that the package is coherent. Figure 6.1 shows an array of features for an orange juice container, Figure 6.2 shows a template, and Figure 6.3 shows what happens when the design features are put together in a template. To the respondents, the package design for orange juice looks like any other package design. Underneath that design, however, is a template and different options that fit in the template.

Experimental Designs

The set of combinations for packages must be designed with a certain amount of art mixed together with statistical considerations. The experimental designs are created by systematic variation of the test stimulus in order to

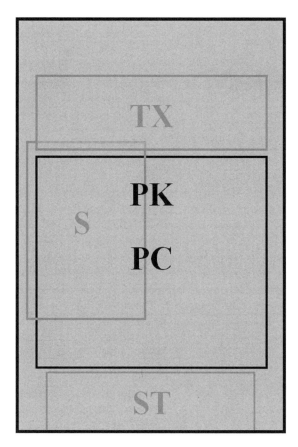

Figure 6.2. Template for the orange juice package that organizes the components into a coherent whole. The abbreviations refer to different parts of the graph: PC, product (e.g., fruit) picture; PK, background picture; S, benefit statement; ST, production method or benefit; and TX, brand name.

Figure 6.3. One of many reconstructed combinations of components for an orange juice package, following the template shown in Figure 6.2.

enable an estimation of each feature's contribution to the response. Unlike concepts, which can be created in such a way that some categories are totally absent for statistical reasons, a package design often has categories that simply *must* appear. The package design must have a shape, so, if shape is a variable, then every graphic package concept must have one of the shapes. There can be no package that lacks a shape, unlike a specific verbal concept that can be developed with the statement of shape missing.

When designing the test stimuli, researchers might wish to address the following three questions:

1. What are the categories of features to be varied?

2. Must each package have one element from a category, or can some categories be entirely absent from a package stimulus and

not be missed? As already noted, if one category is "shape," then that category must appear in every package concept. A package concept simply cannot be created without a shape. In contrast, if the category is "health message," then that category can be absent from some of the packages. Not every package need contain a health message. This distinction between *must be present* and *may or may not be present* has profound implications on the modeling. If the category must be present, then the researcher can estimate the relative value of the different elements in the category but cannot estimate the absolute utility value. This constraint in analysis occurs because the package design must have one element from a particular category. Thus, there is no chance to estimate the effect of absence of the category. The regression model takes care of this issue by leaving out one element from the category, if the category always has an element in the package. This missing element in the regression analysis is called the *reference level*.

3. How many combinations can a respondent evaluate without becoming bored? In concept work, with the respondents recruited to participate for an extended test session lasting 30 minutes, most respondents stop paying attention after having evaluated about 100–120 concepts in one rating question. With concepts the respondent has to read the concept before rating. In contrast, in package design research the respondent need only look at the package and rate the combination without being forced to read the label. This reduced effort allows the respondent to evaluate many more packages than concepts with the same number of concept elements.

The Human Element: Reactions of Package Designers to the Advent of Systematic Research

Package designers welcome feedback. It is the type of feedback that causes some to complain. A sensitive package designer gets a great deal out of focus groups because the designer can see and hear how consumers react to the packages. By presenting different possible packages, the designer can see which ones work and which do not. Focus groups provide the designer with a great deal of feedback, generally in a nonthreatening, nonjudgmental manner (viz., not judged by another professional, although consumers could reject the package). It should come as no wonder that this type of qualitative research has been welcomed by designers, because in essence it reflects how they would intuitively go about obtaining feedback about their creations.

The situation is somewhat different with the experimental design of packages. Like systematic, quantitative concept research, systematic package design research has its critics—some very vocal, some just passively resisting. Many professionals in package design feel very strongly that their efforts are artistic and thus cannot be quantified by research. Advertising agencies feel that same way about concepts: the concept itself cannot be quantified. In package design the professionals may feel that consumers can react to the packages only in an all-or-none manner, like or dislike, and cannot really provide valid direction. Whatever a consumer says, in the minds of these people, could be valid, but could just as easily be invalid. It is left in the purview of the package designer to interpret the consumer feedback.

Case History 1: The Principles Applied to the Graphics Features

Our first case history deals with the direct application of experimental design to the graphics features of a package. The product is frankfurters. The issue was to identify the contributions of the different graphics options to consumer acceptance and to the communication of both health and good taste. Marketing management realized that it might be impossible to communicate both health

and good taste simultaneously and thus wanted to determine which components of the package graphics communicated health and which communicated good taste.

The research approach followed the structure already described for orange juice, but applied to packages for frankfurters (wieners, hot dogs):

1. The artists created the graphics features, generating several alternatives for each category.

2. The graphics features were scanned and made into overlays.

3. Applying the experimental design method generated 48 combinations, with the property that each element appeared an equal number of times for any particular category. In some combinations the category was entirely absent. This absence was made possible by the fact that no single graphics element was absolutely necessary. Even if the graphics element was missing the design would make sense. Figure 6.4 presents an example of a combination with some of the features that were studied in the design.

4. The respondents evaluated all 48 of the combinations, rating each combination on three attributes (interest, communication of good health, and communication of good taste) by using a 9-point rating scale.

5. The ratings for each of the three scales were transformed to percentage top-3 box and analyzed.

6. The results were then arrayed in table form. Table 6.1 shows the utility values for the elements.

7. From the data in Table 6.1 the researcher can immediately identify which particular graphics features drive acceptance, health, and taste, respectively. The analysis simply looks at the elements that score highest in each category to get a sense of winners, and elements that score worst in each category to get a sense of losers.

8. By looking at the entire category (e.g., product name), the researcher quickly determines which category drives acceptance or

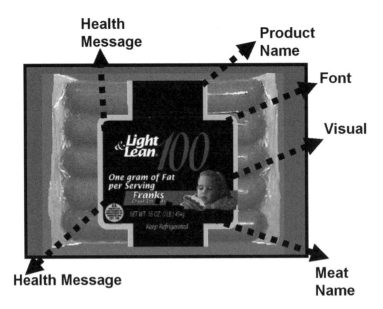

Figure 6.4. Example of a frankfurter/wiener package stimulus, pointing out some of the elements that were systematically varied by experimental design.

communication. Drivers are categories that show high positive or high negative utilities for the elements. For example, product name and design perform most strongly. Just the product name "Light & Lean" itself has a utility value of 19. The basic concept, comprising only this name, would be expected to score 49 (an additive constant of 30 plus a utility value 19 for a sum of 49).

Case History 2: Illustrating the Principles Applied to the Three-dimensional Features of a Package

A second case history on package design illustrates the necessary combination of *high touch* (i.e., client input and insight) and *high tech* (i.e., experimental designs that can be executed in an actual study). The case history presented here for the physical features of a new coffee container describes the approach, test execution, and data analysis (Bernstein and Moskowitz 2003). This case history differs from the foregoing graphics case history with meat because now we deal with actual

product shapes and features, not just with two-dimensional pictures that are superimposed on each other.

Background

The coffee manufacturer periodically reviewed packaging strategy and existing packages. Management at the company recognized that keeping up to date with package design was a critical component of product success, especially in light of the competitive activities of other manufacturers and the demands of the trade.

The goal for the periodic reviews was to understand which packaging features stimulated consumer interest and motivated purchase for both the company's current packaging and for new packaging ideas, as well as for the competitor packaging, respectively. The company wanted to determine the impact and relevance of a variety of packaging options in order to develop the best new coffee package that would appeal to a variety of user groups. These user groups comprised of

Table 6.1. Utility values for the different graphics-design features for frankfurter/wiener packages

	Interest 30	Taste 41	Health 42
Product name/design			
Light & Lean 100	19	9	10
Light & Lean 99	16	7	8
Current Light & Lean 97	15	8	8
Light & Lean 95	13	6	8
97 WHITE	11	3	1
Large Light & Lean, small 97	10	5	6
Light & Lean only	10	5	4
Only 97	7	3	1
Font for features			
Font 5	16	7	6
Font 8	12	6	5
Font 1	11	6	5
Font 4	10	6	5
Font 6	10	6	5
Font 7	10	5	6
Font 2	9	5	5
Font 3	9	7	6
Additional burst			
Great for kids	6	1	3
Hardwood smoked	5	3	1
Meat name			
Hot dogs	5	1	2
Franks	4	2	3
Wieners	1	−1	0
Health message			
97% Fat free	12	4	12
One gram of fat per serving	12	4	12
One gram of fat/one great taste	11	6	11
Extra lean	10	4	7
Fat free	10	2	13
Visual			
Kid eating hot dog	6	5	3
Picture of a heated hot dog	5	5	4
Current runners	4	3	4

The utility values are shown for three rating variables: purchase intent, perceived taste acceptance, perceived level of healthfulness.

a variety of different consumers defined as exhibiting different patterns of product use.

Part of the case history was a desire to understand the feasibility of using the same package, or at least small variations of the package, in multiple countries. This notion of a single base package and local modifications on a country-by-country basis was of particular interest because it combined two key factors: (a) understanding country differences and (b) opportunities for significant cost savings. As one might expect from the practical, business point of view, any commonalities leading to a single or limited number of packages was welcome because it would minimize packaging costs.

In an earlier developmental phase the design team had identified all the viable packaging options. The variables were defined as the package shape, the opening and closing mechanisms, the surface material, and the presence or absence of several value-added features. There were four types of package shape: one type was rigid and three had flexible construction. Since package shape and closing mechanism correlate highly in reality, these characteristics were combined into a single variable to eliminate the correlation (viz., to introduce a new way to describe the features). In total, 42 packaging options were organized into categories of related characteristics. A fractional design specified the composition of 186 test packages. Each of the test packages was realizable (i.e., could be manufactured). It is important to keep in mind that, when companies wish to investigate many design factors, a relatively large number of test combinations must be created to cover this large set of test factors. Often, when faced with the daunting task of creating these combinations, a less committed company or package designer will choose the easier path of creating a few "best guesses" and submitting these to test. Fortunately, in this study the manufacturer stuck with the course, because it was vital to understand truly what the relevant factors were for the coffee container.

A design firm generated photo-realistic depictions of each of the test packages on computer-readable media. Package visuals comprised a main picture of the package on the left side of the visual, with supplementary inserts of key features and descriptive text about these features on the right side of the visual. Text was translated into the appropriate language for the test market (see Figure 6.5).

Orientation

The packaging study was executed in four separate European markets. The respondent sample was constructed so that separate models for contributions of components could be developed in all important user subgroups. Respondents were seated at individual computer workstations. At the beginning of the session, respondents completed a short orientation. Moderators were provided with the orientation script. Each of the various package options was described. The description helped the respondents to understand the stimuli. Then, each of the respondents handled 15 three-dimensional examples. This orientation enabled the respondents to "experience" the various packaging options that they would subsequently evaluate on a two-dimensional screen presented by the computer. It is important to recognize that anything that helps the respondents to understand both the task and the test stimuli adds to the value of the study.

Positioning the Stimulus Through an Orientation Concept

Quite often in concept research the respondents must evaluate the concept against some set of expectations. The expectations are set up by an orientation concept that *sets the scene*. In this study the orientation was achieved by having the respondents read the following concept. Note that the respondents do not rate the setup concept. They simply read it:

You will be looking at a variety of packaging alternatives for the *product category* (actual category in original instructions). Some of the features you see may be familiar and some are new ideas. In any case the package is colored in a neutral color for this test. Do not use the color as a signal for the product quality.

When the computer shows you a package, look at it carefully and read all of the written descriptions. Assume the product you use most often at home is in the package.

Please take your time and read each concept (screen) thoroughly. Now, tell us how you feel about the package by using the questions shown on the bottom of the computer screen. Enter your rating based on the following questions. The entire concept should be rated as a whole.

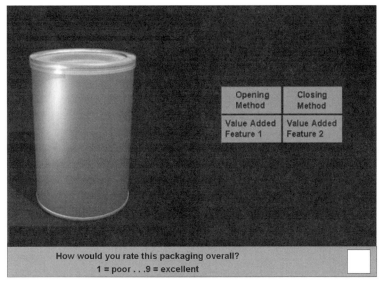

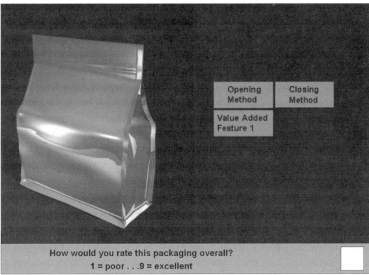

Figure 6.5. Two examples of package stimuli for coffee. **Left:** A photo-realistic rendering of the package. **Right:** A description of the key features.

Modeling the Contributions of the Different Package Features

The test packages can now be treated as though they are package concepts. Each of the test concepts has a specific set of features. By setting up the packages through experimental design, the researcher creates a model relating the presence or absence of the features to the ratings, as is done in conjoint analysis. The model is called *dummy variable* because the predictors (i.e., independent design variables) are either present or absent and thus take on the value 0 when absent and 1 when present. This is called *dummy-variable coding*.

The dummy-variable model works when all the variables are statistically independent. This is not the case for package design, where some features must be present for logical reasons. Knowing the condition of $n - 1$ alternatives in a package automatically determines the condition of n alternatives. This is the case for color or shape. For instance, if there are two colors—A and B—a color must be either A or B. *There are no colorless packages.* If a package is colored A, then it cannot be colored B, and vice versa. Every package has a shape, a closing/opening mechanism, and a material. Because an option from these categories is *always* present, it was necessary to consider one option as the *reference* option. It does not matter which option is chosen to be the reference. Coefficients for these categories estimated through regression represent the contribution *relative to the reference*. For convenience, the current (in-market) packages were assigned as the references.

The analysis of the element contributions provides clear insight into what features of the package drive acceptance. The four countries generated similar models. Within a single category the different elements or options showed similar rank orders. Elements that were highly acceptable in one country were also highly acceptable in the other countries. Table 6.2 lists the detailed results from this study. From a single table, such as Table 6.2, researchers understand the dynamics of the package.

1. *Package shape/closure.* Several type-A packages are liked the most. This finding is particularly intriguing because the different regions have different package histories and different in-market packages.

2. *Opening method.* In every region the same opening method (method A) is preferred.

3. *Material.* Consumers do not show any strong or consistent preference for material. Appeal is only minimally impacted by material.

4. *Package shape.* Package appeal is impacted strongly by shape of the package and how it is opened and closed.

5. *Value-added features.* Both features were designed to increase the usability of the package. Both contribute to consumer liking and seem to be valued by consumers.

The issue of validity of the additive model is often raised for both concepts and packages, but especially for packages because of the belief that package design is an art form. The issue of validity used to be raised frequently among market researchers and marketers new to conjoint measurement when they first became involved in concept research. Now that package design has been brought into the world of experimental design and data-driven development the same questions about data and model validity are raised once again with the same vitality and language as before.

One way to show validity of the package-features equation is to use the model to estimate the likely rating for each package. If possible, one should hold out a few packages from the model to show that the model estimates packages not even used in creating it. However, in any case, a graph such as that shown in Figure 6.6 goes a long way to dispel the designer's and marketer's fears that the model is meaningless statistical representation from a meaningless, self-indulgent statistical exercise. By knowing the features of the package, the researcher can use the model to estimate acceptance, even for combinations not directly tested.

Overview

The migration of experimental design into package development provides an interesting example of how the scientific methods used by researchers have affected the design and creative disciplines. Unlike conventional marketing research of concepts, where empirical data are welcomed, package designers are more cautious and perhaps less than completely welcoming. This diffidence to accept

Table 6.2. Additive model for package liking*

	Total Europe	Country			
		1	2	3	4
Base size	*811*	*204*	*207*	*200*	*200*
Additive constant (k0)	*25*	*22*	*33*	*16*	*41*
Package shape					
Type A-1	21	26	17	29	5
Type A-2	19	27	12	27	3
Type A-3	18	21	10	31	2
Type A-4	16	17	16	27	−1
Type A-5	16	23	9	26	0
Type A-6	14	16	10	26	−2
Type A-7	12	15	6	24	−4
Type A-8	9	9	5	20	−7
Type B-1	6	10	1	19	−11
Type C-1 (reference)	0	1	0	0	0
Type B-2	2	3	−1	12	−14
Type B-3	0	−7	−2	10	−9
Type C-3	−3	−5	−7	2	−7
Type C-4	−3	−8	−3	0	−9
Type C-5	−4	4	−11	6	−21
Type C-6	−5	1	−9	1	−20
Type D-1	−5	−9	−8	3	−13
Type B-4	−5	−5	−11	3	−14
Type D-2	−6	−5	−9	1	−18
Type C-7	−7	−5	−10	0	−22
Type C-8	−8	−7	−12	−1	−17
Type C-9	−10	−14	−15	−5	−13
Type C-10	−11	−12	−15	−4	−17
Type C-11	−11	−11	−18	−1	−21
Type C-12	−12	−14	−17	−4	−21
Type D-3	−12	−13	−17	−4	−23
Type C-13	−13	−15	−18	−6	−19
Type C-14	−13	−16	−15	−8	−20
Type D-4	−16	−6	−20	−14	−31
Type C-15	−22	−16	−28	−19	−30
Opening method					
Method A	8	8	7	9	−8
Method B (reference)	0	0	0	0	0
Material					
Type A (reference)	0	0	0	0	4
Type B	0	1	2	−3	3
Type C	−1	1	−1	0	0
Type D	−1	1	−1	−2	1
Type E	−1	1	−1	−1	0
Value-added feature					
1	7	5	6	11	4
2	7	4	6	14	6

*Numbers are part-worth contribution for packaging components, or the conditional probability that the consumer will say that he or she likes the package concept (top-3 box) if the element is present.

Total European Sample -- Overall Liking of Package
Measured vs Modeled Scores

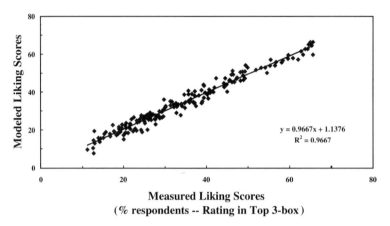

Measured Liking Scores
(% respondents -- Rating in Top 3-box)

Figure 6.6. Validation by means of comparing the measured (actual) versus modeled (estimated) liking scores. The computed regression model accurately describes measured package liking.

concept testing applied to packages comes from the history of research in package design. For many years, researchers provided an evaluation of the package designer's performance. Either the package was acceptable or not acceptable, or somewhere in the middle. The feedback, so important for the designer, was incidental to the grade of pass/fail.

The research approaches espoused in this book, using experimental design, consumers, and the assessment of a broad stimulus range rather than a final single creation, provide a different perspective for designers. The objective of systematic design is to identify what works and what does not, and what communicates and what fails to communicate. Designers welcome this type of information, especially if the objective is to teach, to create a knowledge foundation, and to aid rather than evaluate the design process.

References

Bernstein, R., and Moskowitz, H.R. (2003). The marriage of graphic design & research: experimentally designed packages offer new vistas & opportunities. In: Scott, L.M., and Batra, B., editors. Persuasive Imagery: A Consumer Response Perspective. Mahwah, NJ: Lawrence Erlbaum, pp. 337–348.

Debelak, D. (1998). Package deal: harness the power of packaging to push your product. Business Start-ups, April 1998.

Dettman, D. (2001). The whole package: when it comes to boxing up your product, don't cut any corners. www.entrepreneur.com; 19 March 2001.

Dickson, P., Schneier W., Lawrence, P., and Hytry, R. (1995). Managing design in small high-growth companies. Journal of Product Innovation Management, 12: 406–414.

Duchowski, A.T., Medlin, E., Cournia, N., Murphy, H., Gramopadhye, A., Nair, S., Vorah, J., and Melloy, B. (2002). 3D eye movement analysis. Behavior Research Methods, Instruments, & Computers, 34: 573–591.

Evamy, M. (1994). Call yourself a designer. Design, March: 14–16.

Gitelman, D.R. (2002). ILAB: a program for postexperimental eye movement analysis. Behavior Research Methods, Instruments, & Computers, 34: 605–612.

Gustke, C. (1997). Built to last. Sales and Marketing Management Magazine, August: 78–83.

Hutton, J.G. (2003). REQUEST: packaging & design course. www.marketingpower.com/live/content.php? Item_ID=17189.

Jansen, A.R., Blackwell, A.F., and Marriott, K. (2003). A tool for tracking visual attention: the restricted focus viewer. Behavior Research Methods, Instruments, & Computers, 35: 57–69.

Lingle, R. (2003). Consumer convenience: pouches stand up to retorting [Sidebar]. www.packworld.com.

Lorenz, Ch. (1994). Harnessing design as a strategic resource. Long Range Planning, October: 73–83.

Meyers, H.M., and Lubliner, M.J. (1998). The Marketer's Guide to Successful Package Design. New York: McGraw-Hill.

Pieters, R., and Warlop, L. (1999). Visual attention during brand choice: the impact of time pressure and task motivation. International Journal of Research in Marketing, 16: 1–16.

Ulrich, K.T., and Eppinger, St.D. (2000). Product Design and Development, second edition. Homewood, IL: Irwin/McGraw-Hill.

Young, Brad (2000). Lookin' good: giving your products visual impact. Entrepreneur Magazine, February.

Chapter 7

Segmentation Approaches, Results, and the Differential Importance of Categories

Introduction

Concept testing quantifies the appeal of a product idea to consumers by drawing on company knowledge and experience combined with interviews and focus groups. Concept testing is used to address hypotheses about the benefits and features of a product and about the segments of the general population that will comprise the primary target market. Market segmentation divides consumers into distinct groups, or segments, with similar needs and characteristics (Boyd et al. 1989), who will presumably react differently to the concepts. Well-defined segments are easier to target with specific products and market strategies. Gathering more relevant and predictive lifestyle and demographic information on potential consumers enables the company to develop greater depth of knowledge about the targeted market and consumers. Market segmentation is the selection of groups of people who will be most receptive to a product. The most frequent methods of segmenting include demographic variables, such as age, sex, race, income, occupation, education, household status, and geographic location; psychographic variables, such as lifestyle, activities, interests, and opinions; product-use patterns; and product benefits.

A view into the literature shows six major ways of defining segments (Lappin et al. 1994):

1. *Customer needs:* for example, prestige or convenience (see Lappin et al. 1994)

2. *Product-related behavior:* brand loyalty and frequency of purchase or use

3. *Product use* (Wellner 2000)

4. *General behavioral descriptors:* lifestyle or personality traits, including the propensity to adopt innovative products (Thorson 1989)

5. *Demographics:* for example, age (Wellner 2002), ethnic origin (Gardynhen 2001), and income.

In special cases it is important to distinguish between a concept testing and segmentation for an *existing product* and a *new product*. When testing new products, the physical characteristics important to consumers may not yet be known. To increase market demand for existing products, typical uses of market segmentation studies might be the following:

1. Segment customers on the basis of one or more product-related behavioral variables.

2. Conduct focus groups and surveys on each of the segments, looking for differences in benefits sought, demographics, and lifestyle.

3. When identifying new products or opportunities, the process may be modified to segment customers on the basis of needs or benefits.

4. Conduct focus groups and surveys to identify segments whose needs are not satisfied with current offerings or who could benefit the greatest by new offerings. Further segment those groups by general behavioral and demographic descriptors and, finally, design a product and marketing strategy targeted for identified segments (Lappin et al. 1994).

Despite all of the advantages of segmentation, every researcher has to be aware that in some cases the classifying of consumers could be inaccurate. According to Szmigin (2002), people are not generally acting their age, their class, or their gender. This type of disconnect between segmentation membership and expected behavior regarding product perception and use could be problematic for marketers later when the product has been developed according to the guidance provided by the segmentation research.

What Is Segmentation and Why Do It?

Individual differences in reaction to products and concept are pervasive. During the period up to the 1940s and 1950s there were scarcities of items. Consumers were often happy to purchase what was available. Consequently, the Henry Ford belief that consumers could have a car in any color, as long as it was black, was more common than believed. There was simply too much demand in contrast to too little supply. The notion that one might have to change the features of the product or talk about the product in different ways to appeal to different audiences was simply not recognized as being valid. For each dissatisfied customer who did not like the characteristics of the product, there were three, four, or more individuals who might accept what was available. At this time the famous German college of design *Bauhaus*, under director Hannes Meyer, changed its focus from combining art and production to cost-cutting industrial mass production in order to make products affordable for the mass of the population.

Recently, for North America and Western Europe, the assumption of the mass market no longer holds true for most businesses and product categories. A firm adopts either a mass-market strategy or a market-segmentation strategy. There is no in between (Neal 2003). One variant of market differentiation is *segmented marketing*, the basic principle

of which is to allocate members in the consumer population as clearly as possible to at least two customer groups according to external criteria that can be specified. The bases for segmenting a market are nearly unlimited and can include such factors as product class behaviors, product preferences, brand-selection behavior, demographics, geography, and socioeconomic status. Indeed, the number of factors is almost limitless, although some generate reasonable segments whereas others do not.

Green and Tull (1978) set four basic criteria for market segmentation:

1. The segments must exist in the environment; that is, they are not to be figments of the researcher's imagination.

2. The segments must be identifiable (repeatedly and consistently).

3. The segments must be reasonably stable over time.

4. One must be able to reach segments efficiently through specifically targeted distribution and communication initiatives.

According to Neal (2003), there are, broadly speaking, only two methods for segmenting a market: a priori and post hoc methods. *A priori segmentation* is a procedure whereby a company chooses to break out customer groups by a generally accepted classification procedure assumed to relate to variations in customer purchase or use of the product category. This grouping may be the result of company tradition, recognized industrial groups, or some other external or internal criteria. Examples of a priori segments include such classification schemes as the following:

1. Standard Industrial Classification (SIC) groups (e.g., www.govtsales.com/sics/sicgroups.htm)

2. Geographic regions or sales territories

3. Basic demographic groups (e.g., sex, age, or household composition)

4. Purchase or usage groups (e.g., heavy users, light users, or nonusers)

5. VALS [SRI (Stanford Research Institute)] values and lifestyles classification system)

6. PRIZM or similar geodemographic classification systems (e.g., www.claritas-marketing.com/PRIZMseminar.htm)

Despite the attractiveness of the segmentation criteria, demographic information cannot be used exclusively to determine customer needs. Consider an array of a marketing article that worked with this information about two individuals. Monahan (2003) asked, "Are Grace Slick and Tricia Nixon Cox the same person?" Grace Slick, lead singer for *Jefferson Airplane*, and Tricia Nixon Cox, the preppy daughter of former President Richard Nixon, are demographically indistinguishable. Both women are urban, working women and college graduates, age 25–35, at similar income levels, and are from a household of three, including one child. Demographics could not explain the distinctly nondemographic difference between Grace and Tricia.

More insight emerges about *truer* segmentation through an analysis of an individual's *mind-set*. One of these is the method of "needs segmentation: knowing *why* a customer buys the product," which can do more to help increase customer value than can "studying behavior: knowing what a customer buys" (Neal 2003). The reason is quite simple. The motivation to purchase may be distinct for different customers who are buying the same product. One person buys it for himself or herself, and another one doesn't like the product but buys it for a friend as a gift. Neal (2003) rightly refers to an important caution that should be recognized when using a priori criterion for segmenting a market. Our society is dynamic! This means segmentation studies conducted, or validated, 3 or 4 years ago may not be appropriate today.

Post hoc segmentation is empirically derived from the results of research studies undertaken for the specific purpose of segmenting a market. Segments generated from such a study emerge after aggregating buyers who respond similarly to a set of *basis questions*. *Basis variables* for such a post hoc segmentation study include examples such as product-attribute preferences or product-purchase patterns, as well as benefits preferences or loyalty and socioeconomic status, lifestyles, or self-image (Neal 2003).

Traditional algorithms for conducting traditional post hoc segmentation studies can be found in textbooks (e.g., Kachigan 1991). These are some of the methods:

1. Cluster analysis (Aldenderfer and Blashfield 1984)

2. Correspondence analysis (Benzecri 1992)

3. Q-type factor analysis (Thompson 2000)

4. AID method [automatic interaction detector (Morgan and Sonquist 1963)]

5. Discriminant analysis (Belson 1981)

6. Brand-user profiles (Hammond et al. 1996)

7. Assumption of a mass market (Ehrenberg 1988; Elrod and Keane 1995)

Not all opinions are so positive about segmentation, however. Malcolm Wright, from the Department of Marketing at Massey University in New Zealand, refers to the misapplication of the expression *market segmentation*: in his opinion, any set of competencies or capabilities will be adequate for market segmentation, because the term has come to be void of any meaning (Wright 2003).

In the last 10 years there have been some interesting new developments in market-segmentation research. These are some of those recent advancements:

1. *Multidimensional segmentation* (Bibb 2001). This is a step-by-step process that begins by identifying customer segments from both strategic and tactical perspectives.

2. *Artificial neural networks* (Huston and Klerfors 1998). Using sophisticated and specialized hardware and software, these networks attempt to simulate the multiple layers of simple processing elements called *neurons*. These networks look for patterns that more conventional methods would not be able to discover.

3. *Latent class models* (Vermunt and Magidson 2000). These enable the users to simultaneously optimize a research function and find clusters of cases within that framework.

But if there are many approved methods, why do market-segmentation strategies sometimes fail? Neal (2003) suggests three reasons:

1. A lack of senior management involvement and recognition that market segmentation is a strategy. A strategy must permeate the firm and the way it deals with the marketplace.

2. A lack of understanding of the concept of market segmentation and its need to identify groups that truly exhibit behavioral response differences to variations in the marketing mix.

3. A presumption that all markets can be segmented on bases that are subject to influence by variations in the marketing mix.

For the most part, segmentation by the conventional methods does not reveal many differences in the ratings of concepts or products as assigned by the segments; that is, when a researcher divides the population into different groups of individuals, there appears to be no clear differences among the subgroups in their response to stimuli. What appeals to one group also appeals to the other group. Plotting the concept scores for one group against those of another group shows a 45° line across a large number of different concepts. Although people might differ from one another in ways that we consider dramatic, those differences do not manifest themselves in strong differences in reactions to concepts.

Despite the recognition that segments are usually quite similar in the way they react to concepts, except perhaps in the most unusual cases where the concepts clearly have been designed for one of the segments, researchers continue to dream that the analysis of different subgroups will, in the end, generate different concepts. Users of research continue to demand representation of different groups in the population, trusting and ever-believing that they can find a relation between some conventional measure such as age, or brand used most often, and reaction to the concept. The relation does not appear.

The perplexing lack of relation between concept performance and segment membership developed by conventional means becomes quite obvious when a researcher systematically varies the concept in terms of different types of messages (e.g., health messages vs taste messages). If a person says that he or she likes health products, we might expect that person to respond strongly to concepts that emphasize health. The person might or might not respond strongly to concepts that emphasize taste.

The ability of conventional segmentation can be easily tested through concept screening. The results in Figures 7.1. and 7.2. tell the story. The study, which dealt with vegetarian burgers, comprised eight concepts presented to 120 respondents. Table 7.1 presents the concept summary, the tonality of the specific concept (whether it was more for health versus more for taste), and the top-3 box scores for two *conventionally derived segments*: the "health high" (health more important than taste) and the "health low" (taste more important than health). It is clear from Table 7.1, and especially from Figure 7.1, that the two conventionally derived segments respond almost identically to each other. The concepts appealing to one segment appeal to the other segment, as well. Yet, according to the way these consumers profiled them-

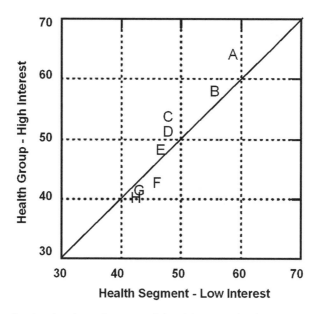

Figure 7.1. Scatter plot showing the performance of the eight vegetarian burger concepts among the two conventionally derived segments. The line is the 45° line showing how identical concept performance would appear.

Table 7.1. Concepts and concept performance for vegetarian burgers, by total panel, and conventional segments (health low, health high)

Concept	Concept Summary	Tonality	Total	Health > taste	Taste ≥ health
				Segment	
				Health higher than taste	Taste higher than health
C	100% organic	Health	46	53	48
F	Endorsed by the American Diabetes Association	Health	41	42	46
H	Recommended by your doctor	Health	38	40	43
A	Delicious, indulgent flavors with ingredients like fire roasted sweet bell peppers and savory sautéed mushrooms	Taste	57	63	59
B	The original, classic vegetarian burger . . . with a great grilled flavor	Taste	52	56	55
D	So moist and juicy	Taste	45	50	48
G	Fills that empty spot in you . . . just when you want it	Taste	38	41	43
E	Healthy eating that tastes great	Taste Health	43	48	46

All numbers are top-3 box acceptance on the 9-point love–hate scale.

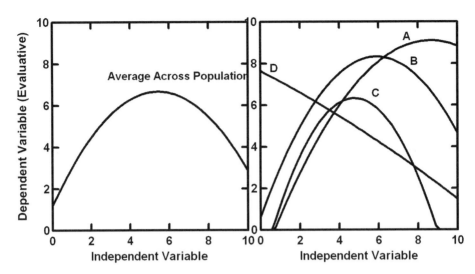

Figure 7.2. The typical relation between sensory intensity/physical stimulus as the independent variable and the hedonic rating as the dependent variable. **Left:** The curve as one might obtain from the average data. **Right:** The curves that might emerge from treating the hedonics data on an individual basis.

selves, they should respond quite differently to the concepts, some of which stressed health benefits and others of which stressed taste benefits. *In the end, the self-profiling of respondents into segments based on their own hierarchy of what is important did not generate appreciable differences in the response to the concepts.*

Response-based Segmentation: Origins in Product Research

Researchers have, for many years, wrestled with the issue of interpanelist variability. Dividing the respondent population into presumed homogeneous groups based on external criteria, or even on the basis of self-profiling of interests and attitudes, appears at first glance to be intuitively the right way to proceed. Often it is not. In the most intractable cases, researchers might discard the person-to-person variability as simply a disturbing, unwanted aspect of data, and merely concentrate on the mean or other measure of central tendency. Such attention to the mean characterizes much of the early

work on direct scaling. In other cases, however, the individual variability is simply too great to ignore as just an inconvenience in the scientific quest. A good example is hedonics of taste and smell in work with actual food products. There is substantial interpersonal variation—so much so, that it cannot be simply explained away as error variability. Such rampant variation is so well known and accepted as to give rise to the aphorism "Of taste, one does not dispute," where *taste* refers to one's liking or disliking. Taste is used here in two ways: in the traditional sensory way and in hedonics. Also, there appear to be no universally liked odors or tastes; one person may dislike what another person likes. This observation has been in the scientific literature for more than 75 years (Kenneth 1927).

The variation is clear for food products and for other products invoking the chemical senses. Research methods have been developed to separate people into segments. Conventional research methods, using self-defined geodemographic variables, suggest no real differences among people on the basis of these variables. Just because a person says that he or she is a resident of a certain market does not mean that the person automatically falls into a particular segment. Furthermore, even brand behavior does not necessarily reveal substantial differences in the reactions to food products. Those people who define themselves as Pepsi drinkers versus Coke drinkers often show similar preferences. There are differences, but these are relatively minor and could in fact be random noise. Finally, even self-defined psychological profiles such as "adventurous eater" versus "timid explorer" do not necessarily drive different sensory preferences.

Contributions to Segmentation from the Psychophysical Viewpoint

Psychophysicists wrestle with the issue of sensory-preference segmentation even in simple systems such as sugar plus water (Ekman and Akesson 1964; Pangborn 1970, 1981). If one adds sucrose at varying concentrations to water and instructs the respondents to rate liking, one quickly finds that some groups like the solution sweeter, whereas others prefer it less sweet. They all taste the solution similarly because their sweetness vs concentration ratings are quite similar to each other. They appear to live in similar sensory worlds, as least insofar as sweetness perception is concerned, but live in rather different sweetness hedonic worlds.

One method for segmentation looks at the sensory-liking curve and separates people by the location on the sensory axis where a person's liking rating peaks. Rather than depending on a person's definition of who he or she is, or what he or she wants, the segmentation approach looks at the pattern of sensory intensity vs liking created by the respondent and defines a person by a single point: the optimum of that pattern. This direction of research into segmentation was developed by the senior author starting in 1981 (in an applied study on responses to coffee) and continues even today (Moskowitz et al. 1985).

The segmentation schemes use the relation between physical stimulus intensity or rated sensory level (independent variable) and the subjective hedonic response (dependent variable). The left panel of Figure 7.2 shows the typical relation between sensory intensity (*x*-axis) and liking (*y*-axis). The right panel of Figure 7.2 shows that this typical relation may result from the combination of data from different individuals, with these individuals showing a variety of patterns. Averaging the individual patterns (or better, the individual ratings) often masks fundamentally different groups of people with different patterns (see Moskowitz 1981, 1986).

Given this single, fundamental value for a single individual (i.e., level of the independent variable where liking maximizes) for either one attribute or many attributes (e.g., liking vs sweetness, or liking vs darkness), the clustering becomes a straightforward statistical operation. The researcher clusters the individuals based on the profiles of their optimal sensory levels. The optimal level is defined as the location on the abscissa (*x*-axis) where liking reaches its peak. In some cases—for example, when the independent variables are highly correlated sensory attributes—one may wish to reduce the set of variables to an orthogonal array to remove redundancy. Otherwise, if the data reduction were not done, the segmentation could be biased in favor of one type of sensory input, such as flavor, to the detriment of another type of sensory input, such as appearance. Redundancy prior to segmentation is reduced by a principal components analysis on the set of variables. Principal components analysis creates a reduced set of factors and locates each individual as factor scores on the reduced set. Clustering then aggregates individuals based on their factor scores. Table 7.2 describes one version of the segmentation algorithm. At each step of the algorithm the researcher can opt to use other statistical techniques or measures.

There are two key benefits to the foregoing segmentation approach:

1. *Patterns drive segmentation.* The segmentation is based on the *pattern* of ratings assigned to an array of stimuli, not just to one stimulus

2. *Scale independence.* The segmentation is independent of the level of liking assigned by a particular individual and depends only on the location of the maximal liking for that individual. This makes the segmentation procedure independent of scale and thus reduces the chance of a statistical artifact that may influence the results.

Applying the Segmentation Algorithm to Concepts or to Concept Elements

There is no reason, however, that the approach be confined to products. With some modifications, researchers can apply the approach to concepts as well. The segmentation approach, originally designed for products,

requires that the evaluative attribute (e.g., overall liking) be related to a descriptive characteristic of the product. This is the sensory attribute. It further requires that the same respondent evaluates several different products, usually at least six, to generate enough data points for a curve. The key to adapting the segmentation approach for concept development is to establish a parallelism, treating concepts or concept elements as if they were products.

The parallelism defines this set of equivalences:

1. *Stimuli.* Originally the stimuli in sensory-preference segmentation constituted complete products. Now the stimuli are either complete concepts or concept elements from a conjoint study.

2. *Dimensions.* Originally the dimensions were sensory profiles of the products, obtained either from expert panelists in their descriptive language or from consumers using consumer terms. For concepts the dimensions are either communication profiles of complete concepts, obtained from respondents in the study, or semantic scale profiles

Table 7.2. One version of the sensory segmentation algorithm used to create subsets of individuals in a population who show different patterns of optimum sensory levels at which liking maximizes

Step 1	Calculate the mean of each sensory attribute for each product. This step generates the set of independent variables (viz., a profile of each product on a set of sensory attributes).
Step 2	For each respondent and for each sensory attribute create a quadratic equation relating that individual's liking rating (e.g., overall liking) to the sensory attribute; the equation: liking = A + B (sensory attribute level) + C (sensory attribute level)^2. The independent variable is a sensory attribute whose values come from a separate expert panel or from the total panel (e.g., average for that attribute and product, computed across all individuals who evaluated the product). The liking ratings come from the products that the respondents evaluated.
Step 3	For each respondent and for each attribute solve the quadratic equation in order to identify the optimal sensory level. Make sure that the sensory optimum lies within the range of the sensory levels tested in the study. Step 3 generates a matrix. The rows of the matrix correspond to respondents, the columns correspond to sensory attributes, and the values inside the matrix correspond to the optimal sensory level for each attribute for each respondent.
Step 4	Many of the sensory attributes correlate with one another. Apply a principal components analysis to the columns (sensory attributes) and rotate the solution by using a *quartimax* rotation method.
Step 5	Cluster the rows (panelists) by using the factor scores by the method of K-means clustering. The clusters represent sensory segments of individuals who differ from one another in their optimal sensory profiles.

Adapted from Moskowitz et al. (1985).

of the individual elements, obtained from a group of dimensionalizers prior to the IdeaMap study (see Chapter 5 on conjoint analysis). For both product and concept research the dimensions are not evaluative but, rather, descriptive. Communication attributes can be scales such as "more for men versus more for women," "for adults versus for children," and "for taste versus for health." These attributes may have a hidden hedonic attribute, but on the surface they appear to be simply descriptions of the concept or of the concept element.

3. *Key evaluative attribute.* For products the key attribute is overall liking. For complete concepts the key attribute is either overall liking or purchase intent, depending on the specific scale used. For concept elements the key attribute is the part-worth utility for liking or for purchase intent. Each concept or concept element has one value for this key attribute for the total panel.

4. *Modeling.* The modeling step remains the same. Modeling relates the key evaluative attribute to the semantic or communication attribute by means of a simple quadratic equation. For example, if the study deals with beverages, and the consumers or the dimensionalizers rate the concept or element on "more for taste versus more for health" (abbreviated as TvH), the equations would be:

$$\text{concept purchase intent} = k_0 + k_1 (\text{TvH}) + k_2 (\text{TvH})^2$$

$$\text{conjoint element utility} = k_0 + k_1 (\text{TvH}) + k_2 (\text{TvH})^2$$

Example of Segmentation: Grapefruit Juice

We can understand the application of concept-response segmentation by a simple example, which comes from the development of grapefruit juice. The original business objective of this study was to identify the product characteristics of a grapefruit juice and, secondarily, to identify the marketing characteristics (Moskowitz 2003). The project had been originally commissioned in order to develop a database of information that could be later used by the manufacturer for a line of different grapefruit juices.

From the scientific perspective the study provides insight about the relative importance of elements and the impact of segmentation on performance of these elements. The actual study was done by using the IdeaMap method (see Chapter 5).

The stimuli comprised 210 concept elements dealing with the different aspects of grapefruit juice. A team of professionals, comprising sensory analysts, market researchers, marketers, and an outside research consultant, developed the concept elements. The elements were created during a 3-hour ideation session and were later refined and edited by the team to create a finished set of concept elements. The text elements comprised 195 stand-alone ideas expressed as simple declarative phrases. The 15 pictures were selected from stock photographs to represent other ideas expressed visually.

The elements were placed into categories. Table 7.3 lists examples of the categories and two representative elements. The scheme of classifying elements into categories is done to facilitate the creation of concepts.

For this study on grapefruit juice, eight respondents [called *dimensionalizers* (see Chapter 5)] profiled each of the 210 concept elements on eight semantic scales. Dimensionalizing in IdeaMap enables the computer algorithm to estimate the utility of untested elements at the individual respondent level. The repeated application of the data imputation algorithm by using the locations of the semantic space to identify *closest neighbors* eventually produces a stable solution.

As part of the setup, the dimensionalizers did their task prior to the study. A sense of the extremes on these eight semantic scales is given in Table 7.4, which shows high-scoring and the low-scoring elements on each semantic scale.

Table 7.3. Examples of categories and two elements from each category for the grapefruit juice study

Category	Element
Aspiration	For a spirited, outgoing person
Aspiration	For the spontaneous fun side of you
Appearance, color	With a ruby red color that's as sassy as its taste
Appearance, color	With a sparkling ruby red color
Health	All the goodness of grapefruit with the taste you will love
Health	Be healthy but have fun doing it
Position	A refreshing break from the ordinary
Position	A unique fruit experience for your mouth and mind
Source	Made with sweet Indian River Ruby fruit
Source	Made with tangy-sweet fruit
Taste, other	A better, not bitter grapefruit taste
Taste, other	A feisty fruit zing
Taste, sweet	A distinctive tangy-sweet taste
Taste	A fresh, tangy-sweet fruit taste
Usage	A cool refresher any time of day
Usage	A healthy alternative to soda
Varieties	Available in Classic Ruby Red and Ruby Red & Tangerine
Varieties	Available in Original Ruby Red and Ruby Red & Tangerine
Visual	Ruby Red product shot
Visual	Woman in robe

Table 7.4. Location of the grapefruit juice elements on the eight semantic scales

Low vs high anchors on the semantic scale	Mean
Relaxed vs energized	
Visual: Woman relaxing with a book	1.3
Visual: Woman mountain biking	8.8
Adventurous vs safe	
The fruit juice that's the "life of the party"	1.8
The healthy refresher	8.0
For taste vs good for you	
Sassy and fun	2.9
Contains 100% of the RDA of vitamin C	8.4
Sweet vs tangy	
Visual: Pink rose	2.9
Made with Florida Indian River grapefruit	7.2
For me vs for others	
Visual: Woman relaxing with book	3.0
Ruby Refreshers juice drinks	7.3
Similar vs different from other products	
Made with citrus fruit	3.6
Visual: Man snorkeling	7.5
Similar to grapefruit vs different from grapefruit	
With the healthy goodness of grapefruit	2.6
A flood of fruit in every gulp	7.2
For fun vs for health	
Visual: Woman in pink making faces	2.3
Contains 100% of the RDA of vitamin C	8.5

Results

The respondents, acceptors of grapefruit juice from a telephone screening, participated in a central location, 45-minute session, during which they evaluated 100 different grapefruit juice concepts. The utility values, as estimated by IdeaMap, are presented in Table 7.5. The segmentation was done by the approach outline previously described, but only using the 59 sensory-relevant elements for the segmentation exercise rather than using the full set of 210 elements. Segmentation is only a statistical technique. The actual names of the segments and the interpretation of their *meaning* (i.e., what they like and what they want) are left to the researcher as an interpretative exercise.

It is clear from Table 7.5 that the winning elements for the three segments differ from one another. The additive constant is the same across the three segments (62–66), meaning that the segments begin with the same base interest in the juice. However, the elements that score best are quite different from one another. The segmentation reveals three nonoverlapping groups responsive to different descriptions of the grapefruit. It is important also that the total result, which comprises a mix of the three segments, shows the effect of *averaging disparate viewpoints*; that is, an element could be a relatively high scorer in one segment (e.g., $+7$ for the visual of a sliced grapefruit), but the lower utility values for the other segments, and occasionally even negative utility values, diminish the total utility value. Segmentation, therefore, brings out the real opportunities in the grapefruit concept.

Table 7.5. Best-performing elements for the three segments, obtained by segmenting respondents on the 59 taste/appearance elements

		Segment		
	Total	1 Taste, nonsweet	2 Color and appearance	3 Sweet seekers
Additive constant	62	62	66	62
Segment S1: Taste, nonsweet				
(Visual) Sliced grapefruit	5	7	3	3
The refreshing tang of grapefruit, without the bitterness or lingering aftertaste	3	7	−3	6
A better, not bitter, grapefruit taste	3	7	1	2
Segment S2: Color, appearance				
With a ruby red color that's as sassy as its taste	2	−1	6	1
With a ruby red color that makes you smile	2	−2	6	4
With a ruby color that brightens up your day	2	2	6	0
Segment S3: Sweet				
As refreshing as lemonade	3	0	1	8
The fruit juice with a refreshing zing like lemonade	1	2	−2	7
Made from tangy-sweet Indian River Ruby fruit	4	3	1	6
A uniquely refreshing tangy-sweet taste	2	−1	2	6
A fresh, tangy-sweet taste	2	−2	1	6
Available in Classic Ruby Red and Ruby Red & Tangerine	3	4	−1	6
A fresh, tangy-sweet taste that really refreshes	3	0	2	6

Numbers in the table body are utility values.

*Practical Considerations
for Segmentation*

A key problem in the foregoing segmentation comes from the up-front effort involved required to dimensionalize the elements. Dimensionalization ensures that the segmentation is scale independent, because the segmentation works at the level of the independent variable, not at the level of the response. The only purpose of the response (e.g., interest, under the control of the respondent) is to help identify where the response reaches its maximum.

This power-culture independence and scale independence come at a price, however. The segmentation requires work ahead of the study, in order to obtain these semantic scales. That prework can take a day or even a week. For research programs that have plenty of time and funding the segmentation approach is fine. One can deal with hundreds of elements in the segmentation analysis because the effort to work with many elements allows both for data imputation and for segmentation (see Chapter 5).

Direct Segmentation by Using the Similarity of Utility Patterns

In some cases, researchers may wish to accelerate the segmentation process and avoid the up-front dimensionalization efforts. For example, in many smaller-scale studies, the researcher can have each respondent evaluate all of the elements embedded in test concepts. In a larger-scale study the researcher would rely on partial set of elements evaluated by each person, followed by data imputation to estimate the utilities of the missing, untested elements for each individual. In this happy circumstance with *complete data* at the outset at the individual level it may be unnecessary to go through the dimensionalization exercise. Rather, one might be able to segment the utilities directly. Direct segmentation using the utilities themselves clusters

the respondents in such a way that the pattern of utilities across respondents within a segment is more similar than the pattern across segments.

The clustering methods can use a variety of measures. One approach uses the utility values themselves or some derived statistic to put people into clusters. Using the utilities themselves may generate a potential artifact: all of the respondents with high utilities fall into one segment, and all of the respondents with low utilities fall into another, even when both segments show exactly the same pattern across the utilities. Thus, using the magnitude of the utilities themselves does not work well all the time. The patterns that drive the segmentation could simply be the magnitudes of the utilities rather than their patterns.

Another way to segment uses the *pattern of utilities, but without dimensionalization.* Patterned-based segmentation is easy to achieve with direct segmentation and clustering programs. The distance between each pair of respondents can be defined by a well-accepted statistic $(1 - R)$, where R is the Pearson correlation between two respondents, based on the utility values. K-means clustering (Systat 1997) separates the respondents into groups that are statistically homogeneous. As in any clustering, the algorithm generates solutions comprising two segments, three segments, four segments, or more. It is the researcher who ultimately decides whether the segmentation itself truly classifies the respondents meaningfully. The clustering approach using the combination of k-means clustering and the $(1 - R)$ statistic is particularly attractive for segmentation using Internet-based interviewing and self-authoring systems, because in those particular situations there is neither time nor budget to allocate to dimensionalization. The only requirement is that each respondent must test all of the elements in the concept, limiting the segmentation to far fewer elements than conventional IdeaMap can handle. We will see ample use of this approach in Chapter 20 dealing with the It! studies.

Appropriate Criteria for Segmentation: Statistical Power versus Interpretable Segmentation

If there are different methods for clustering respondents, and if each of these appears to have its own reasons and justification, then how does a researcher decide which type of clustering to use? Furthermore, and a more profound question, which clustering is correct? We will see in the following that this rather direct and apparently simple question uncovers an array of problems that underlies research and makes us reconsider what we believe to be the truth.

Statistical criteria for goodness of segmentation rely on objective measures, such as the relative variability between or among clusters. To the degree that the statistical system has adequately generated these tight clusters, the researcher can feel confident that the statistical package has done its job. The clustering solutions may be somewhat different, depending on the particular measure to quantify interrespondent distance (i.e., the absolute difference between two respondents across element utilities, or the distance defined by a statistic such as $1 - R$).

The deeper question is not of statistics, but rather of interpretation and research strategy. Today's statistical packages, commonly found deployed in personal computers, can provide any of a number of cluster solutions. The question is better posed in terms of ground rules and interpretation and has at least four aspects:

1. *What is the proper set of variables on which to segment respondents?* Does the set comprise element utilities or does it comprise some external attitudinal profile generated by the respondent, or even some combination of utility value, attitudinal profile, and geodemographics? Indeed, the proper combination may change from study to study. The grape-

fruit juice study could have been run on all of the attributes, on the graphics attributes, on the sensory attributes, and so on. Different results would have been obtained.

2. *What is the proper number of segments to extract, and how are these segments to be recognized?* Like the comparable situation of factor analysis, the more clusters that the researcher exacts, the lower will be the variability across members of the same segment, and the greater will be the variability across clusters. However, one can carry this clustering to its logical, absurd conclusion and make each respondent into a cluster. Statistical considerations and a modicum of intelligence can be brought to bear on the problem so that the researcher does not extract too many clusters. This approach is similar to the corrected R-squared value in regression and the F ratio in analysis of variance. Both statistics correct for the number of independent variables, recognizing that the more independent variables the researcher uses, the higher will be the fit of the model to the data, even at the risk of incorporating nonsense predictors or provincial error variability into the model as explanatory factors.

3. *What is the role of judgment and intuition in the selection of the number of clusters and in their naming?* As previously noted, clustering by itself is a statistical procedure, and one can argue the statistics with a fair degree of objectivity. However, that argument misses the fact that the clustering is done so that the respondent-to-respondent variability becomes meaningful as an aspect of nature rather than as just random variation. Segmentation becomes valuable when married to the researcher's intuition about what the pattern actually means beyond regurgitating the statistics. If the researcher cannot interpret the clusters in a simple way, then the clustering devolves to a meaningless statistical exercise.

4. *How much of the data from a cluster should be used to name the cluster?* When looking at clusters that emerge from clustering

utility values of, say, 150 elements, the senior author generally looks at the elements that score best in the cluster in order to name the cluster. By and large that works. Surprisingly, the poorer-performing elements also can tell a story, but they are rarely consulted. This bias toward strong-performing elements may owe its origin to the difficulty of assimilating all 150 elements when naming a cluster. Looking at the top few elements for each cluster is a much easier task and therefore is the more commonly used.

A Remaining, Nagging Issue: The Assignment Problem

The strategy of segmenting respondents on the basis of the pattern of their responses clearly has attractive aspects. The clusters are often easy to interpret. The results tell a nice story. Most of the time the marketers and product developers will agree with the clustering, often stating publicly that all along they had suspected this type of clustering and segmentation. However, there is no "free lunch." Quite often an analysis of the demographics or product usage underneath these segments fails to reveal a clear pattern; that is, people differ quite clearly on the segments, as can be seen by the results from the study. Yet, for all intents and purposes these segments look identical in their classification. An analysis of demographics for many of these segments suggests that the segments transcend general demographics. *On most other measures besides response to the concepts, segments look similar to each other and indeed are similar.* All too often a disappointed marketer, hoping to find a simple solution, returns to the more conventional ways of segmenting consumers, such as geodemographics, brand usage, or even attitudes. The key learning—that there are differences among groups in response to the stimuli and these could help marketers create new and more powerfully performing products—is abandoned because there is no simple way to

identify these respondents. The unfortunate consequence of that failure to identify these consumers is that the learning from the segmentation is left behind, and marketers revert to what they know best—namely, identifying individuals based on easy criteria—even if that strategy fails to work well in the marketplace.

The solution to this problem is not to run more statistically powerful segmentations with the hope that somehow a segmentation will emerge that lines up with geodemographics or with other information about the respondents. The solution should be to create a predictive model relating geodemographic or usage data to the consumer segments. Currently, a variety of methods can do this, such as multiple discriminant analysis or decision rules such as decision trees. These predictive models based on segmentation provide the hope for a new generation of researchers that powerful segmentation can be married to equally power statisyical analyses. Both look for patterns. The segmentation methods look for patterns in the profile of ratings in order to identify homogeneous groups. The discriminant function and other classification methods look for patterns in information about the respondents that can be used to predict membership in the segments for new respondents.

Category Importance, and Patterns Underlying Segments

Conjoint analysis typically works at the level of the individual concept element. When implementing the study, the category is treated as a unit of bookkeeping. Categories comprise elements. It is important that a concept not have too many elements of a similar type, in order to avoid conflicting information (e.g., two different brand names). The category ensures that a concept will not have contradictory elements of the same type.

At a deeper level, though, we can learn about the way consumers respond to concepts by measuring the importance of a category. Since categories in conjoint measurement comprise related elements, if the elements in a specific category show high utility values, then we might conclude that the category is important to those respondents. In contrast, if the category comprises elements with utility values around 0, then we might conclude that the category is irrelevant. If the category comprises elements with high negative utilities, then again we would conclude that the category is important. The high negative utilities are a signal for us to avoid elements in this category.

An important category is one in which the consumer attends and which sways the judgment, either in the positive or in the negative direction, or perhaps both, depending on the element. An unimportant category is one in which the consumer disregards the information. Does one element in the category do particularly well, but the others in the category do poorly? Or, in the happier case, does a particular element do well because all of the elements in the same category do well? An astute researcher looking at this finding will draw two different conclusions. The strong performance of only one element in the category means that the category itself is not important. Perhaps the element performed well because it was presented in an interesting way. In contrast, if all of the elements do well in a category or, conversely, if all of the elements do poorly, this indicates that it is the basic idea underlying the category that makes a difference and that the particular execution of the element is probably not as critical.

Each element in the category generates its own utility, so it is technically more correct to ask about the importance of single elements than about the importance of a category. An individual element is important when its utility value is either highly positive or highly negative. The utility value for a single element can be operationally defined as the conditional probability that a respondent

will change the rating of a concept from indifferent to interested if the element is present in a concept. A positive utility value means that the concept element adds to the probability of a respondent being interested in the concept, whereas a negative utility value means that the concept element subtracts from the probability.

A category comprises many related elements. Furthermore, the category size is arbitrary in terms of the number of elements, unless the conjoint procedure dictates that all categories comprise the same number of elements. Whatever measure of importance is used, the measure must be independent of the number of concept elements. Thus, the method for computing importance must take into account the potentially very large difference in the sizes of two or more categories in a single study.

The Role of Segmentation

Segmentation as defined in this chapter looks at respondents with either similar *optimal levels* of dimensions or who have similar patterns of utility values. Up to now we have looked at the winning elements, independent of category, to name the segments. We might learn other things about these segments when we look at the importance of the different categories. Since categories comprise related elements, the type of knowledge we obtain is different from the insights that winning elements provide. Concepts comprise many different types of information, whether product features, benefits, heritages, price, availability, usage, etc. Each of these aspects of the concept drives acceptance. When the researcher measures the contribution of the each of the separate components as a driver of interest, then some of the commonalities within a category may become more apparent.

At a higher level, understanding the importance of categories of elements, not just one element, becomes more instructive because it reveals patterns that might be hidden in the array of the individual elements. Understanding the importance of a category of related elements suggests rules or at least generalities. Those rules go beyond the current study to inform future efforts. Thus, to researchers, marketers, and product developers, it is key to *understand* what is important in general—that is, the pattern underlying the win—and not only what specific elements perform well in the particular study.

Operationally Defining Category Importance

Relative importance needs an operational definition that can handle the large number and arbitrary nature of elements and categories. One operational definition of relative importance is some overall measure of the utility values within a category, compared with the utility values of other categories. The definition must be able to accommodate the fact that the categories can have different numbers of elements. The definition should also accommodate the fact that the regression model typically used is dummy-variable regression, so the utilities are in a sense a measure of the absolute strength of the element to sway a respondent toward either accepting a concept or rejecting it.

One possible measure, but by no means the only measure, is the sum of squares of the element utilities *within* a *category* compared with the total sum of squares of all element utilities across all categories. The squaring of the utility value removes the effect of sign. Thus, two elements that have utilities of $+10$ and -10 would become $+100$ each rather than canceling each other out. Other operational definitions could be constructed, such as using the absolute values of the utilities rather than the utilities themselves (Moskowitz et al. 2002). The approach followed here will use the squares of the utilities, but the same type of analysis can be made with other importance measures as well.

The strategy of dealing with sums of squares parallels the statistical framework of analysis of variance. This measure of category importance was suggested to the author by Alain Pioche (personal communication 1994) and can be called the *Pioche RII* (relative importance index) or *RII* for short. The RII is one of a class of formulas that deconstruct the total variation of the concept elements into the components and assign that variation to different categories:

$$\text{importance of a category} = (\text{category SS})/\text{total SS}$$

$$\text{SS} = \text{sum of squares} = \text{sum of squared coefficients}$$

This RII formula using the squares of the utility values (utilities and coefficients) corrects for the fact that the utilities can be both positive and negative. However, the RII is sensitive to the number of elements in a category. The more number of elements there are, the greater is the sum of squares, all other factors held equal. One could further modify the formula by dividing by the number of elements, but that further particularization will not be done here. Thus, the RII formula will be most appropriate to compare two or more segments from a single study where both segments generate utilities for the exact same elements.

Comparison of Relative Importance Across Segments

Conjoint analysis immediately teaches researchers which individual elements do well and which do poorly. The analysis of relative importance presents a different picture. It shows the categories of elements that have large positive or negative swings. Categories with high RII values are those that have more of these strong drivers. Categories with low RII values have utilities that cluster more around 0. Unlike the interpretation of single elements, however, the RII value pertains to a total category. If the category has only one element that performs strongly, then it will have a lower RII value than a category having two or three strong performers.

Looking at the RII values across subgroups, such as concept-response segments, provides a way to understand rapidly the difference between the segments. For example, if one segment shows a high RII value for brands and another does not, then this difference suggests that the former segment responds more strongly to brands, either positively or negatively.

Two Sets of Data About Condiments That Compare the Relative Importance of Categories Across Comparable Segments

The easiest way to understand the RII value is by looking at specific studies. The first study is an Italian condiment. The study was run in Holland, with 420 respondents, and comprised 210 elements. The study was run using the IdeaMap method (see Chapter 5). Two segments emerged, based on segmenting respondents on the pattern of their utilities versus semantic scales. Segment 1 was interested in the heritage of the product (positioning) as well as pricing. Segment 2 was more interested in flavor and product usage. The decisions regarding segment names were made

prior to looking at the relative importance value and are based solely on the elements that "float to the top" for each segment.

The RII values suggest that the two segments are similar in the relative importance of the categories. Table 7.6 shows how easy it is to compare the RII values across comparable segments within a single category. Thus, the value for "flavor" is 15%, meaning that 15% of the total sum of squares can be attributed to the elements in the flavor category. However, that relative value of 15% comes from two different groups. Segment 1, interested in positioning and pricing, shows the flavor category to account for only 9% of the total sum of squares. In contrast, segment 2, interested in flavor, shows flavor to account for 19% of the total sum of squares. The same type of story can be seen for price. For the total panel, 10% of the total sum of squares can be traced to price. The two segments radically differ from each other, however. Segment 1 shows 18% of the total sum of squares traceable to price, whereas segment 2 shows only 6% traceable to price.

The second study deals with a Mexican condiment (salsa product) and was run with 140 respondents (Table 7.7). The two segments show clear differences in relative importance. Segment 1 is interested in the "fresh and natural" elements, whereas segment 2 is

Table 7.6. The relative importance index values for Italian condiment

| | | Segment | |
| | | 1 | 2 |
	Total	Positioning	Flavor and usage
Total sum squares	*5181*	*5720*	*7195*
Positioning	17%	14%	20%
Flavor	15%	9%	19%
Usage	14%	9%	18%
Visual	13%	13%	10%
Pricing	10%	18%	6%
Health statement	7%	9%	5%
Subbrand 1	6%	5%	7%
Subbrand 2	5%	4%	7%
Product line	5%	6%	3%
Light flavor	4%	8%	2%
Brands	4%	5%	3%

Table 7.7. The relative importance index values for Mexican salsa

	Total	Segment 1 Fresh and natural	Segment 2 Mexican imagery
Total sum squares	*2544*	*3999*	*7121*
Fresh/natural	20%	25%	8%
Image	14%	12%	16%
Claims	9%	8%	7%
Texture	9%	6%	6%
Freshness benefits	8%	8%	4%
Flavor/taste	7%	5%	8%
Heritage	6%	4%	3%
Ingredient	6%	4%	11%
Texture benefits	4%	4%	3%
Usage	4%	14%	13%
Flavor strengths	4%	1%	7%
Product types	3%	3%	4%
Visuals	3%	4%	7%
Brand names	2%	3%	4%

interested in "Mexican imagery." The differences in relative importance are quite dramatic, especially for a number of the key categories. For instance, segment 1 shows that 25% of the total sum of squares can be attributed to "fresh and natural" elements, whereas segment 2 shows only 8% of the total sum of squares can be thus attributed.

Self-explication of Importance: A Counterapproach to Derived Importance

Self-explicated importance constitutes a counterapproach. Instead of deriving relative importance from utilities through a computational formula, the idea of self-explication is that the respondent himself or herself knows what is important. Quite often, skeptical researchers argue that the statistical machinations to create indices of relative importance would produce nothing more than might be produced were respondents asked to rate importance. Indeed, in some conjoint methods the respondent is directly asked to identify which elements are relevant and which are not. Only those elements selected as important through direct explication are further analyzed.

This self-explication approach assumes that the respondent knows what is important and what is not. Three issues, however, have to be faced:

1. *No element or category, especially pictures and prices, is really evaluated in isolation.* There must be a framework for evaluation. Price is meaningless by itself outside the framework. The self-explication approach, though, asks respondents to evaluate the importance of each category in isolation.

2. *Dislikes and likes may not be equally known.* Respondents know, in many cases, what they like. They are not quite as attuned to what they dislike.

3. *A category can contain elements that perform in different ways.* Self-explication misses out on the fact that a category can comprise many different types of elements, some of which have strong positive utilities for an individual, some have strong negative utilities, and some have virtually zero utilities. How can a respondent know what is important if the respondent has to take the category as a whole, without experiencing the elements?

References

Aldenderfer, M.S., and Blashfield, R.K. (1984). Cluster Analysis. Thousand Oaks, CA: Sage.

Belson, W.A. (1981). The Design and Understanding of Survey Questions. London: Glower.

Benzecri, J.P. (1992). Correspondence Analysis Handbook. New York: Marcel Dekker.

Bibb, R. (2001). Exploring New Dimensions. www.directmag.com/ar/marketing_exploring_new_dimensions.

Boyd, H.W., Westfall R.L., and Stasch S.F. (1989). Marketing Research Text and Cases. Homewood, IL: Richard D. Irwin.

Ehrenberg, A.S.C. (1988). Repeat Buying: Facts and Applications. New York: Oxford University Press.

Ekman, G., and Akesson, C.A. (1964). Saltiness, sweetness and preference: a study of quantitative relations in individual subjects. Psychological Laboratories, Report 177. Stockholm: University of Stockholm.

Elrod, T., and Keane, M.P. (1995). A factor-analytic probit model for representing the market structure in panel data. Journal of Marketing Research, 32: 1–16.

Gardynhen, R. (2001). Habla English? American Demographics. www.demographics.com.

Green, P., and Tull, D. (1978). Research for Marketing Decisions, 5th edition. Englewood Cliffs, NJ: Prentice Hall.

Hammond, K.A., Ehrenberg, A., and Goodhardt, G.J. (1996). Market segmentation for competitive brands. European Journal of Marketing, 30: 39–49.

Huston, T.L., and Klerfors, D. (1998). Artificial neural networks: what are they? How do they work? In what areas are they used? hem.hj.se/~de96klda/NeuralNetworks.htm.

Kachigan, S.K. (1991). Multivariate Statistical Analysis: A Conceptual Introduction. New York: Radius.

Kenneth, J.H. (1927). An experimental study of affects and associations due to certain odors. Psychological Monograph 37: 164.

Lappin, J., Figoni, P., and Sloan, S. (1994). A Primer on Consumer Marketing Research: Procedures, Methods, and Tools. Cambridge: National Transportation Systems Center.

Monahan, J. (2003). Why customers buy is more important than what. www.marketingpower.com/live/content.php?Item_ID=18414.

Morgan, J.N., and Sonquist, J.A. (1963). Problems in the analysis of survey data and a proposal. JASA, 58: 415–434 [original, AID].

Moskowitz, H.R. (1981). Sensory intensity vs hedonic functions: classical psychophysical approaches. Journal of Food Quality, 5: 109–138.

Moskowitz, H.R. (1986). Sensory segmentation of fragrance preferences. Journal of the Society of Cosmetic Chemistry, 37: 233–247.

Moskowitz, H.R. (2003). Concept-response segmentation for grapefruit juice: what role do sensory statements play as drivers of persuasion and response time? Journal of Sensory Studies, 18: 141–162.

Moskowitz, H.R., Jacobs, B.E., and Lazar, N. (1985). Product response segmentation and the analysis of individual differences in liking. Journal of Food Quality, 8: 168–191.

Moskowitz, H.R., Krieger, B., and Rabino, S. (2002). Element category importance in conjoint analysis: evidence for segment differences. Journal of Targeting, Measurement and Analysis for Marketing, 10: 366–384.

Neal, W.D. (2003). Principles of market segmentation. www.marketingpower.com/live/content.php?Item_ID=1006.

Pangborn, R.M. (1970). Individual variations in affective responses to taste stimuli. Psychonomic Science, 21: 125–128.

Pangborn, R.M. (1981). Individuality in responses to sensory stimuli. In: Solms, J., and Hall, R.L., editors. Criteria of Food Acceptance: How Man Chooses What He Eats. Zurich: Forster, pp. 177–219.

Szmigin, I. (2002). Consumer behaviour.bss2.bham.ac.uk/business/teaching/szmigin/introduction. ppt.

Systat (1997). Systat: the system for statistics. Evanston, IL: Systat Division of SPSS.

Thompson, B. (2000). Q-technique factor analysis: one variation on the two-mode factor analysis of variables. In: Grimm, L.G., and Yarnold, P., editors. Reading and Understanding More Multivariate Statistics. Washington, DC: American Psychological Association.

Thorson, E. (1989). Advertising Age: The Principles of Advertising at Word. Lincolnwood, IL: NTC Business Books.

Vermunt, J.K., and Magidson, J. (2000). Latent class cluster analysis. In: Hagenaars, J.A., and McCutcheon, A.L., editors. Advances in Latent Class Analysis. Cambridge: Cambridge University Press.

Wellner, A. (2000). Brand strategies: profit in loss. American Demographics 2000. www.demographics.com.

Wellner, A. (2002). The female persuasion brand strategies: profit in loss. American Demographics 2002. www.demographics.com.

Wright, M. (2003). Competences/capabilities for market segmentation [Response 3]. www.marketingpower.com/live/content.php?Item_ID=17124.

Chapter 8

International Research and Transnational Segmentation

Introduction: The Practical Side of International Research

Consumer research on concepts began in the United States (Nicosia 1966; Engel et al. 1968; Schiffman and Kanuk 1999). The business use of concept research as a development aid quickly spurred the adoption of concept research internationally. One of the nice features of research for profitable ventures such as products and services is that whatever works in one country is often rapidly adopted in other countries. There will, of course, be the necessary modifications to ensure that the approach works in the different markets, and certainly the egos of the adopters in other countries will ensure a somewhat different format. That is to be expected. The influence of the multinational companies that work worldwide will guarantee, however, that successful research techniques will be adopted in many different markets.

Concept research belongs properly to the domain of marketing research and only later became part of the research and development (R&D) function. Fortunately, marketing research has always had an international aspect to it, where researchers could exchange ideas. The World Society of Marketing Research [formerly European Society of Marketing Research (ESOMAR)], founded shortly after the end of World War II, brought together European and later worldwide marketing researchers to present papers and to discuss research techniques. Concept research was, of course, one of the topics.

Doing the Research Transitionally: Issues of a Practical Level in Research Execution

Researchers raise a variety of issues about conducting concept research outside of one's local borders. To a great degree these issues are nothing more than the unwillingness of many conservative researchers to accept that research can be done worldwide. Rather than stating their case, these conservative naysayers raise "issues" and "problems" as a matter of course whenever presented with new ideas. Typically, the problems come from other researchers, whereas the issues come from nonresearchers in advertising agencies. The terms are nuances of the same condition: "It can't be done other than the way I'm doing it and in the way that makes me feel comfortable."

It is worthwhile addressing these issues head on and, where possible, bringing data to either support the issue or to controvert it and lay the question to rest.

Issues in Accepting New Ideas

It Can't Be Done Because It Never Has Been Done Before Here

Researchers, product developers, and marketers are subject to the syndrome of "It can't be done this way because it never has." All of us, professionals and amateurs alike, fall into this trap. As we seek to make the complex simple, as we attempt to cope with our lives and business problems, by necessity we try to make things rote, machinelike, and thus automatic. This coping mechanism means that we avert our eyes from new methods, once things work. That automatic pilot that helps us deal with the everyday also means that when confronted with new ideas we may not be particularly capable of assimilating them. Researchers are no different from anyone else. When a researcher has a decade or more of experience designing and executing concept studies in a certain way, that experience becomes sacrosanct. Unless the researcher's discomfort is sufficiently high, there is little impetus to learn new methods. Add to that discomfort the differences in country and culture, and the result is the lack of the researcher's experience in concept-research methods other than the ones with which he or she is familiar. This lack of experience means that the researcher in the other country has to cope with a new method for gathering data. Newness, discomfort, and lack of experience lead to problems in design, in the field, and in monitoring the study for problems.

Different Approaches to Analyzing the Same Data

This problem arises when two groups of researchers analyze the same data. Occasionally, the conventions for analyzing data differ by country. Some professionals analyze concepts by looking at the proportion of respondents who rate the concept as "definitely or probably purchase" (percentage top-2 box). Other professionals, often in developing countries, look only at the proportion of respondents who vote "definitely would buy." This is the so-called top-box score. These latter professionals are aware that most of the respondents in their market would say that they would definitely or probably buy the product, and realize therefore that the percentage top-2 box would not discriminate. The problem comes in the attempts to reconcile the two sets of measures. Similar issues arise in simply designing a study. Some researchers want to

look only at the concept evaluated first (pure first monadic), whereas researchers in other countries may wish to look at the ratings for all of the concepts. There are no aspects of right or wrong here—just different conventions for conducting research. When a common structure for analysis cannot be determined, however, the transnational cooperation can break down because conclusions offered by one group could be rejected under the criteria accepted by the other group. It is in cases like these that having standards for interpretation works. Even if the standards do not hold up to 100% scientific scrutiny, they still make the research feasible and enable management to act on the findings rather than forcing management to play the role of umpire.

Different Constituencies as the Clients in the Different Countries

In the United States the market-research function has as its client the marketing department and does not typically work with R&D, except in the smaller companies. The sensory evaluation group works with R&D. In other countries that are neither as large nor have formalized systems for concept development, the sensory specialist and the marketing researcher may be the same person. Research patterns, ways of looking at data, and the ultimate use of the data could differ, depending on the group to which the concept research reports. When a concept researcher does both product and concept research and reports to both marketing and R&D, the tendency is for the concept research to be more practical, executed more efficiently, and analyzed perhaps more superficially, but be more relevant for business. It could be no other way. Research specialists have a group of different constituencies to satisfy and little time and precious few resources with which to work. More specialized researchers can devote a great deal more time to design. These two styles do not necessarily work well together. The jack-of-all-trades, satisfying everyone, working with all constituencies, does not have the time to explore the niceties of the research. The work must get done. His colleague in another country, who is but a single member of a large team, can devote many more hours to any study and consequently may proceed in an agonizingly slow pace. Even within the same multinational company these two types of researchers can be found, depending on the particular market. If one did not know the individuals worked for the same company, or at least for a multinational that has the same name, one might never guess by the work style.

Issues Involving the Mind of the Respondent

The Concepts Cannot Be Understood Because They Don't Make Sense

This is a singularly important and valid issue, but one that can be addressed operationally through a number of good research practices. Researchers tend to be verbally adept in their own language. The experience from this study is that many of the elements must to be translated and retranslated, because otherwise they make no sense. What is idiomatically correct in German, for example, may come out garbled in French, English, and Italian or, what's worse, entirely meaningless. The conventional approach is, therefore, to begin with one set of elements in English or whatever language constitutes the starting study. The elements are finalized and then returned to the countries to be translated. Once translated, the elements are retranslated into English. The translation sequences need not be from English to the other language and back to English. The key is the translation and the retranslation or *back translation*. It is vital that, when the element or concept goes through translation and retranslation, the element maintains its meaning.

Consumers in Other Countries Can't Really Understand Concepts the Way Western Consumers Do

This complaint comes from the researcher who is asked to do concept research in a developing country. Accustomed as the researcher has become to well-executed concept boards and standardized interviewing in the United States and Europe, it is difficult for this researcher to imagine valid research in other than those environments. There may be some substance to this complaint, but it is quite hard to justify it. Perhaps the major justification might be that the mode of presenting concepts (i.e., by computer in a computer-aided personal interview) could be novel for the respondents. Some of the authors' colleagues averred in the mid-1990s that the typical housewife in some European countries did not feel comfortable with the computer keyboard. Time has proven that statement to be quite incorrect. No doubt, over time, consumers in different countries will feel increasingly comfortable with keypads. They already do worldwide, wherever there is an ATM cash machine that they use to get cash with a credit card and a PIN number. It is clear from years of experience that it is perfectly reasonable to conduct valid research with computerized interviewing, as long as the respondents are properly oriented. The respondents do not need to know how to program. The issue, therefore, posed by critics is not so much the problem of the interview as the ability to accept that the field of concept testing specifically and consumer research generally are moving from interviewers to paper, and from paper to computers.

Respondents in different countries attend to different *executional* aspects of the concept, making a valid comparison difficult. *This country-to-country difference is very important for concept research and has major practical implications.* If individuals in one country respond favorably to both white-card concepts and to fully executed concept boards, whereas individuals in another country respond only to the fully executed boards, then we are dealing with possible difficulties in interpreting the results. Since the respondents in the second country do not seem able to respond to white concepts, all of the concepts must be fully executed. This requirement reduces the ability to compare data across countries, because it limits the stimuli to only one type: a format where the concept is fully executed. The problem with this issue, however, is that there are no data to back it up and no sense of the extent of the problem or even if the problem goes beyond one person's experience.

Respondents in a Country Have Their Own Way of Using Numbers

This is a very valid point. For the same concept, some countries (e.g., Scandinavian countries, such as Sweden) tend to down-rate the concepts, assigning them ratings that would indicate less acceptance. In contrast, there are cultures, such as those in Mexico and the Philippines, where it is considered

rude to the interviewer to assign low ratings. Consequently, the respondents in these countries may up-rate the same concepts. The researcher, unaware of the differences in the culture, may think that the concept utterly fails in Scandinavia but does well in the Philippines.

Figure 8.1 presents an example of the distribution of the same concepts in several countries. There were five countries, 240 respondents per country, and 70 systematically varied concepts from a conjoint study on a cereal product that would be eaten in each country. Thus, the distribution of concept scores comes from 16,800 ratings per country, sufficient to show country differences in scale usage. The systematic variation of concepts allowed some concepts to be sparse and perform relatively poorly. The concepts were exactly the same. Figure 8.1 shows that Mexican respondents tend to give higher ratings than do Argentineans and Brazilians, so the scaling behavior is not due to language or geography.

The difference among countries becomes even clearer when we look at the average

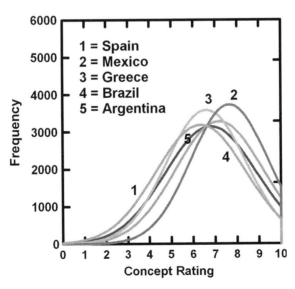

Figure 8.1. Distribution of ratings on concept interest for 70 systematically concepts, tested among 240 respondents in five countries.

ratings across all of the concepts in Table 8.1. On average, Mexican respondents assign the concepts a rating of 1.3 points higher than do Spanish respondents. If we look at the recoded data, we see the differences among countries even more clearly. The 16,800 ratings were transformed into binary values (0,100) in two ways. The first transformation changed ratings of 1–6 to 0, and 7–9 to 100. This transformation, the conventional *top-3 box* discussed at length in the previous chapters, shows that the Mexican respondents use the top part of the scale so frequently that 83% of the Mexican ratings fall into the top-3 box. In contrast, 54% of the Spanish ratings fall into the top-3 box. One strategy is to change the criteria to top-2 box, either for all of the countries or just for some. By changing the criteria for Mexico to the top-2 box as an acceptor, the proportion of Mexican ratings falling into the acceptor group drops from 83% to 68%. However, even with this type of correction factor, one does not know where, when, and how to apply it. Thus, in many cases, interpreting data from other countries involves both quantitative analysis and qualitative, intuitive interpretation that factors in country-specific scaling behavior.

Table 8.1. Mean and top-3 box statistics for the 70 concepts and 240 respondents for the cereal concept

Mexico	Mean	7.66
Brazil	Mean	7.19
Argentina	Mean	6.81
Greece	Mean	6.61
Spain	Mean	6.35
Mexico	Top-3 box	83.
	Top-2 box	68.
Brazil	Top-3 box	72.
	Top-2 box	54.
Argentina	Top-3 box	63.
	Top-2 box	44.
Greece	Top-3 box	60.
	Top-2 box	36.
Spain	Top-3 box	54.
	Top-2 box	34.

Issues Involving the Setup of the Study with Research Partners

A great deal of concept research is run by cooperative groups of research companies and, in recent years, by different offices of the same research company. One might surmise from the globalization of concept research that professionals in different countries have now become proficient in concept testing, have standardized the methods, have recognized best practices, and in general have gotten the process in order.

For new technologies in concept research, such as conjoint analysis, the requestors tend to be in Western Europe or the United States. Quite often the requestors are market-research managers, with the task to answer specific questions posed by their constituencies (e.g., marketing, research, and development). If the requestors have not had experience in foreign research, they typically generalize (albeit incorrectly) from their experiences in a sophisticated environment to the new, less sophisticated environment. Often these managers are unaware of the problems involved in setting up a study in an environment with a limited availability of personal computers and with field services involved primarily in answering simple questions (Rabino and Moskowitz 2001).

Often projects involving less experienced colleagues require a great deal of hand holding, primarily with the immediate client (research user) who requested the project. The client sometimes needs to be educated in the capabilities of the local market field service. Smaller clients, with little experience in international work, often have problems with such hand holding. In contrast, clients working in the large-scale multinational companies are often quite well aware of the field services and have worked with them extensively. This experience, though, is primarily with simplistic interviews done with paper and pencil. For more complex studies, such as computer-based concept testing and optimization, often a PowerPoint presentation helps initially, followed by a good deal of telephone time. Despite the short times involved in concept research, the interaction with foreign field agencies and inexperienced research users often can take weeks and, in some cases, months. Typically, the telephone meeting with the requestor client varies from theory (Why is the method used? How are the data analyzed?) to specifics on field (What exactly happens in an interview?) to process issues (timing and cost). There is always the need to provide the requestor with customized materials that support the approach in theory, analysis, and cost/benefit. These materials typically facilitate the interaction and help move the project forward.

Communication Problems with Different Time Zones and Languages

One of the key problems with international research is the difference in time zones and thus the difference in working hours and availability for discussion. For US researchers there is a 6- to 9-hour difference between the target market in Europe/Middle East and the research company setting up the study. The time difference is even greater for Asia and Australia/New Zealand. This time difference can lead to problems, especially when the research is initiated from the United States, because quite often the difference in time forces US researchers to work from home, without supporting documents and staff.

A second problem is language. Language today is less of a problem than it used to be. Many (but not all) participants in the setup of the research speak English, the apparent lingua franca today. Additionally, and fortunately, the Windows operating system is widespread, providing a de facto standard for computers. With Windows installed on PCs, and with most Eastern European languages being Latin character based, there are no real language problems in the actual (computer-based) interview.

Even though most of the research professionals speak English, there are the ever-present problems in communicating with the local field service and (when appropriate) the local representative in Eastern Europe of the requestor company. Most of the time the language is English rather than the local vernacular. Language capabilities vary greatly among the different people that one encounters. The senior employees of the multinational company typically speak English quite well, whereas the field service actually executing the study often does not. This creates difficulties that must be resolved through written communication (faxes, e-mail).

Issues Involving Field Execution of the Study Among Respondents

Can the Field Service Actually Even Execute the Study?

One of the traditional concerns of market researchers is whether the field service (which actually executes the study with consumers) is actually capable of doing its job. Field services vary in quality in the United States to a great degree and worldwide to a far greater, and often unexpected, degree. Marketing research, as a business with low cost of entry, and based on personal skills, is very subject to staff variability. A market-research field organization can assemble interviewers, go into business, and proclaim itself to be a field service ready to interview consumers in the local country. Multinational companies recognize the quality variation among these different local firms and generally select those that they feel to be most reliable and capable.

For example, shortly after the reunion in Germany the political institute NFO Infratest Germany had a strange problem. When countrywide interviews were conducted by telephone studios in an eastern city like Leipzig, the former Communist Party PDS got higher ranks (9% vs 5%) than if the telephone interviews were done in a western city like Munich or Hamburg. The answer was simple. Respondents in the eastern section felt more at ease with interviewers who spoke in a familiar accent. Thus, they felt more comfortable giving honest answers.

Standards for Quality of Field Execution Vary from Country to Country and Supplier to Supplier

Unlike the lack of experience, which comes from cultural issues, poor fielding comes simply from poor fielding, perhaps with some inexperience thrown in for good measure. Sometimes the fielding is not properly controlled. Other times the interviewers have to execute the study in a limited number of venues. Occasionally, the interviewers in a country have to do things to get responses that we might consider biased. All of these biases are executional or field biases. The worst executional issue, however, comes from the neglect of simple interview requirements, such as screening to ensure that the respondents are who they say they are. Most researchers who work internationally have experience with poor fielding. It is a fact of life, though not necessarily a pleasant one. It is better to admit the problems than to avert one's eyes and pretend that that the study has been adequately fielded when it has not. Simply denying the poor fielding will not make the field execution any better.

Running the Study with Field Services New to Computer Technology May Produce Some Problems in Study Execution

It has come as a pleasant surprise to find that computer-based concept studies can be fielded worldwide in the same way as they are fielded in the United States or Western Europe. We trace this comparability of field execution to the following: most field services can follow explicit directions, have technical knowledge about computers (at least a considerable number of individuals have this knowledge), are motivated to work with high technology, and have good repair shops for faulty equipment. One of the most pleasant surprises, therefore, is the continuing discovery that execution of computer-based interviews is almost flawless, at least at the time of execution. This can be further traced to the enabling and standardization power of the personal computer (and especially that provided by the Windows operating system).

Respondents in Some Countries Are Accustomed to a Personal Interview and Do Not Want to Change

This is not a particularly problematic issue, although it rears its head from time to time. Since research companies in many countries use interview procedures that are less than up to date, there is a feeling that one should continue to do the interviews in the way that they have always been done. The excuse is that the respondents have become accustomed to these types of interviews, generally face to face, and would not feel comfortable interacting with a computer. When all is said and done, these critics aver, quite strongly, that the consumer respondents in their study will never be able to cope with new methods to which they have not become accustomed. Only time will tell whether these critics are right. Consumers often do things that surprise critics and have no problem coming to grips with, and even adopting, new technology. There should be no real reason why consumers in countries other than the most developed ones would have any problems with interview techniques. A lot of the issue is historical in nature, based on the level of development in the particular country.

In the United States and later in Europe the original concept work was conducted *door to door*; that is, an interviewer would go from one home to another, according to a sampling plan devised by the researcher. The interviewer would conduct a personal interview, asking the respondent to evaluate a concept, rating the concept on a variety of characteristics. For a number of years, and with the availability of inexpensive interview labor, door-to-door interviewing was quite popular. Indeed, to many individuals who grew up in the consumer-research field, door-to-door interviewing represented the most scientific controlled approach. The sampling plan guaranteed that the sample interviewed would represent the population. The interviewer was given specific instructions about what to do in the case of refusals by respondents to participate and what to do for incomplete interviews that were terminated. When door-to-door interviewing became too expensive or too dangerous, researchers resorted to other methods, such as telephone interview, mail interviewing, or mall interviewing.

In general, personal interviews constitute the most versatile mode of data collection. They allow for lengthier, more complex, interactive sessions than do telephone or mail surveys. Personal interviews can include product demonstrations and visual presentations (e.g., video loops to engage the attention of passing shoppers). Finally face-to-face mall interviewing is used because it enables interviewers to maintain a face-to-face contact with respondents. Of course, personal interviews remain the most expensive mode of data collection and involve the greatest risk of interviewer bias, because no one feels entirely comfortable with the possibility of covert signals being passed between the interviewer and the respondent (Blankenship and Green 1993).

The Communications Infrastructure Is Not Set Up to Do Internet-based Work Such as Concept Testing

This is again a reasonable criticism, although certainly one destined to become increasingly irrelevant over time. The adoption of the Internet is increasing daily as communications companies recognize there is profit to be made, and as consumers recognize that they can use the Internet for communication, research and, in some cases, actual shopping. What is necessary in the other countries comes down to the infrastructure of the technology. Such infrastructure has a way of emerging even in the most remote, backward areas. For example, cellular telephone service is now widely available. A decade ago, before the cellular network became popular and cellular telephones become cheap and easy to use, the concern was that, for telephony to become popular, one would have to create the infrastructure of exchanges and lines in the way that was done in Europe and the United States. The creation of cellular networks bypassed this need, and now in many countries the cellular phone is the standard communication device. Furthermore, the availability of the Internet by broad-band cellular telephone will allow Internet-based research in a way not possible before. Thus, the infrastructure for Internet-based research is developing in front of our eyes, almost on an hourly basis.

Beyond Research Execution Alone: The Emerging World Market and the Opportunity for Global Segmentation

As this new century begins, the international marketing and research communities announce with increasing frequency a single world of consumers divided by geopolitical boundaries but sharing product and service preferences. Boundaries for products and services exist but transcend the traditional political ones. These preference groups constitute global segments. One can easily recognize political boundaries, but not as easily recognize and then profit from *preference boundaries.*

At a practical level we have not yet reached the stage where we can easily customize a product for a single person, no matter what the business books preach and no matter how much we pay our business pundits and wise men to prophesy. Yet, the time when Henry Ford could dictate that all car colors must be black has also long passed. Marketers now face a dilemma—everyone wants a customized product, but good business sense prevents most motivated marketers from succumbing to the temptation to create and market such a product.

One good alternative creates a set of products to fit a limited group of people. Each different group of people constitutes a segment. The researcher must discover these segments in the population, guide the developer on how to create the product, and counsel the marketer on how to sell it (see Chapter 7). To the degree that these limited groups exist worldwide, albeit in different proportions across countries, the researcher, the developer, and the marketer will have uncovered global segmentation. The next natural step uses this global segmentation to create targeted services and products.

Applying International Research to Global Segmentation and Product Development

Global segmentation reveals groups of like-minded people around the world, with the segments transcending local political boundaries. The assumption is that the same segments exist across countries, albeit in different proportions. Segmentation guides product development and marketing. Proper use of segment information creates world brands targeted to the segments, with the world brands appropriately and moderately particularized to the consumer needs in each particular country. The objective of concept research, in turn, is to identify these segments in terms of what they want to hear about a product or a service, by using wherever possible the same stimuli across many countries. By having the same stimuli, one can be sure that the segments are created using the same building blocks.

Underlying Principles

Segmentation exists at almost every level of our experience. We need only use the Internet to list the specifications of the many variations of product and service offered by each company. We quickly realize that marketers know a lot about segments. But how do product developers and marketers systematically uncover promising *worldwide segments* that are actionable rather than leave such discovery to chance? The following principles provide an *algorithm* or defined set of steps to discover and to use global segmentation:

1. *Develop an organizing principle.* For meaningful global segmentation, not just statistical exercise, the segmentation must embody an organizing principle that is easily understood and intuitively meaningful. At the level of communication the organizing principle comprises preferences for messages (e.g., product features versus price in messages about a product).

2. *Make sure the segments tell a story to the audience, because stories embed themselves in the corporate culture far more easily than do never-ending data tables.* If at the end of the meeting the marketers and product developers talk about the segments as *real people* who can be approached, convinced, and sold to, then the researcher has succeeded. The segmentation has become a real and useful guide. Nothing so frustrates as segmentation that simply demonstrates a researcher's statistical prowess without telling a coherent, useful, actionable story. If the segmentation makes sense, then more than likely it portrays the different consumers in the form of a story or at least a vignette. That story or short description will, no doubt, inspire further new

products and services and form the basis of marketing strategies. Such segmentation quickly embeds itself into the corporate knowledge base.

3. *To make the segmentation actionable, explore a wide range of stimuli and then let the segments emerge from the response patterns to these stimuli.* A lot of global, overarching segmentation works with general statements in questionnaires. After the statistical analysis, the appropriate statements defining each newly uncovered segment demand yet another step: conversion into specific, concrete communications and product formulations. A better way tests many stimuli that represent the end communication or actual product, whether language to incorporate into descriptions of new juice products or offers about QSRs (quick-serve restaurants). Chapter 7 deals with this approach in detail.

4. *Generalities first and then specifics—before looking at a country-by-country segmentation, identify general segments worldwide.* At the start of the research, treat the whole world as one population, cluster the consumers worldwide into the segments independent of country, and afterward determine how these segments occur in the different countries. Start with world principles before getting involved in the details. Then, drill down to the country level. The global segmentation should take precedence, for it presents marketers and developers with the first-order opportunity. Only afterward should one get involved in country details, for they have a way of marring the clarity of the segmentation.

5. *Particularize the service or product to a segment in a country.* Once marketers identify the optimum product or service for a segment, they can then fine-tune or particularize the particular product for an individual country. Particularization takes into account both the global segmentation (efficient development and targeted product) and the local, individual nature of the country.

The International Coffee Study: Concept Research and Worldwide Segments

Coffee, along with water and cola, is an almost universal beverage, albeit in different formats. Coffee preferences vary around the world. It is not unusual for a manufacturer to market the same brands with different formulations in order to suit the consumer taste in a specific market. Hence, coffee concepts are a good example by which to illustrate international research and transnational segmentation. The particular study presented in this section deals with the assessment of coffee concepts created by experimental design and run in ten countries in 1994–1995. The study has been reported in part in a variety of publications. The data from seven of the countries are presented here. Owing to its large-scale nature, in terms of countries, respondents, stimulus elements, and types of stimuli, the study provides an interesting case history by which to illustrate a number of principles involved in conducting concept research worldwide.

Basic Study Design

The study involved 273 concept elements from the categories listed in Table 8.2. The IdeaMap method was used on a personal computer to accommodate the large number of categories and elements. IdeaMap is described in Chapter 5.

The elements for the coffee study were created by marketing research professionals during a set of ideation exercises run in 1993. The 273 concept elements were selected from a set of several hundred elements and then edited to create simple, declarative statements. The coffee elements were then translated into the local language by professionals who were native speakers of the local language. The brand names and prices were individualized to the market, but the remaining elements were identical. Of the 273 elements, 38 were pictures, varying from shots of coffee to shots of situations in which coffee could be consumed. Prior to the study the participating professionals identified which pairs of concept elements were incompatible with one another for logical reasons. These incompatibilities were incorporated into a set of 501 pairwise restrictions. The restrictions ensured that no concept would ever be logically meaningless.

The respondents were prerecruited to participate in a 30-minute session. During the session each respondent evaluated 100 concepts created to comprise 3–5 elements each. Each respondent evaluated concepts set up for an individual design, allowing a model to be created at the individual respondent level. The 100 concepts comprised 80 of the 273 elements. Utilities for the untested elements were estimated at the individual respondent level by the data-imputation procedure discussed in Chapter 5. Each respondent thus evaluated a unique set of combinations, ensuring that no single combination would influence the ratings. This precaution makes the research even stronger because the results are independent of the specific combinations.

The data were made available for analysis, including summarization by gender, age, type of coffee consumed and, finally, for segmentation. The segmentation was done on the responses of the entire set of individuals across the different countries, considered as a single dataset, independent of the country (Moskowitz 1996). Only the brand names and prices were excluded from the segmentation because they varied by country.

The creation of so many concept elements, the implementation of the study, and the analysis provided an opportunity for researchers to get firsthand experience in collaborative research. The following observations apply to the design, execution, and analyses phases:

Collaboration Was Smooth

Among market-research suppliers, there was a surprisingly smooth collaboration in terms of design. Although the researchers came from ten countries with varying degrees of research sophistication, all of the researchers were both positive and able to participate in the research as full members. This may have been the consequence of a self-selection bias, because all of the research participants were members of ESOMAR (European Society of Marketing Research). ESOMAR, as a professional organization of researchers, stands at the forefront of new ideas and technologies in the field of consumer research, so the smooth collaboration should not come as a surprise.

Table 8.2. Categories for the worldwide coffee study and the number of concept elements in each category

Category	Elements	Category	Elements
Aroma	12	Coffee imagery	9
Beans	6	Convenience	6
Country of origin	8	Ways to be served	7
Benefits	16	Ways to be preserved	8
Coffee accompaniment	6	Relaxing	6
Dinnertime	9	Anytime of the day	6
Varieties	23	Morning theme	7
Sociable/friendly	14	Taste	12
Flavors	16	Waking up theme	6
Warm and rewarding	15	Coffee descriptors	7
Roast and grind theme	8	Brands	7
Sophisticated	8	Sizes/prices	7
Associations with coffee	6	Visuals	38

There Was Disagreement About Respondent Task

There were disagreements about the number of concepts that a respondent could evaluate in the prerecruit session. However, this was a minor issue. The number ended up as 100, but the disagreement pointed out the differences among the participants with respect to the length of the interview.

Pictures Reveal Researcher and Cultural Differences More Than Do Text Statements

The choice of the pictures was far different across cultures than was the choice of the text elements. The study called for 30–40 pictures. Each participant offered radically different sets of visuals, from which the final set of 38 was culled. The differences in the type of pictures suggest that the country-to-country differences are clearer when it comes to execution-type stimuli rather than informational stimuli; that is, the pictures, being more executional, were tied more closely to the country than were the text descriptions of coffee. That difference between the nature of

the pictures and the nature of the text was a major surprise in the research.

Countries Differed

The winning elements and the losing elements for the total panel performed differently across countries. Table 8.3 shows the performance of the winning and losing elements, ranked according to performance by the total panel. The winning elements clearly do not do particularly well on a total-panel basis. For example, the element "Show your good taste by serving your guests the very best coffee" scores only a +4. In contrast, the losing elements—most flavor descriptions and novel forms—do quite poorly across the full set of countries. Such ideas as instant espresso and coffee in a tea bag do very poorly (-10).

There Was No Clear Pattern Across Countries

The winning elements differ by country, but the pattern is not precisely clear. Table 8.4 shows three winning elements each from Germany, Hong Kong, Italy, and Norway,

Table 8.3. Winning and losing concept elements for the total panel and performance in the seven countries

Element	Total	Germany	Hong Kong	Italy	Norway	UK	USA	France
Additive constant	*43*	*44*	*44*	*54*	*26*	*38*	*52*	*41*
Show your good taste by serving your guests the very best coffee	4	0	9	4	7	5	5	1
The rich aroma is followed by great taste	4	-1	5	4	6	4	5	6
Delicately roasted beans result in the ultimate drinking experience	4	3	10	4	5	-1	7	3
Pure coffee enjoyment	4	3	7	3	5	4	2	2
Freshly brewed taste	4	-3	6	4	5	6	6	2
Available in coconut	-9	-6	-9	-16	-8	-12	-11	-3
Available in chocolate raspberry	-10	-9	-4	-16	-3	-13	-10	-11
Available in peppermint	-10	-12	-8	-12	3	-12	-15	-11
Coffee in a tea bag	-10	-7	-7	-22	-8	-4	-8	-16
Instant espresso	-10	-8	-5	-22	-4	-5	-17	-13

Table 8.4. Comparing winning elements for Germany, Hong Kong, Italy, and Norway

Winning elements, Germany (GR)	Total	GR	HK	IT	NW	UK	US	FR*
Perfect during working hours	1	5	4	0	−4	−3	2	0
Picks you up when you're feeling down	2	4	5	3	1	1	2	−2
Eye-opener	2	4	2	4	2	−4	4	0
Winning elements, Hong Kong (HK)	Total	GR	HK	IT	NW	UK	US	FR
The coffee you can drink all day	2	0	12	0	−2	−1	1	6
Top quality for the gourmet and connoisseur	2	0	10	2	7	−2	−2	1
So simple to prepare . . . so pleasant to enjoy	−2	−3	10	−6	−2	−1	1	−10
Winning elements, Italy (IT)	Total	GR	HK	IT	NW	UK	US	FR
I know my mind, I know my tastes	4	2	2	9	10	−4	4	4
The way to end the perfect evening	3	1	1	9	4	2	4	0
A carefully blended coffee that will make morning your favorite time of the day	1	0	−1	9	1	−3	1	0
Winning elements, Norway (NW)	Total	GR	HK	IT	NW	UK	US	FR
Enjoyable	3	1	5	2	10	2	5	−2
Slice of cream cake	−1	4	1	−15	10	−4	−1	−2
Specialty flavors	2	−2	5	3	9	−1	−2	0

*FR, France.

respectively. The elements in these countries clearly differ. The German elements tend to be work oriented, those for Hong Kong tend to be relaxation, and those for Italy tend to be self-expressive, whereas those for Norway tend to be appropriate for a *Konditorei* (European coffee and cake shop). The performance of these winning elements in one country may, however, be quite poor in other countries.

There Were No Big Winning Elements

Despite country-to-country differences, there are no massive winners in any country. Table 8.4 shows that the winning elements rarely go above 8–10 on the utility scale.

Segmentation Revealed Clear Groups

One can segment the respondents transnationally, independent of country, to arrive at new segments. Since most of the text elements were identical from one country to the next (except for brands and prices), it was easy to put all of the data pertaining to common elements into a single group of respondents independent of country, segment the respondents, and thus develop new subgroups. The winning elements in each segment define the subgroup. Table 8.5 shows the segmenting solution for a four-segment solution. The number of segments to select is always a matter of judgment, because the statistical procedures simply partition the respondents into different groups.

Table 8.5. Winning elements for the four concept response segments

		Segment			
	Total	1	2	3	4
Additive constant	*43*	*53*	*46*	*51*	*28*
Winning elements, segment 1 (indifferent)					
Cup of coffee from above (visual)	1	0	1	2	2
The perfect way to welcome friends and family	3	−1	0	4	6
Coffee beans (visual)	2	−1	−2	3	5
Coffee and the morning paper	3	−1	5	8	0
Winning elements, segment 2 (relaxer)					
Put your feet up and enjoy a cup of coffee	3	−5	6	6	0
The coffee you can drink all day	2	−3	6	0	4
Hot coffee and a good book	2	−2	6	5	0
Sit by the fireplace with a cup of coffee	3	−2	6	3	5
Winning elements, segment 3 (waker-upper)					
Picks you up when you're feeling down	2	−7	0	14	−3
Wake-up taste	2	−9	1	14	−3
Makes you feel invigorated	1	−13	1	13	−4
Boosts your energy	2	−10	2	13	−4
Winning elements, segment 4 (experiential flavor seeker)					
Available in orange brandy	−6	−11	−12	−17	12
Served in the finest coffee shops	2	−8	−2	1	11
Top quality for the gourmet and connoisseur	2	−11	−4	5	11
Available in chocolate macadamia	−5	−10	−9	−17	11

The four-segment solution is easiest to interpret.

Naming the Segments Is Straightforward

The segments can be labeled by looking at the elements that perform best. Segment 1, which is the *indifferent group*, can be characterized by the phrase "Just give me coffee . . . I don't need to know anything else." Segment 2, which is the *relaxation* segment, responds strongly to elements that deal with taking it easy. This segment can be characterized by the phrase "I drink coffee to relax. Tell me about relaxing and I'm intrigued." Segment 3, which is the *wake-up* segment, responds strongly to elements that deal with waking up in the morning. This segment can be characterized by the phrase "I drink coffee to get energized. Tell me about the morning, and how coffee starts me off." Finally segment 4 is the *flavor seeking* or *flavor adventure* segment, which responds strongly to some of the flavor messages. This segment can be characterized by the phrase "If it's new and different, I think I might just be interested."

The Additive Constant Differs Across Segments

The segments differed in basic response to the concepts. For instance, the indifferent segment shows the highest basic response to coffee concepts (constant = 53), but beyond that high basic response no element can drive acceptance higher. Segment 2, the relaxers, shows lower basic response (additive constant = 46), but the winning elements do drive acceptance higher because they add as much as 6 points to the sum of the utilities. The real effects of segmentation appear for segments 3 and 4, respectively. Segment 3, the waker-uppers, also shows a moderate basic interest (constant = 51), but the right elements can add 13–14 points. Finally segment 4, the flavor seekers, shows the least basic interest in coffee (constant = 28), but

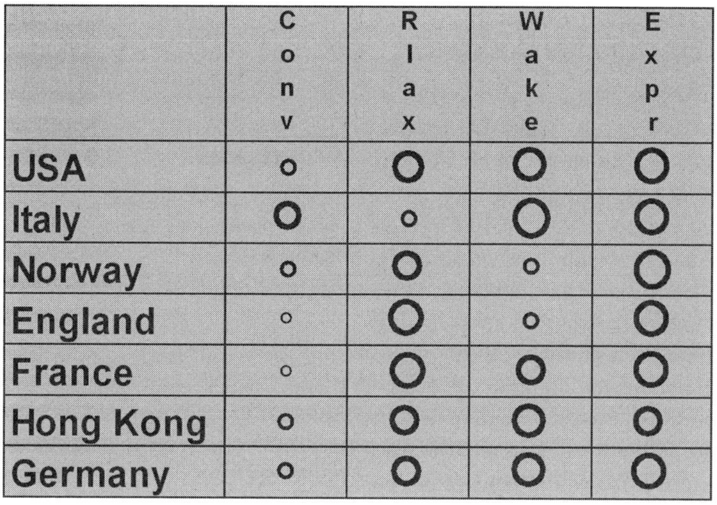

	Conv	Riax	Wake	Expr
USA	∘	O	O	O
Italy	O	∘	O	O
Norway	∘	O	∘	O
England	∘	O	∘	O
France	∘	O	O	O
Hong Kong	∘	O	O	O
Germany	∘	O	O	O

Figure 8.2. Relative distribution of four concept-response segments for coffee among seven countries. Conv, conventional; Expr, experiential, sensory oriented; Rlax, relaxation oriented; and Wake, waker-upper (energizer) oriented.

again the proper flavor statements can add a great deal

The Segments Are Present in All Countries

These four segments appear in different proportions worldwide. Figure 8.2 represents the distribution of the four segments. The size of the circle is proportional to the relative frequency of the segment in the country. What comes across from Figure 8.2 is the presence of the segments in all of the countries, implying that transnational segmentation is possible. As these data and analyses suggest, there is thus the capability to use the concept approach and the ensuing segmentation method to create worldwide coffee concepts.

Creating New Concepts

The final step in transnational segmentation is to create the concepts, based on the global segmentation, but particularized to a country. The concepts should then be appropriate to an

Table 8.6. An example of the world concept for the relaxing coffee

> The two common elements for the relaxing segment
> > Show your good taste by serving your guests the very best coffee
> > A coffee that leaves a warm, relaxed feeling inside
>
> Particularizing concept elements, appropriate for the different countries
> Hong Kong
> > So simple to prepare . . . so pleasant to enjoy
> > (Visual) Picture of a coffee teabag
> Italy
> > The way to end the perfect evening
> > (Visual) Picture of two girlfriends sitting, having coffee
> Norway
> > Bring that special meal to a perfect close with an outstanding cup of coffee
> > (Visual) Picture of a group of friends

identifiable worldwide segment, but particularized to the country by using the information about the country. Table 8.6 gives a sense of the global concept for a relaxing coffee, particularized to the different countries.

Strategies for International Research: Execution and World Concepts

The key messages to take from this exposition of international research can be summarized by the following two points:

1. *Concept development and testing can be done on a worldwide basis, using the same type of platform.* It is important, however, to take proper precautions in executing the study. These precautions include care in translating the elements, creating an understandable questionnaire, and fielding within the constraints of the local culture. It is also important to recognize where constraints on the questionnaire and fielding are irrelevant, compared with where they may have a material effect.

2. *Rather than looking for country-to-country differences, look for organizing principles.* When creating a base concept that is particularized to a country, the elements relevant and unique to that country should then be used to create the finishing touches and thus complement the other elements.

References

Blankenship, A.B., and Breen, G.E. (1993). State-of-the-art Marketing Research. Chicago: NTC Business Books.

Engel, J.F., Kollat, D.T., and Blackwell, R.D. (1968). Consumer Behavior. New York: Holt, Rinehart and Winston.

Moskowitz, H.R. (1996). Segmenting consumers worldwide: an application of multiple media conjoint methods. In: Proceedings of the 49th ESOMAR Congress, Istanbul, pp. 535–552. Amsterdam: ESOMAR [European Society of Marketing Research].

Nicosia, F.M. (1966). Consumer Decision Processes. Englewood Cliffs, NJ: Prentice Hall.

Rabino, S., and Moskowitz, H.R. (2001). International product concept development-research platform and its transfer to a transitional economy. Journal of Euromarketing, 10: 45–63.

Schiffman, L., and Kanuk L. (1999). Consumer Behavior, 7th edition. Upper Saddle River, NJ: Prentice Hall.

Part III

Advanced Analytics

Chapter 9

Believing the Results: Reliability and Validity

Introduction

Business decisions are based on information that must be believable in order to be acted upon, for if the information is doubted, then the conventional alternative is to make an informed, intuitive decision. Although this description of business practice sounds a bit capricious and perhaps somewhat cynical, in actuality it is simply a statement about what typically happens in practice. Unlike science, business makes products, and products make money. Business does not have the luxury to investigate all the alternatives, establish a firm foundation for the decision, and then proceed at a leisurely pace. Managers face the constant challenge to decide quickly and effectively, using only a handful of levers to help build sales and marketing efficiency and market share. This is where *decision models* can play a key role (Kumar and Bohling 2002). Companies may say they follow those steps, but in consumer research the truth hardly confirms this nicely formatted path. Decisions are made using the available information. The information is more often partial than complete, more often somewhat valid rather than really valid, and more often conjecture than hard facts. As a consequence, one must use methods whose reliability and validity are established. It is even more important to build into any research method, where possible, automatic checks that demonstrate reliability and validity. By building in these safety watchdogs, researchers can proceed, generally without too much trepidation.

Although the notions of reliability and validity are used together, occasionally interchangeably, they mean quite different things. A measurement can be reliable but not necessarily valid. However, the measurement must necessarily first be reliable before it can be valid; reliability in itself is not a sole condition for validity. Both reliability and validity are necessary for accurate measurement in a research study. *Reliability* is the degree to which the same result occurs when the experiment is repeated. Another word for reliability is *reproducibility* (Mangin et al. 2002). The results may not be correct, but they can be reproduced. *Validity*, a somewhat more involved concept, is the degree to which the data are *true*. The notion of *truth* can be twisted around. Truth can be the fact that the results seem to make sense on an intuitive level or can be the fact that the data predict performance in another type of test. The problem of validity arises because measurement in the social sciences is, with very few exceptions, indirect. Under such circumstances, researchers are never completely certain that they are measuring the precise property they intend to measure. We explore the different notions of validity in this chapter.

In business, it is important to present clearly reliable and valid data. The business environment is not the university environment, where the results of one's research can be presented to colleagues and published in outside, refereed journals. The scientific endeavor at a university guarantees, albeit perhaps exceedingly slowly, that research methods and research results lacking in reliability and validity will eventually be unmasked and discarded. In contrast, business is fast moving. Decisions are based on the data, and quite often neither the data nor the rationale for the decision last particularly long. One would be amazed to find that, in most companies, even the data for very important decisions are routinely discarded in the corporation's never-ending battle against clutter. The key exceptions, of course, are data recognized as potentially having legal value, such as data underlying patents. For the most part, the humdrum consumer data underlying ideas for new products, and especially the concept research data, are typically filed and forgotten rather quickly after the product is launched. In a matter of 1–2 years, last year's data are overtaken by the tidal wave of additional, new data for the next product and the next launch. Therefore, as already noted, it is ideal if the research methods have embedded within them sufficient information to establish their reliability and validity *at the time of the study execution*.

Reliability

How does one establish reliability in a business environment for consumer data and especially for both methods and results in concept research? The reliability of a specific research tool, the ability to get the same answer again when the study is repeated, is not typically investigated nor established by corporate researchers. Rather, most researchers simply choose a method that they believe to be reliable (as well as valid). When it comes to concept research, all of the involved groups—developers, marketers, and researchers—want to move to the end of the process. They are goal focused, not science focused, or at least their job descriptions talk about goals, not science, per se. Taking the time to establish reliability of a research method is seen as someone else's job and certainly not the job of those who must produce the concept.

Reliability can be established in at least three well-established ways: test–retest reliability, split-half reliability, and statistical tests, respectively.

Test–retest reliability means that the same answers are obtained when the study is repeated under comparable situations. Establishing test–retest reliability involves administering the same instrument at least twice under virtually identical conditions (Fouladi 1999). Of course, no study is ever completely the same from one occasion to another. The respondents may differ from each other and, indeed, in analyses of variance, respondent-to-respondent differences usually far outweigh differences due to concept or replication. The environment may have changed; the test may be done at different times, and during the intervening periods the business situation may have improved or deteriorated. Certainly, the respondents in the different replicate studies are exposed to changing competitive activities, which militates against having a perfect, laboratory-type test of reliability.

With the method of *split-half reliability*, the researchers divide the scale/test into two halves so that the first half forms the first part of the entire test/scale and the second half forms the remaining part. Clearly, a set of items can be divided into two subsets in many different ways: one could use a first-half-last-half split, an odd-even split, random half-split, and so. Which method of achieving split halves of an instrument is best will depend on how the items are organized. The most important consideration for an investigator is to make the two halves as similar as possible, such that they can be thought of as alternate, equivalent, or parallel forms of the same instrument. Both halves are normally of equal length and are designed in such a way that each is an alternate form of the other. Estimation of reliability is based on correlating the results of the two halves of the same test/scale (DeVellis 1991).

Also, statistical tests, such as the Cronbach alpha and the Spearman rho correlation coefficient test, can be used to assess reliability.

Even though we do not test respondents in a tightly controlled laboratory, it is remarkable how reliable concept ratings turn out to be in actual practice. The reliability comes from these three contributing factors:

1. *Large base sizes cancel the noise.* Large base sizes of respondents cancel out the variability due to individuals. Although people differ dramatically from one another, by the time 50 or 100 respondents have evaluated the concept, the variability tends to cancel itself out through averaging. As long as the researcher samples enough respondents with sufficient variation among the respondents, sooner or later the mean shines through. This observation comes both from statistics and from common sense and observation. Statistically, the *standard error of the mean*, or the variability of the mean to be expected across replications, drops with increasing base size. Small base sizes of respondents lead to large standard errors of the mean. In contrast, large base sizes of respondents lead to small standard errors, meaning that, on subsequent replications, one can expect to have means very similar to the mean one is currently observing.

2. *Control of test conditions reduces the noise.* Researchers like to run studies in a controlled environment. Sometimes the control is cosmetic only and has nothing to do with the data itself. However, any type of control that is maintained from study to study, replication to replication, reduces the noise in the system and leads to more reliable data. Unfortunately, however, we do not necessarily know the amount of noise that is reduced by each controlled factor. We do know that the subject variability is very high in consumer research. People vary. Most researchers control this by matching the respondents across replicates so they have similar samples. Other researchers use the exact same instrument, with the same questionnaire, and same interviewer or interview group. In this case the measurement can be reliable but not necessarily valid. The researcher feels safe because the process of checking the data gives a sense of security. However, it might be that the re-

search controls the wrong variables, and the data are ultimately incorrect. The control of test conditions can only add to reliability, but the magnitude of the added reliability is not clear. When the contributory power of these different controls is unknown, prudent researchers often opt for what appears to be the tightest control, such as matching respondents. Matching the externals of the interview, such as field services and questionnaire appearance, is less frequently an issue.

3. *Standardized interpretation against norms.* Researchers always seek norms against which to make their decisions. The norms anchor the results. By looking at the norms for studies, researchers can ensure that the data are reliable, especially if the study contains known factors that should remain stable from year to year. Even if the data move, year to year and market to market, looking at the results in terms of deviations from the norm make the data more reliable. A similar approach that is well known in statistics is *within-subjects designs*. Statisticians recognize that there is enormous variability across people and perhaps across test periods. By using each subject or respondent as his or her own control, statisticians ensure greater reliability. The assumption is that, whatever bias is introduced by the subject or test period, the bias is the same across all test stimuli assessed by the respondents. Thus, the respondent anchors the data by serving as his or her own control. In a within-subjects design the values of the dependent variable for a task or a set of tasks (e.g., tasks involving an online purchase) are compared with the values for another task or another set of tasks (e.g., tasks in which items are added to an online bridal registry) within one participant's data. In other words, within-subjects designs are designs in which the same variable is measured repeatedly on the same participant under different task conditions (Keppel 1993). Finally, interpretations of results from different market countries pose some challenges. Researchers need to be wary of

interpreting results in terms of their own culture and experience and, in particular, of generalizing from experience in industrialized markets to emerging markets (Craig and Douglas 1999).

The Jackknife Strategy: An Emerging Way to Measure Reliability

One way to assess reliability within the context of a business application assesses the degree to which results using subsets of the respondents match results from the total panel. The analytic approach is relevant for business because it shows the base size needed to obtain data that are consistent from trial to trial, and it is relevant from a statistical aspect because it represents an emergent trend in statistical thinking, called the *jackknifing approach* (Perrone and Cooper 1993).

The basic idea behind the jackknife approach is quite simple. The researcher begins with a dataset, knowing beforehand the aver-

age from the total panel. The researcher then creates samples of data by randomly selecting a number of respondents from the dataset and then computes the mean. The sampling can continue with increasingly larger samples from the base population, until at some point the sample size equals the entire population. At some base size the data will begin to stabilize, and one can say that the results become stable. The jackknife strategy thus provides an idea of the number of respondents needed to ensure reliable data.

We can illustrate the jackknife method by looking at data from a large-scale study run with the IdeaMap method on condiments. The study comprised 202 elements and 434 respondents. For each respondent, IdeaMap generates an individual utility model for each of the 202 elements. This very large dataset enables a good use of the jackknife method.

The procedure followed these two simple steps for the condiment dataset:

1. Twelve base sizes were selected, from 10 to 120. For each of these 12 base sizes, five different random samples were randomly

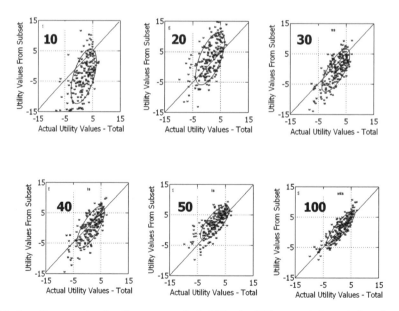

Figure 9.1. A scattergram showing the agreement of utilities for 202 concept elements (condiment), given randomly drawn subsets of different sizes. The *numbers* in the body of the figure (10, 20, etc.) show the base size of the subsample. Only the first subsample (of five) is shown for each base size.

drawn from the full population, without replacement. The utility value for each of the 202 elements was computed by averaging the five replicates.

2. For each base size (10 . . . 120) the 202 average utilities, one average per concept element, were plotted against the 202 utilities computed from for the total panel of 434 individuals. The correlation between the subset and the total panel, based on 202 utilities, provides a measure of reliability. The higher the correlation of the subset with the total panel is across the 202 elements, the more reliable are the data.

The results of this analysis are presented in Figure 9.1, which is a scattergram, with the abscissa comprising the actual utilities from the total panel and the ordinate comprising the utilities from the subset generated by the sampling. As the base size increases, the agreement increases between the utilities for the subsample and the utilities for the total panel. Figure 9.2 shows that the agreement begins to become quite obvious as the base size hits 50 (with the correlation hitting 0.75).

Why Validity?

Consumer researchers are continually challenged, as is every researcher, to demonstrate that their data have validity. Validity in marketing research has often been consigned to validating interviews and specifically establishing that a particular respondent actually participated in interviews. In many consumer research situations the notion of validation has not necessarily gone any further. This is an example of *field level* validation. There are other types of validation, such as the ability of the data to predict future performance. This may be the performance of the test stimulus in a new study and not necessarily the same type of study, or the performance of the product or concept in the marketplace. Internet research in particular is often challenged regarding its validity because the Internet venue

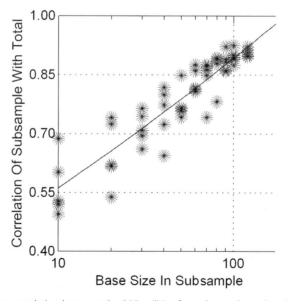

Figure 9.2. The Pearson correlation between the 202 utilities from the total panel and the utilities obtained from smaller subsamples. Each *point* represents one of the subsamples, of a specific base size. As the base size of the subsample increases, the correlation of the utilities shows greater agreement with that from the total panel.

is new to the research community. There are no interviewers who administer the study and who can observe the respondents. The studies are often self-selecting, so the respondents participate when and where they want to participate rather than reacting to a live interviewer.

Another way to validate data checks the results against external criteria that are considered to be more objective. The definition of *objective* is left to the researcher. A good example comes from the field of *product testing*. Product test results can be validated, at least at a basic level, by correlating the ratings of a sensory attribute with the physical level of the ingredient presumed to drive this attribute. Thus, if a beverage manufacturer systematically varies the amount of a sweetener in a set of samples, holding all other variables constant, then the researcher can easily demonstrate basic validity by correlating the concentration of the sweetener known to be in the product with the rating of sweetness. At the aggregate level this type of validity has been demonstrated again and again, primarily in the literature of that branch of psychology known as psychophysics. In more complex cases, when two or more physical variables change systematically, the validation can be shown by models relating the independent variables (ingredients) and occasionally their interactions. This area, which is well known in applied research, is *response surface* methodology. Let us look at the methods to establish validity in concept research.

Types of Validity Available for Concept Research

As noted in the introduction to this chapter, validity is very important in business. Only when the data are believed to be valid will management take a risk and act on the results. Campbell and Stanley (1963) have listed 12 sources of unwanted measurement effects: eight that concern internal validity and four

that may jeopardize external validity. The different validities, their psychological bases in business, and examples are presented in the following sections. Professionals have consistently distinguished between actual validity and face validity (Burns 1996).

Face Validity

A way to define face validity is that "the results look the way they should look, based on my expectation." Face validity is not validity in the technical sense; it refers not to what the test actually measures, but to what it appears superficially to measure. Face validity pertains to whether the test *looks valid* to the examinees who take it, the administrative personnel who decide on its use, and other technically untrained observers (Anastasi 1988). Although, to skeptical readers, the idea of face validity having any role in business makes no sense because such validity is subjective, in reality almost all validity in business is face validity. Face validity can be demonstrated in four ways:

1. *The study looks like it has been correctly executed in the field.* If the study is conducted under supervision and is cosmetically correct (no clear errors and no clear biases), then quite often the approach and thus the data will be accepted as valid, whether the results are valid or not.

2. *The respondents are who they say they are.* Much of consumer research has defined validity as the fact that the respondents actually participated in the study. This definition of validity comes from the heritage of market research in the sociological field rather than in experimental psychology. In experimental psychology, certain defined outcomes need to be satisfied, given the antecedent independent variables, for the research to be considered valid, independent of the actual respondent behavior. In sociology, the validity issue of an individual's behavior is simply that the individual does what he or she ordinarily does. There is no experimentation in

sociology, so the conventional measure of validity against some type of true standard simply does not exist. A person's behavior may not agree with the standard, but it is valid behavior, nonetheless. The only way that the data can be invalid is if they pertain to someone else rather than to the individual whose behavior is being reported.

3. *The results confirm what might be naively expected.* In the business community, especially, where decisions must be made quickly, research may confirm what the researcher, marketer, or developer "knows" to be the truth. The word "knows" is in quotation marks because in reality the marketer and researcher expect things to go a certain way. If the results support their expectations, then all too often they label the research as being valid. In contrast, if the results do not support their ingoing expectations, then one way to reduce the inevitable anxiety produced by the negative news is to label the research as invalid. Whether the research is truly valid is irrelevant. The discomfort from unexpected results leads to a lashing out at the researcher (shooting the messenger). Calling the research invalid is the preferred way to shoot the messenger. When challenged on this, the person often has no reason to call the research invalid, other than it does not confirm expectations.

4. *The results confirm some objective "knowns."* One example of this in concept testing is the response to price. If a respondent lowers his or her rating of interest in a concept with increases in the stated price of the item described, then this is often prima facie evidence that the results are valid. It is not that the researcher absolutely knows that the demand for the item will drop with price, but rather the pattern makes sense. Given this pattern, the researcher will then claim that the data are valid. In a similar way, if the concept describes the product as being "low calorie," and if one of the rating questions is about calorie level, then the concept should score low on questions pertaining to calories.

If it does, then the research is assumed to be valid because the rating reflects what is known to be true.

Construct Validity

This refers to the data conforming to expectations from an antecedent theory. For example, if the construct is that people easily recognize and respond to "value for money," then construct validity would be demonstrated if the concepts with the higher price points would be rated as being lower on "value for money." The rating agrees with the expectations from theory about people being sensitive to money when they evaluate alternatives.

One subtle issue arises with construct validity in business. The validation exercise to demonstrate construct validity should not use the key evaluative measure to demonstrate construct validity. Thus, it is appropriate to validate the concept research by using rated "value for the money," but it is not valid to use overall purchase intent. *There is no direct expectation from the theory that increasing price should yield decreasing interest.* Construct validity and validation are involved whenever a test is to be interpreted as a measure of some attribute or quality that is not operationally defined (Cronbach and Meehl 1955); that is, construct validity requires that there be some external base of knowledge against which the data can be compared and predictions made and either validated or refuted.

In the business environment, construct validity is often glossed over. Even though a concept itself is developed based on constructs, in the heat of analysis it is hard to step back in a detached fashion to ascertain whether the results of the concept study agree with the construct. Often there is not a construct; the goal of the research is to conduct a test, make a measure, and go forward by subjectively interpreting that measure. Too much attention focuses on the immediate business applications of the results, which translates to considering the immediate performance of the concept itself, whether it passes the test, fails, or needs a change.

Predictive Validity

This is generally the most important aspect of validity to business. Business creates concepts with one major purpose in mind: describe the blueprints for a new product, or messages that will be transformed into advertising copy. Unlike sensory analysis that involves products and the science of sensory perception, there is generally no basic body of scientific knowledge about concepts to which the results will be compared. In sensory analysis there are numerous studies of a basic nature, with actual products. The real goal of concept research in business is to pretest the ideas, and get a sense of whether there is a need for this new product.

Content Validity

This means simply showing that the test items are a sample of a universe in which the investigator is interested. Content validity is ordinarily to be established deductively by defining a universe of items and sampling systematically within this universe to establish the test (Cronbach and Meehl 1955).

Empirical Validity

Do the predictions implied by the theory match the data? If they do, then the theory has empirical validity. If they don't, the theory leads one to make false predictions. It doesn't match reality (Torgerson 1965).

Business researchers use a number of criteria for success. These criteria are then wrapped into the assessment of predict validity:

1. *Subsequent consumer tests.* The concept may be tested in subsequent consumer research studies, in which case predictive va-

lidity simply means that the concept that performs well in this study should perform well in future studies.

2. *Product prototype.* The concept may be converted into a physical prototype requiring that concept performance in the current study should predict product or product plus concept performance in future studies. These future studies are often test-market simulators such as bases.

3. *Market launch.* The concept may be converted into an actual product and launched in the marketplace, requiring concept performance in the current study predict in-market performance.

Although the notion of predictive validity appears straightforward and is compelling to businesspeople, the reality of predictive validity is often muddled. Marketers are loath to retest or launch with a concept that scores poorly. Consequently, no one ever validates the performance of poorly scoring concepts with subsequent research, although researchers have done so in some situations. For example, in a study with concepts about military rations, Moskowitz and his colleagues (1994) first identified the components of concepts for military rations and then used these in an IdeaMap study with military personnel. Some elements scored well and some scored poorly. These concept elements, created by conjoint analysis, were then later recombined to produce 25 concepts whose expected values ranged from low to high. When tested among a subsequent and different population of military respondents, the correlation was 0.85 between the predicted performance of the concept and the actual performance of the concept in the later test.

This foregoing validation of data through predictive validity is relatively rare in a business setting, precisely because most businesses refuse to spend money to retest a losing concept. The unspoken goal of business researchers is to save money by weeding out poor performers, even if this eliminates the chance to validate the testing system. With the increasing availability of inexpensive, rapid, low-effort research on the Internet, however, the era of validating the concept stimuli may be at hand. Internet-based research minimizes the cost of the research effort and can be used to determine whether strong-performing elements in an early conjoint screening test do well later when combined into full concepts.

Consistency of Response in Conjoint Measurement: A Blend of Validity and Reliability

Conjoint analysis provides a unique opportunity for researchers to assess validity of the data, even "on the fly" during the study and even in the absence of normative results against which the data can be compared. There is a criterion of *correctness* in conjoint measurement that is at once a measure of data validity at the individual respondent level and also has aspects of face validity.

In conjoint measurement the independent variables are presented in combination to the respondents according to an experimental design. The design ensures that the elements or independent variables are truly independent statistically by being arrayed in a way that ensures they are uncorrelated with one another. For purposes of data analysis, as has been emphasized in this book, the independent variables are coded as either 0 or 1, denoting the absence or presence, respectively, of concept elements in the experimental design. The rating variable is the degree of interest or some other rating attribute. Dummy variable regression analysis relates the presence or absence of the elements to the rating assigned by the respondents. The goodness-of-fit statistic, the Pearson multiple R^2 (henceforth abbreviated as R^2), shows the *proportion of variation in the rating attribute that can be accounted for by the independent variables*. The R^2 value varies from a low value of 0 to a high value of 1.00. Each equation generated by the dummy variable model has a corresponding R^2 statistic. The R^2 statistic does not, however, have any relation to what elements score well versus what elements score poorly.

The R^2 statistic can be introduced as a measure of data quality and possibly an indicator of validity. If a respondent assigns the same number to every concept, then the R^2 value for that respondent's model will have an R^2 of 0 because we cannot trace changes in the dependent variable to any independent variables. If the respondent attends only to a few elements in making a decision, then the rating can capture only some of the variability among the elements; namely, those elements to which the respondent attends. The specific variation captured by the model can be traced to the elements attended to. The remaining variation will constitute error. If the respondent attends to all of the elements consistently, then the R^2 will be quite high. Thus, strategies in which respondents attend to only a limited number of elements cannot yield high R^2 values.

In a set of studies on food, run on the Internet with 3715 respondents, each respondent generated his or her own model relating the presence or absence of each of the 36 concept elements to the rating of interest (craveability; see Chapter 20 on the Crave It! database). Each of the respondents thus generated an R^2 value. The full set of 3715 values generates a density distribution shown in Figure 9.3. The vast majority of the respondents show R^2 values exceeding 0.70 (Moskowitz et al. 2002). The key fact to keep in mind is that at the individual respondent level the R^2 value lets researchers estimate the probability that a respondent is guessing. Only by paying attention to the stimuli and answering honestly can respondents be sufficiently consistent to generate the high R^2 value. Thus, the R^2 value becomes a measure of response consistency and data validity, at least in the case of systematically designed test concepts.

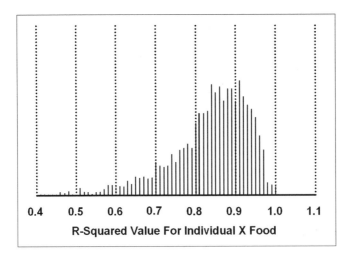

Figure 9.3. The distribution of R^2 values for all respondents participating in the 21 food studies for the Crave It! mega-study. R^2 values greater than 0.65 are significant.

Establishing the Validity of Internet-based Research for Concepts

Sometimes the validity of a research method may be established by demonstrating that the same results will be obtained by two different methods. This definition of validity pervades the market-research community, especially when the issue demanding validity consists of a new way to do the same research. Validity with new field executions simply demands showing that the new method produces the same results, but either gives additional information, is easier to execute, or costs less money to execute.

Migration of research to the Internet provides a wonderful source of such validation studies. As this book is being written (2004) the Internet is rapidly becoming a new venue for concept research. The Internet captured the attention of researchers in the early and mid-1990s but was not considered appropriate for research. By the late 1990s, however, Internet technology had become so visible that market researchers were investigating whether this new medium for interviews would generate valid results. The sequence

of questions and the activities by market researchers are informative:

1. Basic interest: What is this thing called the Internet?

2. What are these new communication technologies? Can we put questionnaires on the Internet? How do we do it?

3. Shall we attend conferences on Internet research? The ESOMAR organization (European Society of Marketing Research) is sponsoring the Net Effects conference? Should we go to learn about the method?

4. Can one demonstrate that I get the same exact answers when I use the Internet as when I use conventional research methods?

5. How can the Internet save me money? I have been to enough conferences and seen enough papers to now accept that the Internet is a valid medium.

Looking at this set of five questions, often posed during the evolutionary period, we find the questions asked repeatedly as researchers migrate their studies to the Internet. The most basic question pertains to the validity of the

Internet-based study, in comparison to the more traditional paper and pencil, or CAPI (computer-aided personal interview) study. The Internet issue poses an interesting case history to show how validity is established among a broad group of research practitioners. It is important to note, at least on a historical basis, that the case history presented here was done in the year 2000, just as the Internet was beginning to achieve very broad exposure and evolving from novelty to serious research venue. Today, 4 years later, such a case history would not even be considered relevant. Market research has moved on.

The case history presented here deals with validating Internet studies by showing that they lead to the same results as conventional CAPI research (Moskowitz et al. 2000). As noted earlier, the validation is of the absolutely simplest kind. Although the data were obtained in 2000 and based on a conjoint study of responses to different political positions by then-presidential candidate Al Gore, the results clearly show the validity of Internet research. The approach can be used as a model on which to pattern other valida-

tions, especially those dealing with research venues.

Study Specifics

The study involved 126 respondents in the New York region who participated in a conjoint study at a central location and 200 additional respondents who participated on the Internet. The respondents comprised a mix of men and women. Both groups of respondents evaluated the same concepts (albeit in different combinations) in the same way with the same screens. The only difference was that the Internet respondents participated at home. The individual utilities were used to create segments. The segment creation was done separately for each group of respondents. Three segments emerged, similar to those found in the Internet study, indicating that the Internet results and central location results led to similar conclusions. Figure 9.4 shows the high agreement between respondents in terms of the concept utilities for the two test methods. The agreement is striking for total panel and for key subgroup (political

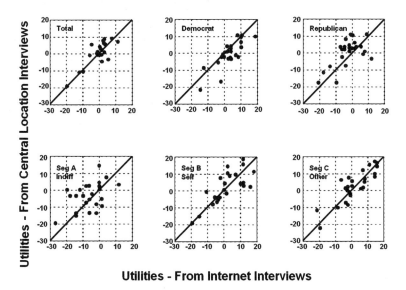

Figure 9.4. The relation between utilities obtained from Internet interviews and utilities obtained from central location interviews.

party and emergent concept-response segment, which were the same for the two test venues). This type of demonstration satisfies most critics because it clearly reveals that the elements tested in the Internet format do similarly as the elements tested in the CAPI format. That type of demonstration satisfies almost all critics.

The Role of Validity in the Business Mind

As this chapter has emphasized, to businesspeople validity is not so much a scientific term as a feeling of comfort with the data. Comfort is critical, especially with concept research, because usually there is no basic body of science against which one can measure one's results. One should never underestimate the importance of emotions in the establishment of validity. Indeed, in the words of S.S. Stevens, late professor of psychophysics at Harvard University, "Validity is a matter of opinion." This viewpoint still rings true, even though Stevens said it to the senior author (H.R.M.), his graduate student at the time, almost 40 years ago.

Despite protestations to the contrary by academics, business is not a university, and the rules of business are not those of the scientific establishment housed in the university. Business is about making products and selling them at a profit. Validity to businesspeople is whether the idea, the product, and the research work. The term *work* has a fuzzy meaning, however; it can mean "makes sense," "does well in other confirmatory tests," or ultimately "does well in the marketplace." When the term *validity* is used in the business world, therefore, it is used with many different meanings. Often businesspeople do not accept the rigorous demonstrations of validity admired and espoused by academics. If the user of the data feels comfortable with them, and if the user has had experience with the research method, then even if the method is wrong it has validity in the user's

mind. Conversely, if the research looks complicated, if the user has had no experience with it, and if the results are hard to interpret, then no matter how much the researcher tries to demonstrate validity (face, construct, predictive, etc.), the businessperson will continue to aver that the approach is not valid. Validity truly, therefore, ultimately remains a matter of opinion.

References

Anastasi, A. (1988). Psychological Testing. New York: Macmillan.

Burns, W. (1996). Content validity, face validity, and quantitative face validity. In: Barrett, R., editor. Fair Employment Strategies in Human Resource Management. Westport, CT: Quorum Books/Greenwood, pp. 38–46.

Campbell, D., and Stanley, J. (1963). Experimental and Quasi-experimental Designs for Research. Boston: Houghton, Mifflin.

Craig, C.S., and Douglas, S.P. (1999). Conducting international marketing research in the 21st century. New York: New York University Stern School of Business.

Cronbach, L.J., and Meehl, P.E. (1955). Construct validity in psychological test. Psychological Bulletin, 52: 281–302.

DeVellis, R.F. (1991). Scale Development: Theory and Applications. Newbury Park, CA: Sage.

Fouladi, R.T. (1999). A Guide to the Methodological Foundations of Quality Oversight and Improvement Process. University of Texas at Austin.

Keppel, G. (1993). Design and Analysis: A Researcher's Handbook. Englewood Cliffs, NJ: Prentice Hall.

Kumar, V., and Bohling, T. (2002). Six steps to better decision models: in the real world, smart decisions depend on accurate predictions. Marketing Research, 14: 8–12.

Mangin, J.F. Poupon, C., Clark, C., and Bloch, I. (2002). Distortion correction and robust tensor estimation for MR diffusion imaging. Medical Image Analysis, 6: 191–198.

Moskowitz, H.R., Beckley, J., Mascuch, T., Adams, J., Sendros, A., and Keeling, C. (2002). Establishing data validity in conjoint: experiences with Internet-based 'mega-studies' Journal of Online Research; http://www.ijor.org/ijor_archives/articles/establishing_data_validity_in_conjoint.pdf. Accessed (size 165810 bytes, 19 September 2002.

Moskowitz, H.R., Gofman, A., Tungaturthy, P., Manchaiah, M., and Cohen, D. (2000). Research, politics and the Internet can mix: considerations, experiences, trials, tribulations in adapting conjoint measurement to optimizing a political platform as if it were a consumer product. In: Proceedings of ESOMAR Conference: Net Effects[3], Dublin. Amsterdam: ESOMAR [European Society of Marketing Research], pp. 109–130.

Moskowitz, H.R., Lesher, L., Cardello, A., and Graves, G. (1994). Computer-based concepts for military ration packaging. Activities Report, R&D Associates, 47: 226–240.

Perrone, M.P., and Cooper, L.N. (1993). When networks disagree: ensemble methods for hybrid neural networks. In: Mammone, R.J., editor. Artificial Neural Networks for Speech and Vision. London: Chapman and Hall, pp. 126–142.

Torgerson, W.S. (1965). Multidimensional scaling of similarity. Psychometrika, 30: 379–393.

Chapter 10

Response Time as a Dependent Variable in Concept Research

Introduction

How fast do we process concept information? Do we process pictures faster than we process text? Do we process information that we *like* faster than we process information that we *dislike*? For more than a century experimental psychologists have dealt with issues such as these using response time instead of rating scales. Up to now we have used rating scales as the key dependent variable against which to measure concept appeal, concept communication, and the contribution of components. By shifting some of the focus to response time as a dependent variable in place of a direct rating we can begin to learn other things about concepts, such as hints about underlying mental processes that may be going on as respondents make their ratings.

Most of the published research on response time (also called latency or reaction time) comes from the scientific literature and very little from the business literature. Response time as measured by tachistoscope has long been a staple research tool in experimental psychology. There is almost a vacuum when it comes to business-oriented research using response times to understand the dynamics of concepts, although there is a substantial knowledge base and practice using response times to understand advertising and packaging. It is not surprising that response time has found a home in advertising testing, to study both reactions to the information and to the mode of presentation of the information (viz., type and layout, etc.) (Taylor 1970; Lancaster and Lomas 1977;

Rhodes et al. 1979; Aaker et al. 1980; McLachlan and Myers 1983). Furthermore, response time fits naturally with packages. The difference between concepts and packages is that packages are searched for, and thus response time is relevant as a measure of "findability" in a crowded environment. This is not the case for concepts. Most concept testing executed for commercial purposes is designed to address a specific marketing issue (e.g., the need to launch a new product or service). There is some, but not a great deal of, written material on concept testing, outside the commercial realm (Haskins 1975).

How Response Time May Be Useful for Concept Testing

Respondents do not react instantaneously to concepts. Rather, they process the information and then respond. The time to process the information may be just as important as the reaction to the concept, because the time to process information may provide more insight about the nature of the stimulus. Response time as an additional measure of reactions to concept provides an entirely new area for research—one that may have practical application for commercial concept testing. Response time provides an objective measure of the time needed to process the information presented by the concept. Two concepts generating equally high ratings of interest and similar communication profiles may actually differ quite substantially in terms of the amount of time it takes for the consumer to react and to assign a rating. Differences between concepts in response time would imply that the consumer needs more time to understand the messages in one concept than to understand the messages in another. In some cases, the response time may be due to other interfering processes, such as an inhibition set up because of the nature of the content in the concept.

Response Time as a Dependent Variable in Conjoint Analysis

Conjoint analysis enables researchers to understand what particular features of a concept drive interest. In most conjoint tasks the respondents rate overall interest or intent to purchase. In other applications they rate the concept on a variety of attributes, not just on purchase intent or likeability. These attributes may include the degree to which a product fits Brand X versus Brand Y, or the degree to which a concept communicates a specific image (e.g., more for men versus more for women).

Response time to concepts can act as the dependent variable. Within the conjoint methods, one can relate the presence/absence of concept elements to response time. Respondents may adopt individual styles in reading and evaluating concepts (e.g., some individuals react quickly in general, whereas others sit and think). The researcher can, in turn, create a model for each individual. The model reveals which elements require a long time to process (i.e., have associated long response times) and which require little time to process (i.e., have associated short response times).

Response Times and Concept Elements: Revisiting the Coffee Study

The coffee study provides good data with which to study response times, because the data come from many elements, across many countries (see Chapter 8). The IdeaMap computer presents the stimulus, records the rating, and also records the number of tenths

of seconds between the presentation of a concept on the computer and a respondent's rating. One can create a model relating the presence/absence of the concept elements to the response time. This model shows the number of tenths of seconds of response time that can be allocated to each element. High values of response time mean that the respondent takes longer to process the information. Low values for response time mean that the respondent processes the information more quickly.

Since the concept elements are arrayed in an experimental design, the researcher can create a model relating the presence/absence of the concept elements to the two dependent variables, interest and response time, respectively.

These two dependent variables for modeling are defined as follows:

1. *Interest* is defined as 0 if the coffee concept is rated as 1–6, and 100 if the concept is rated 7–9. This definition generates the *top-3* box statistic.

2. *Response time* is defined as the number of tenths of seconds from the presentation of the stimulus concept to the rating of interest.

Comparing Two Countries: The United States versus France

We can get a good idea of the differences between countries on both interest and response time. The additive model itself for each country comprises a set of coefficients and an additive constant. Keep in mind that the model can be expressed by the simple additive expression:

$$\text{interest} = k_0 + k_1 \, (\text{element 1}) + k_2 \, (\text{element 2}) \, . \, . \, . \, k_n \, (\text{element n})$$

Table 10.1 shows partial data for aroma elements for the United States compared with France. The additive constant, k_0, shows the estimated interest if all elements are missing from the concept. The additive constant can be interpreted as the basic interest in the concept in the absence of additional information. The additive model is, of course, an estimated value since no concepts appeared without elements.

A similar equation and interpretation hold for response time. The additive constant shows the estimated number of tenths of seconds required by a respondent to rate a concept that has no elements. Again this additive constant is a purely estimated parameter, but it can be used as a baseline value. The coefficients for the elements show the additive number of tenths of seconds in response time required by the presence of a specific concept element.

There are clear country-to-country differences in responses to the elements of coffee concepts, both in terms of interest and response time, respectively:

1. The additive constant for interest is higher for the United States than for France (52 vs 31). This means that the basic interest of US respondents is higher than the basic interest of French respondents.

2. The element utilities for aroma are higher for the US respondents than for the French respondents, meaning that the aroma statements drive concept acceptance even higher.

3. The best single element for the United States, "with that irresistible aroma," combined with the additive constant, generates 58% top-3 box interest. In contrast, the best single element for the French respondents, "the rich aroma is followed by great taste," combined with the additive constant, generates 42% top-3 box interest. This comparison further underlines the disparity between the US and the French respondents.

4. The additive constant for response time for the French and the US respondents is about equal (37 vs 42), corresponding to 3.7 vs 4.2 seconds. This is the *dead time*, or response time that cannot be allocated to the concept elements.

5. The contributions to response time of the elements differ by element and by the respondent's country. For example, the US respondents take the most time (1.4 seconds) to process the message "the rich aroma is followed by great taste." In contrast, the French respondents take the most time (1.1 seconds) to process the message "dark and rich with an invigorating aroma."

Table 10.1. Partial set of coefficients for interest (IN) and response time (RT)*

	IN		RT	
Elements	USA	France	USA	France
Additive constant	*52*	*41*	*37*	*42*
With that irresistible aroma	6	1	7	7
Rich coffee aroma	6	−1	6	8
The rich aroma is followed by great taste	5	6	14	8
Rich, flavorful aroma	4	−1	6	11
Made from the finest coffee beans	4	−1	4	8
Enjoy life—take time to smell the coffee!	3	1	10	6
The wake-up smell of freshly brewed coffee	3	1	6	2
Aroma comes out and spreads over the whole house	3	1	8	3
The aromatic and smooth coffee	2	1	8	9
The wake-up smell of brewing coffee	2	1	4	6
Has a rich, singular aroma	2	0	5	10
Dark and rich with an invigorating aroma	2	0	9	11
Inviting aroma	1	3	5	10

*The elements deal with aroma. Results shown for the United States compared with France. RT units are in 10ths of seconds.

6. There is no clear pattern regarding which types of elements are fastest or slowest to process, although the differences are dramatic. For the United States the difference is 14 vs 4, corresponding to 1.4 vs 0.4 or 1 second for an element. French respondents show a similar range of 11 vs 2 or 1.1 vs 0.2, or almost 1 second.

Fast-to-process versus Slow-to-process Elements

One instructive analysis compares the fastest-processed and the slowest-processed text elements for each country. We saw the beginnings of this analysis in Table 10.1, for France compared with the United States and limited to aroma elements. Keep in mind that these elements were translated into the local vernacular. Table 10.2 presents these initial results, based on the additive model. The fastest processed are those elements with the lowest coefficient (viz., low processing time), whereas the slowest processed are those with the highest coefficient for time (viz., high processing time).

As a first approximation we see the following patterns emerge:

1. The slowest-processed elements appear to have more letters than do the faster-processed elements.

2. Several of the slowest-processed elements require respondents to imagine a situation (e.g., the phrase "Bring that special meal to a perfect close with an outstanding cup of coffee").

3. The slow-to-process elements do not invoke concrete images of products, but rather invoke general, possibly ambiguous situations, forcing respondents perhaps to conjure up the idea.

4. The fast-to-process elements present short, to-the-point phrases, requiring relatively minimal imagination or image. This is not a hard-and-fast rule, but just an observation from the elements that show radically different processing times.

Respondents in Different Countries Require Different Times to Process Text Information

The additive model enables researchers to compare respondents in the seven different countries in terms of how rapidly respondents will react to the text elements. The model for response time enables researchers to compute the total response time for any specific text element. Thus, each element generates an estimated total response-time value. The computation for each element simply comprises the sum of the additive constant and the utility value for response time for that element (element response time = additive constant + response-time coefficient).

Table 10.3 lists the average time and the standard deviation of the average response time for the text elements across the seven countries. For comparison, Table 10.3 also shows the average utility value for the interest. Countries differ. The fastest responders (viz., those with shortest response times) are French, Italians, and Americans. In the middle are the English and the Germans. The slowest responders are Hong Kong Chinese and Norwegians. This is mostly due to the additive constant, but somewhat due to the individual response-time coefficients.

There is a clear difference among countries, in terms of the *spread* of the response times (see Figure 10.1). This difference among groups is due to differences in response times to the various elements:

1. Respondents in Norway, Hong Kong, and Italy show a relatively wide range of response times for text, with some text elements taking far longer than others.

2. Respondents in the remaining countries have a relatively narrow range of response times.

3. In terms of processing patterns for text information, these distributions suggest that respondents in Norway, Hong Kong, and Italy are more affected by the nature of the

Table 10.2. The two fastest-to-process versus slowest-to-process elements, by country: numbers are the coefficients for response time (large numbers = slower responses)

	France	Germany	Hong Kong	Italy	Norway	UK	USA
France							
Colombian coffee	−1	3	3	11	51	10	6
Perfect during working hours	−1	6	22	15	3	7	8
A deep feeling of satisfaction as you consider a job well done	15	14	14	13	31	12	13
Bring that special meal to a perfect close with an outstanding cup of coffee	16	15	19	11	29	11	12
Germany							
Bitter	4	0	9	11	24	12	9
Available in amaretto	2	2	15	1	25	4	9
For the highlights in your life	8	19	24	12	17	9	8
Our flavored coffees are the perfect conversation starters	4	20	18	11	20	12	9
Hong Kong							
Iced coffee	2	9	0	−3	25	3	4
For active people	5	8	2	6	25	3	8
Tastes as good as it smells	10	8	31	11	24	6	11
When you have to be at your best	3	5	34	3	17	12	3
Italy							
Wake-up taste	10	8	13	−4	31	5	3
Iced coffee	2	9	0	−3	25	3	4
Enjoy life—take time to smell the coffee!	6	8	18	17	31	20	10
The way to end the perfect evening	6	11	15	17	0	6	12
Norway							
It's for a relaxing and enjoyable break	6	6	6	7	−1	12	11
Sit by the fireplace with a cup of coffee	14	11	12	9	−1	10	12
A good cup of coffee makes you get more out of the day	8	9	11	5	50	12	11
Colombian Coffee	−1	3	3	11	51	10	6
United Kingdom							
It's ideal for mocha coffee makers	3	4	13	13	21	−2	2
Available in coconut	8	6	4	5	20	1	4
A coffee with a wholesome, well-rounded flavor that will never disappoint you	9	15	19	12	29	18	14
Enjoy life-take time to smell the coffee!	6	8	18	17	31	20	10
United States							
Waker-upper	8	5	6	−1	24	6	1
Available in peppermint	4	2	10	5	23	8	1
For the total, perfect coffee pleasure	8	13	20	9	26	11	15
Coffee—quick pleasure for the way you live today	1	10	3	−1	20	9	16

Table 10.3. Mean and standard deviation (SD) of response times and interest for text elements, computed for all elements, for a given country

Country	Response time Mean	SD	Interest Mean	SD
France	6.63	3.19	−1.30	4.46
Italy	7.41	4.37	−1.91	7.74
United States	7.77	2.87	−0.82	4.65
United Kingdom	9.11	3.36	−0.89	3.93
Germany	9.64	3.69	−0.91	3.00
Hong Kong	13.36	5.92	0.78	4.39
Norway	22.83	8.94	1.42	4.20

elements than are the respondents in the other countries.

4. The difference across countries cannot be accounted for simply by the alphabet, since Norway and other countries share the same Latin alphabet.

Do Respondents Take Longer to Process Elements That Interest Them?

If the response times and interest utilities correlate, then it may turn out that when a respondent finds an interesting element the respondent will stop and read that element more slowly. The simple linear correlations in Table 10.4 reveal that there is a greater likelihood for women than for men to slow the pace of their reading or at least responding when they reach an interesting concept element. This appears to be the case for all countries except Norway and England. Men tend not to slow their reading, again except for respondents in England and Norway. Although the correlations are relatively low, it is important to keep in mind that they are based on 238 points and are statistically significant. Correlations of 0.30 and higher are exceptionally significant and worthy of mention.

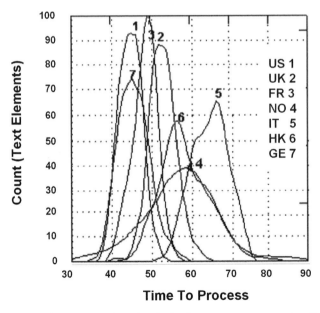

Figure 10.1. Distributions of processing times for single elements, across countries. The distributions are based upon the full set of text elements. The processing time for an element is obtained by this equation: response time = additive constant + coefficient (utility) for the element. US, United States; UK, United Kingdom; FR, France; NO, Norway; IT, Italy; HK, Hong Kong; and GE, Germany.

Table 10.4. Relation between two sets of coefficients—interest and response time—across the text elements*

Country	Total	Men	Women
Italy	0.37	0.11	0.42
England	0.31	0.39	−0.04
United States	0.29	0.05	0.33
Hong Kong	0.19	0.13	0.27
Norway	0.14	0.24	−0.14
Germany	0.13	0.04	0.19
France	0.06	−0.09	0.33

*The correlations are based on total panel, men, and women, respectively, for each country. The numbers in the body of the table are the Pearson R values, i.e., simple correlations.

How Interest in an Element and Text Length Jointly Drive Response Time

Which is more critical to response time: the length of the text (i.e., number of characters in the text element) or the interest? Do either of them matter? Do both factors matter? This type of question goes beyond a simple issue of which element wins and why, but rather looks for a mechanism underlying response time. By relating response time both to interest and to length of text, the researcher begins to uncover rules about how we process concept information.

The foregoing question can be answered by relating response time to a weighted combination of the characters and the interest rating. When both independent variables—number of characters in text and interest—are standardized to remove the distorting effect of scale size, the results show dramatic differences between countries.

The analysis begins with a count of the number of concept elements in each text element. Table 10.5 shows an example of this type of count.

Table 10.5. Four categories of concept elements, four specific concept elements per category, and the number of characters in the concept element (when translated)

Category	Element	France	Germany	Hong Kong	Italy	Norway	USA and England
Flavors	Mysterious flavor	18	31	10	16	12	17
	Delicious flavor	17	24	10	15	12	16
	Lively flavor	24	26	14	12	11	13
	Robust flavor	15	23	10	14	11	13
Aroma	Rich, flavorful aroma	33	33	18	28	19	21
	The rich aroma is followed by great taste	36	52	25	41	53	41
	The wake-up smell of brewing coffee	32	46	24	46	47	34
	The wake-up smell of freshly brewed coffee	58	50	28	46	49	41
Beans	The finest beans, roasted to perfection	45	37	31	43	40	39
	We select only the finest beans	46	41	24	42	33	31
	Hand-selected beans	39	20	20	18	17	19
	Contains gourmet beans	25	22	19	32	24	22
Energy	Boost your energy	22	28	12	15	21	17
	Revitalizing	25	38	8	19	15	12
	To recharge your body	40	35	14	31	30	22
	Wake to the coffee break	21	25	27	23	24	24

The analysis continues by creating an equation. The two predictor variables of the equation are text length and interest, respectively. These two variables are measured on quite different scales. Hence, if we use the text length and interest variables "as is," then we will not be able to discern a pattern very easily, because the different scales distort the effects. To remove the scale factors we standardize the text variable to create a *new* text variable, whose mean is 0 and whose standard deviation is 1.0. We do the same thing for interest in order to create a *new* interest variable.

The analysis using the standard data is done in the same way as all linear regression models are created. Standard ordinary least-squares regression generates this equation:

$$\text{response time} = k_0 + k_1 \text{(interest)} + k_2 \text{(text length)}$$

The coefficients, however, are called *beta coefficient*s because they deal with standardized predictors. Beta coefficients can be directly compared with each other because the effect of the scale has been removed. The additive constant in this case, k_0, will become 0.

We discover the relative impact of interest versus text length by looking at the parameters of the foregoing equation. We see these results in Table 10.6 for the seven countries, which have been ranked by the value of the beta coefficient for number of characters:

1. In all cases the *direction* of the coefficient is correct; that is, response time is positively related to the number of characters in the text element, with one exception (women in Norway). As the number increases, so does the response time. This makes sense; it should take more time to read longer elements.

2. The relative importance of number of characters and interest varies by country. Generally number of characters is more important than is the interest value. Response time tends to be driven more by number of characters than by interest.

3. For Italy, response time is determined only slightly more by number of characters

Table 10.6. Response time as a joint function of the number of characters in the element and interest in the element: the results are shown by subgroup (total and gender) and then by country

	Beta coefficient		
	Characters	Interest	Multiple *R*
Total			
Norway	0.06	0.14	0.16
Hong Kong	0.25	0.13	0.31
France	0.34	0.08	0.36
England	0.36	0.18	0.44
Italy	0.39	0.29	0.53
United States	0.43	0.20	0.51
Germany	0.48	0.08	0.50
Women			
Norway	−0.05	−0.13	0.14
Hong Kong	0.21	0.22	0.34
France	0.27	0.30	0.43
England	0.29	−0.11	0.28
Italy	0.36	0.34	0.54
United States	0.39	0.23	0.50
Germany	0.41	0.15	0.46
Men			
Norway	0.26	0.21	0.35
Hong Kong	0.16	0.07	0.19
France	0.13	−0.11	0.16
England	0.33	0.35	0.51
Italy	0.35	0.05	0.37
United States	0.36	−0.03	0.36
Germany	0.35	−0.01	0.35

than by interest. The Italians appear to vary their response time by the interest value of the concept element.

4. For Germany, response time is determined far more by number of characters than by interest. To the German respondents, the task is the task. Response time is virtually entirely a function of the length of the text (without consideration of its interest value).

5. Genders differ, but not in a consistent way. Typically, response time for women is usually driven by interest as well as by number of characters. In contrast, response time for men is almost always primarily driven by number of characters. This difference between genders means that, in general, women may pay more attention to what they read than do men.

Table 10.7. Percentage of response time ascribable to the elements*

Country	D = constant	P = average*3.5	Total time	Attributable To the constant	Attributable To the elements
Norway	34	81	116	30%	70%
Hong Kong	46	47	93	49%	51%
Germany	36	34	70	52%	48%
England	44	33	76	57%	43%
United States	37	27	64	58%	42%
France	42	24	66	64%	36%
Italy	57	27	85	68%	32%

*The constant, average time per element*3.5, and total time are in tenths of seconds. The table is sorted in terms of the percentage contribution of response time ascribable to the concept elements.

6. Interest in a concept element does not increase response time. Sometimes, interesting text decreases response time (e.g., for women in England and in Norway, and for men in France). This means that an interesting text, as indicated by a high utility value for interest, does not necessarily lead to slower processing of the information.

The Two Components of Response Time in the Assessment of Concepts

Does the total response time show the same structure across different countries; that is, if we look at the structure of response time, does this structure look similar across the countries?

First we have to create a structure or model of how response time can be described. We can create this structure using the parts of the additive model relating response time to the presence/absence of concept elements. The results from this computation are presented in Table 10.7:

1. The average concept in this experimental design comprised 3.5 elements. Although we do not know what specific concepts were presented to each respondent, we do know the average response time for each element.

2. This knowledge enables us to multiply the average response time per element, within a country, by the value *3.5* to estimate the total response time attributable to elements for an average concept comprising 3.5 elements. Call this value P, for processing time.

3. We also know the additive constant, which represents the numbers of tenths of seconds of response time not attributable to elements. Call this value D, for dead time.

4. The total time (D + P) represents the total response time per concept. The total estimated response time for a concept comprising 3.5 elements differs by country, with the United States and France taking the shortest time (6.4 and 6.6 seconds, respectively) and Norway taking the longest (11.6 seconds, or almost 5 seconds longer)

5. The proportion of the response time due to elements compared with dead time can be then computed by country.

6. The structure of response time differs by country. In Italy the elements account for only 32% of the total response time per concept, suggesting that Italians are much more consistent in the way that they respond to different concepts. Response time is less affected by the individual concept elements. In contrast, in Norway 70% of the response time is attributable to the specific elements in the concept. Norwegians thus respond to concepts more individualistically. Response time is far more affected by the individual concept elements.

Continuing the Tradition of Research in Concepts and Advertising

Over the years researchers have searched for methods beyond attitudinal questionnaires to measure communication. There has been an underlying (albeit not verified) belief that more physiological or at least nonverbal, noncognitive measures of consumer reactions to print material may provide more valid data (Stewart and Furse 1982; Klebba 1985). These methods include pupil dilation (e.g., see Stafford et al. 1970), which showed more differentiation power between two commercials than did actual ratings (Krugman 1965). Even the notion of hemispheric dominance as an influencing factor in the types of stimuli shown in concept testing has been offered as a suggestion for understanding differences among respondents (Hecker 1981). Viewed against this history, we see that response-time research provides yet another measure of an aspect of communication that can be associated, in a prima facie manner, with processing speed.

As stated in the introduction, concept testing has been typically relegated to the purview of applied consumer researchers in the business realm. Little in the way of concept testing has appeared in the literature, other than discussions of basic methods, and the application of sophisticated designs for conjoint measurement. A great deal more interest has focused on copy testing for printed advertisements. Thus, it is not surprising to find a paucity of published literature on the results of concept tests. Most of the applications are for specific products, in specific situations, rather than research studies whose goal is to understand the dynamics of consumer behavior. The data presented here suggest that it is quite possible to integrate response time as a measure into the data collection and analysis of responses to concepts. With response time, it becomes possible to partial out the contributory effects of different concept elements (text and visual), by country, and by so doing begin to understand some additional dynamics about how we process concept information.

Application of Response-time Results to Concepts in the Competitive Environment

Practitioners, used to thinking in terms of *good versus bad*, want to know whether a long response time means a good element or a poor element. These data suggest no relation between response time and interest. Response time provides another, orthogonal measure, probably indicating the length of time it takes a respondent to process the information. One may hypothesize that a long response time for an element means that the element may be confusing, or that the element may be very long and thus takes more time to process. A short element (viz., with few characters), but with a long response time, may be an example either of a confusing element or of an element that is complicated intellectually and takes a long time to process.

From an application standpoint, it also now makes sense to optimize concepts on the two aspects of response time and interest. Optimizing on response time generates concepts that are quick to read and to comprehend. Optimizing on interest generates concepts that are highly acceptable. Optimizing on both aspects together (e.g., a weighted combination of the two attributes) generates highly acceptable concepts that are quickly comprehended. One could also optimize concepts to generate combinations wherein the text is quickly comprehended, but where the picture is more slowly processed, and keeps the individual's attention focused on it.

References

Aaker, D.A., Bagozzi, R.P., Carman, J.M., and MacLachlan, J.M. (1980). On using response latency to measure preference. *Journal of Marketing Research*, 17: 237–244.

Haskins, J. (1975). Pretesting interest in messages. Journal of Advertising Research, 15: 31–35.

Hecker, S. (1981). A brain-hemisphere orientation toward concept testing. Journal of Advertising Research, 21: 55–60.

Klebba, J.M. (1985). Physiological measures of research: a review of brain activity, electrodermal response, pupil dilation and voice analysis methods and studies. Current Issues and Research in Advertising, 8: 53–76.

Krugman, H.E. (1965). A comparison of physical and verbal responses to television commercials. Public Opinion Quarterly, 29: 323–330.

Lancaster, G.A., and Lomas, R.A. (1977). Experimental error in T-scope investigation. Journal of Advertising Research, 17: 51–56.

MacLachlan, J., and Myers, J.G. (1983). Using response latency to identify commercials that motivate. Journal of Advertising Research, 23: 51–57.

Rhodes, E.W., Leferman, N.B., Cook, E., and Schwartz, D. (1979). T-scope tests of Yellow Pages advertising. Journal of Advertising Research, 19: 49–52.

Stafford, J.E., Birdwell, A.E., and Van Tassel, C.E. (1970). Integrated advertising: white backlash? Journal of Advertising Research, 10: 15–20.

Stewart, D.W., and Furse, D.H. (1982). Applying psychophysiological to marketing and advertising research problems. In: Leigh, J.H., and Martin, C.R., Jr., editors. Current Issues and Research in Advertising. Ann Arbor: University of Michigan, pp. 1–3.

Taylor, D. (1970). A study of the tachistoscope. British Journal of Marketing, 4: 22–28.

Chapter 11

Children Compared with Adults

Introduction

The cognitive ability of children varies with age and development (Younkin 1989). In contrast, the cognitive ability of teenagers is quite well developed. We do know that the sensory magnitude functions of children are similar to those of adults (Bond and Stevens 1969), at least in a laboratory setting. We also know that the preference patterns for food products for children can be studied to create foods for these children (Birch, 1979). Despite this common knowledge, however, the literature is sparse regarding concept testing, perhaps because, as the previous chapters discussed, concept testing is an application of scientific principles rather than a discipline unto itself with a coherent, accepted body of knowledge (Kroll 1990; Moskowitz et al. 2001).

For the most part, business researchers working with concepts have used adults rather than children, yet there is no reason why children's data cannot be obtained by using systematically varied concepts in a concept task, just like adult data. Many children above a certain age (e.g., ages 5–6) can read and can respond to stimuli. Of course, children may react differently to the test stimuli. This difference in information processing could well show up in the different patterns of utility-score evidence of children versus adults in a conjoint task.

The study of food concepts and foods themselves provides an enormous opportunity to understand how a child reacts to stimuli. Today's children are becoming increasingly sophisticated, are bombarded by commercials, and have learned to interact with computers for extended periods. There is every reason to believe that today's children can participate in a concept evaluation task set up as a game or at least with "bells and whistles" on a computer to amuse the children while they participate in the data acquisition process. Of course, in such a concept evaluation task researchers may not be able to probe more deeply into the individual nature of each child. Nonetheless, if there are differences in the *gross patterns of responses*, then these differences might be quite valuable for hypothesis development about the way children process information. Such information is very valuable for applications, including product development and advertising.

We will apply conjoint measurement to concept research with children because of the richness of the output. From the child's response to the stimuli we will be able to identify what is important, even if the child cannot articulate it. By comparing children of different ages, or children versus adults, we will be able to understand how age affects the way children process information.

When we look at response time, as discussed in Chapter 10, the information will become even richer. It will then be possible to determine whether the *general patterns* shown by adults in processing information are the same among children. For example, do children and adults show the same proportions of response time ascribable to elements versus dead time? Differences between adults and children in the pattern suggest different strategies of processing information and perhaps new insights about creating better concepts.

The information from the data presented here will show that, indeed, children and adults differ rather dramatically in the way they handle concept information.

Case History 1: Response to Systematically Varied Concepts About Toys

This section of the chapter departs from the topic of food to present a comparison of data from adults and children regarding toys. The dataset is valuable because both the mothers and (unrelated) children rated interest for concepts describing a new type of children's toy. Toys are very interesting to children and provide a relevant topic through which to study concepts. The analysis and results shown here may find application to food concepts.

Table 11.1. Examples of elements from the toy study

Packaging elements
>   The product comes in a "try me" packaging so you can try it before buying
>   The packaging is easy to recognize in the store
>   The product comes in a colorful box—you can't wait to open it
>   You can see on the packaging what you get
>   The product comes in a "touch me" packaging so you can touch it before buying
>   The product comes in a big and outstanding gift box
>   The packaging comes in funny shapes and forms
>   The packaging is round, soft, and friendly
>   The packaging is transparent so you can see what's inside
>   The packaging is shaped like a . . .
>   The toy is visible through the packaging to see before purchase

Pricing elements
>   The price for this . . . toy . . . is $1.99
>   The price for this . . . toy . . . is $4.99
>   The price for this . . . toy . . . is $9.99
>   The price for this . . . toy . . . is $14.99
>   The price for this . . . toy . . . is $19.99
>   The price for this . . . toy . . . is $24.99
>   The price for this . . . toy . . . is $29.99
>   The price for this . . . toy . . . is $39.99
>   The price for this . . . toy . . . is $49.99
>   The price for this . . . toy . . . is $69.99
>   The price for this . . . toy . . . is $89.99

The concept study on children's toys addresses three issues relevant to concept testing:

1. *Rating scales.* Do children, compared with adults, *up-rate* the elements of concepts, implying that one has to adjust a child's ratings to account for a false positive, that is, that the product is more acceptable than one might think? We saw the systematic up-rating of concepts by respondents in Mexico, leading to the need to calibrate the data (see Chapter 8).

2. *The processing time.* Do children take longer to rate concepts than do adults, and is the time taken to rate a concept proportional to the nature of the idea in the concept or proportional to the length of the concept elements?

3. *The structure of the response to concepts.* If we deconstruct the rating of a person to a concept into two parts—that contributed by the components or elements of the concept and that contributed by the additive constant or unexplained portion—then how large is the unexplained portion of the concept response for children compared with that for adults?

The study on toy concepts was designed so that the features of the product would be understandable to children as young as age 6; that is, the elements did not comprise abstract, hard-to-visualize characteristics that would elude a child's understanding. The toy manufacturers, experts in this area, provided the elements (which have been modified here to maintain confidentiality). A short pretest done with the elements with six children as test participants showed that children had no problems with the task, nor did they appear to be puzzled when presented with the test stimuli.

Examples of elements are presented in Table 11.1 both for the nature of the packaging and for pricing, respectively. Note that the children and the adults rated the exact same set of elements.

Study Execution

The respondent base comprised 130 adults (mothers) and 56 children (ages 8–12). The respondents were prerecruited to participate and paid for their participation. The mothers and the children were not related to each other, but rather were recruited separately. The respondents were chosen in the United States across three geographically distributed markets:

1. *Adults.* For the adult respondents, an attending interviewer oriented the respondent through a short lecture and afterward took the respondent through the first concept. The adults then completed the evaluation at their own speed, including an extensive classification questionnaire. The adult respondents were paid and then dismissed. This took approximately 30 minutes. The adults participated in groups of 15–25, depending on the particular session. During the interview each respondent saw totally different combinations of the concept elements.

2. *Children.* Groups of 8–10 children participated in each session. The respondents varied in age from 8 to 9, and from 10 to 12. The two ages were kept separate to make the interview more feasible by ensuring similar levels of respondent maturity. An attending moderator took the children through the orientation and through a short practice exercise. The children then evaluated 50 concepts (half the number rated by adults).

3. *Interview control.* The interview, on the computer, comprised built-in delays of 30 seconds between successive concepts, so respondents could not rush through the concepts. During the evaluation senior researchers were present at each test session, whether with adults or with children. A few children attempted to rush through the evaluation, but were stopped by the enforced 30-second delay. For the most part, the children appeared to concentrate on the task, perhaps because of covert social pressure in the test room. There were 7–9 other children, in front of their own computers, all participating (albeit to great or lesser interest levels during the evaluation). In other studies with children the senior author has found that this group evalu-

ation of separate test stimuli sets a serious tone, not unlike that of the school classroom. Furthermore, with interviewers continuing to walk around and check the children, the atmosphere becomes part of an important task. From time to time a child might move his or her chair, wander a little around the computer, and then return to the task. No efforts were made to keep the children in the chair or to rush them through the evaluation. It is important to keep in mind that children often fidget, look around, make faces, and do other things when they are not actively involved. These behaviors can be pretty much ignored.

Results

The dataset generated 186 different equations: one for each of the 56 children, and one for each of the 130 adults. For the attribute of interest, each equation showed the marginal or part-worth contribution of each element (tested by that respondent). For response time (in tenths of seconds), a second dataset yielded 186 different equations.

Table 11.2 shows an example of data, specifically the utilities from the additive model for a few of the elements. In actuality the model comprises 207 different terms: one for the additive constant and one for each of the 206 concept elements.

1. *Additive constant.* The average constant is 52 for children and 57 for adults. These constants mean that 52% of the children would be expected to rate the toy concept positively (7–9 on the 9-point scale) if there were no concept elements present. Both additive constants are modest and quite similar to each other. The children, as compared to adults, do not up-rate the concepts. This is a very important finding that means that children do not like every toy presented to them.

2. *Average utility score and variability for interest.* There is almost no difference between them. This means that, on average, adults and children show the same level of liking. The variability is also the same for

Table 11.2. Average utilities across the 206 elements and representative utilities for individual elements*

	Interest		Response time	
	Adult	Child	Adult	Child
Additive constant	57	52	48	86
Average results across all 206 elements				
Average utility	−0.01	0.06	10.0	1.28
Variability of utility	2.55	2.77	4.8	10.1
Examples of individual elements				
Example: Emotional benefit 2. The toy is				
appropriate for "XYZ"	6.	−2.	14.	2.
Example: Picture B	5.	−5.	12.	−3.
Example: Benefit 4. Dealing with general safety				
of the toy	5.	−3.	8.	−4.
Example: Picture A	4.	2.	12.	9.
Example: Product name "XYZ"	3.	0.	9.	−11.
Example: The price of the toy is "XYZ"	2.	−6.	6.	1.
Example: The toy is acquired at store "XYZ"	1.	−4.	12.	3.
Example: Package 6. The toy comes in				
package "XYZ"	1.	0.	3.	6.
Example: Picture C	0.	−9.	7.	−4.
Example: Picture D	−4.	11.	1.	−23.

*The models are shown for interest and for response time, respectively. Each row corresponds to a specific element (e.g., a different picture or a different benefit). The rows are sorted by the adult utilities for *interest*.

adults and for children (standard deviation of the 206 utility values = 2.77 for children and 2.55 for adults). If the child had been indifferent to the elements, then we would expect the standard deviation to be much lower, with all of the utility values clustering around 0.

3. *Overall similarity in utility values.* Both children and adults show a range of utilities across the elements, indicating that children discriminate among the different elements in this conjoint task (or at least, on an operational level, behave as if they discriminate). There is a positive but relatively low yet significant correlation. The interest utilities correlate +0.24 across the 206 elements, which is highly significant because of the many elements.

4. *Similarity by type of concept element.* The toy study comprised four different types of stimulus elements: names, pictures, text (communication of benefits, packages, age appropriateness, etc.), and pricing. For the four classes of elements the children and adults show highly correlated utility values for

names (Pearson $R = 0.85$), modestly correlated utilities for texts and prices, respectively (Pearson $R = 0.27$ for each, respectively), and uncorrelated utilities for visuals (Pearson $R = 0.13$). Adults and children respond similarly to names. They respond similarly, but less strongly, to texts describing the product and the price of the product. Children and adults differ when it comes to graphics. What a child likes, an adult may dislike. Thus, at a microlevel there are substantial differences between children and adults, but it depends on the nature of the concept element.

5. *Price sensitivity.* Both adults and children show price sensitivity. As the price of a toy increases, the interest level drops.

6. *Additive constant for response time.* With response time the differences between children and adults begin to emerge. The constants differ. The adult additive constant is 48, whereas that of the child is 86. This means that there is a dead time of 48 tenths of seconds for the adults and 86 tenths of seconds for the children. At a substantive level, this

difference can be interpreted as suggesting that, when compared with adults, children take longer to respond to a concept.

7. *Average response time.* These differ from children to adults. Children's average response time across all the concept elements is 1.28 tenths of seconds (with a standard deviation of 10.1). In contrast, the average response time for adults is 10.0 tenths of seconds (with a standard deviation of 4.8). Substantively, these results suggest that more of the adult response time can be related to the elements. Children have more dead time, but react more quickly to each element. Furthermore, children show much greater variability in their response times, depending on the element. In contrast, adults show much greater consistency in response time from element to element.

8. *Two different processing patterns.* The children show more dead time and faster re-

actions to concept elements but more variability from element to element. In contrast, the adults show less dead time and slower reactions to concept elements but less variability in response time from element to element.

9. *Interest and response time to prices.* As price increases, the interest in a toy decreases, as shown by the decreasing utility values for interest with increasing price. This pattern holds for both adults and children. In turn, response time is also sensitive to price. Figure 11.1 shows that, as stated price increases in the concept, the mother's response time dramatically decreases, but the child's response time is not sensitive to price. It appears that, for mothers, acceptance based on low price occurs slowly, whereas rejection based on high price occurs quickly. The range in response time is dramatic, covering a 20-unit range, or 20 tenths of a second, viz., a 2-second difference.

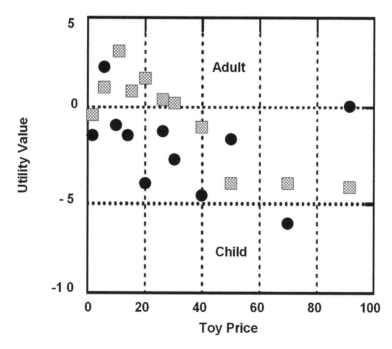

Figure 11.1. The relation between toy price and response time (utility). The utility value for response time is in tenths of seconds. Higher utility or response time values correspond to longer times to respond to the concept.

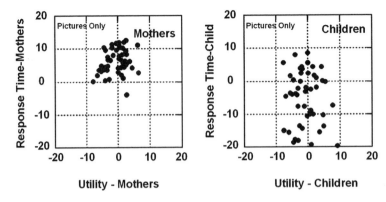

Figure 11.2. The relation between interest utility value and response time, for pictures, for mothers **(left)** and for children **(right).**

10. *Response time to pictures in the concept.* Adults and children differ in their response time to pictures in the concept. Figure 11.2 plots interest in the concept element (i.e., a specific picture) versus response time for that particular picture. The plots are shown separately for mothers and children. For mothers, more interesting pictures generate longer response times. In contrast, for children there is no apparent relation between the utility value of the picture and the response time. These differences suggest that adults, or at least mothers, spend more time attending to the interesting pictures in a concept, whereas children do not attend longer to the more interesting pictures.

11. *Length of text and response time.* Quite often concepts present a great deal of information about which a judgment must be made. This information can get in the way of a respondent's task because the respondent has to sift through the information. We saw a direct relation between response time and length of the concept element in Chapter 10. Length of the concept in text can be important for children. They may react differently because they are more accustomed to serious reading because of school. The text elements in this study are of various lengths, containing

10–95 different letters. It is easy to estimate the total time required for a child or an adult to respond to a concept that has only *one* text element and relate that time to the number of letters in the text. The estimated total response time is the sum of the additive constant and the coefficient for response time for that specific element. Figure 11.3 shows that it takes children longer, *on average*, to respond than adults. We know this already because children have a higher additive constant than do adults. On the other hand, *for children and for adults we see no dramatic pattern relating number of elements to response time.*

12. *Children can do conjoint analysis studies.* Despite the fear of many practitioners that children may not be able to complete a conjoint measurement task, these data suggest that they can do so and discriminate among the elements. The study dealt with toys, a topic of interest to children. We would not know what to expect if the study dealt with a more mundane topic, at least to children. Observations of the children showed them to be quite interested in the task, absorbed with it, and not particularly conscious of interviewers who walked around the test hall to ensure that the respondents were doing their jobs properly.

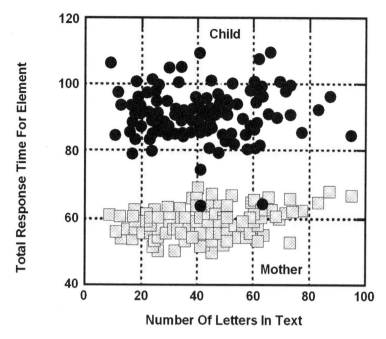

Figure 11.3. How estimated total response time for single concept elements varies with the number of letters in the concept element.

Case History 2: How Children React to Visual Concepts About a New Candy

Overview of the Case History

This study assessed the reaction of children (ages 8–14) to candy colors by using a computer-aided presentation of different bagged candies, both singly and in combination (Moskowitz 2002). The original objective was to identify the optimal combination of the candies. The study is essentially one of graphics concepts, without text (see Chapter 6). The results demonstrated that children could easily discriminate among the different candy options when they were presented as pictures with an underlying, unrevealed experimental design.

Test Stimuli

The stimuli comprised six different circular-shaped candy colors presented on a com-

puter screen. The candies were shown in simulated bowls, comprising either candies of one or two colors, respectively. The six colors were blue, green, red, yellow, purple, and orange, respectively. The shapes were small disks, with an irregular-appearing surface texture. The design was a fractional factorial, comprising all six candies present at 100% and the full set of 15 pairwise combinations of six candies, each at 50% in the respective mixture. Thus, a blue-green combination comprised 50% blue and 50% green candies. The computer program created a simulated bowl, with the actual colored, three-dimensional looking candies appearing as they do in a bowl (viz., random distribution, with some candies covering up parts of other candies).

Panelists

The panelists comprised 242 children and teenagers, ages 8–14, recruited in two markets

Table 11.3. Distribution of ratings on the 9-point scale and in three scale regions (bottom 3, middle 3, and top 3)*

Age	Base	1	2	3	4	5	6	7	8	9	Bottom 3 1–3	Middle 3 4–6	Top 3 7–9
8	61	14%	5%	6%	8%	10%	9%	12%	12%	24%	18%	45%	37%
9	39	14%	6%	7%	5%	9%	11%	11%	12%	24%	21%	43%	37%
10	32	8%	5%	9%	9%	10%	12%	11%	16%	20%	13%	51%	36%
11	34	14%	9%	12%	9%	10%	8%	10%	11%	17%	23%	49%	28%
12	26	10%	8%	9%	12%	11%	14%	11%	8%	18%	18%	56%	26%
13	24	10%	10%	12%	12%	15%	16%	11%	7%	7%	20%	66%	14%
14	26	12%	10%	14%	14%	14%	11%	10%	8%	7%	22%	64%	14%

*Note that the numbers may not add to 100% because of rounding.

(New York City and Chicago). The panel was balanced for gender across the full 242 participants, and the balance was close, but not exactly 50/50, for each age group. The panelists were candy consumers, had eaten this type of candy in the past month, and agreed to participate for a 1-hour session (although in the end the panelists needed only a half-hour to complete the evaluations).

Visual Concept Evaluation

The panelists were oriented in the purpose of the study by an attending moderator and told that the product was a new "bagged candy" that could have different single colors or two-color assortments. The panelists rated the 31 stimuli, one at a time, on an anchored 9-point liking scale (1 = hate to 9 = love). Each panelist evaluated the same 31 stimuli in a randomized order.

Scaling Behavior Differs by Age

Children assigned higher ratings, on average, than did teenagers. The differences followed a simple pattern. Teenagers provided fewer ratings at the top of the scale, but not necessarily an increased proportion of ratings at the bottom of the scale. Teenagers were less positive to the stimuli, but not necessarily more negative (see Table 11.3).

How Acceptance Varies with Single Colors and Color Combinations

The results of this study provide a sense of how children and tweens react to the different colors when candy is presented as a picture concept and also suggest two generalizations for concept work:

1. There are major differences in the acceptance patterns of the different stimuli by age, so that one cannot make blanket statements about acceptance across all age groups.

2. The panelists of different ages demonstrated different scaling behaviors, with younger children showing slight polarization toward using the top of the liking scale, and tweens (young teens, ages 12–14) showing polarization toward the middle points of the liking scale.

3. Single colors are liked more than color pairs. The only exception is yellow.

4. The best-performing color in a two-color combination is blue; the worst performing colors are yellow and orange.

5. Liking of a color combination is not the average of the liking of the component colors.

6. There are greater age-to-age differences in the liking of color combinations than age-to-age differences in liking of single colors.

References

Birch, L.L. (1979). Dimensions of preschool children's food preference. Journal of Nutrition Education, 11: 77–80.

Bond, B., and Stevens, S.S. (1969). Cross modality matches of brightness to loudness by 5 year olds. Perception and Psychophysics, 5: 337–339.

Kroll, B.J. (1990). Evaluating rating scales for sensory testing with children. Food Technology, 44: 78–80, 82, 84, 86.

Moskowitz, H.R. (2002). Children and "tween" acceptance of single candy colors and two-color combinations. Journal of Sensory Studies, 17: 115–120.

Moskowitz, H.R., Hjelleset, T., and Rabino, S. (2001). Children versus adults: on the structure of their responses to systematically varied concepts. Canadian Journal of Marketing Research, 19: 44–55.

Younkin, R.J. (1989). New product development and the kids confectionery panel. In: Wu, L., editor. Product Testing with Consumers for Research Guidance. (STP 1035.) Philadelphia: American Society for Testing and Materials, pp. 85–90.

Chapter 12

Pricing Issues in Early-stage Concept Research

Introduction

Studies of responses to pricing occupy a relatively large proportion of concept research and conjoint analysis in the business community and especially among suppliers. A search on Google for conjoint analysis combined with pricing generated 6500 hits. A search on Google for conjoint analysis alone generated about 22,000 hits. Almost a third of the conjoint analysis articles have something to do with pricing. A further analysis of the nature of these hits shows that many of them are from research companies offering to do conjoint analysis and using pricing as a main topic around which the conjoint study is built.

The issue of pricing is critical in marketing, because price is one of the five key "P's," the others being product, positioning, placement, and promotion, respectively. Pricing is exceptionally effective in driving consumers to purchase a product or to reject it. When it comes to food products, pricing is more important than acceptability as a driver of purchase intent when both price and acceptance are experimentally manipulated (Moskowitz 1995). Most of the pricing studies conducted by market researchers look at the price of the item within the competitive frame. The issues that these researchers address focus on the choice of the item in the competitive marketplace and the expected share to be enjoyed by the item, rather than price as a factor in the development of the product concept.

In this chapter on pricing and concept research, specifically applications in conjoint analysis, we deal with price at the early stage

of development, rather than at the later stages where price is a factor for choice among items. Price becomes relevant at the early stage because, if the respondent is willing to pay a high price for the item, then the item can sustain a lot of costly features. On the other hand, if the respondent is unwilling to pay a high price, then the item cannot sustain those expensive features and must be made so that it can be sold cheaply. These two possibilities put constraints on what the product developer can create.

This chapter deals with price in one of two ways:

1. *Price as simply another type of element,* on a par with the other elements of a concept, such as brand, feature, benefit, and graphics. The issue here is to look at the effects of price as a driver of concept acceptance or rejection.

2. *Price as a dependent variable.* The issue here is to use price as a rating scale, so that the respondent rates concepts in terms of the appropriate price for the item.

Price as an Element in the Concept

The conventional way to explore price is to look at its performance as just another category of concept element. Rather than looking at a single price the researcher can just as easily create a category called *price* and work with several alternative prices. The category can comprise elements that present different specific prices for the entire product, prices for a specific element thus linking element and price, or even statements about price but without prices.

A good example for pricing comes from a study of bagels by a well-known bagel chain. The study was done in the late 1990s, so the elements reflect the then-current prices. The pricing elements were embedded in a large-scale study of other elements for the bagel concept. As in all conjoint studies reported here, respondents evaluated single concepts comprising a variety of statements, including in some cases statements about the price. Some concepts were created deliberately without price statements, allowing the data to be analyzed by ordinary least squares (dummy variable modeling). Thus, the utility values show the conditional probability of a person changing a vote from disinterested in the bagel concept to interested when the particular price element is present.

We see some interesting patterns from the utilities listed in Table 12.1:

1. As the price of a two-pack increases from 59 cents to 89 cents, the utility drops. We would expect this drop because respondents are price sensitive. What makes the results more gratifying is that the respondents assessed combinations of elements, one combination at a time. There is no way that the typical respondent could factor out price alone when the remaining elements were also changing, to all appearances, "randomly," yet the respondent data tracks these changes.

2. The utility is about 0 from the price between 59 and 79 cents. This is the region where respondents feel that the price is "fair," so the price neither adds nor detracts from concept acceptance.

3. The same type of monotonic drop in utility occurs for a six-pack. However, the utility values are greater, both in the positive direction and in the negative direction.

4. We can look at the unit price per bagel by using the same utility values. This will enable us to compare the same bagel incorporated into two different packages. It is clear from Table 12.1 that the total price is more critical than the actual unit price. For example, the element "Pick up a two-pack of bagels for 69 cents," with a unit price of 34.5 cents per bagel, has a utility value of 0; that is, putting this price neither adds nor detracts from purchase intent. In contrast, an offer that has almost the same unit price (33.2 cents per bagel) can be extremely negative. The element "Pick up a six-pack of bagels for $1.99" has a utility value of −9.

5. Large package sizes at low prices can generate very high utility values. It appears that the large package sizes can expand the utility range, perhaps by calling attention to the large range of total prices.

Table 12.1. Utility values and actual unit prices for various offers of bagels

	Utility	Actual unit price
Smaller packages		
Pick up a two-pack of bagels for 59¢	1	29.5
Pick up a two-pack of bagels for 69¢	0	34.5
Pick up a two-pack of bagels for 79¢	−1	39.5
Pick up a two-pack of bagels for 89¢	−9	44.5
Larger packages		
Pick up a six-pack of bagels for $1.49	12	12.4
Pick up a six-pack of bagels for $1.59	6	26.5
Pick up a six-pack of bagels for $1.69	5	28.2
Pick up a six-pack of bagels for $1.79	0	29.8
Pick up a six-pack of bagels for $1.89	0	31.5
Pick up a six-pack of bagels for $1.99	−9	33.2

The utilities were computed on the basis of percent top-2 box purchase intent on a 5-point purchase-intent scale.

Price in the Context of Other Elements and Rated by Different Groups

Another good example of price comes from a study on surimi (Moskowitz and Poretta 2002). The objective of the study was originally to create a lobster surimi and afterward create the messaging for that product both to consumers and to food-service professionals, respectively. The concept research involved different types of elements, including brand name, features, benefits, and prices. The respondents rated the concepts on three key attributes that were relevant to the manufacturer and appropriate to the particular respondent group:

1. *Consumers:* purchase intent, similarity to lobster, and expected good taste

2. *Food-service operator:* purchase for restaurant, expected high quality, and versatility

We can see the results from this study in Table 12.2. The price points were systematically varied, but the respondents never saw two prices together in a single concept, nor did they have a chance to compare concepts, yet among consumers as the price increased the utility values decreased. The price did not have the same effect on food-service operators. Utilities did not decline monotonically as the price increased. Finally, price had no effect on the consumer perception of good taste. The price had a modest effect on the food-service operators' food-service rating of versatility in the restaurant. The highest-priced lobster surimi was judged to be the most versatile, all other factors held constant.

To use these pricing data the manufacturer creates an offering or product concept, without price. The sum of the utilities corresponding to the different elements will be a certain amount:

1. *A strong-performing concept without price.* If the sum of the utilities is high, then this means that there is room to add a high price to the concept. The high price will certainly diminish the final sum of the utilities, but that should not matter because the sum is already high. Respondents are interested in the product concept and show that it can support a high price.

2. *A weak-performing concept without price.* If, however, the sum of the utilities is low, or even negative, then the concept is a poor one, either because the basic idea is no good (i.e., low additive constant) or the elements do not do well. The elements may even have negative values. One way to increase the concept acceptance is by adding a strong positively performing element. This strong element can be a low price. If the price is substantially lower than what the respondent believes is a fair price, then the product looks like a "bargain." The price element will then contribute a strong positive utility, increasing the total utility of the concept.

The statement has already been made that pricing outweighs acceptance. This statement has to be modified a little. Pricing is a tricky variable with which to work. The effect of pricing depends crucially on the level of price. When prices are in the appropriate region, changes in the price do not do much to change the utility value. At regions where the price is clearly too high, the utility value can begin to plummet.

Interindividual Differences: The Relative Importance of the Price Category versus Other Categories

How important is price to respondents? One of the ways to measure the importance of price is to look at the utility values. Chapter 7, on segmentation, discussed the importance of different categories and provided a formula for relative importance. If we apply that to a dataset that has price as a major category, as we do in Table 12.3, we see that price can play different roles, depending on the consumer mind-set. Table 12.3 presents the results for an Italian condiment. The number of

Table 12.2. Utility values for elements of lobster surimi, on different attributes

	Consumer description	Consumer respondents			Food-service description	Food-service operator respondents		
		Buy for self	Similar to lobster	Good taste		Buy for restaurant	High quality	Versatility
	Additive constant	26	41	45	*Additive constant*	10	42	46
	Brand name				Brand name			
B1	Introducing consumer lobster tasties	7	2	2	Introducing food-service lobster tasties	2	3	0
B4	Introducing consumer lobster temptations	4	1	1	Introducing food-service lobster temptations	4	0	−2
B3	Introducing consumer lobsterettes	4	2	2	Introducing food-service lobsterettes	0	0	0
B2	Introducing consumer lobsterimi	0	0	0	Introducing food-service lobsterimi	9	1	−1
	Heritage				Heritage			
H1	From the leader in seafood	4	0	1	From the leader in seafood	−2	4	4
H2	From the finest fisheries in the world	2	2	2	From the finest fisheries in the world	8	2	5
H3	A strong tradition of seafood excellence	2	1	1	A strong tradition of seafood excellence	5	2	−2
H4	From the icy-cold waters of the Pacific	1	2	1	From the icy-cold waters of the Pacific	−2	−1	−1
H5	Quality seafood processors for over 50 years	−5	0	−1	Quality seafood processors for over 50 years	−1	2	−2
	Ingredient composition				Ingredient composition			
I1	All the taste, texture, and flavor of ordinary lobster	15	9	8	All the taste, texture, and flavor of ordinary lobster	13	3	3
I2	Pacific pollock with natural flavors from real lobster	7	4	3	Pacific pollock with natural flavors from real lobster	11	1	−3
I3	Pure ingredients with no preservatives added	3	1	2	Pure ingredients with no preservatives added	9	3	−2
I4	Made with real Alaskan pollock	3	2	1	Made with real Alaskan pollock	3	3	2
I5	A unique blend of Alaskan pollock and other fine ingredients	2	1	2	A unique blend of Alaskan pollock and other fine ingredients	2	6	2
I6	Made with real fish	0	0	0	Made with real fish	0	0	0
	Health and nutrition				Health and nutrition			
N1	The taste of real lobster without the calories, fat, or cholesterol	19	10	7	The taste of real lobster without the calories, fat, or cholesterol	16	7	4

(continued)

Table 12.2. Utility values for elements of lobster surimi, on different attributes *(cont.)*

		Consumer respondents				Food-service operator respondents		
		Buy for self	Similar to lobster	Good taste		Buy for restaurant	High quality	Versatility
N2	99% fat free, 100% delicious	9	1	2	99% fat free, 100% delicious	20	8	5
N3	Low in cholesterol	3	3	1	Low in cholesterol	8	3	1
N4	Very low fat	1	4	2	Very low fat	5	5	5
N5	Good protein source	-3	1	1	Good protein source	8	4	5
N6	No MSG	-4	-4	-2	No MSG	8	2	-2
	Emotion				*Emotion*			
E1	All the flavor of the Captain's finest	10	2	4	All the flavor of the Captain's finest	7	-2	-4
E2	Firm bite and mouth-watering taste	5	0	0	Firm bite and mouth-watering taste	-9	-1	-2
E3	You've never tasted seafood like this	4	-1	0	You've never tasted seafood like this	0	2	-2
E4	Nutritious food for your family	3	2	3	Nutritious, healthy choice for your customers	3	5	2
E6	Seafood made simple without the hassle	3	-2	-2	Seafood made simple without the hassle	0	9	9
E7	Successful seafood at home . . . made easy	1	2	0	Successful seafood . . . made easy	-7	-2	-3
E8	Seafood that satisfies	0	-6	-4	Seafood that satisfies	2	3	7
E9	Healthful and delicious	0	2	3	Healthful and delicious	3	1	-3
	Preparation versatility				*Preparation versatility*			
V1	Works like lobster in any seafood recipe	7	5	6	Works like lobster in any seafood recipe	13	3	5
V2	Use it in your favorite lobster recipes	7	0	1	Use it in your favorite lobster recipes	14	5	4
V3	Fully cooked and ready to use in your recipe	5	5	2	Fully cooked and ready to use in your recipe	-6	1	1
V4	Warm slightly and dip in margarine	3	-1	0	Warm slightly and dip in butter	10	3	7
V5	So versatile, you'll want to pick some up every time you shop	-1	-2	-3	So versatile, you can use it in a variety of recipes	8	6	8
	Usage occasions				*Usage occasions*			
U1	Use as a main-dish ingredient	8	2	4	Use as a main-dish ingredient	12	7	9
U3	Ideal in a variety of dishes—hot or cold	7	5	3	Ideal in a variety of dishes—hot or cold	7	1	6
U4	Perfect as an appetizer	6	1	1	Perfect as an appetizer	4	1	1

(continued)

Table 12.2. Utility values for elements of lobster surimi, on different attributes *(cont.)*

	Consumer respondents					Food-service operator respondents		
	Buy for self	Similar to lobster	Good taste	(description)	(description)	Buy for restaurant	High quality	Versatility
U2	5	0	2	Great for seafood chowder and soups	Great for seafood chowder and soups	18	7	10
U7	4	5	5	Anytime is seafood time—omelet to entree, breakfast, lunch, and dinner	Anytime is seafood time—omelet to entree, breakfast, lunch, and dinner	3	4	11
U5	2	-2	-1	Perfect as a complete light meal	Perfect as a light meal for your customers	15	4	9
U6	0	0	-2	Use in a salad or a main dish	Use in a salad or a main dish	4	-1	0
				Convenience	Convenience			
C3	13	4	3	From refrigerator to table in 10 minutes	From your kitchen to your customers' tables in 10 minutes	12	1	-1
C1	8	2	4	No shells, no cleaning, no cooking, no waste	No shells, no cleaning, no cooking, no waste	29	2	8
C2	4	1	-3	Quick and easy to serve great seafood dishes in less than 10 minutes	Quick and easy to serve great seafood dishes in less than 10 minutes	18	4	8
C4	4	4	1	Easy to prepare and microwaveable	Easy to prepare and microwaveable	10	5	10
C8	3	4	2	Let the Captain make you a seafood chef . . . in minutes	Let the Captain make you a seafood chef . . . in minutes	7	-2	2
C6	0	0	0	Fully cooked and cut for your serving convenience	Fully cooked and cut for your serving convenience	12	4	1
C5	0	-1	-1	Easier to use than ordinary lobster—its fully cooked for you	Easier to use than ordinary lobster—its fully cooked for you	8	1	7
C7	-1	-3	-3	Fully cooked—simply add to your favorite hot and cold recipes	Fully cooked—simply add to your favorite hot and cold recipes	14	2	5
C9	-3	-1	-2	Serve hot or cold	Serve hot or cold	0	-2	0
				Product quality	Product Quality			
Q1	4	-1	0	High in quality and high in fish protein	High in quality and high in fish protein	23	3	4

(continued)

Table 12.2. Utility values for elements of lobster surimi, on different attributes (*cont.*)

		Consumer respondents				Food-service operator respondents		
		Buy for self	Similar to lobster	Good taste		Buy for restaurant	High quality	Versatility
Q3	Made in the USA by America's seafood experts	−1	−1	−1	Made in the USA by America's seafood experts	8	3	4
Q4	Packed under federal inspection	−4	0	−1	Packed under federal inspection for your protection	2	5	3
Q2	Quality seafood for your family to enjoy	−4	0	0	Quality seafood for your customers to enjoy	15	2	1
	Package				Package			
P5	Fully pasteurized and vacuum packaged to maintain superior product freshness	0	1	0	Fully pasteurized and vacuum packaged to maintain superior product freshness	0	7	3
P4	Look for the Captain's logo on our convenient 8-ounce packages	−3	0	−1	Look for the Captain's logo on our convenient 2½-pound packages	14	7	8
P1	Packed in convenient 8-ounce trays	−3	2	−2	Packed in convenient 2½-pound vacuum-sealed packages	2	6	6
P2	Strong, protective packaging to maintain product freshness	−5	0	1	Strong, protective packaging to maintain product freshness	3	7	4
	Price				Price			
M1	An 8-ounce package sells for $2.29	4	0	1	This product sells for $3.05 per pound	1	−2	−4
M2	An 8-ounce package sells for $2.46	5	1	2	This product sells for $3.20 per pound	4	−3	−2
M3	An 8-ounce package sells for $2.63	2	0	0	This product sells for $3.35 per pound	2	−4	−6
M4	An 8-ounce package sells for $2.79	0	0	0	This product sells for $3.50 per pound	0	0	0

Table 12.3. Relative importance: Italian condiment

Category	Total	Segment 1	Segment 2
Total sum squares	*5181*	*5720*	*7195*
Positioning	17%	14%	20%
Flavor	15%	9%	19%
Usage	14%	9%	18%
Visual	13%	13%	10%
Pricing	*10%*	*18%*	*6%*
Health statement	7%	9%	5%
Subbrand 1	6%	5%	7%
Subbrand 2	5%	4%	7%
Product line	5%	6%	3%
Light flavor	4%	8%	2%
Brands	4%	5%	3%

concept elements varied by category, so we can only look across different, comparable subgroups to see the importance of price. There were two segments of respondents for the condiment. Segment 1 is far more responsive to pricing, whereas segment 2 hardly cares at all about pricing. These data suggest interindividual differences in the sensitivity to pricing can be very large with rather different mind-sets.

Pricing as a Dependent Variable

Up to now we have been dealing with price as a category in the concept. What happens when respondents rate the appropriate price of an offering? Rather than rating acceptance, respondents choose a price. This task is fairly easy. All the researcher need do is to change the scale from interest to *fair price*. For example, one could instruct respondents to rate the concept by "selecting the fair price for this offering." The rating scale can be either labeled at each point with a price, or better, anchored at a lowest and a highest price. In either case, the researcher then transforms the rating to a price and runs the regression.

We see the results of this exercise in Table 12.4, which pertains to a frozen-food product, with five categories of elements and four elements per category. Respondents rated

each of 40 concepts on two attributes: the fair price (1 = $1.00 . . . 9 = $3.00) and interest in the product (1 = definitely not interested . . . 9 = definitely interested). In addition, the computer measured the number of tenths of seconds for each respondent to rate the concept. This measure was converted into a response-time measure (see Chapter 10). The data for each respondent was analyzed by dummy variable, ordinary least-squares regression.

We can interpret the results as follows:

1. The additive constant is the estimated fair price, or the price without elements.

2. Each element, whether brand, feature, image statement, quality statement, or picture, generates its own *dollar value*, which can be positive or negative. The dollar value must be added to the constant to yield an estimate of the total dollar value of the concept. Brands have large dollar values, both in the positive and negative directions, whereas the other categories of elements do not.

3. The interest utility value is a different measure. It may or may not covary with the estimated fair price. Indeed, an element may be quite interesting, but not worth much; or, conversely, it may be worth a great deal from a rational viewpoint, but just is not particularly interesting to respondents. For example, the brands contribute a great deal to price. Brands D and C both diminish the estimated fair price because they show negative utility values. However, brands D and C differ in consumer interest. Brand D adds 3% top-3 box to consumer interest, meaning that, although it decreases the fair price, nonetheless it is still interesting. In contrast, brand C is not interesting to respondents and decreases the fair price. Putting brand C into the frozen-food concept diminishes interest by 5% when using the top-3 box utility value as a measure.

4. The response-time model represents a third and independent measure of the concepts. Concept elements that increase perceived price, for example, brands B and A,

Table 12.4. Utility values for fair price, interest, and response time for a frozen-food product

	Estimated fair price	Interest % top-3 box	Response time (10ths of seconds)
Additive constant	*$1.82*	*49*	*45*
Brand			
Brand B	$0.42	0	−1
Brand A	$0.12	4	22
Brand D	−$0.15	3	−1
Brand C	−$0.30	−5	34
Feature			
Feature I	$0.03	2	15
Feature J	−$0.01	0	19
Feature K	−$0.04	1	48
Feature H	−$0.16	−2	19
Image			
Statement O	$0.06	1	−1
Statement P	$0.06	2	18
Statement M	$0.04	−2	9
Statement N	$0.04	2	42
Quality statement			
Statement Q	$0.06	3	44
Statement R	$0.06	1	2
Statement S	$0.03	1	−12
Statement T	$0.03	−1	11
Picture			
FIG3.bmp	$0.09	−8	−8
FIG4.bmp	$0.09	4	45
FIG2.bmp	$0.00	5	−12
FIG1.bmp	−$0.01	8	7

differ from each other in the response time. Brand B is reacted to quite quickly, whereas brand A is reacted to quite slowly.

5. The correlations across the three types of utility values are very low:

a. Price − interest $R = 0.15$
b. Price − response time $R = -0.21$
c. Interest − response time $R = 0.09$

Beyond Price as a Numeric Variable to Price as a Description

A great deal of pricing research deals with price in terms of dollars and cents. One interesting way to look at prices goes beyond statements of money and instead uses statements about the price connected to value for the money, a bargain, or slightly more ex-

pensive. When these elements are used in concepts, their ability to drive interest can be quantified just as the dollar values can be quantified.

We see the results from 30 different studies of both large and small, nonfood items. The data are taken from a large-scale It! Study, in which respondents participated in one of 30 different studies (see Chapter 20 for a full description of the approach). For purposes of this chapter it is only important to know that four of the elements out of 36 dealt with price and in a way that was not numerical but rather descriptive. When embedded in concepts, these elements generated utility values listed in Table 12.5. The data reveal the following insights:

1. Testing for price can be done in the same way as testing other concept elements.

Table 12.5. Utility values for three pricing statements, across 30 products*

Product	Offering a "Great deal" on the suggested retail price	The price is "Just right . . . all of the time"	"Priced a bit more than you would expect—but worth it!"
Average	*4*	*5*	*−9*
Range	*4*	*4*	*2*
Socks	5	6	−14
Dishes	5	6	−12
Refrigerator	5	7	−11
Couch	4	7	−11
Toaster	4	6	−12
Drinking glasses	5	6	−10
Boot	4	6	−12
Bedsheet	3	6	−11
Blender	5	3	−12
Television	5	6	−10
New baby	4	5	−10
Towel	4	4	−10
Pens	3	3	−11
Candles	3	5	−10
Electric drill	4	5	−10
Decorative pillow	3	5	−11
Tires	3	5	−10
Lamp shade	4	6	−8
Exercise	4	3	−8
Bathing suit	5	3	−8
Washer	4	4	−8
Drapes	2	3	−9
Writing paper	3	3	−7
Tablecloth	3	4	−8
Cars	3	3	−9
Lawnmower	3	2	−6
Sunglasses	4	4	−4
Ties	3	3	−6
Business suit	2	4	−6
Sandals	1	3	−6

*The results are taken from Buy It! and from the total panel only. Data courtesy of It! Ventures, Inc.

It is straightforward to test pricing as a statement of value, as well as testing price as a number. Respondents have no problem reacting to these descriptions of pricing, even without a specific number.

2. There are clear differences in the utility values of the price statements across products.

3. Talking about the "price is just right . . . all of the time" and "offering a great deal on the suggested retail price" generate positive utilities, on average.

4. Talking about higher prices such as "priced a little more than you would expect—but worth it" generates negative utilities. This same pattern holds for the entire panel across all 30 products.

5. The positive utility value for the low prices differs across products. The negative utility values also differ across products.

6. The disutility of a high price, verbally expressed, is, in this study, always higher in absolute value than the utility of a low price.

Characterizing Individual Styles in Price Responsivity

We end this chapter on pricing with a deeper look into the minds of respondents who are responding to the price description. Our focus again turns to the Buy It! Study. This time we can go into individual models created from the 4023 respondents who participated in the 30 different studies. The studies averaged 130 respondents, but we will not look at the particular products. Rather, we will look only at the individual response patterns to the concept elements. Keep in mind that these concept elements are phrased in terms of value and cost rather than in terms of dollars.

To better understand the individual respondents, we can classify each respondent into one of three classes for each of the three concept elements. We use the utility value for that individual for the particular element as the foundation on which to classify that individual. We divide the respondents, for each element, as acceptors of the element (utility value of 7 or higher), indifferent responders (utility value of −7 to +7), or rejectors (utility value of −7 or lower). These cutoff values are arbitrary and can be changed to different

numbers. The analysis principles are the same, however. The supporting data are in Table 12.6:

1. *One element: "Offering a* GREAT DEAL *on the suggested retail price."* Let us look at one element, specifically "Offering a GREAT DEAL on the suggested retail price." Let us look at the way any respondent reacts to this particular element, independent of the product in which it was evaluated. Looking at those respondents who really accept the notion of a "great deal" as shown by their utility values in the conjoint study, we see their utility for the element "Offering a GREAT DEAL on the suggested retail price" is +17. In contrast, their *disutility* for "Priced a bit more than you would expect—but worth it!" is −11. *These acceptors of "great deals" show a very wide, 28-point range in the utility values they assign to price. They are really sensitive to price.*

2. *Acceptors versus rejectors of "*JUST RIGHT.*"* The same pattern emerges for those who are defined as acceptors for the element "The price is JUST RIGHT . . . ALL OF THE TIME," based again on their utility values from conjoint analysis. They really like the element "JUST RIGHT," which is no surprise

Table 12.6. Utility values for respondents defined as acceptors, indifferents, or rejectors of the three pricing statements in the Buy It! study*

	Base		Offering a "Great deal" on the suggested retail price	The price is "Just right . . . all of the time"	"Priced a bit more than you would expect—but worth it!"
Offering a "great deal"	1386	Accepts	17	11	−11
on the suggested	1723	Indifferent	2	0	−7
retail price	624	Rejects	−16	−2	−10
The price is "Just	1662	Accepts	8	17	−12
right . . . all of	2013	Indifferent	1	1	−8
the time"	638	Rejects	−3	−16	−8
"Priced a bit more than	560	Accepts	4	3	17
you would expect—	1448	Indifferent	3	3	−1
but worth it!"	2015	Rejects	4	6	−23

*By dividing the respondent population we can see how one's mind-set about prices affects the utilities of price descriptions.

because that is how they were defined. They really hate the element "Priced a bit more than you would expect—but worth it!" Let us see how they react to the utility of the bargain side compared with the disutility of the more expensive, nonbargain side. The bargain side shows a utility of +8, and the nonbargain side shows a utility of −12. They show a 20-point range, and again the disutility of a more expensive price is greater than the utility of a bargain (−12 vs +8, respectively). Furthermore, those who reject the statement "JUST RIGHT" show the same pattern; namely, they dislike the bargain (−3) and they dislike the more expensive price, as well (−8).

3. *Acceptors versus rejectors of "Priced a little bit more."* Following the same logic, but this time for acceptors of the higher-priced element "Priced a bit more than you would expect—but worth it!" we see a different

Table 12.7. Input data for higher-level analysis of utility versus bags of popcorn, bags of seasoning packs, and unit price (in cents)

Element	Bags of popcorn	Seasoning packets	Price	Utility
Each box contains three bags and costs $1.99	3	0	199	2
Each box contains three bags and costs $2.09	3	0	209	3
Each box contains three bags and costs $2.19	3	0	219	2
Each box contains three bags and costs $2.29	3	0	229	−2
Each box contains three bags and costs $2.39	3	0	239	−1
Each box contains three bags and costs $2.49	3	0	249	−1
Each box contains three bags and costs $2.59	3	0	259	−5
Each box contains three bags and costs $2.69	3	0	269	−3
Each box contains two bags of popcorn and two seasoning packets and costs $1.99	2	2	199	1
Each box contains two bags of popcorn and two seasoning packets and costs $2.09	2	2	209	1
Each box contains two bags of popcorn and two seasoning packets and costs $2.19	2	2	219	1
Each box contains two bags of popcorn and two seasoning packets and costs $2.29	2	2	229	1
Each box contains two bags of popcorn and two seasoning packets and costs $2.39	2	2	239	−2
Each box contains two bags of popcorn and two seasoning packets and costs $2.49	2	2	249	−2
Each box contains two bags of popcorn and two seasoning packets and costs $2.59	2	2	259	−2
Each box contains two bags of popcorn and two seasoning packets and costs $2.69	2	2	269	−6
Each box contains three bags of popcorn and three seasoning packets and costs $2.49	3	3	249	4
Each box contains three bags of popcorn and three seasoning packets and costs $2.59	3	3	259	−1
Each box contains three bags of popcorn and three seasoning packets and costs $2.69	3	3	269	−1
Each box contains three bags of popcorn and three seasoning packets and costs $2.79	3	3	279	−4
Each box contains three bags of popcorn and three seasoning packets and costs $2.89	3	3	289	−1
Each box contains three bags of popcorn and three seasoning packets and costs $2.99	3	3	299	−2
Each box contains three bags of popcorn and three seasoning packets and costs $3.09	3	3	309	−3

Table 12.8. Utility for popcorn pricing elements as a function of number of bags, number of seasoning packets, and stated price*

Effect	Coefficient	Standard error	Standard coefficient	t	P (2 tailed)
Additive constant	12.5	2.81	0.0	4.402	0.000
Bags	2.8	0.82	0.540	3.472	0.003
Sea pack	1.2	0.33	0.609	3.576	0.002
Cost	−0.1	0.01	−1.131	−6.444	0.000

*The dependent variable is utility; $n = 23$; multiple $R = 0.830$; squared multiple $R = 0.689$; adjusted squared multiple $R = 0.639$; and standard error of estimate $= 1.503$.

pattern. Their utility is +17 for that element, which is no surprise because this is how they were chosen. They accept the opposite element "GREAT DEAL" (utility = +4) rather than rejecting it.

4. *Bargain seekers versus full-price payers.* These data suggest that a person who likes higher prices will easily accept lower prices and a "great deal." However, people who are price sensitive absolutely abhor higher pricing, even slightly higher prices than they would expect.

Beyond Pricing Utilities to Integrative Models by Using Price as an Independent Variable

Up to now we have been looking at pricing alone. However, pricing can be part of the mix and can either increase or offset the impact of other variables such as number of packages. One way to better and perhaps more profoundly understand pricing is by looking at the utility values for price as themselves new dependent variables and then looking at the nature of the concept elements as providing new independent variables. In Chapter 5, on conjoint analysis, the data for a popcorn product were presented. Some of the elements comprised pricing statements, combined with new features such as seasoning packets.

Let us now create a model relating the features of the new popcorn product to the utility values. The model is straightforward:

1. The independent variables are the number of popcorn packets, the number of seasoning packets, and the stated price

2. The dependent variable is the utility value of the combination, from the conjoint measurement

3. Table 12.7 lists the input data, and Table 12.8 presents the results of the regression analysis.

4. The results show that more bags add to acceptance, far more than more seasoning bags, but that price is again the most important. To counteract price somewhat, one can modify the number of bags of popcorn in the product, although price and number of bags are somewhat linked.

References

Moskowitz, H.R. (1995). The dollar value of product quality: the effect of pricing vs overall liking on consumer stated purchase intent for pizza. Journal of Sensory Studies, 10: 239–248.

Moskowitz, H.R., and Poretta, S. (2002). Contrasting customer and operator concept and product requirements: the case of surimi. Foodservice Technology, 1: 115–130.

Chapter 13

Analyzing a Study: Casual-dining Restaurant

Restaurant Concepts and the Design of a Complex System

Food concepts tend to be quite simple. To illustrate the approach of dealing with many elements, we can go a bit far afield and deal with restaurants. The distinction between the simplicity of a food and the complexity of a restaurant becomes immediate after thinking about what is involved in each. A food comprises physical features: the benefits, pricing, packaging, etc. A restaurant comprises many more aspects, varying from basic theme to the ambiance, service, product, pricing, and emotional statements about how patrons feel. It is no wonder that with the great complexity of restaurants much of the research is not systematic but rather confirmatory. With foods, one knows the range of things to talk about. With restaurants, there are so many options that a disciplined approach may not be easy to design. The researcher, but more often the marketer and general management, must identify which of a variety of concept ideas works best in spite of this complexity. Today's intense competition is forcing food-service professionals and restaurateurs away from the simple "gut feel" and "shoot from the hip" style to fact-based decision making. It is no wonder, therefore, that researchers have only recently begun to deal with concepts that transcend the product itself and to move forward toward the meal and the consumption situation (Meiselman 2000).

Background to the Case History

Most restaurant chains continually monitor the trends and environment for new restaurant concepts. The particular chain, called here O'Steers (not its real name), is well known and has been a key player in the fast-food business. During the course of strategic exercises to define the future, management identified a number of opportunities in the emerging casual-dining segment. This segment lies between the fast-food segment and the local restaurant.

The objective of the concept development exercise was to identify a new restaurant opportunity by having the consumers design the features of the restaurant, ranging from the name to the menu item, to the décor/feel/emotion of the restaurant, and even to policies regarding price and children.

In most concept research the goal is to find concepts that do well. Identifying *why* a concept performs well is of secondary importance. When a researcher runs focus groups or concept tests on these winning concepts the reason is not so much to identify the features that drive the concept but rather to fine-tune the idea. Most work in concepts is either selection (Which of these is the best?) or confirmatory (How big is the opportunity?). Building the restaurant concept from the ground up by identifying which of the particular options drives interest is relatively new but may be expected to become increasingly popular over the next several decades.

An additional objective of this case history was to identify which of the features of the concept made sense from an *information processing* level. In other words, which messages could be understood quickly—an important consideration to an advertising agency for its advertising and to the site committee for its internal and external *signage*. In the highly competitive area of casual and fast-food dining, consumers pay relatively little attention to the messages. The restaurant chain was interested in identifying the types of messages that would take relatively little time and effort to understand. The rationale was that if one could identify both winning messages (high interest) and those easy to understand (short response time), there would be a potentially new concept idea in the intersection of those elements. Since the information could be rapidly processed, the idea made sense. The winning messages, if sufficiently unique, could then provide both the platform for the new restaurant concept and some specific messaging.

Developing the Concept Elements

The restaurant client opted to take a broad look at the different types of options available; that is, rather than following the conventional "beauty contest" approach with its limited scope, the client decided to put all of the options into a hopper and see which ones did well. To this end, the marketing research director ran several brainstorming sessions and conducted a competitive scan of other restaurants. The competitive scan looked at messages from print advertising, television, and the Internet, respectively.

The sessions and the competitive intelligence generated 275 different ideas, culled down from an original 467 elements. Table 13.1 lists examples of the different categories. It is worth emphasizing once again that, when building concepts from the bottom up, the more elements with which one works, the better the data will be, at least on average. One can always discard data that one has. One cannot, however, easily estimate the likely response to elements not in the study at all. If the research method can deal with these elements, then the restaurant client can base the decisions on consumer information rather than on informed intuition.

Table 13.1. Examples of elements and their locations on the semantic profiles

	Lunch	Adult	Casual	Eat-in	Inexpensive	Fast food	Not sophisticated	Low convenience
Dimension 1 =								
Dimension 9 =	Dinner	Child	Formal	Takeout	Expensive	Fine food	Sophisticated	High convenience
Brand								
End the Day at O'Steers	8	5	3	5	4	4	5	3
O'Steers After Dark	9	2	5	5	6	5	6	4
Menu								
Menu that shows each food's nutritional content	5	1	5	3	5	7	6	6
Menu with daily specials	5	2	5	4	5	6	6	5
Main courses								
Filet mignon and other main dishes	8	1	8	1	8	9	9	7
Chicken and steak fajitas, and other main dishes	5	4	4	5	4	4	3	4
Side dishes								
Cole slaw and other side dishes	5	2	3	6	4	3	3	2
Dinner rolls and butter, and other side dishes	5	5	5	5	4	6	5	5

(continued)

Table 13.1. Examples of elements and their locations on the semantic profiles *(cont.)*

Dimension 1 = Dimension 9 =	Lunch Dinner	Adult Child	Casual Formal	Eat-in Takeout	Inexpensive Expensive	Fast food Fine food	Not sophisticated Sophisticated	Low convenience High convenience
Desserts								
Our dessert menu includes all-you-can-eat sundae bar	5	5	4	1	1	3	3	9
Our dessert menu includes Breyer's frozen yogurt	5	4	3	5	3	4	3	5
Atmosphere								
Paper tablecloths with crayons to color on	4	1	1	1	1	3	1	9
Dinner at your own pace	9	5	5	1	5	5	5	9
Service								
All-you-can-eat appetizer bar	5	3	4	1	1	5	7	9
Steak and burgers cooked medium rare or medium well	5	3	5	5	5	7	9	9
Entertainment								
TVs playing all news	9	1	3	1	5	9	9	6
Video game room off to the side	5	7	2	1	2	3	2	5
Children								
Kids eat free Monday through Thursday	5	7	4	1	1	4	6	9
Kids eat free any night with purchase of two adult meals	8	7	4	1	3	5	6	8
Price								
Feed a family of four for under $15	8	2	2	5	1	4	4	9
Feed a family of four for under $20	9	3	3	5	1	4	4	9
Takeout								
Drive-thru service in three minutes	5	5	3	9	2	4	7	8
All items available for takeout	5	5	4	9	3	5	9	9

The IdeaMap study required dimensionalization (see Chapter 5). The elements were located on eight semantic differential scales and restricted so that incompatible elements and redundant elements could not appear together. The semantic differential scales enable researchers to estimate the utility values and response times for elements not tested, are used for segmentation, and permit optimizing concepts that have a specific image profile.

The study was run in four US markets, with 240 respondents, recruited according to standard market-research specifications:

1. *Target population.* The respondents had to be patrons of either fast-food or casual-dining restaurants and interested in

participating. The population comprised two-thirds women and one-third men, all over the age of 18 and approximately equally distributed by age.

2. *Security.* The respondents were not to work in other market-research companies, nor work for an advertising agency. A lot of times it is difficult to exclude individuals. Often people who work for agencies or research companies, or for restaurants, get onto panels and are recruited to participate. Respondents often do not tell the entire truth when participating on a panel.

It is worth emphasizing that, despite this lack of 100% truth, most research works because the security issues are important only for final concepts or final products. When it comes to conjoint measurement tasks like the one described here, security is less important. No respondent really ever sees the final set of concepts. Rather, respondents see systematically varied concepts and, even if they knew how they had responded, nonetheless they do not know the entire set, nor do they know how the elements do in terms of utilities. *That performance information comes only at the end of the study, and only from the aggregation of the individual models.*

Identifying Winning Elements by Inspecting the Entire List of Elements

We can get a sense of what the analyst faces by looking at the results in Table 13.2. These results come from the sorted utility values of the concept elements. Only the winning and the losing utility values are presented in the table, because otherwise it would be unduly long. The response times for these elements are shown as well. Both the utility values for "interest" and the utility value for "response time" (in tenths of seconds) were developed according to the IdeaMap approach (see Chapter 10).

Faced with these data, analysts typically follow these steps to understand the results:

1. Look for potential inconsistencies that could signal trouble. This is generally a preventive measure to ensure that the tables line up correctly!

2. Look for winning elements by sorting all elements, generally by total panel. The data in Table 13.2 suggest that there is no simple, clear pattern, at least for the total sample. The winning elements fall into a variety of categories, including price, main course, and dessert, respectively. Quite often when the initial data are presented to clients they are first befuddled but then delighted as they flip through the data tables, looking for patterns.

3. Look for a common pattern. Is there anything common about the winners or losers that helps one understand the patterns and that can be used for better concepts? We saw some of the search for common patterns in Chapter 7 on segmentation, where category importance helped better explain the meaning of the segments.

4. Look for any other variables that can help explain the winning or losing pattern, such as winners belonging to a specific category or showing a shorter response time.

5. Create new combinations; that is, try out the data as a guide to create new concepts. A winning concept created from Table 13.2 would, without any additional understanding, simply comprise the elements that score well. Once there is understanding, the concepts take on more meaning.

Using Segmentation to Fine-tune the Restaurant Opportunity

In any conjoint-analysis the results always point to winning elements. As Chapter 7 on segmentation suggests, there may be more opportunity if the marketer goes after segments that can be satisfied with a coherent

Table 13.2. Winning and losing elements for the total panel (utility) and their response times (RT): this is the first "cut" at looking at the data and provides a lot of important information and insights

Elements	Utility	RT
Additive constant	*57*	*27.31*
Kids eat for $.99 any night	6	1.82
Family-sized portions of food for family-style eating	6	3.28
Chicken and steak burritos, and other main dishes	6	2.50
Filet mignon and other main dishes	6	1.07
Kids eat free any night	6	1.28
Fresh-baked pizza with premium toppings	6	1.68
Light spending for dinner eating	6	3.12
A menu featuring flame-broiled entrees	6	2.09
All you can eat salad and food bar	6	2.32
A menu that shows each food's nutritional content	6	3.49
Roasted turkey breast with stuffing, and other dishes	6	1.79
Freshly baked pizza in personal and dinner sizes	6	2.42
Buy one pizza get one for half-price	6	3.10
New menu each month with different regional specialties	6	3.29
Garlic bread and other side dishes	6	1.56
Our dessert menu includes cheesecake	6	1.60
Chicken or beef enchilada, and other main dishes	6	2.32
Stir-fried mixed vegetables and other side dishes	6	2.16
TVs playing cartoons	−5	0.64
Kids club characters in costume	−5	0.65
Children's face painting	−5	0.51
Partially cooked dinner items that you finish cooking	−5	2.53
Karaoke sing-along machine	−6	1.36
Clowns/magicians	−7	0.67
Video games in the tables	−8	0.76

strategy rather than going after the entire panel. With the entire panel the odds are very low that the conjoint approach, or indeed any approach, will reveal radically new, exceptionally strong elements. There is a greater opportunity to succeed by working with the more focused segments. It is the researcher's job to identify why winners do well, typically by discerning a pattern among the winning elements.

Often the winning elements in conjoint-analysis, or winning concepts in a regular concept study, come from a variety of different categories or concepts, making it difficult and occasionally impossible to discern the pattern. However, segmentation among the respondents solves this quandary, and more often than not the results provide a very pleasant

surprise to marketing, product development, and the researcher. The data often suggest two or more different groups of respondents with radically different viewpoints. Within each group the winning elements make a great deal of sense. Only when these two different mindsets combine into the undifferentiated *total panel* do the clear segments fade into the mist and what had been so patently clear now becomes clouded and indistinct. The situation is very much like superimposing two patterns on each other, either visually or in music. The patterns by themselves make a great deal of sense and can be easily labeled. When the patterns are superimposed, however, none of the elements do particularly well because they cancel each other. Indeed, occasionally the winning elements comprise mutually contradictory statements. Without knowledge of the segmentation, one cannot really understand why these elements do well. Only when one realizes that the elements come from two superimposed groups that share little in common do the contradictory elements make sense.

Applying the segmentation algorithm revealed two clear groups of respondents. Table 13.3 shows the results from the segmentation, as well as how the concept element scores for the total panel. For restaurants the segmentation is not a case of polar opposites where one segment likes an idea and the other segments hates the same idea. For restaurants the segmentation is rather a matter of focus and degree. This finding is both positive and negative:

1. Positive because there are no dramatic, polar opposites where one person hates what another person loves
2. Negative because the polar opposites are often love–hate, whereas for restaurants

there are only modulated levels of acceptance

The two segments can be named as *upscale* and *convenience*, respectively. These names come from the elements that do best. From Table 13.3, researchers can provide the following three insights to clients in restaurant management:

1. About half the individuals fall into the upscale segment and half fall into the convenience segment. It is not clear whether a person always belongs to the upscale segment and never belongs to the convenience segment, or whether segment membership may be occasion driven. An individual may belong to each segment, but at different times. People frequent different restaurants. The same individual may go to two different types of restaurants, depending on the situation. Thus, the segmentation here is not a segmentation of basic mind-sets (e.g., "I love hot foods" vs "I hate hot foods") but rather segmentation of opportunities.

Table 13.3. Winning elements for the two segments: upscale and convenience

Elements	Total	Upscale	Convenience
Additive constant	*57*	*56*	*57*
Segment-1 winners (upscale)			
A menu that shows each food's nutritional content	6	9	3
Filet mignon and other main dishes	6	9	4
Prime rib and other main dishes	6	9	3
Fresh-baked pizza with premium toppings	6	8	4
Marinated flame-broiled chicken breast, and other dishes	6	8	3
Flame-broiled lobster tails and other main dishes	4	8	0
Chicken Kiev stuffed with herb butter, and other dishes	5	8	3
A menu featuring dishes from around the world	5	8	2
New menu each month with different regional specialties	6	8	4
Segment-2 winners (kids/casual)			
Kids eat for $.99 any night	6	5	8
Kids eat free Monday through Thursday	6	3	8
Order and get your meal in 5 minutes or it's free	5	3	8
Kids eat free weekdays	6	4	8
Kids eat free any night	6	5	8
Buy one pizza get one for half-price	6	4	7
Where kids can be kids and parents can relax	5	1	7
Light spending for dinner eating	6	5	7
All you can eat salad and food bar	6	5	7
Family-sized portions of food for family-style eating	6	6	7

2. The additive constant is the same for each segment, 56, meaning a moderately high degree of basic interest.

3. By inspecting the utilities of the winning elements we see that the segments are not polarized. Segmentation does not always reveal polarizing groups. For foods there is often polarizing segmentation, especially when the segmentation is based on flavor. What one person loves, another person may hate. *When segmentation is based on form or texture, or type of packaging, instead of on flavor, there is usually less polarization.* This finding is important in the development of a restaurant concept, because it suggests that it might be possible to appeal to multiple groups simultaneously. When the segmentation is polarizing, with love versus hate, there is little chance of appealing to both groups at the same time. Perhaps this is the reason for a proliferation of different flavor SKUs (stock-keeping units) for foods and less so for restaurant concepts.

Meta-analyses: Seeing Both the Big Picture and the Small Picture

Conjoint-analysis encourages the dynamic tension between looking at categories (the big picture) and looking at elements (the little picture, or micropicture). Depending on a researcher's predilection, the category may be the key aspect of learning or the element may be the key aspect.

Those Interested in the Big Picture

When *category performance* is key, the elements are simply instances of the category and by themselves are not important. Like a pointillist painting, the individual elements cannot really be appreciated, other than by standing back and seeing how they contribute to the larger picture. It is really when a researcher *understands* category importance that the rules of the game have been discovered, and the researcher can proceed to create

new ideas. Looking at elements alone is simply to jump from one point to the other and never really see the big picture.

Those Interested in the Specific Test Stimuli (Elements)

For those interested in the elements rather than the categories, the *category* is simply a convenient way to organize the elements so that contradictory elements do not encounter each other in a concept, or so that the concept does not repeat the same type of message embedded in two elements. The category is simply a bookkeeping system.

Simultaneous, Parallel Views

Of course, no researcher embraces either position 100%, but within the world of researchers, and indeed within the world of research users, both viewpoints can be found. Many strategists opine that they look for the big picture, never realizing that all strategy must be executed with some specific language and picture. Listening to these ungrounded strategists is listening to general rules that have no concrete substance and, in the end, listening to business cant or jargon. In turn, listening to a person who cares only about specifics is, in ways, listening to someone who lacks the ability to abstract patterns and cannot really see what is going on in the data. There is no rule for this person either, because people need rules and generalities, not just laundry lists.

Measuring Relative Importance

The restaurant study comprised 12 categories, or 14 if we are in a more generous mood to divide the food category into three smaller, more homogeneous categories. These categories comprised different numbers of elements. For this study, as for the other conjoint studies, the categories did not start their life as nice, logical buckets of different ideas. In any ideation session, categories are developed to make the participant's job easier. By using these categories the researcher and the participants in ideation can better cope with the very large number of elements for a restaurant.

Given the different numbers of elements, and the fact that some elements do well whereas other elements do poorly, let us put into practice the analytic tools discussed in previous chapters. We have seen one of the tools—segmentation—reveal to us that the group of respondents is not homogeneous, but rather comprises two groups of different-minded, but not opposite-minded, consumers: the upscale seekers and the convenience seekers. For these groups, as well as for the total panel of consumers and other

subgroups relevant to the study, which of the categories are important and which are not important? We will use the RII (relative importance index) approach with relative sums of squares, discussed more fully in Chapter 7 on segmentation:

The Role of Sum of Squares

The sum of squares is one way to look at the sizes of the elements without respect to the sign. Keep in mind that a positive sign for an element means that the element drives acceptance, whereas a negative sign means that the element reduces acceptance and thus leads to rejection. The groups with the higher sum of squares are those that are more re-

Table 13.4. Relative importance of categories in the restaurant study, indexed by the proportion of the total sum of squares (SS) ascribable to elements in the category

		Gender		Age		
	Total	Male	Female	18–34	35–49	
Total SS	4138	3933	4924	3540	5505	
% Total						
Food (total)	56	57	52	59	52	
Side	26	26	24	27	24	
Main	23	24	21	21	22	
Service	11	13	8	16	7	
Children	10	9	11	8	11	
Price	9	9	9	8	9	
Dessert	8	7	8	11	5	
Entertainment	7	4	12	1	15	
Menu	6	5	7	7	5	
Takeout	4	5	4	6	3	
Atmosphere	4	6	3	4	4	
Brand	1	1	1	1	1	

		Visited		Kids		Segment	
	Total	Yes	No	<12	None	Upscale	Convenience
Total SS	4138	3772	5347	4769	4257	6230	3984
% Total							
Food (total)	56	54	60	48	62	65	37
Side	26	26	25	23	28	29	18
Main	23	21	26	17	28	29	12
Service	11	11	9	10	11	4	18
Children	10	10	10	17	5	4	21
Price	9	9	9	9	9	5	12
Dessert	8	7	9	8	6	7	7
Entertainment	7	8	7	10	5	16	2
Menu	6	6	6	6	6	7	4
Take-out	4	4	5	4	5	1	8
Atmosphere	4	4	4	4	3	2	6
Brand	1	1	1	1	2	2	1

sponsive to the elements. By squaring the utility value we simply look at the power of the element either to drive acceptance or to drive rejection. From Table 13.4 we see that the different subgroups show varying patterns, both in terms of the total sum of squares, a measure of total *driving power of elements*, and in terms of the allocation of the sums of squares to different categories.

Total Sum of Squares as a Measure of General Responsivity to Elements

Women show greater sum of squares than do men, meaning that the elements affect them more. Older respondents, ages 35–49, show greater sums of squares than do the younger respondents, ages 18–34. Nonvisitors of casual and fast-food restaurants show greater sum of squares than do visitors. Having young children under the age of 12 makes no difference compared with not having young children. Finally, being upscale shows greater sum of squares than being in the convenience segment.

The Different Categories Show Different Patterns Across Subgroups

For the total panel, the categories are ranked by percentage of the total sum of squares. Deviations from this ranking by subgroup represent different mind-sets. Three key findings stand out:

1. Most subgroups show similar patterns. These patterns are not identical, but the trends are reasonably similar. There are some notable exceptions, but for specific elements. Two subgroup-to-subgroup differences follow.

2. The age of the child makes a difference. Respondents without younger children are more responsive to the food categories, whereas respondents with younger children under 12 are relatively less responsive to the food category, but as one might expect they are more responsive to statements about children.

3. The concept-response segments differ radically from each other. The upscale segment is far more responsive to elements about food than is the convenience segment. The upscale segment is more responsive, albeit negatively, to the idea of entertainment in the restaurant, whereas the convenience segment is fairly indifferent to entertainment.

Looking at the Different Categories to Understand What Drives Acceptance

Elements are the most important features of the conjoint study. Although the relative importance of categories can identify which subgroups react unusually strongly to categories, it is the elements themselves that do the work. The plethora of data available from a conjoint study makes a systematic approach even more critical. Although the elements are the key aspects of the study, there are 218 elements, making it impossible to do justice to the elements and to the full set of subgroups. Some rational approach has to be used.

Most analysts beginning with the data set up their objectives beforehand and then dive into the results, looking to address the specific objectives. In such cases, serendipity may occur, but more likely any serendipity will reveal itself in an unexpectedly strong or poor performance of single elements that at the start of the study were identified as being important to watch; that is, in the words of Louis Pasteur, "Chance favors the prepared mind." With conjoint analysis and faced with this relative large mass of data, it is more likely that the researcher will discover something if he or she knows what to look for, or at least can frame out a systematic search.

How Semantic Scales Covary with Concept Interest and the Patterns Across Different Subgroups

Recall that prior to the study a group of respondents located the concept elements on a

semantic differential scale. In product research, one learns about the *drivers of liking*
by plotting sensory intensity level on the *x*-
axis and liking on the *y*-axis. The result is a
scatter plot that shows how the sensory level
covaries with liking. Different segments show
different patterns. Segments differ from one
another in terms of the sensory level at which
liking reaches its peak. We can apply the
same functional analysis to concept work,
this time using the semantic scale values in
place of the sensory values (see Table 13.1)
and using the utility values for the subgroup
in place of the liking.

The scatter plot of the semantic profile values (on the *x*-axis) against the utility values (*y*-
axis) reveals no pattern by total panel. An
analysis of the results by concept-response
segment, however, reveals two patterns, as
shown in Figure 13.1. When the independent
variable is the semantic scale "casual vs formal," then the upscale segment shows increasing utility value when the tonality of the concept element becomes more "formal." In
contrast, for the convenience segment there is
no clear relation.

Figure 13.2 shows the same scatter plot,
this time with a fitted quadratic function and

with the size of the individual elements
brought to zero so that they disappear from
the plot. The figure shows the relation quite
clearly. It is from analyses of this type that
one begins to understand the rules underlying
winning versus losing elements. However, as
noted before, the meta-analyses do not deal
with particular concept elements and thus
lack the immediacy of a detailed analysis of
actual elements themselves.

Figure 13.3 shows the relation between
the semantic scale and utility value for respondents who either have children living at
home under the age of 12 or do not. We see
here that the relation is not as clear.

Response Time to Concept Elements Varies by the Interest in the Concept Element

As interest increases, increasing utility values, response time increases. This means
that, at least in the case of a restaurant, when
the element is interesting, respondents will
spend less time reading it, at least in general
(see Table 13.5). It remains for a second
analysis to factor in the different types of
concept elements. This is a major finding.

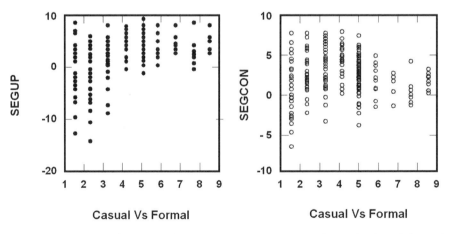

Figure 13.1. How utility value for concept elements covaries with semantic scale *casual vs formal*. The results are
show for the two concept response segments: SegUp, upscale segment; and SegCon, convenience segment.

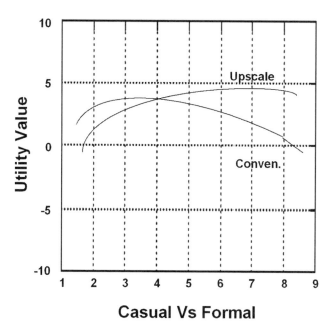

Figure 13.2. How utility value for concept elements covaries with semantic scale *casual vs formal*. The results are shown for the two concept response segments, but this time are presented as fitted quadratic functions in order to clarify the nature of the relation.

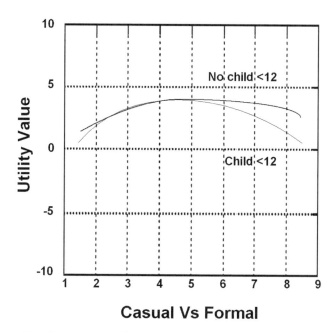

Figure 13.3. How utility value for concept elements covaries with semantic scale *casual vs formal*. The results are shown for respondents who do not have children under age 12 at home and for those who do. The results are presented as fitted quadratic functions.

Table 13.5. How response time (RT) covaries with element interest and element length (number of letters)

Response time vs interest
Multiple $R = 0.3386$
Squared multiple $R = 0.1147$
Adjusted squared multiple $R = 0.1114$
Standard error of estimate $= 0.5781$

Effect	Coefficient	Standard error	Standard coefficient	t	P (2-tailed)
Constant	1.9759	0.05	0.0	39.11	0.0000
Total interest	0.0774	0.01	0.34	94.	0.0000

Response time vs number of letters in the concept element
Multiple $R = 0.3830$
Squared multiple $R = 0.1467$
Adjusted squared multiple $R = 0.1436$
Standard error of estimate $= 0.5675$

Effect	Coefficient	Standard error	Standard coefficient	t	P (2-tailed)
Constant	1.4094	0.1194	0.0	0.0000	
Letters	0.0213	0.0031	0.3830	6.8507	0.0000

Response Time to Concept Elements Increases with Increased Length of the Concept

We saw that before for coffee (Chapter 10). This finding makes intuitive sense because it takes longer to read a long concept than a short one (see Table 13.5)

A Lot More of the Response Time Can Be Explained by Combining Predictors

For example, we can predict 11% of the variability in response time by just knowing the interest value of the utility. We can increase the percentage of variability accounted for by looking instead at the length of the concept element. The multiple R^2 is now 15% rather than 11%. By using both interest and response time we can increase the multiple R^2 to 20%. By knowing the nature of the concept elements—that is, the category from which they come—and by using *dummy variables* to represent the presence of the category in the concept, we increase the predictability of response time to 49% (see Table 13.6).

Respondents Pay Different Amounts of Attention to the Elements

They pay most attention to price (coefficient in response time model = 0.58) and then secondarily to menu items (coefficient = 0.23). Respondents read through those menu elements more. When it comes to children the response time actually decreases (coefficient = −0.33), meaning that as soon as respondents recognize that an element deals with children they *automatically respond*, without much further consideration.

The Analysis Is Constrained

This type of analysis provides an overview of the patterns in the data, but does not yet tell the researcher what particular elements to recommend for the new restaurant concept. The meta-analysis uncovers macrolevel patterns that transcend individual concept elements. From these larger-scale patterns the researcher begins truly to understand the nature of the consumer response, the types of elements that drive interest, how the consumer attends to the different categories, and so on.

Table 13.6. Coefficients of the model relating response time to the utility values (interest), the number of letters, and the different types of categories (coded as dummy variables)

Variable	Coefficient for response time
Additive constant	*1.45*
Variables that increase response time	
Price	0.58
Menu	0.23
Service	0.21
Food: dessert	0.19
Atmosphere	0.12
Interest (utility)	0.05
Brand	0.04
Number of letters	0.02
Variables that decrease response time	
Food: side dish	−0.20
Entertainment	−0.33
Children	−0.33
Food: main item	−0.60

Detailed Element Analyses

In conjoint research a great deal of the learning comes from the analysis of the single elements. The pattern underlying the performance is still left up to the researcher. However, no matter what type of general patterns emerge from an analysis conducted on all of the elements, it is the performance of specific elements that provides the *greatest immediate actionability*. For many practical purposes, though, researchers can do their job merely by reporting on the winning versus losing elements. The general patterns defining what elements win or lose nonetheless immediately allow the marketer or restaurateur to craft a superior concept.

Occasionally, a conjoint study will generate results that are clearly obvious, with elements scoring +10 or higher. The data show immediately that some elements will do superbly. In these simple cases the researcher, the marketer, and the product developer have no problem understanding what to do. The

next steps are simply a matter of selecting the most promising elements.

In other cases, especially when it comes to a restaurant such as that described here, many of the elements perform only modestly. This is not because the study is a failure. With 275 concept elements it is hard to know what else to include. In these situations the usefulness is even greater for a detailed, category-by-category, element-by-element analysis. There are no "white knight" elements that can be used. Rather, the researcher must examine all the available information to identify that set of promising elements, none of which is breakthrough by itself, to create the new concept. One learns more from dealing with these more difficult situations than from dealing with the more spectacular, more anxiety-relieving situations, where a few breakthrough elements steal the show.

Let us look at a few of the categories. We will sort the elements in a specific category by total panel and look at the highest-performing elements by total panel and by key concept-response segment (upscale vs convenience). Patterns may emerge, but the important objective of the analysis is simply to identify promising elements. Table 13.7 lists the data that might be consulted to generate this detailed analysis:

1. *Menu statements.* These are all positive for the total panel and for the most positive part for the two segments. The upscale segment wants menu variety, flame-broiled entrées, and nutritional information. The convenience segment wants pizza and pasta. It is clear that, even if the researcher cannot discern a pattern, the data table will reveal which of the elements is most promising.

2. *Ambiance (atmosphere).* Ambiance is not a particularly positive driver of acceptance. This means that, in the development of the restaurant concept, one has to include a statement about ambiance, but the strategy should be to find something that appeals to both segments. One might communicate

Table 13.7. Utility values for the menu statements

	Total	Upscale	Convenience
General menu			
A menu featuring flame-broiled entrees	6	8	5
A menu that shows each food's nutritional content	6	9	3
Freshly baked pizza in personal and dinner sizes	6	6	6
A new menu each month with different regional specialties	6	8	4
An Italian menu offering everything from pizza to pasta	6	7	5
Ambiance			
Comfortable seating	4	3	5
Unrushed, leisurely dining	4	3	5
Paper tablecloths with crayons to color on	4	1	7
Dinner-sized napkins	4	4	4
Pricing			
Light spending for dinner eating	6	5	7
Buy one pizza get one for half-price	6	4	7
Affordable prices so the whole family can enjoy	6	5	7
5 entrees priced at $3.99 or less	6	7	4
Children			
Kids eat for $.99 any night	6	5	8
Kids eat free any night	6	5	8
Kids eat free Monday through Thursday	6	3	8
Kids eat free weekdays	6	4	8
Service and portion size			
Family-sized portions of food for family-style eating	6	6	7
All you can eat salad and food bar	6	5	7
All you can eat appetizer bar	5	4	7
Order and get your meal in 5 minutes or it's free	5	3	8
Order and get your meal in 10 minutes or it's free	5	4	6
All you can eat salad bar	5	3	7
Steak and burgers cooked medium rare or medium well	5	4	6
Free soft drink refills brought to your table	5	2	7
Serve yourself condiment bar	5	3	6
Friendly host and hostess	5	3	6
Your order is correct or the next one's on us	4	2	6
Free popcorn while you wait	4	1	6

something such as "unrushed, leisurely dining," which is acceptable to the total panel and to both segments.

3. *Pricing.* Generally respondents are interested in low prices, but want to know that they are getting a good meal. Phrases such as "light spending for dinner eating" work well. It is probably best to talk about generally good value for the money rather than to specify a price.

4. *Children.* The segmentation breaks the respondents apart on children. Some of the price statements for the convenience segment

do quite well when mixed with the word "children" or "kid." They include phrases such as "Kids eat for $.99 any night" or "Kids eat free any night." It is important to recognize that bringing in the children to the concept could reduce acceptance. Since the O'Steers restaurant is positioned as casual dining/fast food, bringing in children does not turn off segment-1, upscale respondents, but does reduce their interest. Had the O'Steers restaurant been positioned as a regular restaurant rather than casual dining, then the emphasis on children might have reduced acceptance and indeed made the element a turnoff. As it stands, the elements mentioning children are neutral to slightly positive for the upscale segment.

5. *Service and portion size.* These statements deal with the service, and specials, such as "All you can eat salad bars." These elements do very well with the convenience segment, but only modestly well with the upscale segment. Again we see emphasis on quantity and flexibility. The convenience group perceives the service features as a way to get good food at a reasonable price and thus likes it. The upscale segment looks for something more. They accept the idea of good-sized portions ("Family-size portions of food for family-style eating"). They are not interested in doing anything at the restaurant that requires them to get their own food, such as "Order at the counter and food is brought to your table" or "Get yourself drinks while food is brought to your table."

Transcending the Data: What Are the Next Steps for the Researcher User?

With an embarrassment of data, such as that presented for the restaurant, what is the researcher's real job? How much should the researcher contribute to the analysis? Sometimes researchers feel that they have to do many jobs, such as identify the concept elements, run the study, do the analysis, develop the insights, and then suggest the optimal concepts. Sometimes researchers feel that the only task required of them is to execute the study properly, with the up-front element creation and the downstream interpretation coming from consultants or other experts in the area. Is either of these research requirements on target? Is there a middle ground? Can a researcher really find a comfortable position between the pole of true expert/ superman and data functionary?

The answer to the foregoing question is that both positions, and indeed the intermediate positions between the poles of superman

and data functionary, are valid. The researcher may have been more of a superman when consumer research was a nascent area, and before it became acceptable, professionalized, and the prey of other professionals such as consultants. Since the work flow in concept research was not clearly defined, and there were no professional standards to speak of, nor much experience, most of the senior researchers involved in research were both consultants and field/data functionaries; that is, the same researcher could play a number of different roles. The times were kinder and gentler, and the competition for low price, fast delivery, and canned insights was not as fierce. Many researchers who recount their stories talk about how they were consultants first and data functionaries second. These pioneers would design the study, execute it, and report the results, consulting all the while. Today, a dynamic tension underlies and undercuts the researcher's role. Does the researcher provide just data or act as a consultant, with data acquisition simply one of the duties? Does the report present results or does the report present conclusions, implications, and occasionally fanciful flights of thinking?

Some well-known companies have legislated that researchers may only present results to management, albeit in whatever form is desired. To those companies the notion of *professional analysis* is analysis constrained to lie within a specific presentation format. There is no room to think and hypothesize, except about the actual research process. Rather, researchers have to follow the guidelines in lockstep. Professional research in concept (and product) testing means being able to fit into the format dictated by a rigid, benighted management. Within those tight constraints, researchers are, of course, allowed and even "encouraged" to express themselves. The expression is usually a creative way to analyze and present the data. There is no leeway to become involved in the up-front work or in the later decision making. Fortunately, this tight restriction is not the standard way that market-research analysis is treated. Were it to be so, then restaurant analysis would simply concentrate on the presentation of good data, in the standard method, and possibly some additional data-related insights.

What Happens Next After the Research Has Been Completed?

The thrust of this chapter has been on the actual mechanics of putting together the study, obtaining the data and, in the main, analyzing the results. What happens to the results afterward, once the data have been analyzed? Do the clients who request the data go beyond the report or the tables, does the data stay in some database, or are there other steps beyond the research? All too often those in research are not aware of what happens to their work.

Using the Data Immediately: Work Sessions

A lot of research is commissioned based on immediate problems. In the mind of those who execute the research the task may be completed once the research is reported, either in terms of some brief immediate overview (called a *topline*) or even further in a report. In the author's experience, more and more companies are moving toward sharing the information in a working meeting. This is called a *work session* or *topline presentation*. The objective of this working meeting is to surface the results, answer the questions in an informal way, and share the knowledge of the findings among the key groups who need the data. Often the research provider is asked to prepare a short outline of findings to present to the group at the work session. The session typically begins with a short overview presented by the researcher or the research supplier. Questions that surface during the session are usually key issues. The researcher is asked to answer these, more in terms of col-

laborative give and take rather than being formally examined. The work session can last several hours, but all work sessions have the same objective: namely, to answer the specific issues by going through the data with the individuals most knowledgeable about results. For the restaurant study presented here the work session comprised a meeting with the client (marketing), the advertising agency, and the research supplier. The objective was primarily for the different groups to understand the data and the implications, to refresh their memory of the execution of the study, and to answer key questions about the features of the restaurant. As in most work sessions the atmosphere was collegial and freewheeling, with questions arising and being answered in moments.

Writing the Report

After the heat of the moment, when the study has been digested, it is important to record the study in a formalized document that can be archived. This is the report. For the report the researcher may be asked to include all of the details or to simply provide an outline. The nature of the report, the depth of information, and the structure used are a function of the client demands. In recent years, and as the pace of business has accelerated, the reports have tended to become shorter, more factual, and less detailed. There is little room and little reward in speculating about the causes of behavior, because most clients need the information more than they need the insight.

Storing the Data and Results Long Term

For most concept studies where the objective is to identify which of the concepts win and which lose, the report and data are stored in a corporation's repository. This may be a formalized system, with knowledge-management software, or more often than not, a room where previous reports are stored. Occasionally, the company may have the abstract of the presentation on a form that is available in knowledge-management software. Some companies are moving toward archival storage.

Overview

When researchers do a concept study, and especially a conjoint study, they go beyond the traditional role of providing data that other people can use to make an informed decision. Conventional concept research generally provides a scorecard for a number of concepts, which the researcher then hands over to the marketer or to the product developer. With conjoint measurement, however, the researcher sits in a more powerful position. The conjoint approach lets the researcher into the consumer's mind in a way not previously possible. The data in this chapter represent the typical type of results that one might obtain in a business-driven exercise, and thus the case history is worth studying in detail. There are five key things to keep in mind:

1. The richness of the input
2. The ability to look at *meta-patterns* to discover the drivers of acceptance
3. The understanding of how people process information
4. The existence of segments and the implication for satisfying different groups
5. The ability to drill down into specifics so that the results show both patterns and specific winning elements that can be used immediately to build a concept

Reference

Meiselman, H.L. (2000). Dimensions of the Meal: The Science, Culture, Business and Art of Eating. Gaithersburg, MD: Aspen, pp. 223–244.

Chapter 14

Creating Products from Concepts and Vice Versa

Introduction

Concepts, which are blueprints about a product or a service, can be created in two distinct ways: top down or bottom up, respectively. The first way—*top down* creation—develops fully formed concepts. Conventional concept testing typifies the outcome of a top-down approach. The objective is to measure whether the concept does well or poorly. In contrast, the second way—*bottom up* development—identifies components of concepts, combines them by using experimental design, tests the combinations, and then creates new and better combinations from the winning elements. Bottom-up creation relies on a formal learning approach, identifies the hot buttons of a concept, provides direction, and then develops a database of elements for an ongoing stream of new product ideas. Conjoint analysis typifies a bottom-up approach.

Up to now this book has dealt with the issue of concept development without detailed reference to the product or even to the product development group. The practical question in the mind of a product developer is how to take the learning from the concept phase and translate it into a product. At least four direct questions are involved in using concepts to drive product development:

1. *Direction.* What are the features of the product?

2. *Latitude.* Should the product follow the concept faithfully, or is there wiggle room to change things around, remaining more or less faithful to the general guidelines of the concept, but adjusting the development in light of other realities that the concept cannot address?

3. *Expectations.* Do consumers even know what the product should be, based on a concept, or does this measure called the *product-concept fit* exist solely in the mind of the researchers, marketers, and product developers?

4. *Development strategy.* Finally, is the conventional method for creating products by following the dictates of a concept really the best way to proceed? What about turning the process around 180 degrees and creating concepts based on acceptable products? Would we be any better off?

This chapter presents two case histories intertwining concepts and products. The first case history deals with the very traditional approach of optimizing a concept and then fitting a product to that concept later. The product was a cracker with a Southwestern flavor. The second case history—sweet condiment—deals with conducting the concept and product studies simultaneously and on a post hoc basis creating new products and concepts.

Case History 1: The Southwestern Cracker—A Product to Fit a Concept

The best way to understand the traditional development of products from concepts is by means of a case history. This case history deals with the development of a line of savory snack cracker and chip products that have a decided Southwestern flavor. Keep in mind that there is no single flavor or standard of identity that can be called *Southwestern*, allowing the marketer great latitude in concept and in subsequent product.

Creating the Concept

Top-down development relies on insight, intuition, and just plain luck. To the degree that the principals involved in development and marketing know the market and seize opportunities, they will succeed in top-down development. To the degree that the market becomes competitive, continues to evolve, and sparks unexpected twists and turns, top-down development will fail because it fails to systematize knowledge and insight. In contrast, bottom-up development builds and tests many concepts, identifying particular hot buttons that can be later recombined into newer and better concepts. Bottom-up development is systematic, exhaustive, and potentially more innovative because it can combine concept elements from many different categories to generate wholly new ideas. This difference between the two approaches is worth reiterating here because there is no real Southwestern cracker, per se. There is the notion of a Southwestern product, but nothing definitive in the product to denote it as Southwestern.

The Southwest Concept

The concept study was run with 120 respondents who stated that they would be interested in a Southwestern cracker. The concepts were presented on a personal computer through a self-administered system. The respondent would see a concept and respond, and the computer would go to the next concept. The study comprised 40 concept elements in 60 different combinations. Each respondent evaluated a unique set of 60 combinations created from the same set of 40 elements. The concepts comprised 2–5 concept elements. The respondents rated each of the 60 different combinations (i.e., test concepts) in a randomized order, based on interest in the concept and appropriateness for either special occasions or daily consumption, or some occasion in between. The concepts comprised both text and pictures (see Figure 14.1). The rating on these two attributes provides a mini-snapshot of the performance of product features. Table 14.1 lists the utilities for some attributes from the total panel of respondents.

Nash a Nacho

Barbecue flavor fit for a Texan

Buttery and light

Comes in a can to prevent breakage

> How interested are you in buying this snack?
> 9=very interested 1=not at all interested

Figure 14.1. Example of a concept with text and picture.

Table 14.1. Elements for the Southwestern cracker/chip product and their associated utilities for two rating attributes: interest and special occasions (negative) compared with daily consumption (positive)

	Interest	Special vs daily		Interest	Special vs daily
Additive constant	*21*	*19*	*Additive constant*	*21*	*19*
Names			Visuals		
Cheese Bisquitos	14	−1	P2: Quesadilla	16	8
Nacho Lites	7	2	P1: Nacho—olive, meat	15	13
Taquitos	7	−4	P3: Hot loaf	15	3
Nachitos	3	4	P8: Tortilla, cheese, onion	8	9
Bisquitos	3	1	P7: Family	7	9
Spice Crackers	2	−6	P4: Cracker, cheese, tomato	4	9
Nash a Nacho	−1	1	P5: Tortilla + spinach	−1	3
Crack-o-Jack	−5	−5	P6: Waterfall	−1	−1
Taste/flavor			Package		
Old-fashioned delicious taste	6	9	Comes individually wrapped for		
"Cajun, spicy flavor"	5	2	freshness	8	−1
For a fiesta of flavor	5	−2	Three stacks within each box	4	3
Old West taste	4	−4	Recipes on every box	3	3
Barbecue flavor fit for a Texan	2	−1	Comes in a can to prevent damage	3	1
"Spicy, Tex/Mex flavor"	1	−2	Unique, vacuum-sealed container	1	1
Colorful taste	−4	−3	Comes in a resealable box	1	−2
Mmmm mmmm good	−4	−4	Comes in party-sized bag	0	23
			Easy-to-open can	−3	0
Texture					
Thick and cheesy	13	−4			
Thin and light	9	−1			
Melts in your mouth	5	−5			
Buttery and light	1	−3			
Crisp on the outside, moist on					
the inside	−2	2			
They go "crunch" when you					
bite them	−3	−4			
Crisp all the way through	−4	−6			
Thick and crunchy	−7	−1			

Results

Winning elements for the Southwestern cracker are simply defined as those that score high, preferably 9–10 or higher on interest. Keep in mind that the utility values are the incremental percentage of consumers who would be interested in the concept if the element were present. For instance, in packaging a hot button is "Comes individually wrapped for freshness" (+8). In flavor a modestly performing element is "Cajun, spicy flavor" (+5). There are no clear winning elements, but some elements do reasonably well.

As we have seen previously in a number of datasets, segmentation often reveals subgroups showing much higher utility values, because segments are more homogeneous in their preference patterns (Chapter 7). People differ in what they like. Often the biggest business opportunity comes from identifying smaller groups of consumers with homogeneous preferences. Since the preference patterns within each segment are homogeneous, there is a greater chance that there will be a single set of concept elements that appeal to this segment. Furthermore, the likelihood is greater that this set of concept elements will be consistent with each other. In contrast, when the researcher works with the full panel, competing segments are embedded in the population. These segments become checks and balances. They ensure that no concept element ever becomes exceptionally interesting; that is, when one segment of the consumers in the population may really like a concept element, the other segment may just not like that same element. The result for the element is an average utility value resulting from the two opposing camps.

There are three segments in this population of respondents, although with larger-scale studies more segments may emerge. The segmentation was done using the value $(1 - R)$ as the measure of distance between people, with R as the Pearson correlation coefficient between people, based on the utilities of their 40 elements. Segment 1 is interested in themes dealing with cheese (*Cheesers*), segment 2 is interested in *TexMex* themes, and segment 3 is interested in packaging and storage issues. Segmentation enables marketers to identify new groups of consumers, and developers to identify coherent themes for the creation of new products. Table 14.2 lists the winning elements for each of the three segments.

The final step in concept work creates concepts by recombining elements that were tested into new and potentially more powerful combinations. Optimization may create *close in* or *far out* combinations, depending on the nature of the elements tested. Optimization is not merely painting by numbers, but rather an intelligent use of data to create new combinations; that is, the developer, marketer, and researcher do not simply follow the data blindly but rather use them to select promising elements. Quite often these concepts go on to produce market leaders, simply because they combine two key features: consumer *desirability* (from the data) and *uniqueness* (viz., the concept elements have not been previous used in the particular category). Table 14.3 shows three newly synthesized concepts for the Southwestern biscuit snack.

Table 14.2. Winning elements for three segments: *cheesers, TexMex,* and *package oriented**

| | | Segment | | |
| | | 1 | 2 | 3 |
	Total	Cheesers	TexMex	Package oriented
Segment 1: Cheesers				
P2: Quesadilla	16	13	13	0
Nacho Lites	7	10	1	−4
Thick and cheesy	13	10	1	−1
Segment 2: TexMex				
P1: Nacho—olive, meat	15	9	18	2
P3: Hot loaf	15	6	17	5
"Spicy, Tex/Mex flavor"	1	−7	13	13
Cheese Bisquitos	14	4	12	7
Taquitos	7	−2	11	−7
P8: Tortilla, cheese, onion	8	2	11	4
Segment 3: Package oriented				
Easy to open can	−3	−4	−6	12
Comes in party-sized bag	0	−2	−2	10
Comes in a resealable box	1	−1	3	9

*Segment names were assigned after clustering.

Table 14.3. Three synthesized concepts for the Southwestern biscuit snack*

Most liked	Well liked, scores high on daily use	Well liked, scores high on special and occasional use
Cheese Bisquitos	Nachitos	Spice Crackers
Old-fashioned delicious taste	Old-fashioned delicious taste	Old West taste
Thick and cheesy	Thin and light	Thick and cheesy
P2: Quesadilla	P1: Nacho—olive, meat	P6: Waterfall
Comes individually wrapped for freshness	Comes in party-sized bag	Comes individually wrapped for freshness
Rating: Interest = 78	Rating: Interest = 53	Rating: Interest = 46
Rating: Special vs daily = 31	Rating: Special vs daily = 67	Rating: Special vs daily = 4

*P, picture in the concept.

Fitting a Product to a Concept

The natural second stage to the process creates a physical product to match the Southwestern snack biscuit concept. The product should fit the concept so that the product delivers what the concept promises and should be acceptable so that it will be eaten repeatedly. It does little good to create a product that doesn't fit the concept. However, it is important to create a product that is acceptable, as well, and not to follow the concept too slavishly. Following the concept in minute detail could possibly lead to a product that in the end is really unacceptable.

Homework: It Certainly Beats Guesswork, Although It Takes Longer and Costs More

The speed of business and the competitive environment dictate that for continued success the marketer create a system to produce consumer-acceptable, profit-making products. It is tempting to create products by "rifle shots" consisting of one's own guesses, under the mistaken belief that the marketer, product developer, market researcher, and sensory analyst really understand the consumers. Nothing can be further from the truth. The marketplace is littered with products developed from intuition, researched incorrectly, launched with fanfare, and producing nothing. Homework, homework, homework, and disciplined development must substitute for arrogant guesswork and a belief that speed to market alone suffices to win. Homework becomes even more important when competitors are lurking about, ready to steal one's customers.

One of the key aspects of homework in products is to test many options, not just one best guess. The same advice holds for products as it does for concepts. By testing many products, a developer creates a report card of each product on each attribute, for example, amount of a sensory characteristic such as flavor intensity, degree of liking, and appropriateness for a specific meal occasion. Consumers find it easy to rate products on attributes and can even rate the degree to which a concept fits the product that they are testing.

Category Appraisal: When the Competition Becomes a Developer's Guide

One use of a report card shows developers how liking or image changes with a sensory characteristic. For instance, if a panelist rates the different competitor and even prototype products on liking and use for a specific meal occasion (e.g., to accompany a glass of wine), then data can be plotted to show how the sensory attribute drives the liking and appropriateness rating (Moskowitz 1981a, 1981b). Figure 14.2 presents an example of this type of curve for "amount of visible spices on the surface of the cracker" to "lik-

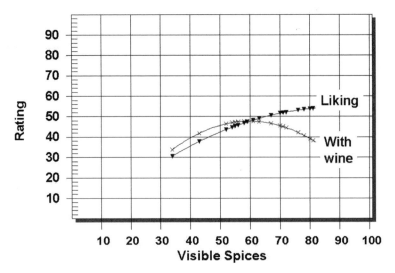

Figure 14.2. How "amount of visible spices" in the cracker prototype drives the rating of overall liking and the rating of "appropriate with wine." Each *triangle* or *x* corresponds to one of the prototypes. The curves are fitted.

ing" (increasing upwards), and "appropriate with wine" (inverted U shaped curve). The curves can even be created for sensory segments (Moskowitz et al. 1985), just as they are created for concepts (see Chapter 7).

Experimental Design of Biscuit Products and Product Optimization

Another approach—experimental design—is gaining favor in product development. When combined with concept–product matching, experimental design generates a concrete roadmap to develop a product that fits a concept. The developer identifies the physical variables under control (e.g., formulation and process) and then creates a set of combinations of these variables, which cover a range of different product alternatives. Unlike concepts, the variables can be continuous or discrete. *Continuous variables* are those that are of the same type (e.g., amount of flavoring). *Discrete variables* are those of different types (e.g., type of flavoring). Experimental design creates the specifications for different formulations, just as experimental design for concept work cre-

ates the specifications for the different combinations of concept elements.

The panelists evaluate the test products, rating them on the sensory, liking, and image characteristics (Moskowitz 1991). The data matrix then can be analyzed by regression in order to develop equations showing how the consumer-rated attributes and other measures, such as cost of goods, change with changes in the independent variables. The equations allow for interactions among ingredients and for nonlinearities. Equations summarize the relation between the independent variables and the dependent variables. Product developers can then identify the optimum combination of the independent variables that, in concert, generates a highly acceptable product within constraints of cost and with specific sensory characteristics. Figure 14.3 shows a schematic of the response-surface *hill*. Optimization finds the highest point on that hill subject to constraints (Khuri and Cornell 1987; Gacula 1993).

The study need not be limited to two variables, even though the graphic representation can only show two variables. The product model may encompass 3–10 variables or

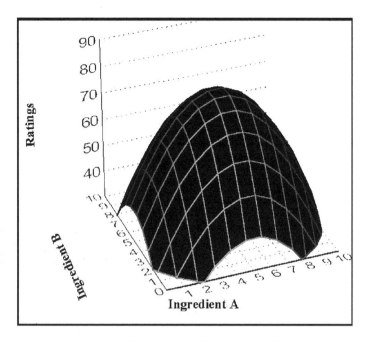

Figure 14.3. Schematic of a model, showing the relation between an attribute rating and two independent (ingredient) variables.

even more (Box et al. 1978; Moskowitz and Krieger 1998). The only requirement is that there be more products tested when the number of variables increases. Table 14.4 lists three product formulations, designed from the same product model and emerging from the same designed study. The researcher simply explores the different combinations by an optimization method, or even by brute force, to identify the formula combinations that satisfy the objectives. The concept to be fit was the maximally acceptable one shown in Table 14.3. Since the developer has done the requisite homework, it is now easy to create new and potentially more acceptable, cost-effective products by simply consulting the product model and imposing specific constraints on the solution.

Five Key Benefits to Fitting a Product to an Existing Concept

Throughout this chapter the emphasis has been on *homework*, viz., the systematic explo-

ration of concepts and products, using a bottom-up approach. There are five key benefits:

1. *Scope.* Marketers and developers are forced to consider a wide range of products, not just a narrow range. The increased scope increases the likelihood of success.

2. *Database.* Traditional development has proceeded in discontinuous steps. The data from one development project may not be easily combined with those from another project. The systematic method creates a database in which one can link and compare different projects.

3. *Corporate learning and memory.* All too often developers in companies do not have access to corporate knowledge because the development and evaluation studies comprise disparate, unconnected pieces of information designed to answer unconnected problems. Experimental design creates a valuable database that promotes ongoing corporate learning. It is not unusual for a company to refer back to, and to use, the results of concept and product studies for a half-decade or more,

Table 14.4. Three optimal products formulated by systematically varying four different ingredients*

Maximize	Liking	Fit to concept	Liking
Constraint	*None*	*None*	*Fit to concept > 67*
Ingredient			
A	1.00	1.95	1.43
B	3.00	1.00	1.00
C	2.42	2.05	2.23
D	1.00	2.16	1.04
Cost of goods	257	259	200
Overall liking	66	58	63
Image			
Older	57	62	61
Sophisticated	51	45	49
For snack	49	55	55
Fits concept	62	71	67
Sensory			
Size	46	51	49
Visible spices	31	32	33
Flavor intensity	35	37	36
Coarseness	51	56	52

*Each combination comprises a specific combination of these ingredients and satisfies several objectives. The sensory objectives come from sensory attributes rated on an anchored 0- to 100-point scale. Cost of goods is on its own scale.

simply because many of the problems can be addressed by looking at new configurations of concepts and products. The corporate learning is forced to be systematic, and the information is warehoused, simply because of the breadth of stimuli that are evaluated and the clear value of the data.

4. *Segments.* Segments emerge from understanding the pattern of responses to concepts and products, respectively. Whereas traditional segmentation is imposed on the data according to information about geodemographics and product usage, segmentation from this type of systematized development study emerges out of the *actual data themselves.* These data, pertaining as they do to the physical concepts and products, provide a valuable basis on which to divide consumers. One need not attempt to make hard-to-intuit bridges nor create abstruse intervening vari-

ables between a generalized segmentation (e.g., according to lifestyle) and immediate subjective responses to the specific, concrete product.

5. *Ongoing applicability of the data for subsequent decisions.* The author's experience is that once managers and developers in the corporation adopt the systematic approach and experimental design, they return to the data again and again. Often the data are used for years afterward, mined to answer new questions and take advantage of new opportunities not even recognized at the time the research was conducted.

Case History 2: A Sweet Condiment—Concept and Product Simultaneously

The Concept May Not Provide the Proper Guidance

The previous case history with Southwestern snack biscuit represents the now-standard approach to product development. It assumes a rational, systematic process for staged development wherein the concept is first developed and becomes in turn a blueprint for the product. This is all well and good. Sometimes it works. Sometimes it doesn't.

From time to time, professionals in the food industry and in other fields may observe to their consternation that consumers often don't know how to describe what they really want in a product, although the consumers do describe something, and to watch the consumers one might conclude that they know exactly what they want. Expressing this observation more concretely, such as what happens in the case of coffee (see Chapter 10), virtually all consumers say they want a rich, robust cup of coffee. The concept of the coffee in their minds is the same from person to person, or at least that's what they say. The conflict between mind and tongue becomes even more obvious when we deal with sensory preference segments. Consumers who differ dramatically in their sensory preferences based on product tests generate similar-looking profiles for their self-defined ideal coffee. The consumers who prefer a weak-flavored coffee in taste tests do not say they want their ideal coffee to be weak. They want a strong, rich, robust flavor, often with a slight nutty note. These observations, repeated time after time, year after year, in category after category suggest a missing connection between what consumers say they want a cup of coffee to taste like and what they actually like in taste tests when hapless product developers attempt to grant their wish.

The discrepancy between the mind and the tongue, the concept and the product, can be further brought home in another set of studies, this time with a condiment sauce. The same respondents evaluated both concepts and products, in different parts of the research sequence, following the standard best practices that market researchers and sensory analysts follow. The concepts suggested preferences for one type of condiment sauce, whereas the products suggested preferences for another type of condiment sauce. The bottom line for this research is that, although the preference patterns for concepts and products are clear, the patterns do not match each other.

On a personal note, it is quite disturbing at first to discover these discrepancies between what people like in a tasting study and what they say they want in a concept study. All too often the researchers involved are ready to throw up their hands in despair, because superficially it seems that one of the two tasks (rating one's sensory experience and responding to concepts) had to yield invalid results.

Business Experience: Creating Products to Meet Descriptions

It does no good to bewail that concept and product preferences quite often do not match each other. On a scientific basis this is important because the mismatch tells researchers that there is no necessary connection between

consumer wants as they describe them and the reality of sensory preferences. The matter becomes considerably more complicated when the task turns to product development. Precisely what should product developers create? Should the development focus on fulfilling the concept or maximizing product acceptance? Fitting expectations means being loyal to the concept; maximizing acceptance means possibly driving repeat purchase. The practical implications are enormous for marketers who face an increasingly competitive environment. What draws consumers to the product (i.e., the concept and the advertising) is not what keeps them coming back. Furthermore, there aren't very many chances to fail. It's hard to come back to management after muffing the opportunity. Thus, the dilemma: what should developers do?

This case history shows a way out of the dilemma. It deals with the simultaneous creation of both systematically varied concepts and systematically varied products. The objective is to identify what to talk about on both a sensory description and a positioning basis, and then what product to create so that it is both acceptable and more or less matches the concept. *What the case history shows is that it may be better to create the product first and then match the concept to the product rather than going in the traditional direction of matching product to concept. This is not the received wisdom, but does represent an alternative, potentially viable strategy.*

The case history deals with a new condiment. The original condiment fell into the category of *savory*, but management recognized an opportunity to extend the condiment to the *sweet* area. The original business objective of this research was to develop a new type of condiment in a fast-growing category. The condiment market in the United States is growing rapidly, especially for condiments that are low in calories.

The manufacturer faced two problems:

1. *What to say about this new condiment.* The condiment was a sweetened version of a

current in-market savory product, but initial qualitative research suggested that the notion of *sweet* itself was not a driver of acceptance; that is, the product concept to guide research and development (R&D) would itself not be acceptable. The product might taste delicious, but the concept was so novel that it might not be as good as the product would be.

2. *What to do physically to the new condiment to make it taste good, but also to accord with the concept.* The sweet taste was a strong driver of acceptance, once the consumers tried the product. However, what was the product developer to do to create a delicious product in light of the concept being novel and thus perhaps not guiding the product in the right direction.

Boiling the issue down to its simplest form puts the marketer and the developer into the dilemma. The concept that would win in this type of study might well guide product development in the wrong direction. It is worth noting that this dilemma occurs more frequently than one would like to admit. The stepwise creation of products from concepts might, in many cases, be the absolutely incorrect approach. The questions can be simply stated: What should a manufacturer say about a product in order to excite interest? And what should a manufacturer do about the product to maintain initial acceptance and repeat purchase? The solution would be easy, except that communication motivators often have little to do with sensory motivators.

Simultaneous Testing and Optimization of Products and Concepts to Address the Dilemma

One approach to dealing with the dilemma has the same respondents evaluate both concepts and products in an extended test session lasting 3 hours. During this time the respondent evaluates the concepts and the products separately. The concepts are rated on interest. The products are rated on acceptance, as well as a variety of both sensory attributes and im-

age attributes. In this way the researcher can build a concept model that shows how the elements of concepts both drive acceptance and fit to various end states or images. Thus, it becomes possible to adjust the concept in a number of different directions, not only in the direction of best liked. We see echoes of this strategy when creating concepts by using first principles (Chapter 18) or when creating concepts that maximize acceptance, but also deliver against a semantic profile. The researcher can also test systematically varied products so it becomes possible to engineer the product to be highly acceptable, but at the same time fit an image profile. The image profile is provided by the ratings on a set of image attributes, with the product rating on these image attributes.

Concept Stimuli

The concept stimuli comprised 118 different elements (Table 14.5). Each respondent rated only a limited set of elements, embedded in 100 concepts. The concepts were set up according to the IdeaMap system (Chapter 5).

Product Stimuli

The products comprised 27 systematically varied condiment formulations, comprising variations of nine ingredient variables, following a fractional factorial design. Given the great number of independent variables, a three-level, Plackett–Burman screening design had to be used (Plackett and Burman 1946). This design enables researchers to explore many different combinations and thus produce a variety of prototypes with different sensory properties. The design ensures that ingredients appear statistically independently of one another, and each ingredient appears equally often at every level with every other ingredient. The experimental design enables researchers to create a model relating the level of a physical formulation to attribute ratings, be these liking, sensory, or image at-

Table 14.5. An example of concept elements for the sweet condiment product

Party setting (visual)
Lifestyle: drinks (visual)
Gourmet chips (visual)
Deliciously different
A tangy new taste
A fresh new taste
Sweet and spicy all in one bite
Great mix of sweet and heat
Sweet with a kick
Chunky
Thick 'n rich
Thick 'n chunky
Hearty chunks
Great with fish
Brings chicken to life
Great for snacks
Only the best
Highest-quality ingredients
Quality ingredients
Only the finest ingredients
Made with fresh ingredients
A more nutritious way to snack
Low in sodium
For those who like to try something different
For the gourmet in you
For the whole family

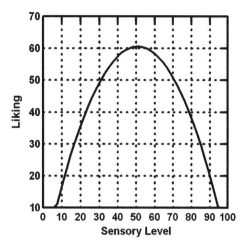

Figure 14.4. Prototypical curve showing how the sensory-attribute level drives liking.

tributes. The model is created by conventional regression modeling, using linear, quadratic, and significant cross-terms (Box et al. 1978; Khuri and Cornell 1987).

Drivers of Sensory Liking

Scientific research over the past 30 years suggests that, as a sensory attribute increases, liking first increases, then peaks, and then declines (Moskowitz 1981a, 1981b). The typical function relating sensory attribute to liking follows an inverted U-shaped curve of the form shown in Figure 14.4. The nature of the curve depends on the particular product, the respondent, and the sensory attribute being investigated. Figure 14.5 shows the data for three different sensory attributes. We saw these patterns, as well, in Figure 14.2, when we dealt with the Southwest biscuits.

The Issues in Creating a Product Based on Liking versus Creating a Product Based on a Concept

The presupposition in this chapter is that the development process is systematic and that in most cases the concept development precedes the product development. When the products and the concepts are evaluated at the same time, though, the concept cannot serve as the goal against which the product is formulated. As we shall see, this is both a problem and a major opportunity:

1. *Problem.* The problem is that there is no immediate goal against which to direct product development. This lack of a goal can cause great discomfort, especially among those who need a structure.

2. *Opportunity.* The opportunity, however, is great, as well. Sometimes the development goal as provided by the concept can be too limiting. Sometimes the concept can be too narrowly defined because that is what consumers are accustomed to. The narrow concept definition (i.e., the narrow range of sensory attributes that respondents want, based on responses to communications) is translated into a correspondingly narrow

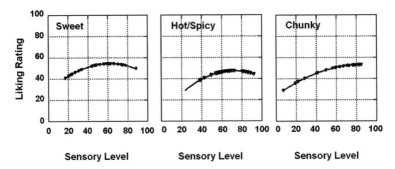

Figure 14.5. Relation between sensory-attribute level (*abscissa*) and liking rating (*ordinate*) for a sweet condiment. The curves are fitted, and all points are brought to the curve.

sensory range in the test product. In some cases this narrow concept range can mislead and make the product developer overly conservative when it comes to formulating a product.

Development Direction from Tasting versus Development Direction from Reading

We can compare the direction from the product and the concept portions of the study for three specific attributes key to product acceptance: sweetness, spiciness, and chunkiness. These three attributes are presented in Figure 14.5. We first deal with the directionality from the product portion of the study and then with the directionality from the concept portion of the study. The question we want to keep in mind is whether the two types of directionality given to the developer agree with each other.

Basic Development Direction from the Product Evaluation

Figure 14.4 suggests that the optimal sensory level should be a middle level. This is generally the case, keeping in mind that the figure is schematic and represents the common sensory attributes. Consumers might say that they want a very high or a very low sensory level, for example, when the profile

is their self-designed ideal, but quite often the most acceptable sensory level is midrange for many attributes. This optimal level can be identified by looking at the relation between liking and sensory-attribute level on an attribute-by-attribute basis (Figure 14.5). When it comes to the sweet condiment we see a similar inverted U-shaped curve for sweetness versus liking, a somewhat inverted U-shaped curve for heat/spiciness versus liking, and a very chunky/thick product for chunkiness versus liking.

Product Sensory Segments and More Precise Development Direction

Let us go one step further in our analysis of product data and look at sensory segments. Substantial variability exists in what consumers like. Some of this variability is simply random differences from consumer to consumer. Like the concept-response segments that we have seen in Chapter 7, there are sensory preference segments in products (Moskowitz et al. 1985). An increasing body of evidence suggests that preference patterns in the population may fall into a limited number of classes [e.g., likers of spice versus haters of spice (Moskowitz and Bernstein 2000)]. In at least one case (pickles, by the US manufacturer Vlasic), sensory segmentation has been profitably applied to the creation of a line of condiments to appeal to

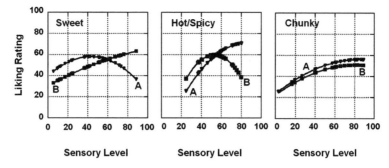

Figure 14.6. Relation between sensory-attribute level (*abscissa*) and segment liking rating (*ordinate*) for a sweet condiment. The curves are fitted, and all points are brought to the curve. Segment A, spicy seekers; and segment B, sweet seekers.

three sensory segments identified through product evaluation.

Using the sensory ratings and the ratings of liking, segmentation algorithm revealed three distinct groups. Segment A likes the hot/spicy taste and shows the highest liking at high spiciness. Segment B likes sweet taste and shows the highest liking at the highest sweetness. Segment C shows no clear pattern of liking versus sensory attributes. We won't deal with segment C any more. Figure 14.6 shows how the sensory-liking curves differ for the two key segments—spicy seekers versus sweet seekers—and thus how the optimal sensory level differs.

The sensory segments provide two product directions: a strong, sweet tasting but moderately spicy condiment versus a strong, spicy, moderately sweet-tasting condiment. It is clear from Figure 14.6 that these two segments differ radically in their sensory preferences, and most likely the product accepted by one segment would differ dramatically from the product accepted by the other.

Development Direction from Responses of These Sensory Segments to Condiment Concepts Rather Than Products

A nagging question for the researchers is whether these sensory segments, with such clearly different taste preferences, would

show those same polarizing differences in their responses to concepts? We see from Figure 14.6 how different they are when confronted with a food. Do these same differences show up as clearly at the concept level? Since the same respondents evaluated both concept and product, we can look at how these two radically different sensory preference segments, obtained from the pattern of product liking, respond to concepts that also present sensory characteristics. The logic is straightforward:

1. We know from the sensory-liking curves the approximate sensory level (low, medium, or high) that respondents from a specific segment would like most.

2. We also know, for the same sensory segment, how they respond to descriptions of the sensory characteristics.

3. Thus, it becomes possible to discover what a sensory segment likes from *tasting* versus what a sensory segment likes from *reading* about the product.

4. It turns out that the sensory segment likes a strong product from tasting, but likes descriptions of medium intensity, not strong intensity; that is, people may like stronger products from tasting, but don't want to read about the strength of what they like to taste.

5. Thus, what appeals to them in a description of the product may suggest a weaker product than what they actually like.

This would imply that the product direction is "less correct" based on concepts than on actual tasting.

We can see the support for this logic from the empirical data. The point has already been raised that people tend to describe their ideal product in pretty much the same way, especially for coffee, where everyone wants a robust coffee, whether they like mild or bitter coffee. Let us continue the analysis with the condiment data in order to show how that truism and product development paradox applies here.

We begin with what people say that they want from concept research, based on the concept data. Table 14.6 shows that what people say they want from concept research tends to underrepresent their real preferences when they are actually confronted with a food. For example, when it comes to the actual spicy taste, the sweet-seeking segment behaviorally dislikes the strong spice when they evaluate the product and prefer the weaker spice taste (column D). Yet, when it comes to direction from the concept (column E), these sweet seekers react modestly positively to statements about spice. From the concept data we would conclude that the sweet seekers want a fair amount of spice in their condiment, whereas from actual tasting we see that they do not want this spicy taste. By looking at the direction for product development in Table 14.6, comparing what one might conclude from the concept research versus the product research, we see some notable differences.

To sum up the issue, one gets a sense that the respondents are more *centrist* when direction is taken from their response to concepts, and more *extremist* when product development is taken from their response to actual products. The bottom line is that the product-based or so-called sensory preference segments show much stronger desire for impactful products, with clear high or level sensory levels, than one might think from simply looking at their responses to con-

cepts. The concept data generate a much more restrained description of the product than the developer might deduce from looking at their reaction to actual products.

Implications for Commercial Product Design, Prototype Development, and Advertising

In most companies product design proceeds first from the concept stage and onward to actual formulation. Concept work may involve focus groups, ideation of new product configurations, concept evaluations, and the like. The concepts are then tested to determine whether they have potential. Finally, the products are created from the winning concept, as R&D labors to deliver what the concept promises. This traditional approach may have to be reexamined in the light of differences in development direction emerging from tasting as compared with concept evaluation.

There are five implications from this case history for the conventional process that comprises product definition followed by formulation and finally advertising the product:

1. *Concepts may understate the magnitude of a sensory characteristic needed to drive acceptance.* Directionality from concepts is more conservative than directionality from products. We see the conservatism when we compare the relatively muted levels of winning concept elements describing sensory properties with more intense sensory levels emerging from the tasting of actual prototypes. This conservatism shows up very strongly in the data from the sensory preference segments.

2. *Guidance in product development should come from the product first, not the concept first.* It is tempting to stage the development process in an orderly way, with concepts preceding products. That could lead to disaster. Consumers should taste the product. From the pattern of their responses (i.e., the sensory-liking curve), the developer should

Table 14.6. Comparison of development direction obtained from product tasting and from concept reading

Sensory segment		Interest		Segment	
A	Direction from B	Concept-based direction D	Product-based direction E	Concept-based direction F	Product-based direction G
Sensory category	Sensory attribute	Utility value	Target sensory level	Utility value	Sensory level
Chunky	Thick 'n rich	1 = Low	High	6 = Intermediate	High
	Thick 'n chunky	−1 = Very low	High	5 = Intermediate	High
	Rich 'n chunky	4 = Intermediate	High	7 = Intermediate	High
	Hearty chunks	4 = Intermediate	High	5 = Intermediate	High
	Chunky	0 = Intermediate	High	3 = Low	High
	Average	*2 = Intermediate*	*High*	*5 = Intermediate*	*High*
Spice	A zesty, lively condiment experience	5 = Intermediate	Intermediate	10 = High	High
	A hot and spicy condiment	2 = Low	Intermediate	8 = High	High
	A livelier taste in condiment	1 = Low	Intermediate	7 = High	High
	A big, bold taste in condiment	8 = High	Intermediate	4 = Intermediate	High
	Zesty blend of fruit and spice	2 = Low	Intermediate	7 = High	High
	A tangy new taste	8 = High	Intermediate	2 = Low	High
	Average	*4 = Intermediate*	*Intermediate*	*6 = Intermediate*	*High*
Sweet spicy	Perfect balance of sweet and spicy	3 = Low	High	5 = Intermediate	Intermediate
	Sweet with a kick	4 = Intermediate	High	8 = High	Intermediate
	Sweet and spicy all in one bite	2 = Low	High	6 = Intermediate	Intermediate
	Sweet and spicy together	1 = Low	High	4 = Intermediate	Intermediate
	The sweetness of fruit combined with the spice of condiment	2 = Low	High	5 = Intermediate	Intermediate
	Great mix of sweet and heat	2 = Low	High	3 = Low	Intermediate
	A sweet and hot fruit condiment	−1 = Very low	High	5 = Intermediate	Intermediate
	Naturally sweet and spicy	2 = Low	High	3 = Low	Intermediate
	Average	*2 = Low*	*High*	*5 = Intermediate*	*Intermediate*

determine what to do so as to optimize the product. Describing what one wants, or picking out the description from a set of concepts will not do, or may in the worst case misguide the developer.

3. *The concept development process should use the same sensory ideas expressed in different ways through many elements, not just one element.* It is tempting to screen the concepts down to a promising few, select the winner,

and then begin the product development step. It is more productive to cast a wide net in the concept elements and find sets of winners from this wide array of contenders. The multiplicity of *similar* elements as winners is more comforting than one winning element alone. For example, there are many ways to talk about spicy, bold, and high impact. Some of these concept elements work better than others in exciting consumers and thus giving the product developer direction. Some of them do not work as well. Rather than limiting oneself to the study of one or two different concepts or elements to guide development, the researcher should explore many dozens of different ways to talk about spiciness, dozens of ways to talk about texture, and so on. It is only from the pattern of the responses to these many different elements that the researcher can really identify what sensory characteristics the product should have. A wide scope, not a narrow scope, in concept development is to be sought.

4. *A within-subjects design, with the same people evaluating products and concepts, is preferable to the staged approach that uses different respondents for concept testing versus product testing.* The most efficient development design calls for the same respondent to evaluate both product and concept in order to identify what the product should be from tasting and at the same time in order to identify what the product positioning should talk about from concepts. By following the within-subjects paradigm, the developer and the researcher can be sure of obtaining both sensory direction and positioning/conceptual design direction. The potential discrepancies between what people say they want and what they actually want can be immediately identified rather than being allowed to misguide development.

5. *Strategy for advertising.* One of the roles of advertising is to predispose consumers to purchase a product by presenting a *word picture* of the sensory experience to be enjoyed. We see that sensory preferences are strongly segmented. It is in the conservative nature of manufacturers not to take advantage of these segmented preferences in advertising, but rather to appeal to the great middle group, where most of the consumers are thought to reside. Rather than striking out for the concept elements that turn on some consumers, but turn off others, the typically conservative manufacturer selects those sensory phrases that are acceptable to all parties. This strategy would be effective if the concept elements were to be as powerful to consumers as are actual product experiences. However, this is not the case. First, the range of acceptance of concept elements is narrower than the range of acceptance of products. Consequently, there is an initial loss of potential acceptance by beginning with the concept and moving on to the product. Second, given the conservative nature of manufacturers, there is the tendency to regress to the middle range of acceptance, where no segment of consumers is turned off. The manufacturer has thus twice constrained the potential of the advertising to convince consumers. The first constraint occurs because most sensory phrases do not properly convey the potential for sensory satisfaction. This is the problem of concept promise as compared with product delivery. The second constraint occurs because most manufacturers opt for a conservative middle ground and choose the weaker, nonpolarizing concept elements. This is the problem of overly conservative marketing strategy.

References

Box, G.E.P., Hunter, J., and Hunter, S. (1978). Statistics for Experimenters. New York: John Wiley and Sons.

Gacula, M.C., Jr. (1993). Design and Analysis of Sensory Optimization. Trumbull, CT: Food and Nutrition.

Khuri, A.J., and Cornell, J.A. (1987). Response Surfaces. New York: Marcel Dekker.

Moskowitz, H.R. (1981a). Relative importance of perceptual factors to consumer acceptance: linear versus quadratic analysis. Journal of Food Science, 46: 244–248.

Moskowitz, H.R. (1981b). Sensory intensity vs hedonic functions: classical psychophysical approaches. Journal of Food Quality, 5: 109–138.

Moskowitz, H.R. (1991). Optimizing product acceptability and perceived sensory quality. In: Graf, E., and

Saguy, I.S., editors. Food Product Development. New York: Avi, pp. 157–188.

Moskowitz, H.R., and Bernstein, R. (2000). Variability in hedonics: indications of world-wide sensory and cognitive preference segmentation. Journal of Sensory Studies, 15: 263–285.

Moskowitz, H.R., Jacobs, B.E., and Lazar, N. (1985). Product response segmentation and the analysis of individual differences in liking. Journal of Food Quality, 8: 168–191.

Moskowitz, H.R., and Krieger, B. (1998). International product optimization: a case history. Food Quality and Preference, 9: 443–454.

Plackett, R.L., and Burman, J.D. (1946). The design of optimum multifactorial experiments. Biometrika, 33: 305–325.

Chapter 15

Exploratory Modeling and Mapping, Simulating New Combinations, and Data Mining

General Introduction

Concept research provides a fertile area for advanced analytic methods. Over the past several decades, interested researchers and computer programmers have created a variety of different methods by which to analyze the concept data. Some of these methods involve *mapping*, or showing stimuli (concept and product) on a geometric map such that those stimuli lying close together are qualitatively similar. Other methods, following a third direction, use the data to synthesize newer and better concepts or to estimate how respondents would react to new concept ideas. This is called *simulating*. Still other methods, going in a third direction, involve relating response profiles to other aspects of the respondent. This is called *data mining*. These three topics each could have a book devoted to them because of their popularity. We will touch on each of these only in passing, to introduce them as they relate to concept development and testing.

Part 1: Mapping and Optimizing

The Role of Mapping in the Analysis of the Geometric Representation of Concepts

Mapping—that is locating test stimuli in a geometric space—constitutes one of today's most popular ways to represent data and, by so doing, uncover hitherto hidden relations among stimuli. People are naturally visual, responding to pictures faster than to words and perceiving relations that could not have been as clearly expressed by words. It is easy to show stimuli as points in space, with the distance between the points reflecting subjective dissimilarity. Those who look at the geometric representation often feel compelled to label the axes. Once they have created the labels, they then look for underlying patterns and rationales that make intuitive sense. Occasionally, powerful insights emerge. Even without insights the mapping exercise can summarize relations efficiently among the stimuli.

Mapping is often done in conjunction with physical stimuli, on the one hand, or with brand names, on the other. Thus, the researcher might map actual products, whether these be commercially available products or test stimuli. The respondents might directly judge the perceived dissimilarity between pairs of brand names on a white card or, the converse, perceived similarity. Other types of mapping use *attribute profiles* of the products or the brands, from which distances are developed and maps created. When it comes to brand names, these are treated like actual products. Both products and brand names are *gestalts*, or "wholes," that are not deconstructed to their components. For the most part, looking at the history of mapping one gets a feeling of a strategy to understand stimuli that researchers could not otherwise deconstruct into components and then vary.

Although mapping is a very powerful method, increasingly popular among the researchers in the product testing community, the question is "What is the place for mapping in concepts?" Mapping assumes that the products themselves are of great interest and that one can learn more about products by showing their juxtaposition on a map or their projects on some axes in the map (Greenhoff and MacFie 1994; Vigneau et al. 2001). Furthermore, one can optimize products by using the mapping (McEwan 1996). Concepts, however, for the most part are not of such great interest unless these are final concepts reflecting different products in the market. In this happy situation the concepts stand in for the products. In most concept research, however, there is no real need for the concepts until they have been passed by research so that they become blueprints for new products and services. Until then, the concepts, like prototypes, are simply test stimuli. If the prototypes are just best-guess "rifle shots," then they, like test concepts, are not typically mapped. There is little to learn from these prototype products to test concepts, because they have no clear business validity since they are not in the market.

Going a Step Further: Applying the Mapping Exercise to Concept Elements and Their Systematically Varied Combinations

Systematically varied concepts transcend single concepts because the concepts have been created according to a structure. The structure itself provides insight. When subjected to mapping, these systematically created concepts may generate insights, especially when there are two pieces of information about the concept to use in the mapping:

1. Basic interest level, as measured by utility values

2. Profile of each element on a set of semantic dimensions to provide a *signature* of the element

The particular study to be mapped comes from a set of studies on beverages called It! studies, described more fully in Chapter 20. The data to be mapped concerns red wine. The study itself comprised 36 elements, spanning a wide range of alternative statements about

product features, emotions, brands, etc. The study generated data from 220 respondents.

At the time of the Drink It! study, the elements were created and combined into concepts, and the concepts were tested in a conjoint evaluation over the Internet. There was no mapping involved. The elements were not profiled on any semantic scales to enable such mapping. The original objective of the Drink It! study was to identify the drivers of acceptance at the concept level. Since the elements were systematically combined into concepts, there was no need to use any additional information to understand the performance of these elements on concepts. The elements by themselves would do, because one could assess the performance of each element.

Let us go beyond the systematic experimentation that characterized experimental design and combine it with mapping to create a different class of insights. To explore what additional information can be gained from mapping, one first needs to profile the elements from the red wine study on semantic scales. We first dimensionalize the wine elements, following the approach discussed for IdeaMap (Chapter 5). For the elements a small group of 12 respondents profiled each of the 36 wine elements on a set of six attributes. Although in IdeaMap the dimensionalization had been done prior to the actual field evaluation, when every respondent evaluates every attribute the dimensionalization can be done at any time. The dimensions will be used for mapping rather than for estimating the utilities of untested elements. Therefore, the dimensionalization is simply an additional analysis done any time before, during, or after the fieldwork.

The dimensionalization provides some additional insight about the concept elements because in a sense the exercise generates a signature of each element, albeit on only six scales. There could have been more scales used to create a more detailed profile of each concept element, but the six chosen will suf-

Table 15.1. The 36 concept elements for red wine, their locations on six semantic scales, their utility values, and location on two factors derived from the semantic* scales

A	B	C	D	E	F	G	H	I	J	
	1 = 9 =	Men Women	Alone Together	Cheap Expensive	Common Unique	Young Old	Morning Night	Utility Total	Factor FA	FB
E01	Wine with a pale ruby color and a fruity taste	6.9	5.5	5.3	4.6	3.4	5.2	2	1.00	3.01
E02	Pale pink colored wine with fizz and the flavor of fruit	8.0	6.0	3.9	4.6	3.1	5.2	−11	0.61	3.84
E03	Deep red wine with a sweet complex fruity flavor and delicate flowery fruit aroma	7.1	5.4	6.9	6.3	4.8	6.0	6	2.64	3.25
E04	Ruby red color with a black currant flavor and a hint of oak	4.2	4.8	6.3	6.0	6.3	7.4	6	3.55	1.04
E05	A heavy red wine with a warm plumy fruit flavor and a little bit of dryness	4.3	5.6	5.9	5.7	7.3	7.0	12	3.17	1.41
E06	A pale red wine with a light flavor of raspberries . . . so soft and sweet	7.5	5.8	5.2	5.5	3.9	4.9	−3	1.29	3.34
E07	A chilled light bodied red wine with a fresh fruity flavor	7.2	4.7	4.6	4.7	4.5	5.3	3	1.12	2.91
E08	Red wine with added antioxidants	6.3	4.2	5.1	6.1	6.8	6.7	−3	2.67	1.96
E09	Bubbly fresh and sweet with a hint of raspberry tart flavor	7.1	5.2	3.6	4.3	3.9	5.3	−6	0.68	2.91
E10	Hits the spot on a hot summer day	4.4	4.6	3.5	4.2	3.4	4.3	−1	0.53	1.19
E11	Goes down smooth and easy	5.3	4.5	6.7	5.1	6.0	5.8	3	2.21	1.83
E12	Made in the tradition of the greatest wine producers all over the world	4.7	5.9	8.1	6.8	7.5	6.6	5	3.73	1.96
E13	Premium quality	5.0	5.7	8.4	7.4	7.3	6.9	2	4.08	1.90
E14	So refreshing you want to savor how it makes you feel	5.3	3.7	6.2	5.3	5.8	6.1	1	2.40	1.40
E15	100% natural	5.7	5.2	6.5	5.9	6.0	5.9	3	2.44	2.11
E16	Dry, Sweet, Semi Sweet . . . whatever you're looking for	4.2	5.9	4.5	4.2	4.9	5.7	4	1.56	1.61
E17	You can imagine the taste even before you drink it	4.6	3.7	6.8	5.7	4.4	6.1	0	2.21	.01

(continued)

Table 15.1. The 36 concept elements for red wine, their locations on six semantic scales, their utility values, and location on two factors derived from the semantic* scales *(cont.)*

A	B	C	D	E	F	G	H	I	J	
	1 = 9 =	Men Women	Alone Together	Cheap Expensive	Common Unique	Young Old	Morning Night	Utility Total	Factor FA	FB
E18	So refreshing . . . you have to drink some more	5.0	4.9	5.6	5.8	4.3	5.7	0	1.81	1.54
E19	Quick and fun . . . ready to drink, on the shelf, no bartender required	3.7	7.2	3.1	3.1	2.7	5.9	−3	0.62	1.95
E20	When you think about it, you have to have it . . . and after you have it, you can't stop drinking it	3.7	3.4	5.7	5.2	4.1	6.6	−3	1.95	0.28
E21	Simply the best	5.3	6.1	7.8	5.7	5.7	6.3	4	2.76	2.27
E22	Relaxes you after a busy day	4.7	2.3	4.9	4.5	6.4	6.8	3	2.11	0.02
E23	A great way to celebrate special occasions	5.7	8.3	7.0	5.3	5.9	6.5	6	2.89	3.55
E24	Looks great, smells great, tastes delicious	4.9	5.0	5.9	5.1	4.6	6.2	2	2.05	1.68
E25	A wonderful experience . . . shared with family and friends	6.3	8.3	5.7	6.1	6.8	6.9	3	3.09	3.94
E26	Pure satisfaction	5.0	3.6	6.9	6.0	5.3	6.7	2	2.67	1.08
E27	It quenches THE THIRST	4.1	3.9	4.3	4.7	3.3	4.4	−3	0.85	0.86
E28	From Northern California	5.3	5.4	6.3	4.1	5.6	6.0	2	1.91	1.92
E29	From Gallo	4.8	4.8	4.9	3.1	4.7	5.5	−2	1.09	1.46
E30	From Kendall Jackson	5.1	5.9	6.0	5.1	5.5	6.5	2	2.43	1.90
E31	Imported from France	5.7	5.8	8.0	6.3	7.2	6.4	2	3.48	2.34
E32	Tastes great . . . without the taste of alcohol	7.3	6.1	4.3	5.6	5.6	5.8	−8	1.38	3.46
E33	Multi serve containers . . . so you always have enough!	4.3	7.3	3.5	4.9	3.1	6.0	−5	1.27	2.52
E34	Icy cold	4.7	5.7	4.7	3.9	4.1	4.8	−8	0.77	1.79
E35	Resealable single serve container . . . to take with you on the go	4.0	3.1	2.7	5.0	2.6	5.1	−5	0.53	0.18
E36	With the safety, care and quality that makes you trust it all the more	6.3	5.6	6.1	5.5	6.9	5.9	0	2.44	2.57

*FA and FB are the two factors emerging from principal components factor analysis of the semantic profiles done on the concept elements for wine.

fice for the illustration. The concept elements were profiled on anchored 9-point scales. Nonetheless, there is additional information about the *tonality* of the concept elements that one obtains simply by looking at the profile for each element and by comparing all elements on one dimension. For the mapping, we will use the scale results. Table 15.1 presents an example of the dimensionalized elements. The dataset comes from the evaluation, on the Internet, of systematically varied concepts by a panel of 220 respondents.

Representing Elements on the Map: Results and Limitations

Mapping at its simplest level comprises the geometric representation of the data along a limited number of dimensions, preferably orthogonal or independent ones. Limits to the ability to visualize data drive the creation of maps having 1–3 dimensions. The most frequently presented maps have two dimensions. It's not that two dimensions work so much, but rather two dimensions are simply the most convenient representation. Four-dimensional representations can-

not be readily visualized, three-dimensional representations are hard to visualize, and one-dimensional maps are simply not particularly satisfying to researchers and to their audience.

We can get an idea of the outcome of mapping from a simple representation of the data in two dimensions. We could choose any of the two dimensions as the coordinates of the map and draw the plot. Figure 15.1 shows the 36 elements for red wine mapped on a two-dimensional scatter plot. The size of the rectangle is proportional to the utility value of the element for the total panel. In many cases this type of mapping is all that a researcher presents, without explanation, without rationale and, in Shakespeare's words, "sans everything." Yet, at some level the mapping satisfies the viewers because there is a sense of seeing the data (Moskowitz 2002).

Rethinking Mapping: Doing Exploratory Data Analysis First

One of the issues with Figure 15.1 is that it doesn't really teach much. Certainly, we see

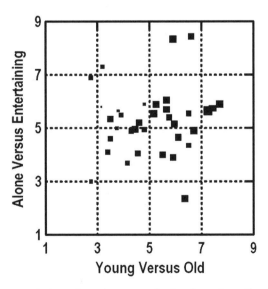

Figure 15.1. Plot of the concept elements on two nonevaluative dimensions. The size of the *square* is proportional to the utility value from the total panel.

the different points, we see that the utilities appear to rise when the element is rated as being "old" rather than "young" and "for entertaining" rather than being alone. However, something is missing. We lack some type of rule or generalization from the mapping. It's time to go back to square one and rethink the mapping so that it could be more useful. The abscissa (*x*-axis) is the semantic scale value, the ordinate (*y*-axis) is the utility value, and the scatter plot shows the relation. The points are scattered, as they should be because this is actual data. Statistical programs can plot a *smooth curve* through the data to reveal the underlying relation. When we do that plotting we find that there is a clear relation of utility to the semantic scale "young vs old," with those elements profiling "old" having a high utility. In contrast, there is no clear relation between utility value and the semantic scale "alone vs for entertaining."

What is the outcome of this exercise? There are at least four things to keep in mind:

1. Mapping alone on semantic dimensions does not tell us much.

2. Adding a depiction of utility, e.g., by the size of the square, tells us a little more.

3. Univariate plots of utility versus semantic scale tells us a lot more, even though we are dealing with one dimension at a time (Figure 15.2).

4. Mapping in the conventional way simply shows us where holes may be in the map, with no elements present. There is little additional information that can be extracted from the map. *Thus, mapping by itself has moderate, not exceptional, value.* It is a good way to portray the data, but beyond that, the additional information must be teased out separately.

Segmentation, Exploratory Data Analysis, and Univariate Mapping

If we continue with the univariate mapping exercise of semantic scale versus utility value, we learn more from segmentation. The segmentation was done on an individual by individual level by using the relation between semantic scale value and utility value (see Chapter 7).

Segmentation uncovered two different groups of consumers with respect to the utilities of red wines:

1. Sensory oriented. Segment 1 comprises those individuals who react strongly to

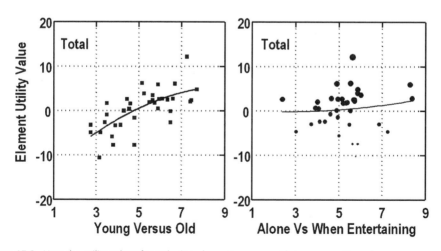

Figure 15.2. How the utility values for red wine elements covary with two semantic scales: young versus old *and* alone versus when entertaining.

the description of the wine. Winning elements for segment 1 include

    a. A heavy red wine with a warm plumy fruit flavor and a little bit of dryness

    b. Ruby red color with a black currant flavor and a hint of oak

    c. Deep-red wine with a sweet complex fruity flavor and delicate flowery fruit aroma

    d. A pale red wine with a light flavor of raspberries . . . so soft and sweet

    2. *Imaginers.* Segment 2 comprises those individuals who react modestly to promises of quality and some brand names:

    a. Made in the tradition of the greatest wine producers all over the world

    b. Dry, Sweet, Semi Sweet . . . whatever you're looking for

    c. Simply the best

    d. A wonderful experience . . . shared with family and friends,

    e. With the safety, care and quality that makes you trust it all the more

The segmentation by itself simply shows two different groups of respondents with different mind-sets. If we now plot the semantic scale on the abscissa versus the element utility value, we see radically different patterns (see Figure 15.3). Segment 1, the sensory-oriented respondents, shows relatively little sensitivity of utility value to the semantic scale "young versus old." In contrast, segment 2, the imaginers, shows increasing utility with the increasing level of "old." Concept elements that are perceived as being appropriate for older people are of more interest to segment 2 respondents.

Is There Anything More to Be Learned from Conventional Mapping If We Use Segments?

The initial foray into mapping showed that it was perfectly straightforward to map products as locations in a two-dimensional space, but that there was little to learn from the mapping (see Figure 15.1). We can follow the same logic and map the elements once again on the two dimensions ("young vs old" and "alone vs entertaining"), only this time create two plots, one plot per concept-response segment. This pair of plots is presented in Figure 15.4, with the left panel devoted to segment 1 (sensory-oriented respondents) and the right panel devoted to segment 2 (imaginers). Again the size of the square represents the utility value. Again we have a hard time learning more from this

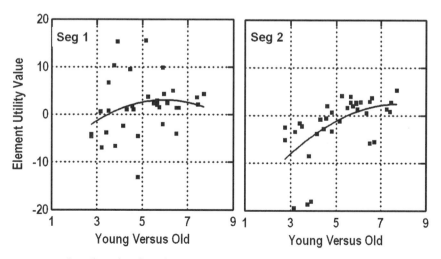

Figure 15.3. How the utility values for red wine elements covary with young versus old. The results are shown for the two segments.

two-dimensional map than we did prior to segmentation. In the way they have been presented, the scatter plots in Figures 15.2 and 15.3 teach us far more than do the two-dimensional plots.

The key message to take away is that simple mapping of the concept elements on semantic scales adds little insight. There is some modest learning to be obtained by locating the elements on the semantic profile. There is a little more to be learned by looking at the relation between semantic scale value and utility value, especially when it comes to the segments. We learn only the relations between elements, but have a hard time putting that learning into direct application.

Beyond Simple Mapping of Concept Elements to Mapping Plus Modeling

During the past decade, the senior author has experimented with mapping as a prelude to modeling (Moskowitz 1994). As already discussed, mapping simply represents multiple points in a geometric space and, as such, is really not much more than a way to represent data. The insights that emerge come more from the knowledge of the researcher than from the contribution of the mapping itself.

The mapping can be made far more valuable, however, if the mapping is combined with modeling (i.e., equations). The mapping–modeling combination follows these simple steps, which we will illustrate with the wine data:

1. Create a factor-based map, to ensure that the coordinates are parsimonious (few) and orthogonal (statistically independent) to each other. This first step involves the principal components analysis of the six semantic scales on which the 36 red-wine elements had been dimensionalized. *Principal components analysis* (Sharma 1996) is a standard, well-known method for reducing the complexity of the data to a set of orthogonal, parsimonious factors. To some researchers, these factors represent fundamental dimensions (e.g., of perception). For the purposes of this analysis, we simply use the principal components analysis to locate the 36 elements in a small-dimensional geometric space. Each of the wine elements has a specified location in that space.

2. The principal components analysis of the six semantic scales generate two factors. These factors are presented in Table 15.1 to the far right (FA and FB), both made positive by the addition of the number 2 to each factor

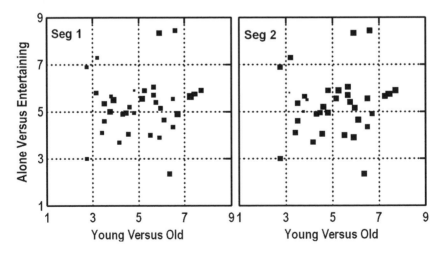

Figure 15.4. Plot of the concept elements on two nonevaluative dimensions. The size of the *rectangle* is proportional to the utility value from the two concept-response segments.

score). Each of the 36 elements becomes a single point in the two-dimensional factor space, defined by the factor scores, just as it had been a point in the maps based on the semantic differential scales. The utility for the total panel can be used to determine the size of the point. Higher utility values are bigger points (or squares), whereas lower utility values are smaller points (or squares). This mapping exercise is no different from the mapping in Figures 15.1 and 15.4, except that the factors (i.e., the coordinates of the map) were derived from all of the semantic scales.

3. Make the map far more useful by creating equations that incorporate the locations in the map. One way to increase the usefulness of the map creates a model or equation relating the location of the element on the map as an independent variable, either to semantic scale profile or utility value as a dependent variable. Such optimization problems have become standard issue in product research, wherein the products are mapped on a set of independent coordinates, with these coordinates in turn derived from principal components analysis of the sensory attributes used to profile the products. The creation of an equation is straightforward:

a. The independent variables correspond to the location on the map; that is, the two-factor score of the elements. By design, the coordinates of the factor map were created through principal components to be statistically independent of each other. Therefore, researchers can treat the coordinates as independent variables.

b. Each of the 36 elements has two coordinates and therefore has values for each of two independent variables. These are FA and FB, listed in column J in Table 15.1.

c. The dependent variables are the utility values (column J) and the six semantic scales (columns C–I). Each semantic scale generates its own equation, as does the utility value. Each subgroup for which the utility value is estimated from the conjoint analysis also generates its own equation. For

example, the two segments generate separate equations.

d. Table 15.2 lists the equations for factor scores versus one semantic scale (men vs women) and for the utility values for total panel. The factor scores, serving as the independent variables, are constructed from all of the six semantic scales, so there is no reason to assume that the factor scores would generate a model that fit each semantic scale perfectly. There is also absolutely no a priori reason why the factor scores should fit the utility values, since the factor scores were not obtained from the utility values. Still, the equations fit the data, albeit only modestly.

4. *The concept-element model.* The two equations in Table 15.2 and the remaining five equations, one for of the other five semantic scales, constitute the *concept-mapping model.* Figure 15.5 gives a sense of the way the map looks in two dimensions. In essence, we treat the map as a coordinate system.

Table 15.2. Parameters of two equations relating one semantic scale (men vs women) and the utility value for the total panel as dependent variables versus the location on the factor map as independent variables*

Dependent variable: Men vs women	
Squared multiple *R*	0.77
Constant	3.04
FA	0.96
FB	0.72
FA*FA	−0.12
FB*FB	0.16
FA*FB	−0.29
Dependent variable: Utility value, total panel	
Squared multiple *R*	0.55
Constant	−6.46
FA	4.62
FB	0.40
FA*FA	−0.67
FB*FB	−0.49
FA*FB	0.62

*The factor scores have been adjusted to be positive by adding 2.0 to each factor score. FA and FB are the two factors emerging from principal components factor analysis of the semantic profiles done on the concept elements for wine.

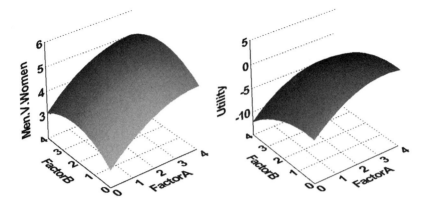

Figure 15.5. Two dimensional maps (response surface), plotting the utility value or the semantic scale (men vs women) as a function of the two factors: FA and FB. The equations underlying the response surface come from Table 15.2.

5. The equations allow us to search through the map, looking for a combination of coordinates that satisfy certain objectives. One objective might be the highest utility value, subject to the constraint that we remain within the range of FA and FB (the two factors) that were actually achieved in the study. Another objective might be to maximize the utility of a segment.

6. With the equations we now search for the location on the map that corresponds to a specific objective, such as most highly acceptable location or most highly acceptable location subject to a constraint. Once we locate the coordinates, FA and FB, we return to the set of equations and estimate the full semantic profile corresponding to these coordinates. We now know the *semantic profile* of the optimum, allowing us to get a better sense of what is the tonality of this new concept element. We do not know the element, per se. Rather, we know the nature of this element and how it is perceived. It is left for the creatives and the researchers to develop new elements with this type of semantic profile.

7. *An example of the approach.* Tables 15.3 and 15.4 give one a sense of what might be done with mapping and modeling combined. Table 15.3 shows the locations in the factor map (values of FA and FB) correspon-

ding to the optimum utilities for the total panel and the two segments, respectively. Table 15.3 also shows the expected semantic profiles of these optimum elements. Note that the elements do not exist. Rather, the approach synthesizes the location of the elements and estimates their likely semantic profiles.

8. *Getting help from the existing elements.* Table 15.3 shows the semantic profile (labeled *semantic goal*) corresponding to the optimized elements. We know the tonality of the concepts, but we also need to have some concrete exemplars to guide the development of new concept elements having those semantic tonalities. One good way to get that guidance searches for elements that have the requisite semantic profile. By looking at these *exemplar elements* the creatives and the researchers can begin to get a sense of what the new element should be like. Thus, the final task identifies those individual concept elements that generate a semantic profile as close as possible to the newly synthesized semantic goal. Each of the 36 red wine elements generated its own semantic profile. All we need do is compute the absolute distance between the semantic profile of an element and the semantic goal. This computation gives us an idea of what elements lie closest

Table 15.3. Locations in the factor map corresponding to optimized utilities for total panel and for segments 1 and 2*

Location	Total	Segment	
		1 Sensory oriented	2 Imaginers
FA	4.0	4.0	4.0
FB	3.0	0.8	3.7
Semantic profile			
Men vs women	5.1	4.7	5.6
Alone vs entertaining	7.6	3.8	8.5
Cheap vs expensive	7.8	7.5	7.6
Common vs unique	6.9	6.7	7.1
Young vs old	7.6	7.6	7.4
Morning vs night	6.7	7.4	6.7
Expected utilities			
Total panel	5.6	3.3	5.3
Segment 1	6.4	16.2	4.6
Segment 2	4.9	−6.7	5.9

*Data from red wine. FA and FB are the two factors emerging from principal components factor analysis of the semantic profiles done on the concept elements for wine. Source: 2002 Drink It! study. Data courtesy of It! Ventures, Inc.

to the goal and what elements lie further from the goal. The absolute distance is simply the sum of the absolute differences on the six dimensions. Each element has its own absolute distance. Table 15.4 ranks the order of the elements from low to high on the basis of the absolute distances. The *close* elements are those that have a semantic profile similar to what we are searching for and might themselves be part of the new concept. The *far* elements are those that have a semantic profile dissimilar to what we are searching for and probably should not be included; nor should elements similar to them in tonality.

9. *Validation: Estimating the utility values, and validating those derived utility values against actually observed utility values.* If the mapping is to go beyond simply an exercise in statistical data handling, then we must demonstrate that the utility values obtained from the conjoint analysis can be validly related to the distances between the semantic goal and the semantic profile of the elements; that is, those elements that are close to the semantic profile that is *optimum* should have high utility values, whereas

those elements far away from the semantic profile that is optimum should have low utility values. If we demonstrate the relation between distance from optimum and utility value, then we have a mechanism by which to show the usefulness of the map as a heuristic to identify promising areas. The logic is straightforward:

a. Semantic profiles generate factors. We saw that earlier, in the exercise to generate principal components, with the concept elements being factor scores on these principal components.

b. Factors are used as independent variables in equations to predict the values for both utilities and semantic scales, one semantic scale per equation.

c. One of the equations is optimized, the utility value for total panel or for a segment, while the factor scores must remain within the range tested. Furthermore, the estimated semantic profiles of the optimum must also remain within the range tested, so that we don't end up with an estimated semantic profile outside the range in which we are working. All semantic levels of the optimum must

Table 15.4. Elements for red wine ranked in terms of absolute distance of their semantic profile to the semantic goal*

	Distance	Semantic/bottom Semantic/top Semantic goal:	Men Women 5.1	Alone Entertain 7.6	Cheap Expensive 7.8	Common Unique 6.9	Young Old 7.6	Morning Night 6.7	Utility
Close elements									
E13	3	Premium quality	4.9	5.8	8.3	7.1	7.5	6.8	2
E12	3	Made in the tradition of the greatest wine producers all over the world	4.8	5.9	8.0	6.5	7.7	6.3	5
E31	4	Imported from France	5.7	5.8	7.7	5.8	7.4	6.7	2
E23	6	A great way to celebrate special occasions	5.6	8.4	7.0	5.1	5.9	6.6	6
E25	6	A wonderful experience . . . shared with family and friends	6.2	8.5	6.0	5.8	6.6	6.8	3
E21	6	Simply the best	5.3	6.1	7.3	5.4	5.7	6.2	4
Distant elements									
E16	12	Dry, Sweet, Semi Sweet . . . whatever you're looking for	4.4	5.9	4.8	4.1	5.3	5.5	4
E33	12	Multiserve containers . . . so you always have enough!	4.8	7.3	3.6	5.1	3.2	5.7	-5
E06	14	A pale red wine with a light flavor of raspberries . . . so soft and sweet	7.3	5.5	5.3	5.2	3.9	4.9	-3
E34	15	Icy cold	4.8	5.7	4.4	3.7	3.8	4.8	-8
E07	15	A chilled light bodied red wine with a fresh fruity flavor	7.2	5.0	4.8	4.4	4.5	5.1	3
E01	15	Wine with a pale ruby color and a fruity taste	7.0	5.4	5.0	4.3	3.5	5.2	2
E19	16	Quick and fun . . . ready to drink, on the shelf, no bartender required	4.3	6.9	3.3	3.5	2.8	5.6	-3
E27	16	It quenches THE THIRST	4.4	4.1	4.4	4.8	3.4	4.5	-3
E10	17	Hits the spot on a hot summer day	4.6	4.6	3.7	4.5	3.5	4.2	-1
E09	17	Bubbly fresh and sweet with a hint of raspberry tart flavor	7.2	5.0	3.8	4.2	3.8	5.0	-6
E02	17	Pale pink colored wine with fizz and the flavor of fruit	8.0	5.8	3.9	4.3	3.2	5.0	-11
E35	19	Resealable single serve container . . . to take with you on the go	4.3	3.0	2.7	5.3	2.8	4.8	-5

*Representative close and far elements are shown. The semantic goal is the semantic profile corresponding to the optimum utility for the total panel.

have been achieved by the elements or in the range achieved by the elements.

d. The optimization yields a location on the map where acceptance is high.

e. The location, a pair of factor scores, is used to estimate the likely semantic profile of the location simply by solving each of the equations in turn relating a semantic scale to the location of the factor scores. We thus generate an expected semantic profile.

f. This semantic profile is the reference. We can compute the geometric distance between this semantic profile and the semantic profile of each of the 36 elements. This is simply an exercise in computation. Each of the 36 elements thus has a distance value (from the profile of the optimum) as well as an associated utility value that the element had achieved in the actual study.

g. The distance to a goal semantic profile should be inversely related to the utility value; thus, an element that has a very different semantic profile is expected to have a low utility value. Figure 15.6 shows that this ex-

pectation is confirmed. Close elements have high utilities, whereas distant elements have low utilities.

10. *Implication 1.* If we know the semantic profile of a new element, previously untested but whose semantic profile is comparable to the semantic profile of the test elements, then we should be able to estimate its likely utility value. Thus, it does not matter where the semantic profile comes from . . . as long as we have a comparable semantic profile, we can calculate its distance from the optimum and have a good guess as to its likely utility value. This makes the concept development system open to elements not directly tested, but rather retrofitted after the fact.

11. *Implication 2.* If we are looking for new elements, we can look at the semantic profiles *near the optimum* to get a sense of the nature of the new element. We cannot synthesize the language of the element (as we might synthesize the formulation of the product), but we know the element tonality.

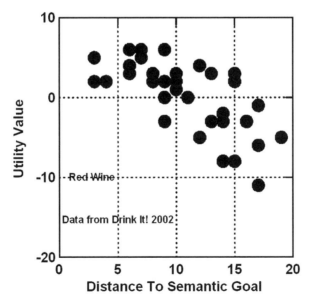

Figure 15.6. Relation between the utility value of an element and its absolute distance from the semantic profile of the optimum concept element estimated to have the highest expected utility. The relation is an inverse one, as expected. Data courtesy of It! Ventures, Inc.

Overview to Mapping and to Mapping Plus Modeling

The popularity of mapping as a heuristic to show the relation among stimuli often stops at simple representation, whereas it could go much farther. The approach shown here, adapted from product research, enables researchers to go from representation to discovery. What is necessary, however, is a different viewpoint from that usually involved in mapping. The map provides the location of a set of stimuli in a geometric space. Analytic geometry and regression modeling do the rest, enabling researchers to create the equations relating position in the space to attribute rating (semantic scaling) and acceptance rating (utility value). Optimization and reverse engineering complete the picture, enabling researchers to find a desired location in the map corresponding to some objective, find the corresponding semantic profile to that space, and then determine the set of concept elements that deliver that semantic profile.

Part 2: Synthesizing New Concepts

Introduction

This section goes beyond modeling to synthesizing. Modeling in the previous section identified the opportunity for new concepts, based on the utility values and semantic scales. We can go much further with the conjoint analysis results, to synthesize new concepts directly. We need, however, the following information:

1. The set of elements (raw materials)

2. The utility value of each element, for each key subgroup

3. Semantic profiles for each element (optional), which can direct the synthesis

4. A list of pairwise, incompatible elements (constraints)

A simulator for concept research is, in very simplest terms, a computer program that does one of two operations:

1. *Synthesizes new concepts subject to constraints.* Identifies combinations of concept elements that either maximize/minimize some criteria or match a profile. This could be maximizing the purchase intent of one subgroup, while at the same time satisfying other subgroups or fitting some type of communication profile.

2. *Estimates the likely proportion of respondents to exhibit some specific behavior.* For example, if one has the individual respondent data for a study and one has a specific concept, one might ask about the distribution of interest ratings for the particular concept. Another computation might be made for two such created concepts to determine the proportion of respondents who would choose one concept over another. This type of simulation is generally done at the end of the development cycle, where the interest focuses on how many people will be interested in the concept.

This chapter concentrates on the first use of simulation—that is, the synthesizing of new concepts—and deals with the second use (share of choice) in passing. The underlying reason for this treatment of the topic is that in the early stages of concept development, where this book concentrates, the developer and marketer are interested in *what is best*. They are not necessarily interested in market share, per se. The author's experience in development has, therefore, focused on the maximization of some variable, such as purchase intent, by the selection of the appropriate concept elements with which to go forward.

Simulators are important in business for the very simple reason that they reduce cost and time, making the results of a study easy to apply to practical issues and useful for a long time afterward. If a researcher tests one or a few concepts to answer a momentary problem, then the usefulness of the data lasts about as long as the problem lasts. Once the problem is solved, the data leading to the solution are filed away because they are intimately linked to the problem. If, however, the data are incorporated into a simulator with alternative scenarios, then there is a greater likelihood for subsequent use. The simulator can be used later when new situations arise and when managers want to guesstimate the likely strategy or outcome for that situation. Often managers turn first to the simulator for data that they already have acquired rather than commissioning a new study.

Not all studies lend themselves to simulation, however, so the decision to create a simulator has to be made in light of the type of research conducted and the nature of the data obtained. Often research conducted in business answers one question and has no other use (e.g., How well does a particular concept score?). The stimulus input is limited, the consumer response is equally limited, and the study answers the limited question with a limited answer. In this case there is nothing to simulate. The stimuli cannot be recombined into new combinations and, even if one were to do so, there is nothing on which to base the simulation in order to answer the new question. Quite a lot of business research is done for such limited objectives. Many concept tests are executed whose sole objective is simply to measure the performance of a concept for some next steps; for example, launch, rework, and abandon.

Revisiting the Coffee Dataset to Illustrate the Development and Use of the Simulator

We visit again the international coffee study to look at the creation and use of the simulator. The final coffee study comprised responses from more than 10 countries, each respondent evaluating a unique set of 125 concepts comprising 100 elements. Data will be shown here from seven of the countries. The concepts were created by experimental design, following the IdeaMap method for conjoint analysis (Chapter 8). This chapter deals with the creation of two simulators based on those data from the seven countries: one to optimize concepts and the other to assess proportion of times a concept would be chosen in an absolute sense or relative to a competitor. The simulator is presented by three steps in each section:

1. Explanation of principle behind the activity

2. Screen shot

3. Discussion of the step in the context of the coffee study

Background to the Concept Simulator

The *concept optimizer*, which mixes and matches concept elements to maximize interest in a concept or to match a specific goal profile, works with the aggregate data from the total panel, key subgroups (e.g., gender, country, and concept-response segments), and the set of dimensions used originally for data imputation and for segmentation. The mathematics behind the method is known as *integer optimization*. Fundamentally, the objective is to maximize or minimize the utility equation by selecting the appropriate elements. There are constraints on the optimization, such as certain pairwise combinations cannot occur or the combination must comprise a fixed number of elements.

Rather than go through the theory, this chapter shows the steps involved, the objectives, and then screen shots of the results. The actual approach, however, can be used in any of the conjoint analysis studies where the utilities of the elements have been established.

Preparing and Specifying the Basic Dataset

The input data for the concept optimizer comprises a spreadsheet matrix. The rows of the matrix correspond to the concept elements, and the columns of the matrix correspond to the key subgroups. An additional set of columns corresponds to the semantic differential values used for dimensionalization (Chapter 5). The only other information that the system needs is the organization of the elements (e.g., how many of the elements belong to each category). An example of the data input is presented in Table 15.5. The same type of data is presented in Figure 15.7, which shows the way the data look in the concept optimizer itself.

Study Restrictions

In concept research, quite often combinations of concept elements do not make sense. Sometimes the irrational combination comes from knowledge in the marketer's or the de-

veloper's mind; the combination is infeasible, although the respondent would never know that. Examples include specific features with prices. Other restrictions come from mutually contradictory statements that would make no sense if present together in a concept. Examples include one element that talks about strong aroma and another that talks about mild taste. The two elements are perfectly reasonable, but the combination makes no sense to the consumers reading this for the first time.

The coffee study generated 502 original pairwise restrictions. This number of restrictions sounds massive, but when we deal with hundreds of elements the number of pairs of elements that do not go together can grow larger. Figure 15.8 shows how these pairwise restrictions are identified for the simulator. The element on the far left is the *base* element, which cannot go with any element to its right. It is important to know that, when the concepts are developed for the experimental design, the restrictions are obeyed, as they are when the concepts are optimized.

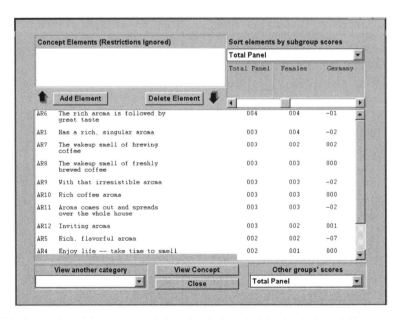

Figure 15.7. Screen shot of the concept optimizer data (only part of the data is shown). The screen shows the utilities of the aroma concept elements by total panel, women, and German respondents.

Table 15.5. Excel matrix elements and utilities (rows) × key subgroups (columns)

A	B	C	D	E	F	G
		Subgroup 1	Subgroup 2	Country	Dimension 1 Local vs international	Dimension 2 Weak vs strong
Element	Text	Men (MALE)	Women (FEM)	Germany (GER)	(INTL)	(STRG)
AR1	Has a rich, singular aroma	1	4	−2	4.75	4.94
AR2	The aromatic and smooth coffee	−1	0	2	4.88	4.69
AR3	Dark and rich with an invigorating aroma	2	1	3	6.13	6.31
AR4	Enjoy life—take time to smell the coffee!	3	1	0	4.88	5.00
AR5	Rich, flavorful aroma	2	2	−7	5.06	6.19
AR6	The rich aroma is followed by great taste	4	4	−1	4.94	6.56
AR7	The wake-up smell of brewing coffee	4	2	2	4.00	6.50
AR8	The wake-up smell of freshly brewed coffee	2	3	0	3.63	7.06

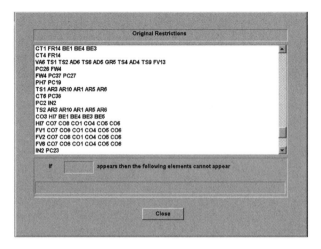

Figure 15.8. Partial set of pairwise restrictions. Each element is identified by a category code and a one- or two-digit number code showing its location within the category.

Imposing a Structure on the New Concept: Forcing in Categories

In any simulator, one can vary the amount of structure forced on the concept. For example, one might wish to optimize the concept without any constraints, other than those imposed by the original 502 pairwise restrictions. More typically, however, the goal is to create a concept following a structure that makes business sense. For example, in the coffee study we chose to force in the following categories: pictures (category PC), energizing statements (BS, benefits), roast and grind statements (GR), and taste promise (TS). The elements themselves are allowed to vary and must conform to the pairwise restrictions, but the categories must appear. Figure 15.9 shows the user interface, and Figure 15.10 shows the selection of the categories.

Selecting Subgroups (or Total) to Optimize

The optimization algorithm must be told which group to optimize. To *optimize* means to identify either the highest-scoring or the lowest-scoring combination of elements within

the constraints. For purposes of this demonstration we optimize the interest from the total panel by using a four-element concept.

To provide business value the simulator must generate a concept that can be tested, as well as a prediction about the performance. It is this specific output that differentiates the simulator and its progenitor conjoint measurement from conventional concept research. Concept testing is akin to a report card. Insights come from the experience and intuition of the researcher, not from the data. In contrast, conjoint analysis generates a database that can be interrogated at will to identify a new combination of elements that presumably should score better. The actual magnitude of improvement needs to be tested, however.

A key benefit of the simulator is that, once it identifies the concept corresponding to the objectives, it can then estimate the performance of that concept for all subgroups and for all semantic profiles. This capability is a direct outcome of the nature of the simulator. Since the simulator is privy to both the elements and their utility values for all subgroups and semantic scale attributes, it can immediately calculate the profiles by substituting the appropriate utilities into the equations.

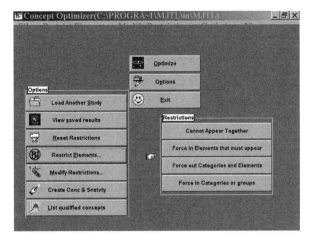

Figure 15.9. User interface to impose constraints on the optimized concept. These constraints are in addition to the pairwise constraints set up at the start of the study.

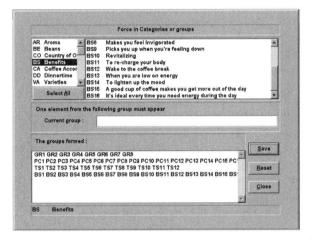

Figure 15.10. Example of categories chosen (constrained to appear in the concept). The groups are the categories. Exactly one element from each category must appear in the optimized concept.

Figure 15.11 shows the optimized concepts, subject to the 502 original pairwise constraints. One element from each of the four categories has been forced into the concept. Since the optimized concept was created for the total panel, the key interest should focus on the performance among the total panel only. Figure 15.12 shows the expected performance of this newly synthesized concept. We see the additive constant, the utility values for the four elements, and their sum. The top of Figure 15.12 shows the sum. By default, two other groups are shown, but they are not necessary to the results. They just appear there because the simulator was written to show three groups at a time. The same type of expected profile can be inspected for any other key subgroup(s).

Figure 15.11. Optimum, four-element concept for the total panel as created by the simulator.

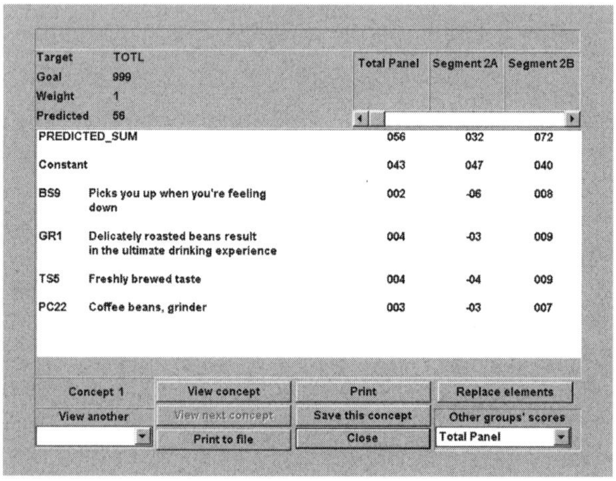

Target	TOTL		Total Panel	Segment 2A	Segment 2B
Goal	999				
Weight	1				
Predicted	56				
PREDICTED_SUM			056	032	072
Constant			043	047	040
BS9	Picks you up when you're feeling down		002	-06	008
GR1	Delicately roasted beans result in the ultimate drinking experience		004	-03	009
TS5	Freshly brewed taste		004	-04	009
PC22	Coffee beans, grinder		003	-03	007

Figure 15.12. Expected performance of the optimum concept, showing the utility values of the four elements and the additive constant, as well as the sum.

Performance of the Optimized Concepts on the Semantic Scales

One of the nice aspects of the conjoint method used in conjunction with the simulator is the ability to intermix optimization based on utilities with optimization based on the semantic profile. This dual capability generates concepts that are simultaneously highly acceptable by maximizing the utility and at the same time have a specific tonality created by maximizing or minimizing specific semantic scales.

The approach to simultaneous maximization of utilities and semantic scales is fairly straightforward. Recall that each of the coffee elements was originally profiled on a set of eight semantic differential scales, in the dimensionalization step, prior to the fieldwork. These semantic scales are not evaluative, but rather descriptive only; that is, the scales describe where an element lies in terms of meaning. The original scale comprised nine points. Thus, the scale for "regular versus gourmet" has been transformed from an original 1–9 scale (1 = regular . . . 9 = gourmet) to a −10 to +10 scale (−10 = absolutely regular . . . +10 = absolutely gourmet points) and can be transformed to a scale similar to the utility scale. Thus, the picture of coffee bean and grinder is given a value of +4 on the semantic scale. The transformation is quite simple:

1 is transformed to −10
5 is transformed to 0
9 is transformed to +10

The additive constant is arbitrarily set to 50. This approach creates a utility value for each concept element and commensurate utility values for the semantic scales.

Once this convention is adopted, all of the elements now have commensurate utility values for interest and for semantic scales. Following this logic, the concept optimizer computes the expected score of the optimal concept for subgroups and for semantic scales (see Figure 15.13). One can add the additive constant (arbitrarily set at 50) to the locations of the elements on the semantic scale to generate an expected magnitude. The optimized

concept has an expected semantic level of 62 on the regular vs gourmet scale (0 = regular, 50 = equal, and 100 = pure gourmet). On scale of inexpensive (0) vs expensive (100), the optimized concept scores 57. By inserting the semantic scale as another variable on which to optimize, just as on the utilities, and by giving it a relative weight, one can create concepts that both interest consumers and have the appropriate messaging.

Simultaneously Optimizing Several Groups/Semantic Scales

There is no reason why the optimization must be limited to one subgroup or one semantic scale. Rather than optimizing one group, therefore, the researcher, the developer, or the marketer can optimize the *weighted average* of several subgroups and the semantic scales. The result is, therefore, some compromise combination of elements that attempts to satisfy multiple goals simultaneously. Figure 15.14 shows an example of this approach. Figure 15.15 shows the resulting concept, and Figure 15.16 shows the performance of the concept for the key groups. The top left of the screen shot shows these estimated scores.

Target	TOTL		Total Panel	egular/Gourm Blend	ensive/Expe
Goal	999				
Weight	1				
Predicted	56		◄		►
PREDICTED_SUM			056	062	057
Constant			043	050	050
BS9	Picks you up when you're feeling down		002	001	-01
GR1	Delicately roasted beans result in the ultimate drinking experience		004	004	004
TS5	Freshly brewed taste		004	002	001
PC22	Coffee beans, grinder		003	004	003

Concept 1   View concept   Print   Replace elements
View another   View next concept   Save this concept   Other groups' scores
   Print to file   Close   Total Panel

Figure 15.13. Expected performance of the optimized concept on two semantic profiles: regular vs gourmet blend *and* inexpensive vs expensive.

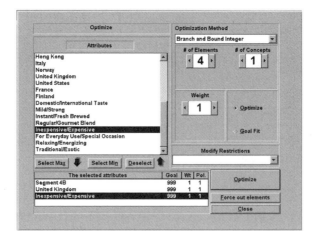

Figure 15.14. Optimizing three variables at the same time, with all of the variables having the same weight. The *dependent variable* for optimization comprises equal parts of two subgroups [segment 4B or the relaxer segment, and the respondent from the United Kingdom (UK)] and one semantic profile that introduces *tonality* (expensive).

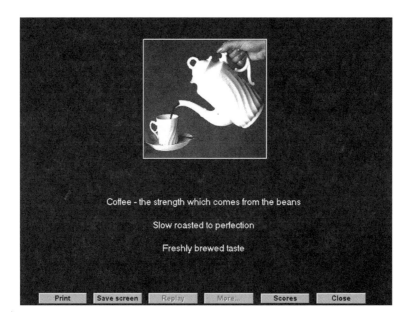

Figure 15.15. The optimized concept generated from the instructions shown in Figure 15.14.

One of the issues in optimization comes from the fact that one may be able to optimize some attributes, but as a consequence other attributes will be lower than expected. For example, Figure 15.16 shows that segment 4B (the relaxers) will not be satisfied with this optimal concept. Even though segment 4B was part of the objective, the solution did not identify a winning combination. This failure to optimize comes from the fact that, when all of the objectives are given equal weight, sometimes one or more of

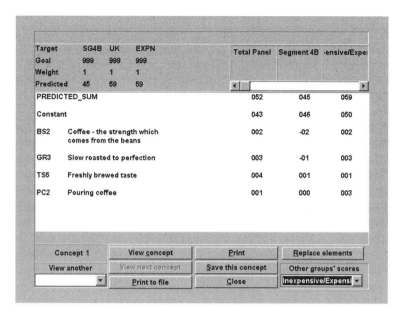

Target	SG4B	UK	EXPN		Total Panel	Segment 4B	ensive/Exper
Goal	999	999	999				
Weight	1	1	1				
Predicted	45	59	59	◄			►
PREDICTED_SUM					052	045	059
Constant					043	046	050
BS2	Coffee - the strength which comes from the beans				002	-02	002
GR3	Slow roasted to perfection				003	-01	003
TS5	Freshly brewed taste				004	001	001
PC2	Pouring coffee				001	000	003

Concept 1 | View concept | Print | Replace elements
View another | View next concept | Save this concept | Other groups' scores
▾ | Print to file | Close | Inexpensive/Expensi ▾

Figure 15.16. The expected performance of the optimized concept. All objectives (subgroups and semantic profile) are given equal weight.

them is overshadowed. Other strategies must be followed to ensure that all objectives are met. The next section deals with that issue.

Working with Unequal Weights in Order to Satisfy All Objectives

An alternative concept can be developed by differentially weighting the dependent variables in order to compensate for weak-performing subgroups. This weighting ensures that the weak-performing dependent variables will increase their acceptance, albeit at the expense of the strong-performing ones. Figure 15.16 shows that the results of optimizing all of the dependent variables may not result in a desirable concept. Segment 4B scores only a 45, whereas UK (British) respondents score a 59 and the semantic profile of inexpensive vs expensive scores a 59. The objective was to get as high as possible for all of them.

A different strategy might be to weight the segment 4B respondents more in order to lift their low score, while at the same time trying to keep the UK acceptance high and the se-

mantic profile also high, toward the *expensive* side. This new objective is accomplished by giving the segment 4B subgroup a weight of 3, the UK subgroup a weight of 2, and the inexpensive vs expensive semantic profile a weight of 1. The goal is still to maximize each, but the weights ensure that segment 4B's acceptance will be higher because it is more important. The three objectives move closer together, but of course it still is impossible to satisfy all of them. Figure 15.17 shows the new elements from this revised set of objectives.

Estimating Choice

Another use of simulators is to estimate choice among alternatives. For example, based on the expected sums of utilities, what is the proportion of choice between concepts? This issue of choice is very important for estimating market share when the concepts are complete and the marketing plan is created. Choice is less important at the developmental level, where one wants to create new concepts. In practical terms, the estimation of choice has

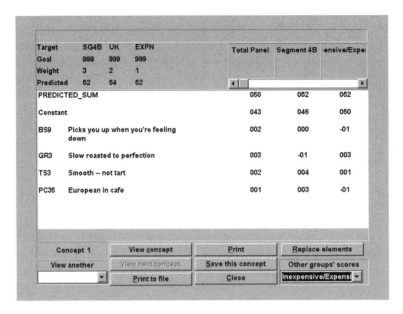

Target	SG4B	UK	EXPN		Total Panel	Segment 4B	ensive/Expe
Goal	999	999	999				
Weight	3	2	1				
Predicted	52	54	52	◄ ||			►
PREDICTED_SUM					050	052	052
Constant					043	046	050
BS9	Picks you up when you're feeling down				002	000	-01
GR3	Slow roasted to perfection				003	-01	003
TS3	Smooth -- not tart				002	004	001
PC35	European in cafe				001	003	-01

Concept 1	View concept	Print	Replace elements
View another	View next concept	Save this concept	Other groups' scores
▼	Print to file	Close	Inexpensive/Expensi ▼

Figure 15.17. The expected performance of the optimized concept. The objectives (subgroups and semantic profile) are given different weights to maximize all objectives at the same time.

been used by market researchers and other interested individuals for situations in which a concept has already been developed, and the real issue is to estimate how successful it will be. *Choice analysis based on utility values is rarely used at the early development stage when dealing with situations with large numbers of concept elements, simply because the issue of choice is not yet relevant.* Acceptance, not choice, is relevant at this early stage. Nonetheless, we can illustrate the approach by looking further at the coffee data, specifically for data from the United States.

The issue of utility value versus choice has been dealt with by a variety of researchers (see Green and Srinivasan 1978), resulting in two different general rules:

1. *Winner takes all.* In this rule, the researcher compares the utility values for two or more concepts, across many people. For any single person, the concept having the highest utility is assumed to be chosen 100% of the time. By looking across all of the people and assigning each person an all or none choice, the researcher can estimate the pro-

portion of respondents in a population who will choose each of a set of concepts.

2. *Choice is proportional to utility.* Concepts are chosen in proportion to their relative utility values. In this model the all-or-none rule is replaced by a proportion rule. The researcher adds the utility values of the individual concepts and computes the ratio of utility for each concept to this total utility. There are a variety of ways of expressing the relative utility, including simple proportion, and proportion of utilities after the utilities have been *exponentiated.* In any event, this approach, called the Bradley, Terry, Luce (BTL) model, is more flexible (Bradley and Terry 1952).

Figures 15.18 and 15.19 show the composition of two concepts, with known utility values for total panel and for subgroups. Indeed, with the large array of concept elements and utilities available to researchers, it is straightforward to create an array of concepts whose components have known utility values and therefore whose total utility values are known on a respondent-by-

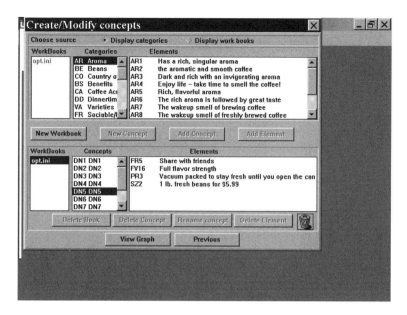

Figure 15.18. Composing concept DN5 by combining elements, each of which has a utility value. The final concept is shown. The total utility value is known on a respondent-by-respondent basis because the database comprises the utility of each element by each respondent.

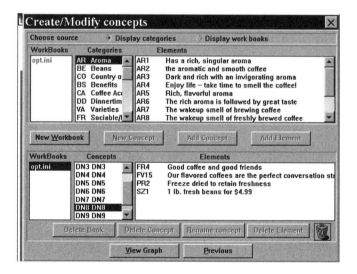

Figure 15.19. Composition of a second concept by combining elements with known utilities.

respondent basis. Applying the BTL rule of choice, one can estimate the likely proportion of times that each respondent will choose one concept over the other, based on the relative magnitudes of the total utility values for the two concepts. By averaging these choice proportions across respondents, researchers can estimate the net allocation of choice across the two concepts. This choice is presented in Figure 15.20. Finally, if a third concept or a fourth concept is introduced, composed from the same

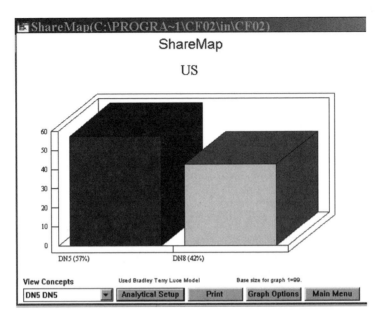

Figure 15.20. Estimation of choice between two concepts, based on their relative utility values, and the Bradley, Terry, Luce model for choice among alternatives with known utility values.

grand set of elements, the analysis can go further to estimate the likely choice of each of the four alternatives.

The Business Value of Concept Simulators

Simulators are important for decision making. They add value to the conjoint analysis results, primarily because they allow the answer to various "what if" questions. The conjoint database provides, as it were, the alphabet of the consumer mind. The simulators combine the letters of those alphabets into different concepts and estimate the likely response to the concept. These foregoing simulators are based on two different development and marketing needs:

Simulation to Identify Features

This simulation, discussed at length in the first part of this chapter, finds its main use in the creation of new alternatives. When there are many alternatives, simulation to identify features can be viewed as a sophisticated version of a sorting program. The algorithm searches

through alternatives until it comes up with a combination that fits specific objectives, such as maximizing a certain set of criteria. At that point the development job is finished, because the researcher has identified the particular combination that satisfies the objectives.

Simulation to Estimate Choice

This simulation, discussed in passing at the end of the chapter, deals with the expected performance of the combinations when they are pitted against each other. The major goal here is to see how well a combination will perform in a competitive array. This type of simulation makes little sense in a developmental mode when there are many alternatives to choose from and literally thousands or tens of thousands of combinations to sort through.

Some Recurrent Issues: Choice Analysis at the Early Development Stage

Quite often novice researchers believe that choice analysis (i.e., the second type of simulation) can be used to answer the question about what the optimal set of features is (i.e., the first type of simulation). Certainly, choice simulations run for many hundreds of alternative concepts can solve the problem, but it is inefficient. The same type of misapplication of method occurs in product tests. Developmental product research with many alternatives usually requires monadic ratings of each product prototype, the creation of a model relating the features of the prototypes to acceptance, and then the optimization of the product based on the model. All too often, however, researchers begin with paired comparison testing between two products, which is used mainly as a measure of preference. In the hope of identifying the optimum product, they then try to extend this paired-comparison approach to dozens of different prototypes. The same inefficiencies occur, and incorrect results emerge as occur with concepts. Paired comparisons of products and choice modeling with concepts both deal with estimated share and selection within a narrow world of finalized alternatives. It is inappropriate to use the paired method to create the products from among many alternatives. The strategy is inefficient, incorrect, and generally will not work. Rather than providing better answers because there is choice involved, the paired-comparison modeling or the choice behavior ends up constraining the number of test alternatives to a limited number. The field portion of the study, sensitive as it is to respondent tolerance to the task, ends up dictating more of the research than might be desired.

Part 3: Linking Results to External Variables

Introduction

Researchers are often requested to link performance of the test stimuli to *exogenous* or external information about the respondent, going beyond the response to the concept(s) itself. For example, if a researcher tests only

one concept and divides the respondents into those who accept versus those who reject the concept, then is there anything else about the respondent that covaries with the rating of the concept? For instance, do respondents who accept the concept differ in any clear way from those who reject the concept? If the researcher discovers a relation between accept/reject ratings and an external, respondent-based characteristic, this information helps the marketer and the developer. The marketer can target his efforts toward those who accept the concept. The developer can further work with the concept acceptors to perfect the product. For both the marketer and the developer, the knowledge about aspects of who specifically accepts or rejects the concept makes their job easier. Indeed, one of the first questions to come out of a segmentation study is "Who are in the segments?" This is not strictly a theoretical question. It is a very cogent business question.

Market researchers and, to a growing extent, sensory analysts are recognizing that they must learn a lot more about their respondents in order to link concept performance to external factors. The ideal situation would be some clearly defined *double dissociation*. This term means that there would be a clear differentiation in performance that could be easily traced to some clear external factor. For example, it would be ideal that older respondents, or some other easily defined group, accept the concept, whereas younger respondents reject the concept. More often than not, though, such simplicity simply does not exist. It may frustrate the researcher and his research user. Indeed, the more typical situation is that no clear defining characteristic, at least from easily measured demographics, ties together respondents who accept the concept and differentiates them from respondents who reject the concept. The frustrated marketer or developer might, in the end, opt for some type of soft direction, such as "a slightly greater than 50% of the older respondents accepted the concept."

Table 15.6. The winning concept for chocolate comprising four elements

Dense chocolate with swirls of dark chocolate and chocolate sprinkles on the surface
Premium quality . . . that great classic taste, like it used to be
A joy for your senses . . . seeing, smelling, tasting
Simply the best chocolate in the whole wide world

A sense of the frustration that might be felt by the marketer can be appreciated from data tables that a researcher might present to the marketer. Consider the winning concept for chocolate shown in Table 15.6.

Table 15.7 lists sample geodemographics for responses to the foregoing concept. The data come from the first 13 respondents and are sorted by response to concept. A *1* signifies that the respondent accepts the concept (rated the concept as 7–9), whereas a *0* signifies that the respondent rejects the concept (rates the concept as 1–6). The data show that a simple pattern for responses to this concept is hard to identify. Even if there seems to be an underlying pattern, all too often the pattern is not statistically robust, and also it is hard to understand why it should be this way in the first place. There is simply no rational organizing principle.

Types of Ancillary Information

Inspection of Table 15.7 fails to reveal any simple pattern underlying acceptors or rejectors. The same lack of pattern pervades the entire table, with all of the respondent data. Respondents can be easily characterized by geodemographic information, but for the most part the acceptance data show little relation to this standard type of information about respondents. Table 15.8, for example, shows that information about respondents assigning ratings 7–9 (percentage top-3 box acceptance) bears little relation to the conventional data of market, age, and income. Occasionally, researchers might opt to purchase additional geodemographic information about the respondents from companies that specialize in such information. The covert, generally unproved assumption is that somewhere in the midst of mountains of information about the respondents there might lurk that magic key that will predict acceptance or at least covary with acceptance in a statistically significant fashion. Such magic keys do not necessarily exist, except in the most unusual, and generally most obvious, of situations.

Marketers and developers have had to come to grips with the relatively low usefulness of the geodemographic data. The data

Table 15.7. Example of data for the chocolate concept, showing acceptors versus rejectors and ancillary information about the respondent from a self-profiling classification

Respondent	Accept = 1 Reject = 0	Gender	Hunger	FACT scale*	Age	Market
1	0	Female	4	8	4	2
2	0	Female	2	2	4	2
3	0	Male	1	6	4	1
4	0	Female	1	5	7	4
5	0	Female	1	7	6	3
6	0	Female	2	8	6	2
7	0	Female	1	6	4	5
8	0	Male	2	6	7	1
9	1	Female	3	7	3	5
10	1	Female	4	8	3	5
11	1	Female	4	8	3	6
12	1	Male	4	8	6	6
13	1	Female	4	6	3	2

*FACT scale, food action scale.

Table 15.8. Summary table relating concept rejection/acceptance to conventional geodemographic information: numbers in the reject and accept columns are percentages

	Reject concept (%)	Accept concept (%)	Base size
Gender			
Male	63	38	72
Female	58	42	445
Age			
Under 13*	0	100	1
13–17	80	20	5
18–30	61	39	132
31–40	58	42	139
41–50	53	47	130
41–60	61	39	75
61–75	71	29	34
75+	0	100	1
Market			
Northwest	60	40	55
Southwest	66	34	79
Mountain	79	21	14
Midwest	67	33	112
Northeast	49	51	117
Southeast	55	45	103
AL/HW	0	100	1
Canada*	63	37	19
Mexico*	41	59	27

*Data from respondents who are outside the appropriate age range or live outside the United States.

describe respondents and are useful for reaching consumers because the information is set up in the way marketers have been accustomed to think about it. At the same time, however, when this type of information is collected about respondents there is no expectation that the data will be used for anything other than a quick check to identify exceptions; that is, there is no longer an ingoing assumption that there will be meaningful differences in response to concepts based on the conventional geodemographics. Most of the time, good research practice dictates that these data be collected, but experience leads the researcher to search for light in other corners.

If the conventional geodemographics do not work, then perhaps other breaks in the data will work. What specifically should the researcher look for? Should the researcher instruct the respondents to profile themselves on interest in chocolate products? Questions of this nature appear, at the surface, to be more relevant. One does not necessarily assume that age, income, gender, market, and so on, would covary with response to the concept, no matter how much one would like that to be the case. One does assume, however, that attitudes toward the product might covary with response to the concept. The link of general attitudes with consumer responses to test stimuli has led to overarching segmentation approaches (e.g., see Wells 1975). There is the feeling that at least attitudes should predict concept responses because attitudes somehow inhere in the mind as do the concept responses.

Without going into overarching segmentation we can look at attitudes toward chocolate. This focus is a bit more concrete. Table 15.9 tabulates the proportion of respondents interested in the chocolate product based on concept, compared with the individual's predisposition to the chocolate product based on the FACT (food action) scale (Schutz 1964) from the self-profiling classification that the respondents completed at the end of the concept evaluation. There is no clear evidence from Table 15.9 that one's predisposition to chocolate products covaries with one's rating of the chocolate concept.

If Geodemographics or Predispositions Don't Work, Then What?

Tables 15.8 and 15.9 suggest that, at least in the case of chocolate candy, it is difficult to predict response to a concept from either standard geodemographic data about a respondent or from predisposition to the candy from direct rating scales. If this is true, then what should the researcher do? Perhaps one should segment the respondents on the basis of their responses to several different self-profiling questions in the classification questionnaire.

Table 15.9. How predisposition to the chocolate product [FACT scale (Schutz 1964)] covaries with response to the chocolate confection concept

Scale description	FACT scale*	Reject concept	Accept concept	Base size
Hardly ever eat	2	100%	0%	4
Don't like—eat on occasion	3	50%	50%	2
Eat if available	4	52%	48%	31
Eat now and then	5	71%	29%	93
Frequently eat	6	55%	45%	140
Eat very often	7	63%	37%	142
Eat at every opportunity	8	48%	52%	105

*FACT scale, food action scale.

We saw that this strategy really doesn't work when we use conventional scales about predisposition to predict concept acceptance (i.e., the FACT scale).

One possibility is to look at the pattern of responses to several questions that describe the respondent rather than simply confining oneself to a search for the magic "single" question. It is important to stress here that the goal is to identify what features of a respondent covary with a specific rating: accept or reject. There is only one test stimulus: the concept. There are, however, a number of variables that possibly covary with the response. These variables are assumed to be linked to the concept response because they deal, in another way, with the topic of chocolate candy.

The additional information about respondents comes from the classification questionnaire that they filled out after the concept evaluation. The classification is patterned after a standard classification questionnaire in the Crave-It mega-study (see Chapter 21), which deals with the drivers of product acceptance. The respondents completed a large self-profiling questionnaire and were instructed to check the reasons for craving a candy, including why, when, and where.

One way to understand more about the acceptors and rejectors is by listing the averages. We saw clearly that this did not work (see Table 15.8). Other methods have been developed to predict membership in a class, such as concept acceptor (versus concept re-

jector), by looking at a composite predictor from many of these questions rather than one question alone. One of the most important and widely used methods—*discriminant function analysis* (DFA)—can be applied here quite easily. DFA is similar to regression in that it creates a weighted combination of a limited number of independent variables. Rather than predicting the magnitude of a dependent variable, however, DFA tries to assign weights to the independent variables so that one can predict whether a respondent falls into one category or the other (Cooley and Lohnes 1971; Klecka 1980; Dunteman 1984).

DFA enables researchers to scan through a long list of predictors and, in so doing, identify the particular predictors that differentiate the acceptor and rejector groups. The steps are straightforward:

1. By setting criteria for including or excluding predictors, the researcher gives the DFA program criteria to evaluate the different predictors. DFA rapidly identifies these potential predictors. Table 15.10 shows the candidate set of predictors (left side) and their respective *F* ratios (right side). DFA includes only those predictor variables that are significant. Otherwise, the equation could be filled with irrelevant variables.

2. Once the predictors are identified, DFA then creates a set of equations that, when applied to the predictors, each produce a value. With two categories (accept and reject), DFA works with two equations: one pertaining to

Table 15.10. Variables from discriminant function analysis that significantly covary with concept acceptance or concept rejection*

Variable number	Name	F ratio
70	Mood	5.62
60	Aroma	4.35
81	Alone	3.54
76	After dinner	3.06
50	Superstores	2.99
67	Packaging	2.65

*The variables come from the self-profiling and are ordered by the ability to predict membership (F ratio).

Table 15.12. Performance of the discriminant function analysis, classifying respondents as concept rejectors (0) or acceptors (1), based on the six criteria listed in Table 15.5, along with the discriminant function

Classification matrix (cases in row categories classified into columns)			
		Expected reject 0	Expected accept 1	% Correct
0	Actual reject	175	129	58
1	Actual accept	84	129	61
Total		259	258	59

concept acceptance and one pertaining to concept rejection (Table 15.11). Each equation generates its own estimated value. The two values are used to estimate the membership of a new individual in the two groups.

3. DFA then estimates the proportion of respondents correctly and incorrectly classified (Table 15.12).

4. A given respondent's profile on these six key variables can be submitted to the two equations. The goal is to classify this respondent as belonging to one group or another. The approach uses the weighted distance of a person's set of six numbers from the center of the acceptance cluster or from the centroid of the rejector cluster. Thus, there are two distances for each respondent in the study and

Table 15.11. Discriminant functions (equations) that, when applied to the specific attributes in the self-profiling questionnaire, generate discriminant function values*

	Reject	Accept
Constant	*−2.4616*	*−2.3968*
Superstores	1.9858	1.6676
Aroma	1.5917	2.0508
Packaging	1.9888	1.1824
Mood	2.5434	2.0978
After dinner	1.8544	2.1871
Alone	1.2158	1.7568

*Depending on the relative values of the two numbers the researcher can assign a respondent to the acceptor group or rejector group, with either greater confidence or less confidence.

for any new respondent who can be profiled on the six questions listed in Table 15.11. Whichever of the two distances is smaller (i.e., from the respondent to the centroid) defines the cluster to which the respondent belongs. Clearly, this is a purely statistical classification, and it will be wrong some percentage of the time, as Table 15.12 shows.

5. Table 15.13 presents data from eight respondents: four from the rejector group and four from the acceptor group. We know their membership because we have it empirically from the study. We then use the DFA equations to compute the weighted distance of each of the respondents from each of the two centroids and pick the smallest distance. Thus, for the first respondent (number 9), the distance is 4.1 from the rejector centroid and 4.2 from the acceptor centroid. We define respondent 9 as a rejector. We do the same for each current respondent, generating some correct assignments and some incorrect assignments. Table 15.12 indicates that we are right about 59% of the time. We could increase the proportion of times we are right by adding more predictor variables to the DFA. This strategy would increase the number of times we are right, but at the risk of achieving that improvement by incorporating variables that may be spurious predictors of membership. It is always a good idea to be parsimonious, using the fewest number of significant predictors possible, while maintaining a good "hit rate."

Table 15.13. Weighted distances of a respondent to the centroid of the rejector or acceptor segment (so-called Mahalnobis distance) and the probability of being correct for each respondent

	Distance from rejector	Probability	Distance from acceptor	Probability	Decision	Correct?
Actual rejector						
Pan 11	4.1	0.51	4.2	0.49	Reject	Yes
Pan 19	2.7	0.59	3.4	0.41	Reject	Yes
Pan 07	9.9	0.46	9.6	0.54	Accept	No
Pan 08	3.5	0.48	3.4	0.52	Accept	No
Actual acceptor						
Pan 03	4.0	0.48	3.8	0.52	Accept	Yes
Pan 04	6.2	0.45	5.8	0.55	Accept	Yes
Pan 11	2.7	0.67	4.1	0.33	Reject	No
Pan 13	2.7	0.59	3.4	0.41	Reject	No
Centroid						
Superstore	0.5000		0.4225			
Aroma	0.1941		0.2958			
Mood	0.5230		0.3931			
After dinner	0.3717		0.4460			
Packaging	0.0461		0.0235			
Alone	0.1020		0.1455			

Overview

During the past decade, and with the emergence of high-powered computation tools, researchers have begun to deal more effectively with the assignment problem. The *assignment problem* is the assignment into a segment of a newly encountered person. The reason for that assignment is quite straightforward. If the marketer or the product developer knows the preferences of that segment, then the individual can be more easily satisfied because the developer can generate the correct product and the marketer can choose the right message to interest the individual.

Experience shows, however, that the assignment problem cannot be easily solved. It is quite straightforward to measure concept acceptance/rejection. It is also straightforward to identify the salient characteristics about a respondent through the classification questionnaire, whether these characteristics consist of standard geodemographics or psychographics (attitudes). Matters become difficult when the researcher attempts to marry these two

sources of information. The marriage typically does not work. One can use ad hoc methods, such as DFA, to create a predictive model, but for the most part the descriptive model does not provide much insight about the underlying reasons, the *why*.

Newly emerging, more complex methods, such as data-mining procedures, with a variety of algorithms can help. Some algorithms attempt to construct human-interpretable representations of the derived patterns, such as decision trees or rule sets. Other algorithms focus more heavily on a statistical characterization of the patterns, such as Bayesian networks or hidden Markov models (Rabiner and Juang 1986; Berry 1996). Still other algorithms avoid the issue of interpretability altogether, opting for powerful methods for capturing patterns with even a certain degree of nonlinearity, such as multi-layer neural networks (Bengio and Bengio 2000). These are more powerful descendents of the DFA methods. Data-mining combines statistical techniques and knowledge-based methods to extract meaningful patterns from large datasets. These procedures, commonly used in financial services to "score" individuals for credit risk (e.g., mortgages), might work for food concepts, but there are not enough case histories to warrant any conclusions yet. One of the complicating factors is that, unlike financial services, there is no real accounting for product preferences. One might make the reasonable conjecture in financial services that an individual's credit history, pattern of payments, the salary, type of job, education, gender, and marital status might all be relevant. These variables are not necessarily relevant to credit risk, per se, but they are probably involved in an individual's financial behavior. Married individuals have different patterns of purchasing than single individuals. Education is important in the type of job and the type of income that a person has. Patterns of payment probably remain a characteristic of a person and could transcend what is being paid for (unless there is a severe penalty for delayed payments).

The problem with data mining, as with DFA and the family of assignment algorithms, is the lack of a fundamental set of rules that would lead us to expect a certain pattern. There is no clear-cut *why* and *rationale*. There is only data fitting. There is no such accounting for food preferences. Knowing about an individual does not predict how that individual will do in terms of accepting or rejecting a concept, or whether a person belongs to a specific concept-response segment. Solving the assignment problem *with basic knowledge, and not just with computational algorithms*, remains one of the first-order problems for future research in concepts. Knowing that a person belongs to a segment enables one to satisfy that person. One knows a lot about that segment from prior research. Knowing to which segment any *new* individual belongs is the answer to the $64,000 question. Knowing why any new individual belongs to the segment is, in turn, the pearl without price, representing an entirely new order of knowledge and insight.

References

Bengio, S., and Bengio Y. (2000). Modeling high-dimensional discrete data with multi-layer neural networks. In: Solla, S.A., Leen, T.K., and Müller, K.-R., editors. Advances in Neural Information Processing Systems. Boston: MIT Press, 12: 400–406.

Berry, D.A. (1996). Statistics: A Bayesian Perspective. Belmont, CA: Duxbury.

Bradley, R., and Terry, M. (1952). Rank analysis of incomplete block designs, I: The method of paired comparisons. Biometrics, 324–345.

Cooley, W.W., and Lohnes, P.R. (1971). Multivariate Data Analysis. New York: John Wiley and Sons.

Dunteman, G.H. (1984). Introduction to Multivariate Analysis. Thousand Oaks, CA: Sage.

Green, P.E., and Srinivasan, V. (1978). Conjoint analysis in consumer research issues and outlook. Journal of Consumer Research, 5: 103–123.

Greenhoff, K., and MacFie, H. (1994). Preference mapping in practice. In: MacFie, H.J.H., and Thomson, D.M.H., editors. Measurement of Food Preferences. London: Blackie Academic and Professional, pp. 137–166.

Klecka, W.R. (1980). Discriminant analysis. In: Quantitative Applications in the Social Sciences Series. Thousand Oaks, CA: Sage, 19: 7–19.

McEwan, J.A. (1996). Preference mapping for product optimization. In: Naes, T., and Risvik, E., editors. Multivariate Analysis of Data in Sensory Science. New York: Elsevier Applied Science.

Moskowitz, H.R. (1994). Product testing 2: modeling versus mapping and their integration. Journal of Sensory Studies, 9: 323–336.

Moskowitz, H.R. (2002). Mapping in product testing and sensory science: a well lit path or a dark statistical labyrinth? Journal of Sensory Studies, 17: 207–213.

Rabiner, L.R., and Juang, B.H. (1986). An introduction to hidden Markov models. Journal IEEE Acoustics, Speech & Signal Processing Magazine, 3: 4–16.

Schutz, H.G. (1964). A food action rating scale for measuring food acceptance. Journal of Food Science, 30: 202–213.

Sharma, S. (1996). Applied Multivariate Techniques. New York: John Wiley and Sons.

Vigneau, E., Qannari, E.M., Punter, P.H., and Knoops, P. (2001). Segmentation of a panel of consumers using clustering of variables around latent directions of preference. Food Quality and Preference, 12: 359–363.

Wells, W.D. (1975). Psychographics: a critical review. Journal of Marketing Research, 12: 196–213.

Part IV

Putting the Approaches to Work

Chapter 16

Developing from the Ground Up: Self-authoring Systems for Text and Package Concepts

Introduction

Today's manufacturers and service providers are beset by increasing competition in all aspects of their businesses. In many companies the search is on for ways to better understand what consumers and customers want. In the race to maintain a competitive advantage, knowledge and insight provided by researchers have assumed an increasingly valuable role as a strategic weapon to maintain the competitive edge. As a consequence, research buyers have become increasingly open to developments in survey research, especially those that provide insight and actionability (see Chapter 1 for details).

The Distinction Between Developmental and Confirmatory Research

We should first revisit the distinction between early-stage developmental research with many options and late-stage confirmatory research where the data are used for a go/no-go decision. This book deals with the early-stage developmental research. Late-stage research typically confines itself to one concept, with the goal to measure the acceptance of that concept and perhaps to estimate market share.

Conjoint analysis is often used in the early development stages, where there are many options from which to choose and where the objective of the research is to understand the lay of the land. A version of conjoint analysis called *discrete choice analysis* is used for confirmatory research, where the goal is to measure potential market share.

A problem that traditionally plagues conjoint analysis is the issue of *complexity-driven delay*. The typical conjoint study is executed only after a great deal of time has passed, with a lot of different, often conflicting, inputs obtained during that time. There are inevitable project delays. Some arise because historically conjoint analysis was used for the very important projects. Other delays arise because the approach involves *homework*—that is, the creation of multiple concept elements. All too often these delays reduce the business usefulness of conjoint analysis.

Time and complexity are generally important as a determinant of whether researchers use conjoint measurement instead of, say, a focus group or a simple concept test. Most clients want to glean as much information as possible from a single test. This knowledge hunger leads to studies that have many elements and thus require long setup times. The sheer desire to pack many elements into a single study inevitably leads to delays because everyone involved in the design of the study feels compelled to offer a few elements to the project and then to act as an editor of the full set of elements. No one in these larger-scale studies wants to assume the risk of implementing the study without having thoroughly investigated the inputs, editing them, and then proceeding, armed with the group consensus.

Speed and well-thought-out quantitative research often appear to be mutually contradictory. In the custom research world a great deal of pontificating is about the need to be precise in terms of the research design. Researchers involved in quantitative research, with large-scale studies, discourage rapid, low-cost research methods because they feel that the results may be compromised by the speed of execution and analysis. At some unconscious level the results can be valid only when they have been massaged, fully digested, and presented in detail with the necessary insights. When it comes to the combination of conjoint analysis and rapid, easy research, the combination is even more disconcerting. Thus, from the senior author's (H.R.M.) viewpoint such discouragement of speed and ease promotes risk avoidance in favor of statistical correctness, probity, and propriety. It is better to work with arcane but powerful methods than with simple, perhaps transparent, ones. What is lost is the power and spontaneity of discovery at the early stage (Wheelwright and Clark 1992).

The concentration of quantitative research on *probity* leads to an inevitable outcome or at least to an outcome that seems to be a natural consequence. The outcome is the growth of the qualitative at the expense of the quantitative. A great deal of quantitative research has been and continues to be scrapped in favor of the more rapid qualitative research approach. Myriad are excuses given, such as a lack of time, a lack of budget, and a need to get close to the consumer. Focus groups and depth interviews, while not providing the necessary quantitative information, are still perceived as being more responsive to time pressures. Consequently, qualitative research has boomed worldwide, being perceived to be the only way to do rapid developmental work in primary research in order to understand customer needs.

Iterative Concept Research at the Development Stage

Besides speed to market, a factor is the need for success. This is a truism, but the need for true success is often overlooked in the mad rush to get to market. The senior author (H.R.M.) has found in almost 30 years of

consulting and research that the more times a nagging problem is tackled in an iterative fashion, the greater is the likelihood of success. When this notion of *multiple, repeated, structured* research efforts comes to customer research, the results are even more striking. Although companies may not use sophisticated research tools, the more frequently they return to the problem in an iterative fashion, the more likely they are to be successful.

The word *iteration* as used in this chapter does not refer to an interminable project or to a desultory visit to the problem once every year or several times per year to learn in a programmatic, rote way, whether anything has emerged that can solve the problem. The notion of iteration means a consistent, systematic, disciplined approach. The iterative approach comprises a sequence of studies. The results of one study are added to the results of the others, and the results of an early study are used to modify the product or positioning prior to the next iteration (Moskowitz and Ewald 2001). The approach redefines research as a series of transactions with the customer where information is communicated and acted on equally rapidly (Saguy and Moskowitz 1999). Iterative research is best used as a tool for early-stage development where there are many options (Pawle and Cooper 2001).

Early Attempts at Making Conjoint Analysis an Iterative, Personal Computer–based Tool

The increasing importance of research, or more accurately the advantages conferred by the knowledge that research provides, has not been lost on survey researchers. The increasing use of computers, the time and quality pressures on survey research, and the obvious need for actionable information all have contributed to the growth of new research tools. Technology continues to appear, enabling survey researchers to provide new types of knowledge and insight. This first section focuses on an early version of self-authored conjoint analysis that could be executed rapidly. The chapter then moves onto the Internet, where the technology and vision were better realized.

The personal computer (PC)-based research tool, *IdeaMap.Wizard* (abbreviated *Wizard*), was conceived as a *do-it-yourself*, multimedia conjoint measurement tool and originally designed 8 years ago before the power of personal computers and Internet communication had dramatically grown (Moskowitz and Martin 1993). The objective of the self-authoring system was to bring the power of conjoint analysis to researchers as a complement to conventional focus group and other low-cost procedures.

Rather than designing the self-authoring system and then marketing it, the system was designed with user feedback obtained in a study. The objective was to identify customer requirements for the system, much as one might run a similar type of study for the features of a product, based on consumer reactions. The goal was to identify what the different constituencies want from early-stage development research. The topics covered the plethora of different possibilities, ranging from what the technology can do to what business and process benefits it provides. The research tool was conjoint analysis. The three client constituencies were marketing research, marketing, and product research and development (R&D). All three use early-stage research extensively to identify the features of products and services. All constituencies understand the importance of getting the right information early in the development cycle.

Basic Interest in the Self-authoring System

The basic interest level is shown by the additive constant (Table 16.1). For instance, when we compare different client user groups (marketing research, marketing, and R&D)

we see a great enthusiasm for the approach from R&D (constant = 53), a moderate level of enthusiasm from marketing research (34), and absolute disinterest by marketing (−7). The negative constant is possible because the constant is an estimated parameter that never stands by itself. Marketing is not interested in research technology by itself, whereas R&D is very interested. These quantitative results confirm the authors' observations that R&D is much more hands on than is marketing research and that marketing is primarily interested in results rather than in method.

What Features in a Self-authoring System Are Most Interesting to Prospective Users?

The additive model shows those particular elements that are very strong (viz., high positive numbers) compared with those that are very weak (viz., high negative numbers). Rules of thumb for this type of research are that utilities (coefficients) above +15 are extremely strong, above +10 are very strong, above +5 are meaningful, and below −5 are negative and to be avoided.

On a substantive basis, the key hot button from the total panel is the general description: "Wizard is an easy step-by-step guided approach to technology-based concept development and screening."

This general statement contains little in the way of specific benefits, specific end uses, etc. As we will see, the different constituencies have different viewpoints, so only the most general statement about the approach appeals to everyone. The other three elements that pass must also provide end benefits without specifics:

"Wizard is an innovative self-contained research technology that mathematically sorts and evaluates your ideas."

"One system allows a wide variety of testing methods: from conjoint to concept testing to package screening."

Table 16.1. Base size, additive constant, and winning elements for the model relating interest to presence or absence of elements*

	Total	MRD	Marketing	R&D
Base size	*32*	*17*	*4*	*8*
Additive constant	*32*	*34*	*−7*	*53*
Market research				
Can handle a wide range of variables and help you understand independent utilities for each piece of stimuli	17	27	17	9
Wizard is an easy step-by-step approach to evaluate or create a variety of concepts	18	18	29	6
Rapidly test your ideas versus new competitive products	11	14	3	5
Lets you test the full creative array using consumers to identify what "turns them on"	6	13	2	−11
Identify powerful ideas and how to express them most effectively	13	11	25	9
Can be applied to any category, any topic, for consumer research and beyond	11	10	30	0
Marketing				
Can be used for concept screening, package designs with internal or external consumers	14	9	45	6
Wizard is an easy step-by-step guided approach to technology-based concept development and screening	9	8	40	−16
Wizard is a do-it-yourself innovative multimedia expert research system	11	9	32	2
Can be applied to any category, any topic, for consumer research and beyond	11	10	30	0
One system allows a wide variety of testing methods: from conjoint to concept testing to package screening	7	8	30	−10
Wizard is an easy step-by-step approach to evaluate or create a variety of concepts	18	18	29	6
Identify powerful ideas and how to express them most effectively	13	11	25	9
Explore many alternatives—do not limit your possibilities too early on	11	3	25	15
For you if you know research and have access to respondents	0	−4	24	−4
A new tool for customer satisfaction studies	6	7	22	4
Wizard is an innovative self-contained research technology that mathematically sorts and evaluates your ideas	7	9	20	3
Get it right the first time—use this expert research partner to make knowledgeable decisions	2	0	18	0
For start-up companies	−17	−21	17	−36
Can handle a wide range of variables and help you understand independent utilities for each piece of stimuli	17	27	17	9
Complete with setup, interviewing, and reporting software	8	6	17	9
A nicely balanced combination of powerful insight and ease of use	7	5	17	−1
For consultants who want an added-value tool	−10	−13	15	−19
For the Market Research Department	−5	−1	15	−23

(continued)

Table 16.1. Base size, additive constant, and winning elements for the model relating interest to presence or absence of elements* *(cont.)*

	Total	MRD	Marketing	R&D
Wizard is PowerPoint for concept development	−3	−9	15	1
Wizard is an easy step-by-step advanced concept development technology	4	4	15	−5
Wizard is your link into our high tech research technology	0	0	13	0
For retailers for on-the-spot research	−13	−22	12	−16
Your own market research "guru" that sits in your computer and guides you through the concept development maze	6	0	10	11
Test, analyze, modify and reset in one session	3	4	10	−3
Gives a clear answer to complex issues	4	3	10	−2
R&D quality control				
So fast you can modify the test while it is fielding	11	8	9	26
Explore many alternatives—do not limit your possibilities too early on	11	3	25	15
Your own market research "guru" that sits in your computer and guides you through the concept development maze	6	0	10	11

*MRD, market research department; and R&D, research and development.

"A nicely balanced combination of powerful insight and ease of use."

What Features of a Self-authoring System Appeal to Each Constituency?

The three groups of respondents come in with very different predispositions to the technology, as evidenced by their additive constants. Table 16.1 further shows the different winning elements:

1. *Market researchers.* The key themes are those of *process*, including ability to handle many variables, testing large arrays of elements, and flexibility. Market researchers pay attention to process, not to the direct economic benefits (e.g., increased market share and competitive dominance). To market researchers the benefit lies in the expanded number of stimuli and the process itself. The additive constant (34) means that, without any elements, market researchers are moderately interested in the product. The top element "Can handle a wide range of variables and help you understand independent utili-

ties for each piece of the stimuli" when used alone in a simple concept generates an expected interest level of 61; that is, 61% of the market researchers would find this proposition interesting.

2. *Marketers.* They are interested in the power of the approach—what it does for them. They are interested in the business aspect rather than in the process aspect. One quite telling benefit is that "For retailers for on the spot research" the utility value is positive (+12), whereas it is negative for the marketing researchers and for R&D professionals.

3. *R&D professionals.* The self-authoring system is basically interesting to R&D professionals (additive constant of 53), but the features and benefits that really break through are the ability to modify the test while it is fielding and the ability to explore many alternatives. These benefits are consistent with the way that R&D professionals approach their job.

It seems clear now in retrospect that researchers and research users look at the same technology with radically different viewpoints. Some individuals look at the technology from the viewpoint of process: what does it do and how does it do it? Others look at the technology from a more global approach: what does it accomplish for the business? The lesson from this exercise is simple: what to researchers is a unified, single technology cannot be positioned or sold to prospective users in a single way. Rather, we have to put the technology into the customer's mind in such a way that it taps his or her own particular hot buttons. Each group of clients comes in with a different set of needs.

Self-authoring Concept Development: Porting the Technology to the Internet

The Internet offers a particularly attractive environment for concept testing or new product market research. First, the cost of creating and testing virtual prototypes is considerably lower than that for physical prototypes, and more product concepts can be tested within the same market-research budget. Second, the Internet allows efficient and expedient access to respondents. Third, the new technologies, such VRML (virtual reality markup language), streaming video, and interactive sensory peripherals, enable visual, auditory, and tactile information to be disseminated and retrieved in rather powerful ways if it is needed (Mosley-Matchett 1998).

Trial and error over the past 5 years, following the early efforts with the PC Wizard, and ongoing discussions with clients generated nine considerations that moved the self-authoring system from a dream to a reality. It is important to note that this structure follows the approach that one might use for a food product, except that the research deals with the user-driven specifications of software, specifically a consumer research tool:

1. *Democratize: Allow anyone access to conjoint measurement, no matter what the experience level.* The most favorable response and some of the most novel ideas come from individuals at the R&D and consultant level. R&D creates physical proto-

types. They tend to be interested in anything that helps them to do so. Consultants are also important in this regard. Many consultants earn their fees by creating knowledge that leads to action. They tend to be interested in anything that adds to their research toolbox.

2. *Scope: Allow the conjoint task to handle a sizeable, yet manageable, number of elements.* The value of early-stage conjoint research lies in the ability to explore many elements. However, there is a problem: Experience with the self-authoring system shows that there is an initial hesitation of users when they are confronted with the opportunity to assess a large number of elements. By the time users have had a second or third experience with the technology these same individuals push for many more elements and larger designs. The number of alternatives must be nonintimidating and manageable.

3. *Guidance: Make the system "bulletproof" by using templates to reduce human error.* A *template* allows the timid and forces the adventurous users to follow a set of screens and type the stimulus material into those screens. The template provides additional, automatic error checking before permitting the user to move on. Over time it was clear that the template significantly reduced the error rate and accelerated the learning. A key discovery was that richness of alternatives was a negative instead of a positive. Offering users too many different templates was counterproductive. It took more time to explain the templates than actually to perform the setup and the study. Offering less flexibility, fewer designs, and fewer options forced users to think about the problem to be answered rather than to spend a lot of time understanding the system. The right number of features in the program made the program easy to use rather than difficult and inviting rather than daunting.

4. *Link users to enable collaboration across functions and geographies.* In many small studies run with the Internet-based system there was a strong spirit of collaboration

between different parties who were located in different states in the USA and, in some cases, in different countries. One person could save the editorial comments, and afterward another individual in a completely different location could immediately access the comments and make additional changes.

5. *Create a context-sensitive help system.* Guide users from the start to the end of the study by a set of context-sensitive pop-up screens. Thus, at the start of the project, a user can see the entire process outlined in a series of *screen shots*, similar to what the individual would see while following the template. During the setup, the user can press the help button for an explanation of the step and advice about what to do.

6. *Simplify fielding.* Although many users appeared to be able to *create* usable questionnaires, they felt paralyzed when it came to *fielding* the studies with actual consumers. Many potential client users felt uncomfortable dealing with services that sent out e-mail invitations. Over time the users became increasingly more comfortable with the field operation, but development of a comfort level required far more time than we expected.

7. *Make the data themselves "bulletproof": assure high-quality data by inserting automatic traps for data that are suspect.* One of the key issues with users of self-authoring systems (and indeed any research) is the fear that the results will be of poor quality and thus not usable. In projects handled by professionals rather than through self-authoring, many of these problems either disappear entirely or at least remain hidden. In contrast, when an individual is left to do the study itself, the flaws in the study execution become obvious. We recognized that data from respondents on the Internet could not be controlled. Therefore, the quality-control system would have to rely on features inherent in the actual data themselves. Each respondent evaluates concepts that are arrayed by experimental design. We created a system that measured

the *consistency* of an individual's ratings, based on the goodness of fit of the individual's model. This goodness of fit was indexed by the multiple R^2 of the individual's equation. The quality-control system identified which respondents had low R^2 values. Those data could be prevented from coming into the system at the time of data collection or flagged for later consideration.

8. *Simplify reporting.* An early version of the self-authoring system downloaded the data in Excel format, requiring the users to create tables. Subsequently, users requested easier-to-read, simpler formats with fewer numbers and more formatting. The simplified reporting created a more user-friendly system. One of the key findings here was that user-friendliness was far more important, at least observationally, than were statistical power and programming prowess.

9. *Create a cadre of strategic partners to help users throughout the entire process.* Many individuals who were interested in using the system felt uncomfortable. Most of these individuals felt uncomfortable creating the *raw materials* (viz., elements) for the conjoint study. Consequently, we developed a network of strategic partners who could help facilitate the early stages of the process, as well as handle entire projects when the need arose. The use of strategic partners also helped the acceptance of self-authoring procedures because it made the technology less intimidating.

Examples of the Self-authoring System: Screen Shots and Their Explanation

The best way to understand the self-authoring approach to concept development is to use a few screen shots to show the steps that a researcher follows in setting up a study, followed by a short presentation of the types of data. It is important to keep in mind that the self-authoring approach represents a simple type of computer application known as an

ASP (application service provider); that is, the technology resides on a central server that can be located anywhere worldwide. Furthermore, the self-authoring system is scalable for three reasons:

1. *Research users, not a central expert, do the work.* The users do the setup, invite the respondents, and inspect the results. Thus, there is no limit placed on the number of different users or respondents who can participate in the system. The general benefit is the distribution of effort and the empowerment of many people. This general benefit allows many researchers to do high-level research almost instantly, cost-effectively, and virtually effortlessly (except, of course, for the thinking that is involved in setting up the elements).

2. *Upgrades are inexpensive, off-the-shelf technology.* As the system expands in scope, all that need be added is greater channel capacity, with either faster communication and/or more processing systems. It is hardware, rather than human brainpower, that becomes a limiting factor. Hardware becomes cheaper all the time. People do not.

3. *New technology is presented in a format and structure that facilitates learning.* Any new technology or modification can be instantly inserted into the system. The modification can be at the level of design, rating question, statistical analysis, or reporting. All research participants in the system can avail themselves of the new ideas instantly. The training can also be downloaded as a Power-Point *show* or as a set of exercises. The new technology follows the approach of Windows, which has created a standard format into which new applications can be easily slotted. The users need not learn an entirely new language. Rather, the applications fit into what users already know.

The self-authoring system described next has been used since 2001 (Moskowitz et al. 2001). It was based on the PC version, but with numerous modifications based on user

responses to the original IdeaMap.Wizard. These are the two key modifications:

1. *Make the system easier to use.* The PC IdeaMap.Wizard proved unduly cumbersome to use. The key difficulties encountered were primarily setting up of the study on a PC, collection of data into a central repository after the study was completed, and subsequent processing. The users had to be technically proficient, which was not often the case. In all but a few cases there were glitches along the way, some more serious, some less serious. It was critical to repair this. Ease of use, as already noted, beyond anything else was critical to success.

2. *Make the setup completely transparent to the users and "bulletproof."* Researchers, like other individuals, are perennially seduced by power, by *feature creep*, that is, by the ability to do more advanced types of data collection and analysis. There is the nagging feeling that, unless one keeps up with the latest technology, one's perceived technological prowess is called into question. The experience with the IdeaMap.Wizard disabused the developers of that fallacy. Indeed, the most important aspect of the system was the ability through engineering beforehand to create a system that prevented errors. Power of the system was never challenged or debated. Ease of use was always a response to the question about why the system was useful.

The Process

The process follows these steps. Note that each step is driven by a template so that the user need not remember the steps:

Step 1: Log-in (Figure 16.1). This step simply requires an account and a minute's training.

Step 2: (Figure 16.2). Follow the navigation page with a simple point and click.

Step 3: (Figure 16.3). Define the study. Study definition includes the name of the

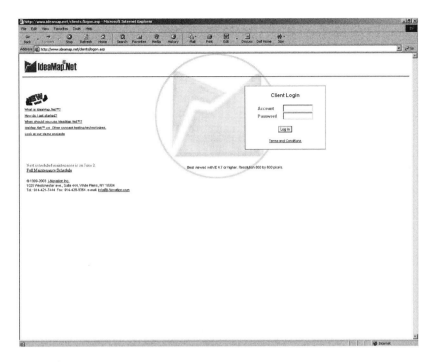

Figure 16.1. Log-in page.

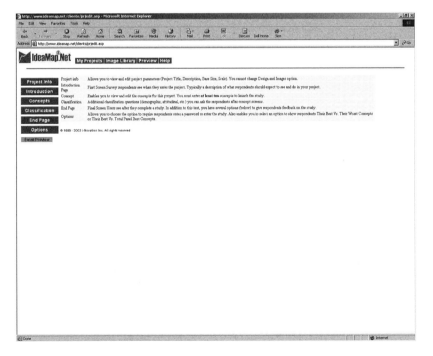

Figure 16.2. Navigation page.

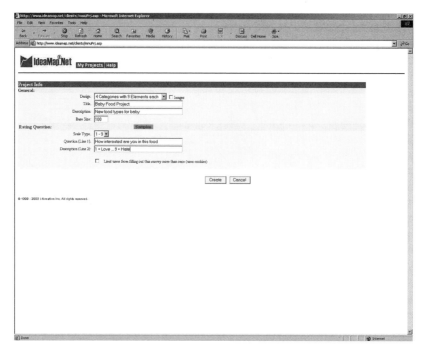

Figure 16.3. Study design.

study, the experimental design structure, the rating question, and the type of feedback that the respondents should get (none, their best combination versus worst combination, and their best combination versus the best combination created from the previous study participants).

Step 4: (Figure 16.4). Using a standard word-processing program with HTML instructions, type up the introductory page that defines the study. The typing is fairly effortless, and the instructions merely show the spacing, carriage returns, and the like. Some word-processing programs automatically convert the text to HTML, so the users need not even know HTML.

Step 5: (Figure 16.5). Following an Excel format, type in the elements, edit them, and so on. This can include pictures as a category. Most people are familiar with spreadsheets, so the effort becomes focused on the creativity rather than the computer technology, per se.

Step 6: (Figure 16.6). Select different classification questions (single answer, mul-

tiple answer, rating scale, open end), and type them in.

Step 7: Type in the end page (Figure 16.7). The end page thanks the respondents for participating. The end page may also lead them to a new questionnaire or back to the survey host company.

Step 8: Launch the study (Figure 16.8). Launching the study on the server means obtaining a link that can be distributed to the respondents, either by e-mail or by pop-ups from a Web site.

Step 9: Monitor the study as the data come in (Figure 16.9). The self-authoring system is fitted with a screen by which the researcher can check the results as they come in. Each respondent evaluates a full set of concept elements sufficient to create an individual model. Since each respondent generates an individual model, the parameters of the model are stored for that individual in a database, along with the responses to the classification questionnaire. These results are accessed in real time through a screen that enables the researcher to analyze the utility values across all respondents, rank the elements, and then display the elements on the screen. Monitoring can be a mesmerizing activity at first, as the data rush in, but with repeated studies the monitoring is soon abandoned by everyone but the most anxious of users.

Researchers are by their very nature curious. One of the interesting aspects of self-authoring systems is the ability to obtain data relatively quickly and easily. This leads naturally to the involvement of nonresearchers who originally commissioned the work. Researchers have "trained" their clients to expect data in a slow, measured way, with the data and the implications predigested by the researcher and presented by the researcher to his client. Internet-facilitated research speeds up the process, making the ultimate user impatient. Furthermore, self-authoring systems that place the research process under the user's control lead further to the diminution of patience. One consequence is that the person or group that commissions the research also wants to know how the work is going. It is no longer sufficient to placate a research user by an appeal regarding the process in the way it used to be, when one could say that the "data were still in the field" or the results "still at the tab house." Today's research user is savvier and wants data instantly.

A few words are in order about the reaction to this information by novice users of self-authoring systems, compared with experienced users. Conjoint analysis provides a very rich database of reactions to concept elements. A simple topline analysis provided to the user reveals which features win and which lose. One consequence of this instant information is the interest that novice users show in the topline data. Since the data can be obtained instantaneously, novice users often accesses the results again and again for the few hours of the study. The results typically stabilize with 50 or so respondents. At that point the utility values no longer change, since each

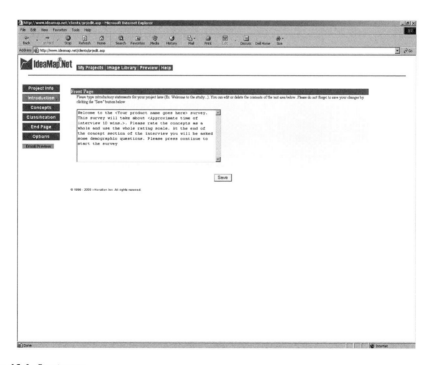

Figure 16.4. Front page.

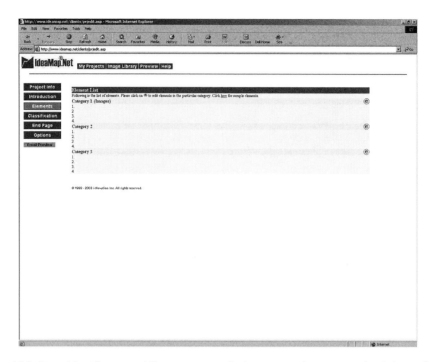

Figure 16.5. Spreadsheet format enabling users to type in the concept elements or upload pictures into the study.

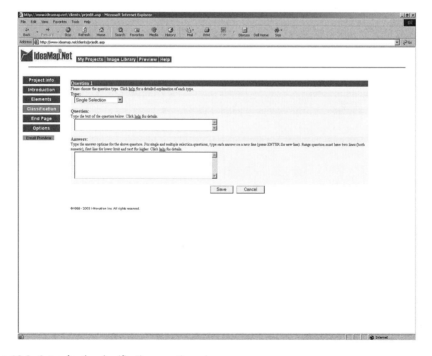

Figure 16.6. Setup for the classification questionnaire.

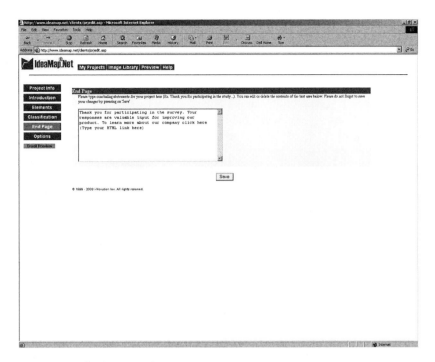

Figure 16.7. End page for the respondents.

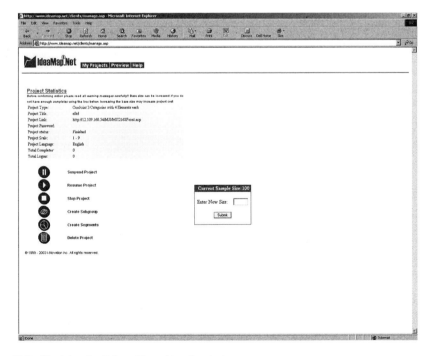

Figure 16.8. Obtaining the link and launching the study.

respondent contributes less and less to the mean. After a while, novice users become bored with the topline data and move on.

Self-authoring and the computer-based interview and analysis provide the researcher with instant feedback in terms of the number of respondents. Since every respondent generates his or her own model, and since the model data are saved in a file, one can examine the data on an aggregate basis in real time. Usually, the data converge after the first 50 respondents, at least for the total panel.

Step 10: Download the data (Figure 16.10). The data are downloaded after being processed. The analysis creates a set of Excel files that comprises the raw data and the summarized data. The data comprise total panel, key subgroups (user defined), and automatic segmentation into two-, three-, four-, and (optionally) five- and six-segment solutions. The

data are in the form of a set of zipped files for further analysis or for immediate use.

Step 11: Opening the summarized data (Figure 16.11). The data are summarized by total panel and key subgroups. The results are created for the models both before transformation to a binary scale (top-3 box) and after binary transformation.

Extending the Self-authoring System to Package Design: Visual Conjoint Analysis

For the self-authoring notion to be applicable to package design, a step beyond text concepts, the following in-going issues are worth considering:

1. *More combinations are better than fewer combinations.* Each respondent should evaluate many combinations. The more

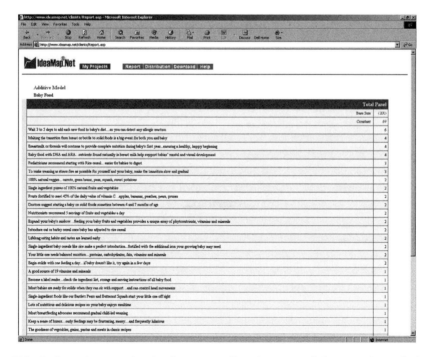

Figure 16.9. Study-monitor page showing the current utility values across all the respondents who have participated.

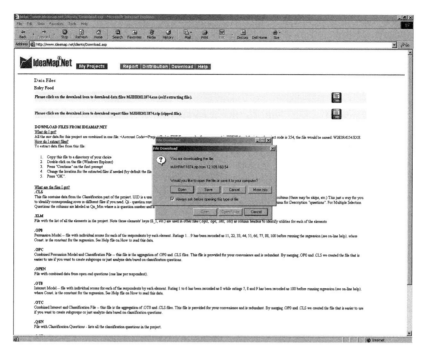

Figure 16.10. Downloading the results.

	Additive Model Table For: MJIHRM11874											
		Total San	Q1_1 chil	Q1_2-3 cl	Q1_4 or n	Q2_0-3 m	Q2_4-6 m	Q2_7-9 m	Q2_10-12	Q2_13-18	Q2_19-24	C
	Base Size	200	66	111	23	14	18	16	23	38	32	5
	Constant	68.7	76.7	66.3	57.5	59.7	94.5	73.5	64.4	57.3	65.1	7
E1	Single ingredient purees of	2.4	2.9	1.4	5.2	6.1	-8.6	0.6	-3.4	9.3	1.4	3
E2	Certified organic baby food	-1.9	-3.5	-1.7	1.3	2.1	-7.1	-1	-11.6	0.2	2.1	-
E3	Fruits fortified to meet 45%	2.4	2.6	1.7	4.7	-6.5	-0.6	-2.4	1.1	3.3	7.6	3
E4	100% natural veggies…carr	2.5	2	1.1	10.4	3.2	-2.3	3.7	0.7	4.6	3.4	2
E5	Single-ingredient foods like	1	2.5	0.7	-1.6	-5.2	-5.6	3.2	8.6	2.1	-2.3	2
E6	Pear Berry Medley and yum	-1	-0.7	0.2	-7.4	2.3	-2.9	3.6	-0.6	0.9	-3.7	-
E7	The goodness of vegetable	0.7	-1.3	1	4.6	-1.8	-4.6	7.5	-4.5	5	1.3	-
E8	A variety of organic baby fo	-0.9	-0.1	0.4	-9.6	-1.7	-4.3	-0.5	-10.1	0.7	3	0
E9	Apples & Bananas, Pears &	-0.3	-4.2	1.6	1.5	10.8	-4.4	5.3	-8.1	4.1	-3.4	-
E10	Lots of nutritious and delici	0.9	2.5	1.1	-5	-1.8	-3.7	3.5	-0.7	3.2	3.9	-
E11	Lifelong eating habits and t	1.9	2.7	0.9	4.4	-4.9	1.7	-2.6	2.4	5.9	5.4	0
E12	Expand your baby's rainbo	2	0.3	3.3	0.4	2.6	1.4	2.7	-0.8	4	5.1	-
E13	Expose your baby to a vari	0.3	-3	3.1	-3.9	-9.4	1	9.3	-1.6	1.7	-0.9	0
E14	Baby food with DHA and AR	3.6	2.1	5.2	-0.5	6.1	3.8	10.1	0.1	7.9	5.7	-
E15	Your little one needs balanc	1.8	4.4	1.3	-3.1	-6.3	-0.3	-1	4.6	0.6	4.9	3
E16	Most babies are ready for s	1.4	1.2	2.5	-3.7	-2.3	-0.2	5.1	10.9	4.8	-6.7	0
E17	Become a label reader…ch	1.4	1.1	1.4	2	-1.5	4.6	6.3	2	4.1	2	-
E18	When feeding solids, choos	-1	-0.8	-1.1	-1.3	-14.4	-0.9	-12.5	2.9	4.8	-2.5	1
E19	Nutritionists recommend 5	2.3	2.2	1.8	4.7	7.3	-4.6	3.5	-3	9.2	-1.7	3
E20	To make weaning as stress	3.3	-0.1	5.9	0.2	3.4	2.9	9.7	-1.2	6.1	4.4	0
E21	A good source of 19 vitamir	1.4	1.3	1.1	3.3	7	0.2	9.3	0.2	1.1	-0.6	0
E22	Breastmilk or formula will c	4	1.1	5.4	5.4	11.8	1.4	-1	-0.6	8.2	5.8	2
E23	Talk to your baby's doctor	0.4	3.4	-2.4	4.9	10.7	-4	-5.7	-0.1	-1.4	4.7	-
E24	Most breastfeeding advoca	0.8	-3.4	3.3	0.8	15.7	-4.5	-9.4	2.8	7.9	2.7	-
E25	The American Academy of	0.1	-4.1	2.4	1.4	9.1	-3.7	-3.3	-6.9	3	2.2	-
E26	Pediatricians recommend s	3.5	4.8	3.2	1	12.2	2.2	3.3	1.8	2.9	7.4	0
E27	Doctors suggest starting a	2.3	2.7	2.4	0.4	16.4	-2	2.8	6.3	-1.7	0.4	1
E28	Wait 3 to 5 days to add each	5.8	3.1	8.1	1.9	6.5	5.9	4.2	18.5	7.3	2.3	2
E29	Making the transition from	4.2	4.8	5.4	-3	7.2	1.6	-2.7	10.2	6.9	0.6	4
E30	Keep a sense of humor…ea	0.7	-3.9	4.7	-5.1	-1.4	-4	-1.8	1.2	9.5	0.9	-
E31	Introduce oat or barley cere	1.9	2.5	2.5	-2.4	-3.1	0.3	-1.3	10.8	0.3	3.3	1
E32	Begin solids with one feedi	1.5	-3	5.1	-2.8	5.7	3	-3.7	-5.1	3.8	0.1	3

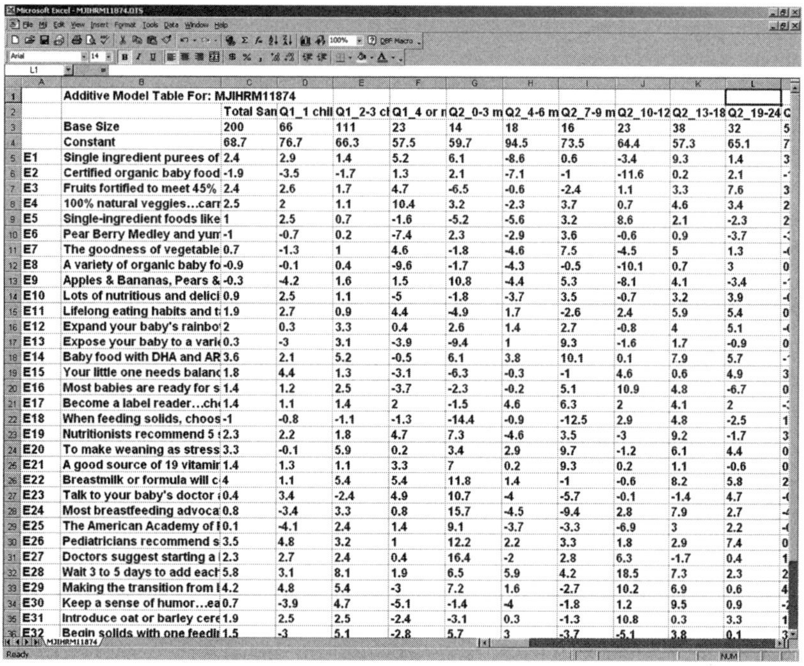

Figure 16.11. Results of the study, which are summarized in a spreadsheet file. The *rows* are the elements, and the *columns* are the subgroups.

combinations that a respondent evaluates, the more tightly will one be able to estimate the utility values for the different visual elements. Of course, the respondents are not machines and have limits to how many combinations they will test. With package design, the respondents need not do the reading in the same way that they read concepts. Packages can be scanned quickly and reacted to quickly. Hence, the respondents might well do 50–100 package evaluations in the time it takes to read 25 concepts.

2. *Let machines create the combinations because they do a better job than humans do.* A machine can create these combinations by putting together text or, later, for package design superimposing graphics transparencies. The machine can produce hundreds of these combinations by following an experimental design, which would put stress on the researcher to create the combinations beforehand. It is one thing to create 5–10 text combinations of the type shown in Figure 16.12. It is another to create 50–60 of these combinations. The issue gets worse with pure graphics, such as those shown in Figure 16.13. It becomes very difficult to create many different combinations by hand. By the time 40–60 combinations are created, even the most patient researcher may lose interest.

3. *Each respondent should rate different sets of combinations.* One of the benefits of a basic design structure is the ability to permute the structure to make many different sets of combinations. The permutation of the design creates *isomorphic designs* that are equivalent to one another. As a consequence there is far less chance that the data will be affected by one or two unusually strong-performing or poor-performing combinations. Furthermore, by allowing each respondent to rate different sets of combinations, the researcher lays the groundwork for studying interactions among variables, which could not be studied heretofore without creating the specific combinations wherein the interactions were thought to exist.

The Six Steps for the Self-authored Package-concept System

Applying conjoint analysis, or, more correctly, experimental design principles, to the problem reveals the rules by which the package components drive the response. Conjoint analysis introduces a structured approach to the problem of understanding drivers of consumer response. One way to apply conjoint approach to package-design research might follow these six steps in an almost algorithmic fashion (see Moskowitz et al. 2004 and Chapter 6):

1. *Template.* Create a template for the package being studied. The template is the basic design. Within its body are the locations where the package-design features are placed. Thus, the template is the outline or blueprint. The template itself is a means to organize the stimuli.

2. *Features and elements.* Once the features are identified, create a series of alternative executions or elements for each feature. Thus, if there is a brand name, create several alternative brand names. In the simplest of cases, create two options: the brand name and a blank to denote no brand name. The key here is the notion of alternatives. Conjoint analysis looks at the contribution of many alternatives for each feature, trying to discover the *utility* or contributing power of each alternative.

3. *Combine features by experimental design.* Experimental design provides a schematic whereby the researcher creates a number of alternative package stimuli to test, such that each of the stimuli contains different executions of the features. The package designs look different, but they are all connected by the experimental design.

4. *Test the combinations among consumers.* Each consumer can evaluate all or just some of the package combinations, on one or several rating scales. In consumer research the conventional rating scales deal

with one or another form of evaluation, good or bad, but there can be other rating scales, as well.

5. *Model the results.* Relate the presence/absence of the different package-design elements to the response. Usually, this analysis is done by ordinary least squares, although a variety of other regression methods are commonly used, as well, such as logit (Batchelor 2001).

6. *Interpret the data.* The coefficients of the least-squares model show the part-worth utilities or *driving* capability of the elements.

Specific Issues Involving Visual-based Conjoint Analysis

When the test stimuli are all visual rather than text, several practical issues arise. These involve the nature of the test stimulus, the modeling, and others that touch on the *artistic involvement* of the following:

1. *The nature of the test stimulus.* In conventional *full profile* or *partial profile* conjoint analysis the researcher puts together phrases and/or pictures systematically, such as the test stimuli shown in Figure 16.12. The concepts in Figure 16.12 show the effect of

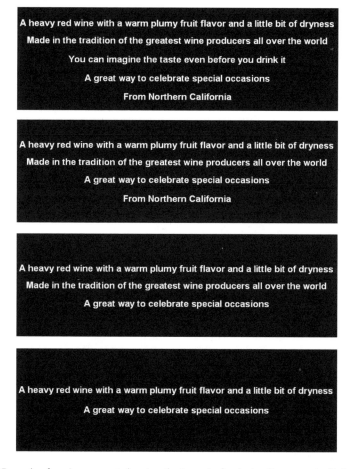

Figure 16.12. Example of a wine concept showing the impact of reducing the amount of information.

reducing the amount of information. When it comes to verbal concepts the respondent often fills in the missing information. Concepts can be incomplete, and the respondent will still react to the concept with little discomfort. There may be missing information, but the information being missing is not disturbing. In contrast, for visual stimuli, reduction of information may be bothersome, as shown by the concept in Figure 16.13. Reducing the amount of information leads to a disquieting feeling that there may be a lot missing from the package. Thus, the integrative and mental fill-in that occurs with text concepts does not necessarily exist for visual concepts.

2. *Modeling issues.* Conventional conjoint analysis uses regression analysis, which itself comes in a variety of types. The most direct type of regression is OLS (ordinary least squares). One very useful form of OLS is known as *dummy variable* because the elements take on one of two values: 0 if the element is absent from the concept, and 1 if the element is present in the concept. For the concept work in this book, OLS has been used extensively because of its simplicity of analysis and the intuitive meaning. By arraying the combinations of concept elements in a specified experimental design with *true zeros* (i.e., with some combinations entirely missing a category of elements), the researcher can estimate the absolute values of the utilities (see Table 16.2, left panel). In other situations, often encountered by researchers, there is the demand by the research user that each concept have one element from each category (see Table 16.2, right panel). In this case the absolute utility value of an element is impossible to estimate. Rather, the utility values are estimated relative to a *reference* value. The reference element is one of the concept elements. One cannot compare the utilities of elements across different categories of concept elements. One can only compare the utilities of elements within the same category. When it comes to verbal concepts the use of true zeros is generally no problem. If the concept lacks a category, the respondent will mentally fill in the missing connectives and easily

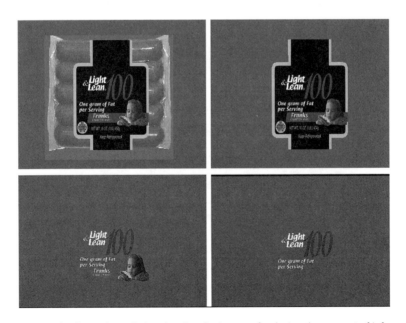

Figure 16.13. Example of a package design showing the impact of reducing the amount of information.

Table 16.2. Comparison of two types of designs*

| | A concept can comprise 2–3 components, enabling estimation of absolute utilities | | | A concept *must* comprise 3 components, enabling estimates of relative utilities only | | |
| | Category | | | Category | | |
Combination	1	2	3	1	2	3
1	Absent	3	1	1	1	1
2	4	Absent	3	1	2	2
3	1	4	Absent	1	3	3
4	1	1	4	1	4	4
5	2	1	1	2	1	2
6	1	2	1	2	2	3
7	Absent	1	2	2	3	4
8	3	Absent	1	2	4	1
9	2	3	Absent	3	1	3
10	2	2	3	3	2	4
11	4	2	2	3	3	1
12	2	4	2	3	4	2
13	Absent	2	4	4	1	4
14	1	Absent	2	4	2	1
15	4	1	Absent	4	3	2
16	4	4	1	4	4	3
17	3	4	4			
18	4	3	4			
19	Absent	4	3			
20	2	Absent	4			
21	3	2	Absent			
22	3	3	2			
23	1	3	3			
24	3	1	3			

*The leftmost design presents combinations with categories absent from some concepts, allowing for an estimation of the absolute magnitude of utilities for all concepts. The right-hand design shows the combinations wherein every concept always has one element from each category, requiring estimation of the relative values of utilities.

make a judgment. The OLS, dummy-variable regression analysis will then generate an estimate of the absolute utility values. With package-design concepts, as Figure 16.13 suggests, sometimes the visual concept may not make sense if too many elements are missing. Furthermore, when it comes to visuals, those individuals who commission the research feel strongly that each visual stimulus must comprise one element from each category. We will deal with approaches to answer this problem, trading off the rational appearance for the powerful analysis. They do not go hand in hand with visual stimuli.

3. *Artistic involvement.* In conventional concept research using only phrases, often one can get an intuitive feel about the order of the categories in the concept. In visual research there is far more left to interpretation. Even the template may vary, so some investigators feel that the label should be at the top and others feel it should be at the bottom. Unlike text-based concepts, there is no right or wrong order for visual concepts, except for the most obvious cases.

Considerations and Issues for Self-authoring, Visual Conjoint Analysis

Experimentally designed visual stimuli represent an emerging area for self-authoring systems. Once the self-authoring approach for text-based concepts (text plus picture) is accepted, the self-authoring system for visual conjoint analysis becomes a technical challenge rather than a research challenge. The objective is to develop the authoring system in such a way that it is easy and intuitive to use by novices.

A Case History and Some Considerations for the Visual Conjoint Analysis

The remainder of this chapter presents a case history about visual conjoint analysis involving a tea package. Unlike concept research, package-design research on the Internet brings with it a host of even more complex issues, problems, and controversies. In addition to the substantive issues, there are procedural issues, as well, and presentation of the process from the user and respondent viewpoints.

The Test Stimulus: What It Is and How It Is Created

The issues involved in visual conjoint analysis are more difficult than those involved in conventional text conjoint. Since the respondents do not necessarily fill in the blanks to generate a meaningful combination, we explored different types of designs that would allow for dummy-variable regression modeling, but also generate a relatively large proportion of complete concepts.

Design the Template

The template for this first project was a simple package design for tea bags. The objective was to create a system that gave full power to the respondent. The user first creates a template showing the location of the different categories. For this particular study, there were five different categories and thus five locations in the template. The template itself was a simple box.

Figure 16.14 shows an example of the first attempt to locate the different categories. The assumption is that the researcher is not an artist and cannot easily conceptualize the template, but can use trial and error to put the template together. The user identifies the nature of the categories. A category might be a picture, a colored background, a splash, a brand name, etc. The user selects the categories, drags and drops them to the location, and then manipulates the final outline and categories until the appearance is acceptable. In the more general case, users may wish to create their own template. This can be accomplished by having available to the user a set of predrawn silhouettes.

Figure 16.15 shows the finalized *trial* template, which comprises the categories located in their proper places. The key to Figure 16.15 is that the user does not know about templates at all, but, when seeing a test figure, can move the categories around until things look right.

Figure 16.16 shows the actual template that is created by the user's efforts. The template will then be used for the actual stimulus presentation.

Select Elements for Each Category

Experimental designs require a specific number of elements for each category. In conventional conjoint analysis, these are phrases that can be typed in. With self-authoring visual conjoint analysis, the phrases are replaced by pictures (jpg, gif, etc.) of relatively small size. Issues that needed addressing included the size of the pictures in terms of file size (small to facilitate Internet transmission) and size of the picture itself in terms of the template (fixed, to facilitate drawing on the template). Once the outline silhouette and template structure have been decided, the user selects different visuals from a library of visuals. One can add visuals at will to the library. Figure 16.17 shows the five different categories, four elements per category. The

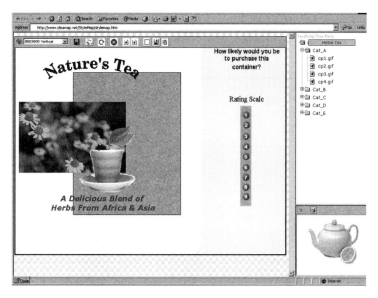

Figure 16.14. First attempt to create the template by locating concept elements (one per category) on a trial template.

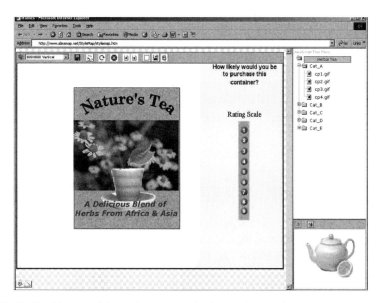

Figure 16.15. Finalized format of the package design, leading to the template.

up-front preparatory work for the visuals takes place prior to the development of the template and is an entirely different operation from the actual research itself. Visuals can be prepared automatically by a set of utilities that *grab* visuals from existing pictures and adjust them by shrinking, contracting, and changing their file formats.

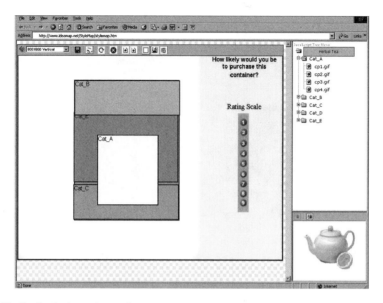

Figure 16.16. Finalized schematic template.

The Rating Question and Scale for the Entire Concept

This is very straightforward. The question is typically a 5-point or 9-point category scale. The scale appears below the figure. For the tea study, the question was "How interested are you in this herbal tea package?" The rating scale was 1 = "Not at all interested" to 9 = "Very interested."

How to Obtain and Then Orient the Respondents

Figure 16.18 shows the invitation screen that the respondents received by e-mail and to which they respond online. Figure 16.19 shows the orientation page.

A Respondent-oriented Feedback Mechanism to Improve the Quality of the Interview

A good way to ensure participation provides respondents with feedback about the interview progress, as well as about individual performance. The feedback mechanism comprises several parts. One part consists of a counter that shows the number of screens completed and the total number of screens. Nothing so irritates a respondent as having to go through many screens while not knowing how many are left. It is not clear as to the number of respondents who continue to work when they have this information; that question is worth a methodological study in itself. It is clear, however, to those that the author has interviewed that this type of progress bar helps the respondents to feel less frustrated. Figure 16.20 presents an example of the feedback, which comprises a counter at the top of the screen. In Figure 16.20 the counter says "28 of 43," meaning that the respondent has completed 28 of the 43 screens.

Another feedback procedure presents the respondents with their best combination of pictures. Showing a respondent what they like most reinforces them. Furthermore, it is possible to have them rate that newly emergent combination on attributes. Finally this feedback enables the user to determine whether the respondent feels that his or her answers correspond to this best product. They should, because the best concept picture is assembled from the data provided by

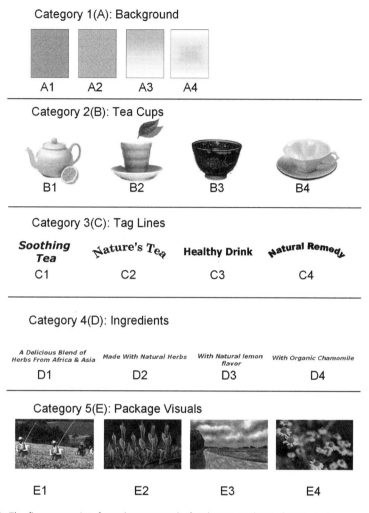

Figure 16.17. The five categories, four elements each, for the tea package-design study.

the respondent after the ratings have been analyzed by regression analysis. The best combination comprises the winning elements.

Results from the Visual Conjoint Analysis for Package Design

We will deal with several different types of issues when we look at the results:

1. Interest in the experience: how the respondents rated their impressions of this new type of interview

2. Statistical issues about design

3. What the results showed

4. Statistical analyses afforded by the self-authoring system

5. Approaching the knotty problem of pairwise interactions, now made feasible by the permuting method of the self-authoring system

Measuring the Pleasantness of the Experience

Is the package-design interview pleasant, even though some of the concepts were visually

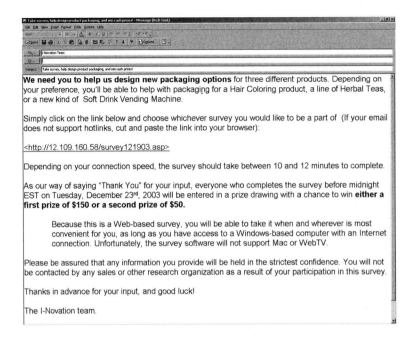

Figure 16.18. Invitation page.

Figure 16.19. Orientation page for the package-design study

incomplete (i.e., missing some categories, as dictated by the experimental design)? As stated previously, respondents tend to fill in the blanks with text-based concepts, but often people say that respondents do not fill in the blanks with visual stimuli. Do we see the respondents recognizing the stimuli as being somewhat meaningless?

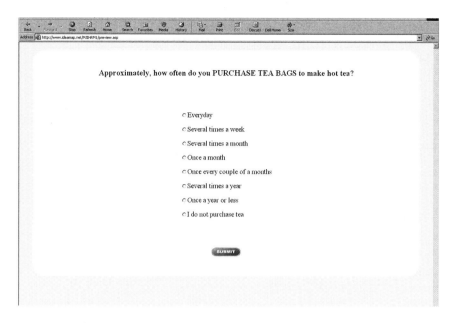

Figure 16.20. Classification questionnaire showing feedback about the number of completed screens (28) versus the total number of screens (43).

In this study of tea packaging, two of the screens for the classification questionnaire asked for positives and negatives about the experience. The response is submitted automatically after the respondent assigns the rating. Altogether, 95% of respondents provided at least one answer. Table 16.3 summarizes the most frequent responses. The three key responses were interesting package designs, easy to use, fast and simple. The ease of use category included the understandability of the survey and the simplicity created by need for simply clicking the responses without using the *submit* button. The fast speed of the survey was related to the quick download time. A verbatim summarization of this thought: "I liked the speed that the photos of products changed—not a lot of time loading each photo. I also liked the way the answer bubbles were set up, easy to click on." A large majority of the respondents used the words "easy," "fast," "enjoyable," and "simple" to describe their experience during the interview. The key learning

Table 16.3. Summary of responses to the interview by using visual conjoint analyis

Theme	% Respondents
Likes about the interview	
Pack designs	34
Easy	29
Fast	27
Easy and fast	14
Their opinion influences pack designs which will come onto the market	11
Simple	8
Fun	4
Dislikes about the interview	
No dislikes	49
Difficult to differentiate between packages	13
Did not like the tea visuals/ samples shown	8
Packages were too plain/ bland	6
Difficulty with computer (downloading, scrolling)	5
Boring	4
Open-ended questions	4
Too long	3
Miscellaneous	9

that emerged is the need for fun and speed. By and large the experience is pleasant, perhaps because the stimuli are attractive, because they are easy to process and judge because only visual inspection is required rather than reading, and because respondents need not tap the *enter* key to submit the response. Once the answer is pushed, the next visual screen comes up rapidly.

Statistical and Methodological Issues

Experimental Designs

The objective of conjoint measurement is to identify the part-worth contributions or utilities of the component elements. For this visual implementation the ideal solution is to estimate the absolute utility values by using an incomplete design. The basic design comprised combinations listed in Table 16.4. This is a very simple design called a Plackett-Burman

screening design (Plackett and Burman 1946), which is the same underlying design structure that was used for IdeaMap (see Chapter 5). Other designs for more complex concepts are necessary for bigger issues. It is important to develop experimental designs that can handle 4–12 variables or categories, each with multiple levels, because of the nature of visual stimuli. Those stimuli comprise many types of information and are by nature more complex than simple verbal concepts that cannot convey this amount of information. Simple verbal concepts with too much information would be impossible to process; visual design concepts with detailed information are more common.

Dummy-variable Models, Missing Elements, and the Appropriate Nature of the Test Stimulus

The experimental design shown in Table 16.4 calls for concepts in which, and on some oc-

Table 16.4. The actual experimental design used in this study*

Concept	A = backgrounds	B = teacups	C = names	D = ingredients	E = visuals
1	1	4	2	4	4
2	Absent	1	4	2	4
3	2	Absent	1	4	2
4	2	2	Absent	1	4
5	3	2	2	Absent	1
6	2	3	2	2	Absent
7	1	2	3	2	2
8	4	1	2	3	2
9	3	4	1	2	3
10	3	3	4	1	2
11	Absent	3	3	4	1
12	3	Absent	3	3	4
13	1	3	Absent	3	3
14	2	1	3	Absent	3
15	Absent	2	1	3	Absent
16	Absent	Absent	2	1	3
17	4	Absent	Absent	2	1
18	Absent	4	Absent	Absent	2
19	1	Absent	4	Absent	Absent
20	3	1	Absent	4	Absent
21	4	3	1	Absent	4
22	4	4	3	1	Absent
23	2	4	4	3	1
24	4	2	4	4	3
25	1	1	1	1	1

*The design for this simple study is the five-level Plackett-Burman screen design, with one level 5 missing from the concept.

casion, one or more categories are absent. The purpose of missing elements is to enable a valid estimation of utility values by using the dummy-variable approach for ordinary least squares. In a conventional concept study the absence might pose no problem because the respondent would fill in the missing elements, as previously noted and suggested by Figure 16.12. In the visual conjoint analysis for package design, however, the issue of missing categories is more serious and must be addressed via experiment. This issue was raised earlier after considering the nature of the package concept with various elements missing (see Figure 16.13).

The issue of responses to concepts with missing categories was addressed by doing three parallel studies, all with the same Plackett-Burman design shown schematically in Table 16.4. The objective of the additional experiments was to determine whether respondents reacted any differently when they evaluated concepts that had missing elements versus reacting to concepts that always had five elements; that is, there may be some degradation of the concept. Did this degradation affect the respondents' attitude toward the study, and did this degradation affect their ratings? Keep in mind that there are two opposing forces. Degrading the visual concept enables a better estimation of utilities because of the power of dummy-variable analysis. Degrading the visual concept also means presenting respondents with a less full, less complete concept.

The strategy to address missing categories followed these steps:

Run Parallel Studies, All of Which Conform to the Design Shown in Table 16.4

Follow the design completely, with some combinations comprising two, three, or four rather than all five categories. This approach follows the method used for text-based conjoint analysis:

Study 1: Absent categories are really absent from the concept. This is called the *no*

compensation condition. We see from the experimental design in Table 16.4 that many concepts call for the category to be absent. The study was set up so that a concept category could be legitimately absent from the concept. This first study is the *real study* to be analyzed here, because it provides true estimates of the utility value for each of the 20 concept elements.

Study 2: Only two of the categories can be absent from a concept. This is called the *partial compensation* condition. This strategy produces a concept that is not as incomplete. Two of the categories (B and D) were physically absent from the concept *when the experimental design called for their absence.* The remaining three categories (A, C, and E) were also supposed to be absent from certain concepts according to the experimental design. When a concept to be created and presented to the respondent with either category A, C, or E missing, the computer randomly selected an element from the same category and put the element in the concept to *compensate* for the missing element. In this way the respondents evaluated fewer degraded concepts.

Study 3. Whenever a category was absent from the concept, it was compensated for by a random selection of a concept element from that category, to produce a complete concept. This is called the *full compensation* condition. At the level of visual appearance the objective was to produce a concept that looked complete. Figure 16.21 compares two concepts: the first with all five categories present, and the second with three of the five categories present and the remaining two categories absent.

Random Assignment to Studies

Respondents were invited to participate. A respondent was guided to one of the three studies and participated without knowing the nature of the test stimuli, other than by visual inspections. The computer program assigned respondents to the studies in a rotating fashion in order to equalize the base size of respondents. Altogether, 102 respondents participated in each of the studies. A respondent participated in only one of the three studies.

Respondents Down-rate the Incomplete Concepts, But Do Not Reject Them

The in-going assumption was that the respondents would severely down-rate the degraded, incomplete concepts having few elements, because elements were missing and the visual appearance was incomplete. As Figure 16.22 shows, the distributions of the ratings for the concepts are not particularly different for the matched group of 102 respondents who participated in each study; that is, after separating the concepts into the three sets (no compensation to all categories compensated) and looking across all of the concept ratings we find a moderate tendency to down-rate concepts because they are missing elements. The means on the 9-point interest scale of the three studies are 4.39 for no compensation, 4.77 for partial compensation (3 of 5 categories), and 5.06 for compensating all five categories. The equivalent top-3 box statistics are 20.7%, 26.2%, and 30.7%, respectively.

The average scores and top-3 box statistics are higher when the missing categories are visually compensated for by the insertion of a random element so the concept is complete, with the maximum effect being approximately two-thirds of a scale point and about a 10% increase in top-3 box statistic.

An Analysis of Dropout Rates to Measure Boredom with the Interview

The proportion of respondents who drop out during the interview (i.e., log in but fail to complete the interview) is an indirect, inverse measure of interest. If the interview is boring we would expect respondents to drop out more frequently than if it is interesting. The proportion of dropouts is 24% for concepts that can be missing any of the categories so that the concept can comprise 2–5 elements, 16% for concepts that always comprise 4–5 categories, and 25% for concepts that always comprise 5 categories. It may be that concepts lacking some information, but not a lot, maintain respondent interest a lot more than do concepts that are always complete. This issue is worthy of further investigation.

All Five Categories   **Three of Five Categories**

Figure 16.21. Comparison of appearance of two concepts, one having all five categories present **(left)** and the other having only three of five categories present and two categories missing **(right)**.

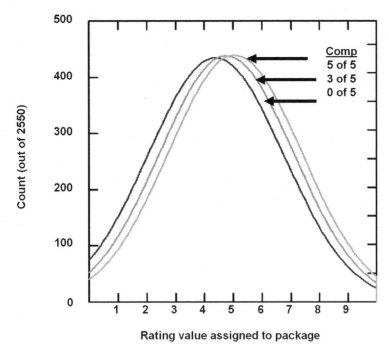

Figure 16.22. Distribution of ratings on the 9-point scale for the first 102 respondents participating in each of the three studies: 0 of 5 corresponds to concepts that could lack 2 of 5 categories (no compensation); 3 of 5 corresponds to concepts in which three specific categories always had to be present (i.e., were compensated for when the designed called for their absence in a concept); and 5 of 5 corresponds to concepts in which all 5 elements had to be present in a concept even though the designed called for their absence (maximal compensation).

The Individual-level Model

We create the individual-level model for the respondents from study 1, which used incomplete concepts and did not compensate for missing concepts. Following the approach discussed in Chapter 5, the ratings from an individual respondent are then subject to dummy-variable regression analysis, first, after a simple multiplicative transformation ("persuasion," multiplied by 11) and, second, after the individual's ratings were transformed to a binary scale ("interest," 1–6 transformed to 0, and 7–9 transformed to 100). The individual data show high R^2 values, a measure of the goodness of fit of the individual model to the data (see Chapter 9 and Figure 16.23). These results mean that, even with degraded visual concepts that are

missing elements, the respondents are consistent in their ratings.

The results are listed in Table 16.5 in a form already well known to the readers. Each of the 20 package-design elements generates its own utility value, just as was done for the conventional package-design research prior to self-authoring (see Chapter 6). Table 16.5 presents the average utility values across all of the respondents in study 1 (no compensation for missing elements in a package concept).

Concept-response Segmentation Based on the Utilities of the Visual Elements

The respondents were *automatically* segmented by the pattern of utility values (except for the additive constant), as discussed previously in this chapter. The two segments

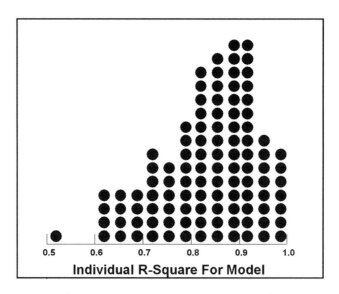

Figure 16.23. Individual-level R^2 values for the additive model. The median R^2 is 0.87.

that were most easily interpreted are best defined by the elements that appeal to them:

Segment 1: Nature seekers. Respondents in this segment react strongly to the big pictures (flowers, plants, meadow scene, and hikers). They also react strongly to the names Soothing Tea, Nature's Tea, and Natural Remedy. They accept, but not strongly, the name Healthy Tea, so it is not health that attracts them. These respondents appear to want a package that speaks to *nature* and *natural*. Their strong response to the large pictures confirms that. One of the interesting speculations is whether in the population there exist groups of individuals with this type of strong reaction to nature for many products.

Segment 2: Ingredient seekers. Respondents in this segment react strongly to novel ingredients such as "A delicious blend of herbs from Africa and Asia." They respond strongly to visual messages about tea, such as a teapot or an oriental bowl from which tea is sipped.

Modeling Response Time

A key benefit of conjoint analysis is the ability to identify the nature of response time to the total concept or package and then to partition that response time across the different ele-

ments (see Chapter 10). On the Internet, response time can be tricky to measure, because the response time must take into account the transit time over the Internet. The self-authoring system was created so that all of the visuals load onto a respondent's computer at the start of the study. This strategy of using *cached* visuals significantly reduces the time it takes to move from one complete visual concept to the next, making response time more meaningful. Response time was analyzed by using ordinary least-squares regression, *but without an additive constant.* The logic is that the response time should be 0 without any elements in a fully graphic concept. The response time, in tenths of seconds, appears to be longest for the pictures of tea and tea-related items and secondarily to the large pictures (see Table 16.5). The response time is clearly shortest for the background colors.

Regression Methods: Logit and Probit Methods Compared with Ordinary Least Squares

Different approaches for estimating utilities have been suggested by researchers, especially since the independent variables are binary (present/absent). Do these different approaches generate the same decisions? If

Table 16.5. Utility models for the 20 package design elements for total panel and for two package-response segments: the elements shown in descending order of interest by total panel

		Persuasion total	Top-3 box interest total	Response time	Interest Segment 1	Interest Segment 2
	Additive constant	23	−1	NA	−1	−2
	Background					
A4	White/green center	1	3	7	4	0
A2	Blue	1	2	7	3	−1
A1	Salmon	1	1	5	1	0
A3	Fuchsia/white	1	1	5	−2	3
	Teacups					
B1	Teapot	10	11	16	9	12
B3	Oriental bowl	8	8	16	5	10
B4	Teacup and saucer	3	3	20	0	6
B2	Tea leaves	3	1	19	−1	4
	Names					
C1	Soothing Tea	9	10	13	8	11
C2	Nature's Tea	11	10	15	7	13
C4	Natural Remedy	7	8	12	6	9
C3	Healthy Tea	5	5	15	3	7
	Ingredients					
D3	With natural lemon flavor	6	6	14	3	9
D1	A delicious blend of herbs from Africa and Asia	6	5	16	−2	14
D2	Made with natural herbs	6	5	15	1	8
D4	With organic chamomile	6	5	15	0	11
	Pictures					
E4	Flowers	14	12	16	18	1
E2	Plants	14	10	16	15	2
E3	Meadow/mountain scene	10	9	16	13	1
E1	Hikers	6	2	15	10	−11

so, then it does not matter which of the valid estimation methods are used. Ordinary least squares (OLS) is the easiest to interpret, but is good only if it generates the same conclusions as the more conventional logit or probit models. The raw data from this study—2500 cases or data from the first 100 respondents in study 1—were subject to three regression analyses that generate estimates of coefficients, which cannot be directly compared because they mean very different things. The *t* ratio and the corresponding *F* ratio for each term, for each method of regression, can be compared meaningfully. Figure 16.24 shows a scatter-plot matrix for the 20 different *t* values, with one *t* value for

each of the 20 visual elements. The *t* values are almost identical, and the relation is a straight line. The *t* values are almost identical across the three regression methods. This means that no matter what estimation method is used the significance level of each of the 20 utilities is identical. Thus, the same decision would be made about the contribution of the element to the acceptance of the package. From a practical viewpoint the virtually identical results mean that OLS methods can be used to estimate the utility values. Furthermore, OLS has the advantage of being far easier to explain, both to oneself at an intuitive level and to potential users of the data who need to

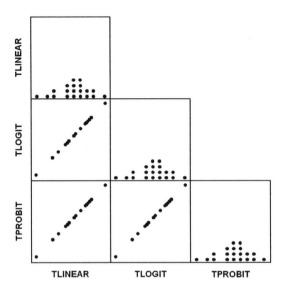

Figure 16.24. Scatter plot matrix of the *t* values for the 20 elements after they have been estimated by three regression methods: ordinary least squares (*TLinear*), logit (*TLogit*), and probit (*TProbit*).

understand what the utilities mean. OLS results show the intuitively simple notion of the additive conditional probability of a concept being *interesting* if the element is present in the picture or concept.

Uncovering Pairwise Interactions Among Elements

Interactions among variables remain a recurrent theme in consumer research and especially so in conjoint analysis. Those who aver that researchers cannot adequately understand drivers of acceptance often base their argument on the combination of the parts being greater or occasionally less than the whole. According to these skeptics, the contributions of components to concepts cannot really be measured because the measurement does not incorporate within it an adequate treatment of interactions.

In traditional conjoint analysis these skeptics may be hard to refute. The analysis works with a limited set of combinations or *cards*, even though these cards may be ported to a computer. Most researchers work in the mode of testing a fixed set of combinations that are designed to be efficient so that the ratio of concept elements to concepts is as close to 1.00 as possible; that is, for regression purposes it is necessary to have more combinations or *cases* than elements or *variables*. At the same time, however, it is also vital to have as few combinations as possible. The goal is to have enough combinations to create a valid model. Finally the traditional methods use the same test combinations among a large number of respondents to create a *solid base size* behind each concept element.

The self-authoring approach presented for both text and visual stimuli is set up in a way that enables researchers to identify pairwise interactions, even if these interactions were not previously specified. This ability to model interactions comes from one of the features of the self-authoring system: the systematic permutation of combinations in a basic design; that is, there is a single, relatively simple design structure for a given study. However, the actual element corresponding

Table 16.6. Three permutations of the basic experimental design: the design structure remains the same, but the assignment of elements changes

Background			
White/green center	A1	A4	A4
Blue	A2	A2	A3
Salmon	A3	A1	A2
Fuchsia/white	A4	A3	A1
Teacups			
Teapot	B1	B1	B4
Oriental bowl	B2	B3	B3
Teacup and saucer	B3	B4	B2
Tea leaves	B4	B2	B1
Names			
Soothing Tea	C1	C1	C4
Nature's Tea	C2	C2	C3
Natural Remedy	C3	C4	C2
Healthy Tea	C4	C3	C1
Ingredients			
With natural lemon flavor	D1	D3	D4
A delicious blend of herbs from Africa and Asia	D2	D1	D3
Made with natural herbs	D3	D2	D2
With organic chamomile	D4	D4	D1
Pictures			
Flowers	E1	E4	E4
Plants	E2	E2	E3
Meadow/mountain scene	E3	E3	E2
Hikers	E4	E1	E1

to a design element varies. Thus, Table 16.6 shows three possible permutations. With this type of strategy there are 4! (4 factorial) or 24 assignments possible for each of five categories. Altogether, therefore, there are $(24)^5$ or 7,962,624 different permutations of the same design. One need create only about 100 of these permutations in order to ensure that no particular combination of elements unduly affects the ratings. The foregoing strategy produces an unexpected benefit: it allows new pairwise combinations to emerge that would not have emerged had the researcher remained with a single set of combinations. This happy result leads to the unexpected ability to identify synergistic combinations, as well, as we see in the next section.

The Strategy to Uncover and Measure Significant Interactions Above and Beyond the Linear Contributions of the Elements

A continuing issue in conjoint measurement is the search for interactions among pairs of elements. Most conjoint studies have to build in pairwise interactions beforehand, so the search for interactions is actually the evaluation of the magnitude of such interactions. By building in these interactions the researcher assumes that they exist.

What happens, however, when there are many elements and when these elements are not continuous variables (e.g., they are not different levels of one variable, such as number of calories), but instead are discrete, uncon-nected elements, such as the different ingredi-ents in this study (e.g., "natural lemon flavor" and "a delicious blend of herbs from Africa and Asian")? The following six-step approach provides one way to identify the significant in-teractions between pairs of concept elements, beyond the simple contributions of the ele-ments alone [The authors have filed for a US patent to cover this as a business process]:

1. Work with a relatively generous set of elements. In this study we deal with only 20 elements, but there is no reason why the study cannot deal with 100 elements.

2. Create the set of permutations of the basic design. For each respondent, permute the combinations, keeping the same design structure.

3. The permutations create different com-binations for each respondent. The net effect of these permutations is to create many more observations than elements. Whereas in the design structure used here (the Plackett-Burman five-level design) there are 20 ele-ments and 25 combinations, the number of combinations among 100 respondents is re-ally 2500.

4. If we move away from looking at data at the individual-respondent level to looking at the data from the entire panel, we now have 20 independent variables and 2500 combinations (25 combinations, 100 respon-dents = 2500 combinations). There are many more unique combinations than predictors and far more unique combinations across these 2500 concepts than the original 25.

5. Using a spreadsheet or a transforma-tion program, create all meaningful pair-wise combinations of the elements by multi-plying the two elements in the pair. Creating combinations of two elements from the same category is not meaningful. Creating combinations from two separate categories is meaningful. For any pair of categories, each of which comprises four elements, there are 16 such pairs. Ten unique pairs can be created from five categories of elements.

Thus, there are 160 pairs of elements and 20 single elements, or 180 predictors, for the 2500 cases. The 180 independent variables take on the value 1 if present in the concept and 0 if absent from the concept. By the na-ture of the design, there will be far more sin-gle elements than combinations.

6. Analyze this dataset in a single regres-sion. Using stepwise regression, force in the 20 linear terms, one term for each of the 20 concept elements. Afterward, use stepwise regression to add new combinations if these combinations are statistically significant and add predictability to the model.

We see the results of this analysis in Table 16.7. The comparison of the utilities should be made only within the context of Table 16.5. (The utilities will differ here from the those obtained by averaging the utilities for the individual models.) The dependent vari-able again was the top-3 box, which took on a value of 100 if the rating were 7–9 and a value of 0 if the rating were 1–6.

There are 12 significant interaction terms out of the additional 160 terms. The equation with the interaction terms fits the data better, as it should, since we are adding more pre-dictors to the same dataset.

The real question, however, is what the magnitude is of the effect. That is, does iden-tifying interactions materially impact the de-cision that we would make? Are there some combinations to be sought after that are truly synergistic so that the combination is far greater than the expected linear sum? Are there some combinations to be avoided so that the combination is far lower than the ex-pected linear sum? One way to answer the question regarding magnitude of effect esti-mates the utilities for the various pairs of elements that show significant interactions. We do the summing in two ways. First, we assume that there is no significant interac-tion. We look at the utilities of the *interacting elements* by using the model that ignores in-teractions (column B). Each element has a utility value, so we can compute the sum of

Table 16.7. Coefficients for a large-scale regression analysis relating the presence or absence of the concept elements to respondent ratings*

A	Linear with no interactions			Linear + significant interaction terms			Diff
A	B	C	D	E	F	G	H
Multiple *R*	0.11			0.19			
SE estimate	30.63			40.21			
Effect	Coefficient	*t* value	*P* (2-tailed)	Coefficient	*t* value	*P* (2-tailed)	
Additive							
constant	*12.01*	*3.29*	*0.00*	*14.86*	*3.99*	*0.00*	−2.85
A1	−1.13	−0.53	0.60	4.08	1.57	0.12	−5.21
A2	−1.25	−0.58	0.56	−6.15	−2.37	0.02	4.90
A3	−2.36	−1.09	0.27	−2.65	−1.23	0.22	0.29
A4	−4.83	−2.23	0.03	−6.64	−2.84	0.00	1.81
B1	8.53	3.79	0.00	3.93	1.51	0.13	4.60
B2	3.00	1.31	0.19	−3.19	−1.16	0.25	6.19
B3	5.40	2.39	0.02	7.17	2.94	0.00	−1.77
B4	3.85	1.67	0.09	5.22	2.11	0.03	−1.37
C1	3.81	1.72	0.09	0.59	0.25	0.81	3.22
C2	3.94	1.78	0.08	3.53	1.60	0.11	0.41
C3	2.12	0.94	0.35	0.64	0.29	0.78	1.48
C4	1.81	0.82	0.41	2.92	1.22	0.22	−1.11
D1	0.72	0.33	0.74	−2.31	−0.96	0.34	3.03
D2	3.68	1.68	0.09	3.47	1.60	0.11	0.21
D3	4.34	2.01	0.04	1.27	0.42	0.68	3.07
D4	0.24	0.11	0.91	2.01	0.83	0.41	−1.77
E1	3.30	1.31	0.19	0.56	0.18	0.86	2.74
E2	5.50	2.17	0.03	5.65	2.25	0.02	−0.15
E3	4.69	1.86	0.06	5.95	2.21	0.03	−1.26
E4	1.11	0.44	0.66	0.00	0.00	1.00	1.11
A1B4				−11.75	−2.23	0.03	
A4C1				10.33	2.01	0.04	
A1D4				−10.38	−2.16	0.03	
A2D1				11.21	2.31	0.02	
A2D3				20.05	3.80	0.00	
B1D3				11.91	2.12	0.03	
B2D3				13.96	2.57	0.01	
B1E1				11.99	2.36	0.02	
B2E1				15.33	3.00	0.00	
B3E3				−12.38	−2.49	0.01	
C4D3				−11.36	−2.23	0.03	
D3E1				−17.67	−3.21	0.00	
Source	Sum of squares	*df*	Mean square	Sum of squares	*df*	Mean square	
Regression	55786.51	20.	2789.33	168008.9	32.	5250.28	
Residual	4423513.	2679.	1651.18	4311291.	2667.	1616.53	
F ratio			1.69			3.25	

*All the respondent data are combined in this single model. The left side shows the model without interactions, and the right side shows the model with interactions. Diff, the difference in the coefficients, which shows how the utility value of an element changes when interactions are introduced; SE, standard error.

the two utilities. We then look at the utilities of the interacting elements by using the model that accounts for pairwise interactions (column E). Each element has a utility value, but the pairwise interaction also has a utility value. We compute the sum of the three utilities (two linear terms and one interaction term). Finally we compare the two sums and then look at their difference. The difference of *sum with interactions* minus *sum without interactions* tells us what the effect is of accounting for interactions. The results presented in Table 16.8 suggest very little real synergism, such as A1 and D4 or A1 and B4. The combination adds an additional 4%–5% to the top-3 box. On the other hand, we find more suppressions, so the combination of elements generates a *lower utility value* than we would have expected based on the utilities computed with the linear model.

The additive constant differs for the model with synergisms (constant = 14.86) compared with the linear model without synergisms (constant = 12.01). We might wish to correct the sum of the utilities by incorporating the different additive constants. In that case we still see negative effects or *suppression* outweighing positive effects or *synergism*, but the magnitude of highest suppression is about equal to the range of highest synergism (−6.2 for highest suppression versus +7.7 for highest synergism).

Overall, therefore, there are synergisms and suppressions, but these data suggest more suppression than synergism. Furthermore, with 160 pairwise combinations, only 12 reached statistical significance. Finally it appears perfectly feasible to identify the individual pairs of concept elements that exhibit synergy or suppression, as long as the researcher permutes the design to create these combinations. It appears virtually impossible to identify these combinations any other way, unless the researcher has an idea of what interacts before the start of the experiment and can build in these combinations as part of the stimulus set.

The Six Contributions of the Self-authoring System, Especially to Packages

The approach provides the following six key benefits:

1. *To the company:* better insight, obtained from consumer research, about consumer responses to existing concepts and packages

2. *To the company:* more rapid concept package design, obtainable in a low-risk, iterative fashion

3. *To the researcher:* a more powerful research tool for applied concept and package-design development

4. *To the researcher:* a more powerful scientific tool to understand the subjective response to concept packages by discovering patterns in response to test stimuli

5. *To the researcher:* experience with a new type of Internet research

6. *To the creative and the package designer:* an ability to probe the mind of consumers and get responses to package ideas on an ongoing basis

Better Information About Packages

Conventional package research has usually involved evaluation of single packages, either alone or on the shelf. For the most part, researchers have been content to work with a relatively small number of test stimuli and ask a lot of questions. From these questions, posed either in focus groups or in quantitative surveys, the researchers have identified what particular features appeal to consumers versus what do not. Rules about drivers of acceptance and rejection generally do not emerge from such limited sets of stimuli, nor can they, because rules generally emerge from the pattern of response to many stimuli rather than from the patterns of many responses to a few stimuli.

More Rapid, Iterative Package Design, Generating Better Learning and a Competitive Advantage

Iterative, small-scale research represents a major opportunity for consumer researchers. Such research is not unknown—indeed, the enormous popularity of qualitative research speaks to the recognition that low-risk research at the early development stages is both desirable and feasible. One of the issues in qualitative research is the limited number of stimuli. With more stimuli there is a greater chance that this early-stage research can drive changes that represent significant improvements in the package. One way to do this iterative research has the designer sitting either among the respondents or in the back room sketching away as the respondents voice their opinions. The designer becomes an active participant, feeding back in the package what he or she hears the consumers say they want.

The self-authoring approach to package design brings this iterative approach to a different stage. The consumers who participate do so by voicing their response to a set of stimuli, with immediate results. The designer can either change the template or change the elements on the spot or wait for large-scale quantitative results the next morning, look at the segments, and decide about the next steps. With either approach the package design becomes an iterative exercise generating quantitative data at each stage to guide development. In the most active case the designer either actively creates new alternatives (categories) on the spot or new components of the current alternatives. Once these are created and uploaded to the server, the respondents can immediately evaluate combinations, have their results tallied, and the development is moved forward to the next step. The self-authoring system combines with the designer and the respondents to create an iterative, low-cost, easy-to-implement loop comprising development-feedback modification.

At the more relaxed pace, the designer might have the luxury of 24 hours between iterations. Within the 24 hours there might be as many as 300–500 respondents who evaluate the new alternatives through Web evaluation. Within this time frame the data provide the designer both with quantitative responses to design alternatives and with segmentation. The segmentation, in turn, may provide additional insights regarding the importance of the different types of graphic features. For instance, the segmentation in this study revealed two mind-sets interested in different types of graphics because the segments are interested in different aspects of the tea package. Their mind-sets predispose these segments to attend to different package features. Knowing this type of information is invaluable to the designer, who now goes beyond measurement of the limited stimulus set into understanding the rules or at least emergent generalizations underlying consumer attitudes.

More Powerful Tools for the Consumer Researcher

For many years, consumer research relied on the intellectual power of practitioners to identify patterns in data. Certainly, advances in statistics and in research methods helped the field to progress. Self-authoring technologies for higher-level research problems, however, may provide an even greater benefit to the field. Package design in particular may benefit from the technologies. Until now, package-design research has been done principally in an evaluative mode. The researcher has been called in after the fact as the voice of the consumer. Some designers use consumer research up front, before the package is developed, but for the most part the research is evaluative.

The self-authoring technology for package design makes design research at the early stages far easier, more flexible, and more cost effective. Whereas previously the designs were created, now there are no real designs, per se, but only alternative options from which the design emerges. The consumer respondent integrates the alternatives, reacting to combinations by liking some and disliking others. The ability to disaggregate the response into the contribution of the components shows the "algebra of the consumer mind." Once the researcher and designer know what is important to the consumers, it is straightforward to use this knowledge to guide development of newer and better package designs. The researcher now has become part of the development team, playing a role early on when there are many options from which to choose rather than acting as the final arbiter of the consumer's voice.

Discovering Patterns in Test Stimuli

Ask any experienced researcher about trends in data and associated consumer learning, and one will be shown a wealth of opinion and information. It is hard in the consumer research industry to practice one's craft without observing certain repeating behaviors and patterns. These observations are not necessarily idiosyncratic either, because they are often shared among researchers, and the level of agreement is far too great to ascribe to mere politeness.

At the same time there are no hard-and-fast rules. For the most part, researchers in the business work on the basis of their experience, insight, and intuition. It is the nature of ad hoc consumer research to be dealing with different problems all the time. The learning and pattern recognition emerging from these varied efforts are not systematic. There are no formal rules of patterns, but rather simply repeating observations.

By itself, conventional package-evaluation research, like the science of astronomy, can be a scientifically valid endeavor, but in general it is not experimental. Both evaluation research and astronomy use science to construct frameworks of knowledge about natural phe-nomena, construct and test hypotheses, and arrive at an understanding about how the world works. Neither, however, admits systematic variation of the physical stimulus, so the nature of the information is limited to observations about what exists. By systematically varying the test stimuli, researchers go beyond simply polling the population about packages to discover how nature works by seeing *cause and effect* relations.

Researchers interested in *rules* about the perception of package designs can learn a great deal by varying the nature of the package stimulus and measuring consumer reactions. The stimulus variations afforded by self-authoring methods include different type fonts, different types and locations of the graphics, and different amounts of information. Consumer responses can range from interest to ratings of appropriateness or communication and even to response time. In the end, self-authoring systems rapidly provide experimenters with an infrastructure by which to manipulate stimuli in an experimentally rigorous nature, measure responses, and draw inferences.

A New Type of Research: Self-authoring Systems, the Internet, and the Growth and Maturation of Design Research

A great deal of research is migrating to the Internet. Much of this research deals with concepts. As companies find it less expensive and faster, the Internet will become an even more popular venue for concept research. There is no reason why package-design evaluations should lag behind. One of the key benefits of the Internet is its ability to provide a lot of data about a lot of stimuli in a very short time. Consequently, as package-design research migrates to the Web we may expect to see a great deal more of this research being done, just as there is a great deal more concept research afforded by the Internet. Such growth in research with design means that the field will mature. Simply by being able to do cheap experiments, researchers will understand a great deal more about how consumers respond to packages. What may begin as desultory studies in a cheap, easy research venue will, in turn, become the foundation of a new science and a new understanding.

Designers, Research, and the Development of Stimulus-based Insights

Package designers are in the process of becoming accustomed to consumer insights. For many years, package design as a branch of marketing services used consumer insights from focus groups, in-depth interviews, and trends analysis. However, one might surmise that the personality of the package designer differs from the personality of the researcher. The work product of the former is a package—a visual and artistic expression—whereas the work product of the latter is a report. Both use insight, but work in radically different ways. Traditionally, those who aver that they are artists have been loathe to use formal statistical data because all too often those data were used to evaluate the work product rather than to aid it. A similar type of response held sway among the creatives at advertising agencies, who were loathe to work with researchers because they felt that the research was judging them. Indeed, in both cases—package designer and agency creative—the research did indeed judge because it worked with one stimulus and in the end gave a good or bad rating (no matter how strongly the researcher attempted to disguise that fact).

Self-authoring research—an easy, quick, private technology—presents package designers with a radically different capability. Self-authoring conjoint measurement invites designers to play, to experience, to try out ideas. Certainly, the results are quantitative, because quantification is the very soul of conjoint measurement. Yet, at a different level, the quantification is nothing more than feedback about how the different notions and ideas perform. Designers are curious about what works and what doesn't, but like any artists they want that information on their own terms, in their own way, and in the privacy of their own creative space. Like any artist they are curious about their creations, about the medium in which they work. Witness the number of sketches that artists and writers make before their oeuvre is started. The metaphor of sketches applies just as well to self-authoring conjoint measurement. For package design these are sketches, albeit sketches in a new knowledge medium (conjoint measurement) and in a new venue (the Internet) with new materials (namely, the elements of package design and the consumer audience).

References

Batchelor, R. (2001). *Logit Regression and Neural Networks.* London: City University Business School.

Moskowitz, H.R., and Ewald, J. (2001). Always on: bringing market research down to the development engineer, closer to the customer, and into the vortex of product development. In: *Proceedings of the 54th ESOMAR Congress, Rome.* Amsterdam: ESOMAR [European Society of Marketing Research].

Moskowitz, H.R., Gofman, A., Katz. R., Itty, B., Manchaiah, M., and Ma, Z. (2001). Rapid, inexpensive, actionable concept generation & optimization: The use and promise of self-authoring conjoint analysis for the foodservice industry. *Foodservice Technology,* 1: 149–168.

Moskowitz, H.R., Gofman, A., Manchaiah, M., Ma, Z., and Katz, R. (2004). Dynamic package design & optimization in the Internet era. In: *Proceedings of ESOMAR Conference. Technovate2, Barcelona.* Amsterdam: ESOMAR [European Society of Marketing Research].

Moskowitz, H.R., and Martin, D.G. (1993). How computer aided design and presentation of concepts speeds up the product development process. In: *Proceedings of the 46th ESOMAR Congress, Copenhagen, Denmark.* Amsterdam: ESOMAR [European Society of Marketing Research], pp. 405–419.

Mosley-Matchett, J.D. (1998). Leverage the Internet's research capabilities. *Marketing News* 32: 13.

Pawle, J.S., and Cooper, P. (2001). Using Web research technology to accelerate innovation. In: Proceedings of Net Effects[4]. Barcelona: European Society of Market Research, pp. 11–30.

Plackett, R.L., and Burman, J.D. (1946). The design of optimum multifactorial experiments. Biometrika, 33: 305–325.

Saguy, I.S., and Moskowitz, H.R. (1999). Integrating the consumer in new product development. Food Technology, 53: 68–73.

Wheelwright, S.C., and Clark, K.B. (1992). Revolutionizing Product Development: Quantum Leaps in Speed, Efficiency and Quality. New York: Free Press.

Chapter 17

Deconstruction and Competitive Intelligence

Introduction

Deconstruction refers to the analysis of competitive messaging in concept research. The deconstruction exercise identifies what specific elements in the competitive frame drive consumer acceptance or communication and what do not. The knowledge obtained from such an exercise is exceptionally valuable because it shows what specific features in the communication work and what do not

Competitive intelligence is an old concept in research and in business. The Society of Competitive Intelligence Professionals (SCIP) was founded to bring together the individuals in the business and to provide a center locus for developing procedures. Competitive intelligence, as typically practiced, looks at stimuli in the environment, such as advertisements and spending patterns. From the pattern of stimuli a competitive intelligence analyst can deduce what might be going on in the mind of a competitor. The pattern gives the competitor away.

Another form of competitive intelligence actually evaluates the competitor offerings. One may profile the food products offered by competitors in order to identify level of performance and, where possible, the sensory characteristics of the product that drive acceptance and image. Moskowitz (1984) presented the systematized approach for such experimentally oriented competitive analysis, first for health and beauty aids, and then later for foods. The approach, called *category appraisal*, is now widely used (Munoz et al. 1996). In many cases, category appraisal has supplanted the more traditional paired-comparison method wherein the researcher simply tests one competitive product without considering the profiles of the other products in the competitive frame.

Experimentally, varied test stimuli to understand the competitor mind do not make either intuitive or business sense when the research must be executed by using finished food or beverage samples. It is simply too difficult to create products that simulate all of the competitors by using the different ingredients by competitors. Thus, there is little in the way of systematic analysis of the components of finished products and beverages, except in the most unusual case where the financial rewards make this analysis worthwhile. In contrast, however, the competitive analysis using concepts is far more feasible, and systematic variation of the competitor elements makes a lot of sense. With competitor concepts or advertisements, the concepts can be assessed as *gestalts* or "wholes" and/or be broken up into components to discover how each component performs.

From this competitive analysis using actual in-market products expressed as concepts, the researcher uncovers specific patterns that reveal aspects of the competition, such as the following:

1. *Performance.* Which concepts are accepted and which concepts are rejected?

2. *Communications.* What do the concepts present in the nature of information? What is the style of the communication (e.g., serious vs fun)?

3. *Brand value.* If respondents evaluate two sets of concepts that are identical except for brand names, then what is the effect of

the brand name on the rating? The effect is estimated by subtracting the *blind* rating (without brand name) from the *branded* rating (with brand name). The difference is a measure of brand value.

At a further level (not shown here) the competitive analysis of concepts can "drill down" to more fine-grained aspects of the data.

4. *Predictors of responses.* Do differences exist in response to the concepts as a function of some aspect of the respondent (e.g., male vs female)? Market researchers, sensory analysts, marketers, and the like look for these differences both as guides to improving the concept for the target audience and for hints about who will be most receptive to the concept when it goes into the market.

5. *Segmentation.* What segments exist in the population that respond differently to the concepts? Throughout this book we have seen the issue of segmentation to be very important because the population is not homogeneous with respect to statements about products, benefits, brands, etc. To the degree that the researchers can uncover segments in the population by showing how the respondents exhibit different patterns of responses to the same stimuli, it will be possible to learn more about the population from responses to the array of current communications. We have already seen the power of segmentation (see Chapter 7).

The strategy of deconstruction into components is known by other names, such as content analysis. *Content analysis*, which looks at the types of messages in communications and attempts to identify trends, often works at a gross, morphological level, counting the frequencies of specific communications. Through an analysis of different media, one can identify the spread of information, the tonality of the way the topic is presented, and similar general analyses that deal with metapatterns. The content analyst is akin to a sociologist looking at groups of individuals. Instead of getting into the mind of the individual and identifying like-minded clusters of people, sociologists look at the interaction of masses of individuals in order to identify emergent trends of a social, interpersonal nature.

Taking the Approach One Step Further: Systematizing Deconstruction

The steps involved in deconstruction are straightforward. Table 17.1 lists one sequence of seven steps that represents a systematized approach. It is important to keep in mind that the discipline of deconstruction can be as valuable as the actual exercise itself, even if the researcher stops there and does not move forward with any empirical analyses using concept testing. Deconstruction via experimental design uses the current set of messages broadcast by the competitive frame rather than concentrating on new ideas. There is room, however, for new ideas in the analysis. These new ideas may be added into the element set, depending on a researcher's objectives.

A Worked Example: New Methods for Preserving Food

The case history data presented here come from research on new ways to preserve food (Moskowitz et al. 2000). During the past three decades, companies in the food industry have investigated a variety of preservation methods beyond the conventional ones. Some of these methods include irradiation and electrical stimulation. The senior author was involved in the consumer evaluation of products preserved by irradiation when the research was gearing up in the early 1970s at the US Army Laboratories in Natick, Massachusetts. At that time, research focused on identifying the treatment conditions that would generate acceptable products after they had been irradiated. Often the product would have an off-flavor, conven-

Table 17.1. Seven steps to deconstruct the competitive frame and model the results to identify what works and what does not: the approach is geared to Internet sites

1.	*Gather* communications for ideas in the topic area. This step entails visiting competitor Internet sites, gathering their advertising for print media, and recording their language. The competitive frame may be very narrow or very wide. The elements may come from the Internet site, from point of purchase materials, from articles in the media talking about the competitive frame, and so on. All types of communication are fair game, but the stimulus elements must be capable of being reduced to print for subsequent research.
2.	*Deconstruct* communications to elements. Deconstruction entails parsing the messages in order to create simple, stand-alone declarative statements. The editing may create an element by slightly modifying a sentence or phrase in a paragraph. Deconstruction generates elements that will be later used as components of concepts, so the editing of the deconstructed elements is important. The editing must leave intact both the idea and the way the idea is expressed. However, the editor must realize that the elements have to be stand *alone* and make sense by themselves in a variety of different concepts.
3.	*Classify* elements into categories by using conjoint analysis (see Chapter 5). The classification is a mere bookkeeping device. It has no effect on the ultimate statistical analysis of the data. However, classification ensures that the two elements of the same type (e.g., method of sterilization), but presenting different messages, do not appear in the same concept; that is, classification prevents elements from bumping into one another, creating an irrational, meaningless concept. One could go one step further and create a system for pairwise restrictions.
4.	*Dimensionalize* the elements by locating them on a series of simple bipolar semantic scales (e.g., more for selling versus more for education). Dimensionalization is optional when there are few elements, but important when there are many elements and each respondent can evaluate only a subset of these elements.
5.	*Create the test combination* by experimental design, run the study, and create the model relating the presence/absence of elements to the consumer response. Again this follows the conjoint analysis paradigm.
6.	*Segment.* Cluster the respondents into homogeneous groups (see Chapter 7 and Moskowitz 1996).
7.	*Look for patterns.* The utility values can be cross-tabulated easily for a variety of different subgroups, including gender, age, self-defined concerns with specific issues, and even the newly emergent segments from step 6.

tionally termed the *wet-dog taste*. As the years went on, however, and as the technology improved, the focus shifted more toward the promotion of these new processing and preservation ideas beyond simple technical virtuosity and the obvious immediate benefits.

New methods for preserving food make an excellent introduction to deconstruction of the competitive frame, because several companies have developed methods and are advertising their products. The processing technology is sufficiently novel that consumers have not yet had time to form strong associations with, and attachment to, technological methods for processing and preservation. Furthermore, the new methods for preservation are not well known, making it impossi-

ble to dismiss the discipline of deconstruction cavalierly as an exercise that will not reveal much new. In the world of business Realpolitik, it helps to introduce new ideas by using case histories that cannot be dismissed on the basis of "We already know all there is to know about that topic."

Steps 1 and 2: Gather Communications and Deconstruct Them into Elements

The first two steps assemble and analyze messages from different sources about the new technologies. For this case history the messages were obtained in early 2001 from the Internet. An example of the Internet site is presented in Figure 17.1. Table 17.2 has a partial

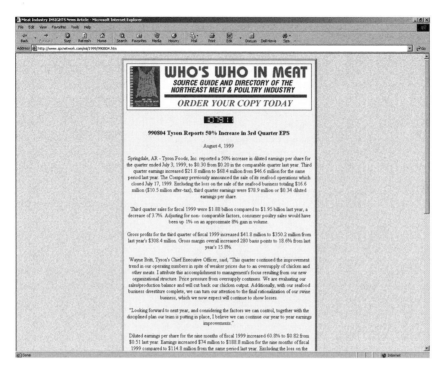

Figure 17.1. Example of a page from a Internet site dealing with irradiated foods.

list of messages and the Internet source from which these messages were obtained. It is important to keep in mind that, in a deconstruction exercise, one can come across many hundreds or even thousands of messages to study. A prudent researcher need not conduct an exhaustive search of all Internet sites and other advertising material, but should attempt to sample a representative number of sites and extract a reasonable number of messages.

In this study, as in other deconstruction studies, readers should keep in mind that some of these sites provide information, whereas others provide sales material for foods processed by these new methods. Furthermore, some of the sites were positive to new methods, whereas others were critical. Some sites were clearly commercial, whereas others were created by groups that want to present contrary information in the "public interest."

The 11 sites generated 90 final elements. One should keep in mind that some amount of

editing of the competitive messages is necessary to create the stand-alone concept elements. The editing may be at the selection stage, where implicit judgments are made about what is relevant. Often these judgments are rapid, undisciplined, and subconscious. Other editing changes the language where relevant to create short, easily understood phrases. This editing is explicit and conscious. The field of text and content analysis deals with these issues all the time, and interested readers can find a world of alternative approaches simply by looking up the term *text analysis* in an Internet-based search engine.

Step 3: Classify the Elements into Categories or Buckets

This is mainly an editorial job. Since the research presents the elements in combination to the respondents, it is important that two elements of the same type, which may have con-

Table 17.2. Competitive analysis by deconstruction: example of elements and their Internet source

Element	Source
The USDA approved use of irradiation on refrigerated and frozen meats in December 1999.	Bbq.about.com/food/bbq/library/weekly/ aa012200b.htm
Food irradiation does not prevent contamination that may occur during storage or preparation.	saferfoods.com/ada.htm
X-rays can be used to treat fruit when it is ripe.	surebeamsafe.com/nonflash/index.htm
The FDA recommends labeling for informational purposes, not as warning.	www.bostonherald.Internetpoint.com/ food/shirrad.htm
The significant factors favoring irradiated food are superior quality and safety.	www.foodsafety.ufl.edu/comsumer/sf/sf077.htm
Irradiation is used on ground-beef products to guard against harmful bacteria such as *Escherichia coli*.	www.foxmarketwire.com/wires/ 1117/f_ap_1117_47.sml
The new method uses ordinary electricity to pasteurize frozen beef patties after they have been processed and packaged.	www.huiskenmenats.com/press.htm
Consumers readily choose irradiated foods in commercial marketing and tests.	www.iaea.or.at/worldatom/inforesource/ other/food/status.html
Most irradiated foods will now be labeled to allow consumers to make an informed choice.	www.nwamorningnews.com/1999/september/ 21/news/0921martin2.shtml

Table 17.3. The ten elements falling into the category of consumer-relevant health claims

HC01	Consumers should still be careful in handling meat so as to avoid contamination.
HC02	Food isn't being steamed for 4 hours, killing all those vitamins.
HC03	No more environmental threats are posed.
HC04	Food irradiation does not prevent contamination that may occur during storage or preparation.
HC05	Changes in food caused by irradiation have all been found to be benign.
HC06	Multigenerational animal studies have shown no toxic effects from eating irradiated foods.
HC07	Human volunteers consuming up to 100% of their diets as irradiated food have shown no ill-effect.
HC08	Irradiated foods are healthier and fresher.
HC09	Irradiated foods can prevent food poisoning.
HC10	Irradiated food does not appear to cause cancer.

flicting messages, never appear together. The classification does this. Table 17.3 shows an example of the ten elements found on the Internet sites that deal with health claims and health issues. It is important to keep in mind that the classification, like the initial editing, is a subjective task. The *categories* (or *buckets*), can neither be too large, because the elements are then really unrelated to one another, nor too small, because then there will be too many categories and insufficient numbers of elements per category. Thus, in Table 17.3 one might find qualitative differences among the elements that could lead to a disagreement as to whether the classification is accurate. This definition of categories in content analysis is

perennial, especially when the goal is to map many stimuli to a few categories.

Step 4: Dimensionalize the Elements on Nonevaluative Semantic Differential Scales

As discussed in Chapter 5 on IdeaMap, one objective is to create individual-level models, with the models presenting the utility values for all of the concept elements. The objective becomes even more interesting and tantalizing when the elements are chosen from the competitive frame. The IdeaMap method enables researchers to create the individual-level model for all of the

90 elements developed from in-market Internet-based communications.

Dimensionalization, a key feature of IdeaMap, locates the 90 concept elements on semantic scales. One newly emergent benefit of the dimensionalization exercise is its ability to provide a signature of the concept elements, and thus current communications and communicators, on the nonevaluative scales. By profiling the communications of the different in-market companies and products, the researcher rapidly understands the way the different communication elements are perceived, which companies are communicating specific types of messages, and whether any interesting *tonalities* of communication exist that are not being exploited by companies. For example, if one of the semantic scales is *safety vs quality*, and if everyone is communicating on the *safety* dimension, then communications of *quality* might provide an opportunity (see Table 17.4).

Step 5: Experimental Design, Study Execution, and Modeling

Design

The experimental design comprised easy-to-read combinations of 2–4 concept elements. Each respondent evaluated a unique set of 36 of the 90 elements, combined into 60 concepts. The utilities of the remaining 54 elements not tested by a particular respondent were estimated for that respondent by the data-imputation method used in IdeaMap (see Chapter 5).

Invitation

Each respondent was sent an e-mail invitation. Figure 17.2 shows the text. For studies of this type it is important to grab the respondent's attention when the invitation is first received and to provide some type of prize. A sweepstakes often works. Offering the re-

Table 17.4. Elements on four semantic differential scales were located as part of the dimensionalization exercise: the results show the pair of elements that lie at the extremes for this specific study

1 = 9 =	Prevent Treat	Safety Quality	Consumer Manufacturer	To buy To teach
Text				
Prevent vs treat				
Irradition prevents stored potatoes, onions, and garlic from sprouting.	2.4	6.3	4.3	4.7
Irradiation kills microbes, insects, and parasites in spices, pork, and poultry.	7.6	2.4	2.9	5.4
Safety vs quality				
Cold pasteurization irradiation has been scientifically proven to be a safe process.	3.2	1.9	3.8	5.4
The process of irradiation has no effect on the taste or quality of food.	5.7	7.1	2.6	3.6
For consumer vs manufacturer				
Cold pasteurization irradiation has been scientifically proven to be a safe process.	4.4	3.4	1.7	4.3
Spices or foods that are merely ingredients do not have to be labeled.	4.6	5.0	7.1	6.8
Buy vs teach				
Irradiated foods are healthier and fresher.	3.9	5.2	2.1	2.4
Radiation doses allowed by the US Food and Drug Administration (FDA) are the most restrictive of all countries that use irradiation.	4.6	3.7	5.1	7.8

Dear Consumer,

I-Novation, an independent online research company, has been asked to learn more about how consumers like you feel about new technologies currently being developed for processing and preserving foods.

Here's your chance to tell us. It's easy to participate and your responses will be completely confidential. Just click on the following link to access the survey (If your email does not support hotlinks, just cut and paste the link into your browser.):

http://64.23.174.214/pr09/

Tell us your thoughts and opinions by participating in this important online survey today. As a thank you for your prompt response, everyone who completes the survey before midnight on [xxx] will be entered in a draw featuring a first prize of $400, and 4 second prizes of $50.

Thanks in advance for your time. Your feedback is appreciated.

Regards,

The I-Novation team   24

Figure 17.2. Invitation to participate.

spondents the chance to win a prize entices them. The cost of administering the sweepstakes is fairly low, and the return on the responses makes it a good investment in the research.

Orientation to the Task

Once the respondents have clicked on the link, they are sent to the study site and begin the interview. The first screen is an orientation, which introduces the respondent to the goals of the study (see Figure 17.3). Quite often researchers make a big fuss about the absolutely proper wording to be used in an orientation screen and feel that the data are meaningless unless the respondents have been properly oriented. From a theoretical viewpoint, and for novice researchers, this meticulous attention to detail is worthy and should be commended. However, the reality of the interview is far different for five reasons:

1. Most respondents don't pay attention to instructions.

2. Respondents read part of the interview script and then proceed.

3. Respondents may look at the question occasionally as they proceed through the interview.

4. For the most part the respondents "get it" fairly quickly and really look at the bigger picture—what they must do.

5. The researcher's effort should go into the construction of the stimuli more than into the construction of the perfect orientation screen.

Results: Who Responded?

One of the key benefits of Internet-based research is the ability to source respondents all over the country (or, for that matter, all over the world). The deconstruction data comprise the responses from 999 individuals (see Table 17.5). The acquisition of these 999 respondents took a little less than 3 days, attesting to the power of the Internet to provide data. The classification questionnaire shows

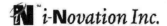

Figure 17.3. Respondent orientation.

a good spread in gender and age (items 1 and 2 below). The additive constant suggests significant, positive interest among consumers in these new methods of food preservation (item 3 below).

1. *The genders are well represented.* Quite often in research studies the respondents comprise a preponderance of women. In this study the breakout was 60% women and 40% men. Previous studies have shown instances where there are as many as 75% or more women (e.g., chocolate candy) or as few as 55% women (e.g., steak).

2. *There is a good representation of age, market, and education.* Quite often critics of Internet research argue that only poor respondents with low incomes participate. Certainly, the respondent demographics are weighted toward those with a lower income, since 75% of the respondents have incomes less than $50,000. On the other hand, with nearly 1000 respondents, a sample of 256 individuals with higher incomes is certainly a reasonable sample.

3. *Even without additional information in the concepts, the high additive constant sug-* *gests many of the respondents were interested.* We know nothing, though, about the interest of respondents who chose not to participate.

The regression model generates an equation of the form:

$$\text{utility} = k_0 + k_1$$
$$(\text{element 1}) \dots k_{90} \text{ (element 90)}$$

The additive constant, k_0, shows the expected percentage of respondents who would be interested in these new methods of food preservation. The additive constant is approximately 48–50, depending on the subgroup. This constant is a high midrange, compared with other additive constants for such products/services as credit cards and computers ($k_0 \sim 20$) all the way up to pizza ($k_0 \sim 65$).

Winning and Losing Elements

The key outcome from a competitive analysis is, of course, the performance of the competitors. Anyone who has seen the report of a deconstruction analysis immediately becomes

Table 17.5. Results from the 999 respondents who participated in the deconstruction study: the key subgroup, number of respondents, and the additive constant for the particular subgroup are presented

	Subgroup	Base	Additive constant
	Total sample	*999*	*49*
Gender	Men	401	50
	Women	598	48
Age	Under age 21–35	218	51
	Age 36–50	464	48
	Over age 51	317	48
Market	Live in Northeast	272	52
	Live in Southeast	212	47
	Live in Northwest	114	48
	Live in Midwest	252	46
	Live in Southwest	149	49
Education	HS diploma or less	316	50
	Attended/completed college	545	49
	Grad school or above	60	47
	Technical/nursing schools	78	39
Income	Under $25k–$49k	743	49
	$50k or more	256	46
Food concern	Extremely concerned about types of food consumed	375	57
	Very concerned about types of food consumed	443	48
	Somewhat/not concerned about types of food consumed	181	33
Organic food	Consumes organic food when possible	74	57
	Consumes organic/nonorganic food	374	50
	No difference what types of food consumed	499	47
	Organic food prices (greatly/somewhat) affects purchase decision	727	50
	Organic food prices (slightly/does not) affect purchase decision	272	46
Food safety	Experienced food poisoning	445	50
	Have not experienced food poisoning	554	48
Food information	Obtain food information from supermarket	833	50
	Obtain food information from media (TV, radio, magazines, and newspapers)	864	49
	Obtain food information from internet	472	54
	Obtain food information from friends/relatives	540	50
	Obtain food information from doctor/nutritionist	510	49
Food technology	Obtain food technology information from supermarket	419	52
	Obtain food technology information from media (TV, radio, magazines, and newspapers)	855	48
	Obtain food technology information from Internet	510	53
	Obtain food technology information from friends/relatives	340	51
	Obtain food technology information from doctor/nutritionist	316	52
Concept segments	Segment 1: Basically indifferent	366	47
	Segment 2: Irradiation works, and short communications work	199	68
	Segment 3: Irradiation prevents bad things from happening	303	43
	Segment 4: Technology is interesting	130	35

well aware of the keen interest by marketers and research and development (R&D) in the performance of the products that directly compete with their own. The competitive nature of marketers pushes people to compare their performance with the performance of relevant competitors, whether these are the messages or even brand names.

Table 17.6 presents the possibly disappointing finding that no elements perform particularly well. Nor, in fact, do any elements perform particularly poorly. One might surmise from these mediocre scores that the majority of the elements have no impact at all. Yet, anyone who has participated in the creation of advertising knows full well many talents are applied to the task and that much of what is created does quite well initially, at least in the opinion of advertising professionals. If that is the case, why then do these elements perform in such a muted, even poor, fashion? Why are there no elements scoring

+10 or higher? It is worth noting that the same "mediocre" performance of in-market competitive elements appears in other studies about products (e.g., toothpaste, cars, and health-oriented foods) and public policy (Moskowitz et al. 2002b). One reason may be that the self-policing of communications by companies generates elements that are relatively weak. These elements neither infuriate any clearly defined group of respondents nor inspire any respondents either.

Segmenting Respondents and Uncovering the Mixed Strategies Adopted by the Competition

As we have seen before with segmentation (see Chapter 7) and international research (see Chapter 8), segmenting respondents often reveals patterns that were hitherto invisible within in a larger dataset from the total

Table 17.6. Winning and losing elements for the total panel for the new food-preservation methods

Element	Source	Utility
Winning elements		
Irradiated foods can prevent food poisoning.	www.saferfoods.com/ca.htm	3
Irradiated foods are healthier and fresher.	www.saferfoods.com/ca.htm	3
If given information about irradiation, half or more of consumers choose irradiated foods.	www.saferfoods.com/ca.htm	3
Irradiation doesn't make food radioactive.	www.whyfiles.org/054irradfood/index.html	3
Irradiated foods last longer in your refrigerator.	www.saferfoods.com/ca.htm	3
The new process will give consumers an added measure of confidence and safety for their families.	www.huiskenmenats.com/press.htm	3
Irradiated food can safely be consumed by everyone.	www.saferfoods.com/iifs.htm	3
Irradiation can kill viruses, bacteria, fungi, and insects.	www.whyfiles.org/054irradfood/index.html	3
The process of irradiation has no effect on the taste or quality of food.	www.spcnetwork.com/mii/991076.htm	3
Losing elements		
Irradiation is legal although seldom used for fruits, vegetables, poultry, and pork in the United States.	www.whyfiles.org/054irradfood/index.html	−3
Radiation doses allowed by the US Food and Drug Administration (FDA) are the most restrictive of all countries that use irradiation.	www.saferfoods.com/ada.htm	−3
Irradiation exposes food to gamma rays from radioactive material, such as cobalt 60.	www.spcnetwork.com/mii/991076.htm	−4

panel. This discovery of segments, and the subordination of some strongly performing elements to the will of the majority (i.e., the total panel), makes a great deal of sense for the following three reasons:

1. *Suppression of the contrasting and possibly outlandish.* Competitive analysis deals with stimuli that are already in the marketplace. Wild new ideas generally are not featured in advertisements simply because these ideas depart so much from the norm.

2. *Legal constraints in the public domain.* A great deal of what is in the current messaging comes from legal considerations that transcend what will sell and deal with what can be legitimately claimed and protected and will not be harmful to the corporation. A major goal is to avoid potential lawsuits resulting from one's miscommunication. When lawyers become involved in communication, as they must in today's litigious society, they must act to protect the corporation at all costs. This protection leads to conservatism. In the end and despite protestations to the contrary, conservatism leads to safe, middle-of-the-road, nonintrusive communications, especially with respect to the actual content of the message being communicated. Boring wins because boring is safe. The execution of the message, not the message content, must carry the day.

3. *Multiple, competing constituencies.* Current messaging must appeal to a number of different groups simultaneously. This naturally leads to countervailing forces, which minimizes the appeal of the concept to any particular group. Such strong appeal to one group would reduce appeal to the other groups in the consumer population. The marketer, ever hesitant to reduce the total base of interested respondents, attempts to satisfy each group. The result, as one might expect, constitutes a middle-of-the-road communication that offends nobody but has no chance to excite anyone either. This strategy of appealing to different constituencies leads to average scores, not to very high scores. It also leads to the winning scores, however high they may be, having different messages. There is no pat-

tern because the winning elements appeal to different groups. One winning element may appeal to a segment interested in product descriptions of a strongly factual nature. Another winning element may appeal to yet another segment, interested in the health benefit, whereas a third winning element may appeal to a segment interested in price.

The rich base size of 999 respondents allows for segments of substantial size to emerge. Internet-based research for segmentation, using utilities from conjoint analysis, combines the benefit of segmentation, the large base size afforded by the Internet, and the clarity of the results afforded by conjoint utilities.

Four segments emerged from the segmentation analysis, which was based on the methods described by Moskowitz (1996). It is interesting to look at the elements that appeal to each segment, but also at the total number of elements that appeal to each segment. If we stand back for a second and look at the segments that are interested in many of these elements, we can get a sense of the "mind of the company." At least three aspects from these data add to competitive intelligence:

1. *Companies have constituencies with different levels of business importance.* Companies try to satisfy their constituencies. If they satisfy one constituency more than others, this indicates that, from the company's viewpoint, that constituency is the most important.

2. *Companies have given up on constituencies.* The company realizes that there are other important constituencies, but it may be legally impossibly to satisfy them, no matter what the company would like to do.

3. *Companies do not realize the nature of constituencies.* The company does not realize there are other constituencies to satisfy.

These are the four segments:

Segment 1: Those who are basically indifferent. They are indifferent to communications. They are basically interested in food preservation, but most of what is said they do not like. What they barely tolerate is a description of general benefits only. They want nothing to do with scientific explanations.

Segment 2: Those who believe in irradiation and prefer short communications. This segment likes new methods of food preservation, but fundamentally respond to short, simple, nonintellectual messages. The less said to them the better, and whatever is said should be in short, declarative statements. They start out with a very high positive response to these new methods of food processing (additive constant = 68).

Segment 3: The preventors. This segment wants to know and be reassured that the processing will do something specific to the food, but do not want to know about science. They like to hear messages about food processing as doing something to prevent bad things from happening.

Segment 4: The technology junkies. This segment likes new technologies and wants to hear how the processes work. They begin with a low basic interest in the idea (additive constant = 35), but there are plenty of elements that can dramatically influence their opinion.

Segments differ, often dramatically. What one segment likes, the other segment is often indifferent to or actively dislikes. The same message does not appeal to multiple groups, except by chance.

A sense of the difference between the segments emerges from looking at the winning elements for each segment (see Table 17.7). What becomes very clear is that the four segments show radically different patterns of what existing communications appeal to them. Segment 2 is fundamentally predisposed to the entire new preservation system, and communications don't do much to add to that positive predisposition. Segments 3 and 4 are less interested at a basic level, but the proper communication can do wonders to excite these respondents.

A Second Example: Deconstructing Toothpaste, an Overly Advertised and Overly Analyzed Product Category

As a short follow-up to the previous case study on food preservation, which is a new technology and Internet based, one can contrast the case of toothpaste (Moskowitz et al. 2002a). Like soaps and detergents, toothpaste falls into the category of the overly advertised. Toothpaste has been around for more than 100 years. Found in all households, commonly accepted, the question is whether the principles emerging from deconstructing new food preservation methods apply to toothpaste. It is one thing to have segments coming out of high-involvement categories such as food preservation, where attention focuses on safety and technology and brings in one's latent feelings about social policy. It is another thing to identify segments for toothpaste, which is a low-involvement staple, health and beauty aid product.

Rather than going into the entire case history for the remainder of the chapter, let us visit the highlights of a toothpaste study conducted in early 2001. The stimulus elements were obtained from print advertising and from materials distributed at the point of purchase in stores. Toothpastes are common, messaging combines old and new stories, and the information is generally not revolutionary. Unlike new methods of food preservation, there is no need to overcome neophobias about novel, possibly off-putting technology. Toothpaste advertising can be banal, the messages can be hackneyed, the brands are well known, and the efforts are focused on finding new things to communicate. Nothing can differ so much from food irradiation as toothpaste (or perhaps laundry detergent, which is too far afield to discuss in a book on food concepts).

The study with 408 respondents again used only the in-market communications. The same principles of mix and match were adopted. The study itself used 258 different

Table 17.7. Winning elements for the four segments for new methods of food preservation

	Total	Segment			
		1	2	3	4
Base	*999*	*366*	*199*	*303*	*130*
Additive constant	*49*	*47*	*68*	*43*	*35*
Segment 1: Basically indifferent					
Human volunteers consuming up to 100% of their diets as irradiated food have shown no ill-effect.	2	−1	−3	7	7
The process of irradiation can legally be used in the United States for killing insects in grains, flour, fruits, and vegetables.	1	−1	−3	4	10
Treating foods with gamma rays offers benefits to consumers, retailers, and food manufacturers.	2	−1	−5	8	5
Segment 2: Irradiation works and short communications work					
Irradiated foods are healthier and fresher.	3	−2	2	10	4
Irradiated foods last longer in your refrigerator.	3	−2	1	8	6
Irradiated foods can prevent food poisoning.	3	−2	1	10	6
Segment 3: Irradiation prevents bad things from happening					
Irradiation dramatically reduces harmful bacteria in food products including poultry.	2	−2	−4	11	1
Irradiation doesn't make food radioactive.	3	−2	−1	10	5
Irradiated foods can prevent food poisoning.	3	−2	1	10	6
Irradiated food can safely be consumed by everyone.	3	−2	0	10	2
The process of irradiation has no effect on the taste or quality of food.	3	−3	−1	10	5
Irradiated foods are healthier and fresher.	3	−2	2	10	4
Segment 4: Technology is interesting					
The new method uses ordinary electricity to pasteurize frozen beef patties after they have been processed and packaged.	0	−3	−12	1	18
The increase in cost for irradiated foods is estimated at 2–3 cents per pound for fruits and vegetables and 3–5 cents per pound for meat products.	−1	−3	−12	0	18
X-rays double or triple the shelf life of fruit.	1	−3	−10	5	18
Irradiation has been used for years on limited amounts of produce, spices, poultry, and other foods.	−1	−3	−14	2	17
The US facilities currently in operation process spices, citrus fruits, tropical fruits, strawberries, tomatoes, mushrooms, potatoes, onions, and poultry.	1	−3	−11	5	17
Spices or foods that are merely ingredients do not have to be labeled.	−2	−3	−15	0	17
X-rays can be used to treat fruit when it is ripe.	−2	−4	−14	−1	17
There's one facility in the United States that uses x-rays on fruit.	−1	−3	−11	0	17
Electronic pasteurization does not lengthen a food's shelf life.	−1	−2	−12	1	17
Irradiation prevents stored potatoes, onions, and garlic from sprouting.	2	−2	−6	7	16
Consumers have indicated that they would pay a premium price for irradiated ground beef.	0	−2	−11	2	16

(continued)

Table 17.7. Winning elements for the four segments for new methods of food preservation *(cont.)*

	Total	Segment 1	Segment 2	Segment 3	Segment 4
The USDA approved use of irradiation on refrigerated and frozen meats in December 1999.	0	−2	−9	0	16
One process uses electricity to process meat, vegetables, fruits, and other foods.	0	−2	−8	2	15
X-rays have replaced a controversial earlier proposal of radioactive cobalt 60 to treat fruit.	−1	−3	−12	0	15
Consumers consistently rate irradiated fruit as equal or better than nonirradiated fruits in appearance, freshness, and taste.	2	−3	−5	8	13
Cold pasteurization can add valuable days to shelf life of refrigerated products.	2	−3	−6	7	13
Only primary products such as meats and vegetables have been treated by irradiation require labeling.	−2	−3	−11	−1	13
Irradiation is legal although seldom used for fruits, vegetables, poultry, and pork in the United States.	−3	−3	−15	−1	12
Irradiation does not change the taste of ground beef.	2	−3	−3	8	11
Irradiation can kill viruses, bacteria, fungi, and insects.	3	−2	−4	9	11
The process of irradiation can legally be used in the United States for killing insects in grains, flour, fruits, and vegetables.	1	−1	−3	4	10
Only foods of good hygienic quality are irradiated.	2	−2	−3	6	10
Food isn't being steamed for 4 hours, killing all those vitamins.	2	−2	−3	7	10
Handling of foods processed by irradiation are governed by the same food safety precautions as all other foods.	2	−1	−5	7	10
Irradiation has been approved by health and safety authorities for about 40 different types of food.	1	−1	−5	4	10

communications chosen from the 12 first-tier and second-tier brands and four lesser-known brands. One of the benefits of using familiar products is that the messages are often well known, recognizable, and can be tied back to specific brands. Furthermore, the brand names are well known.

Table 17.8 lists the utility values of the 12 brand names. The brand names were treated as elements, independent of the messages with which they were paired. Each of 408 respondents was instructed to check the brand used most often in the classification questionnaire. This generated a set of four different subgroups with readable samples. What becomes very interesting to the marketer is the degree to which brand names drive the utilities and whether a person's favorite brand (so-called brand used most often)

achieves a higher rating. Table 17.8 answers those questions:

1. The brand name is not a particularly strong concept element, at least for the total panel. The strongest brand name is Mentadent. The weakest is Tom's of Maine.

2. The utility values for the four groups (defined as used most often) generate more interesting results. For example, the Mentadent most-often user shows a very high utility value for the brand name Mentadent, compared with the utility value from the total panel (14 vs 6). Indeed, one could actually estimate the utility value for the Mentadent name among the non-Mentadent users by a simple algebraic substitution:

$$(49/408) * (14) + (359/40) * x = 6 \ or \ x = 0.48$$

Table 17.8. Utilities for toothpaste brands in the deconstruction study

		Brand used most				
		Total	Mentadent	Aquafresh	Crest	Colgate
	Base size	*408*	*49*	*34*	*131*	*119*
	Additive constant	*22*	*20*	*22*	*26*	*19*
NM06	From Mentadent	6	14	8	8	5
NM02	From Crest	6	8	6	8	4
NM07	From Rembrandt	5	8	7	6	5
NM01	From Colgate	5	4	10	5	5
NM12	From Gleem	4	6	10	3	4
NM10	From Pepsodent	4	3	10	5	3
NM03	From Aquafresh	3	4	8	2	4
NM08	From Listerine	3	5	3	3	4
NM09	From Arm & Hammer	3	4	7	3	1
NM05	From Close-Up	2	3	6	1	3
NM04	From Aim	2	2	3	4	1
NM11	From Tom's of Maine	1	2	7	0	1

3. In contrast, the brand name Colgate does no better among those who say they use Colgate most often than among the other respondents. The brand name Colgate, therefore, does not carry with it either a very strong brand *equity* among its users or a very weak brand equity among its nonusers.

Performance of the Messaging

Unlike messaging for the unusual, new food technologies discussed earlier, toothpaste messaging appears to be relatively ubiquitous, if not necessarily attended to. Marketers strive to create messages to which consumers will attend, recognizing full well that consumers are inundated with toothpaste messages. Most consumers, in fact, tend to tune out these messages, making the marketer's job all the more difficult. How well do the competitive messages perform, given this quite different competitive and communications environment?

In a deconstruction study the messaging elements are set up so that the elements are completely free agents; that is, any brand can appear with any message, so both are independent, which means that the respondent does not really know which particular toothpaste manufacturer is responsible for a spe-

cific message, unless the respondent is personally familiar with that message. Thus, the deconstruction exercise becomes even more interesting, because the deconstruction measures the degree to which the marketer and the advertising agency have been successful.

We can see the performance of the different elements from five major brands in Figure 17.4. The data suggest quite clearly that the elements currently being communicated to the public score in the intermediate to low range. One of the nice things about deconstruction is the ability to trace back each message to a specific manufacturer. The array of element utilities achieved by a specific toothpaste manufacturer quickly reveals whether any one of the well-known manufacturers is doing *messaging* better than the others. Such superior performance would show up in higher utility values, which would in turn be impressive simply because the utilities correspond to the performance of elements that, in turn, do not have brand names to back them up. None of the major toothpaste brands provides elements that perform particularly better than do elements from other brands.

A deeper analysis of the results reveals that the winning elements from the competitive frame show a narrow range, and indeed far narrower than one might expect. This

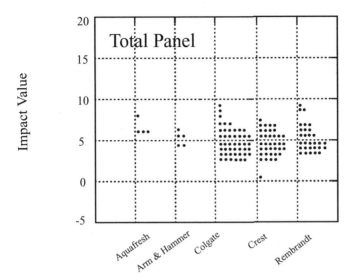

Figure 17.4. Utility values for the toothpaste elements, arrayed by manufacturer. The *impact value* is another name for the utility value obtained from conjoint analysis.

narrowness of the winners in terms of utility value or impact value is telling. It suggests that marketers, legal counsel, and advertising agencies opt for a narrow band of communications aimed at satisfying legal requirements and appealing to the mainstream of consumers. We see three additional patterns from Figure 17.4:

1. *High-performing elements come from many manufacturers, not just one.* What is furthermore interesting is that those messages that attract the general population come from a variety of brands, not from one brand alone. This means that the toothpaste marketers know what the key elements are, as might be expected given their long history of marketing. The particular success or failure of the communications probably arises from the choice of who to market to rather than knowledge of what communications really sell. Unlike the new methods of food preservation, where there is an absence of knowledge yet a plethora of marketers, toothpaste has both a plethora of knowledge and an abundance of well-known brands (see Table 17.9).

2. *Consumers do not necessarily respond strongly to communications from their own brands.* In many cases they may be just as strongly attracted to messages communicated by competing brands.

3. *No breakthrough elements emerge for the total panel.* We see little evidence of breakthrough elements whose utilities score +15 or more for the total panel or even for the respondents who identify themselves as users of a specific brand. That strong responsiveness will be seen later to be a function of segmentation, as it was for the first case history described in this chapter.

Segmentation Again Reveals the Key Differences Across the Groups

The segmentation analysis for toothpaste, like the segmentation for food preservation, reveals the true structure of the category in terms of consumer mind-sets. Applying the segmentation algorithm (see Chapter 7) generated five clear segments with radically different responses to the in-market communications. One of the nice things about Internet-based studies is the possibility of working with large numbers of respondents so that even small segments comprise large

Table 17.9. Relatively strong-performing elements and utilities selected from total panel results*

Brand		Total	Utilities among respondents who say that they use this brand of toothpaste			
			Colgate	Crest	Mentadent	Aquafresh
	Base size	*408*	*119*	*131*	*49*	*34*
	Additive constant	*22*	*19*	*26*	*20*	*22*
Aquafresh	Restore your teeth to their natural whiteness.	8	9	7	7	6
Arm & Hammer	For cleaner, whiter teeth and fresher breath.	6	7	5	8	10
Colgate	Whitening toothpaste helps remove daily surface stains, whitens teeth, and helps prevent tartar buildup, working to leave your teeth sparkling white.	9	9	7	13	11
	Whitening toothpaste with baking soda and peroxide.	8	9	7	11	11
	Tirelessly fights plaque, tartar and cavities, even gingivitis and bad breath. All day.	7	7	8	9	9
Crest	Multicare formula-loaded with all the powerful protection you want against cavities and visible tartar.	7	8	9	10	6
	Formulas for the entire family: multicare advanced cleaning, multicare, extra whitening, cavity protection, tartar protection, baking soda, and peroxide. Whitening, sensitivity protection, gum care, and kid's cavity protection.	7	7	5	11	10
	Multicare advanced cleaning—the best toothpaste ever at repelling stains, tartar buildup, and that filmy feeling.	6	7	7	6	7
	Fluoride formulation starts to penetrate on contact to help strengthen weak spots and reverse the early stage of tooth decay.	5	8	3	2	6
	Clinically proven to help get your gums healthier and reduce gingivitis associated with plaque.	5	7	3	2	8
Rembrandt	Helps remove the debris that leads to gum inflammation.	9	9	8	10	9
	Age-defying formula penetrates the tooth's surface to oxidize stains and reverse discoloration.	8	9	8	8	7
	Safely whiten teeth up to 5 shades.	8	9	8	10	8
	Complete protection for your teeth and mouth.	6	8	6	6	8
	Safe, maximum-strength peroxide whitening power.	6	7	6	3	11

*The panelists identified themselves as users of four popular toothpastes. The elements are sorted first by brand and then within brand by utility value for the total panel.

numbers of respondents. In that way, even small segments emerge, with reasonable numbers of respondents. Was the dataset to have fewer respondents, one might not see these segments emerge or, if they did emerge, the pattern might not be as clear.

We can easily understand the nature of these segments from the winning elements

Table 17.10. Top-five scoring toothpaste elements for the total panel and for the five segments that emerged in the deconstruction study

Brand		Total	Segment				
			1	2	3	4	5
	Base size	*408*	*71*	*35*	*69*	*81*	*152*
	Additive constant	*22*	*38*	*39*	*11*	*22*	*15*
	Segment 1: Disinterested						
Rembrandt	Complete protection for your teeth and mouth.	6	1	13	15	8	3
Colgate	The only toothpaste to receive the American Dental Association Seal of Acceptance for protection against plaque, gingivitis, and cavities.	6	1	13	14	7	3
	Leaves your mouth feeling baking-soda clean and fresh, and has a great-tasting natural mint flavor.	4	1	0	12	7	2
Crest	All this packed into one tube.	5	1	5	12	7	1
	Available in clean and easy to use stand-up tubes.	5	1	9	12	6	3
	Segment 2: Care oriented						
Rembrandt	Complete protection for your teeth and mouth.	6	1	13	15	8	3
Crest	Multicare formula-loaded with all the powerful protection you want against cavities and visible tartar.	7	−1	13	7	10	9
Colgate	The only toothpaste to receive the American Dental Association Seal of Acceptance for protection against plaque, gingivitis, and cavities.	6	1	13	14	7	3
Crest	Formulas for the entire family: multicare advanced cleaning, multicare, extra whitening, cavity protection, tartar protection, baking-soda and peroxide whitening, sensitivity protection, gum care, and kid's cavity protection.	7	0	12	12	10	4
	Every time you squeeze out toothpaste, the "Neat Squeeze" tube pulls back what you don't use.	6	1	11	14	9	3
	Segment 3: Authority oriented						
Rembrandt	Complete protection for your teeth and mouth.	6	1	13	15	8	3
Crest	From Crest.	6	0	4	14	10	3
Colgate	Accepted by the American Dental Association.	6	0	8	14	9	2
Crest	Every time you squeeze out toothpaste, the "Neat Squeeze" tube pulls back what you don't use.	6	1	11	14	9	3
Colgate	The only toothpaste to receive the American Dental Association Seal of Acceptance for protection against plaque, gingivitis, and cavities.	6	1	13	14	7	3

(continued)

Table 17.10. Top-five scoring toothpaste elements for the total panel and for the five segments that emerged in the deconstruction study *(cont.)*

Brand		Total	Segment 1	Segment 2	Segment 3	Segment 4	Segment 5
	Segment 4: Whitening						
Aquafresh	Restore your teeth to their natural whiteness.	8	−2	2	5	12	12
Crest	The first toothpaste that the ADA accepted for whitening by polishing away surface stains.	6	−1	10	10	11	4
Colgate	With microcleaning crystals.	6	−1	3	6	11	7
Rembrandt	Helps remove the debris that leads to gum inflammation.	9	−2	5	4	10	15
Crest	Formulas for the entire family: multicare advanced cleaning, multicare, extra whitening, cavity protection, tartar protection, baking-soda and peroxide whitening, sensitivity protection, gum care, and kid's cavity protection.	7	0	12	12	10	4
	Segment 5: Techie						
Rembrandt	Safely whiten teeth up to 5 shades.	8	−2	3	1	10	16
Colgate	Whitening toothpaste helps remove daily surface stains, whitens teeth, and helps prevent tartar buildup, working to leave your teeth sparkling white.	9	−1	2	5	9	16
Rembrandt	Helps remove the debris that leads to gum inflammation.	9	−2	5	4	10	15
	Age-defying formula penetrates the tooth's surface to oxidize stains and reverse discoloration.	8	−1	−2	5	7	15
Colgate	Whitening toothpaste with baking soda and peroxide.	8	−2	2	5	10	15

for each. Table 17.10 lists the top five scoring elements for each of these segments:

Segment 1: Disinterested. This segment shows no responsiveness to the elements. The low additive constant (22) indicates that, at the very basic level, these respondents are not interested in toothpaste. No elements interest them either. To them toothpaste is a commodity. Their highest utility values are +1. Furthermore, the elements to which these individuals respond vary in meaning. We conclude that this group of consumers is disinterested and that the elements driving acceptance do so at a very minimal level.

Segment 2: Care oriented. This group shows a higher additive constant (38) in contrast to the disinterested segment. They respond very strongly to messages about total care (utility, +13).

Segment 3: Authority oriented. This group shows a very low additive constant (+11), meaning that for them the elements do all the work. The most effective elements are those that talk about the imprimatur of some authority.

Segment 4: Whitening. This group responds very strongly to statements about whitening, but again they are indifferent to the basic idea of toothpaste (additive constant = 11). To drive their interest the concept must talk about whitening, and it is that promise that increases their interest.

5. *Segment 5: Techie.* This group again has a very low constant (15) and responds very strongly to technical word pictures about how toothpaste works.

Segments Vary in Size

The segments are of different sizes, with the techie group comprising almost 40% of the respondents and the care-oriented comprising less than 10% of the respondents. Thus, these segments do not distribute equally in the respondent population. Since toothpaste is a popular topic, the deconstruction analysis provides some indication about the possible distribution of basic mind-sets in the population.

Lack of Segment Focus

No manufacturer appeals strongly to only one segment. The manufacturers present a variety of different messages spanning different segments. This means respondents in each of the segments will be attracted to messages from different manufacturers.

Overview

Analysis of the competitive frame by deconstructing the communications provides marketers and developers with insights about what works in the category. Whereas most research efforts in early-stage development begin with ideas, there is a world of learning to be had by looking at what competitors are doing and how well they are doing it.

The deconstruction approach presented in the two case histories shows that the full benefits of conjoint analysis can be enjoyed, even if one cannot create new ideas. The exercise of gathering the competitor communications is itself quite valuable. Even more valuable, however, is the discipline of sys-

tematic deconstruction. By parsing the communications the alternatives in a category can be understood even before entering it. By dimensionalizing the elements on attributes and looking at the dimensions, elements, and manufacturer, it becomes possible to identify some of the strategies that competitors follow, because the dimensionalization itself creates a database of element × manufacturer by dimension. Quite a great deal can be obtained from the analysis of this material. Finally, by working with large numbers of respondents, preferably on the Internet, the deconstruction can reveal which, if any, of the in-market communications work and, even further, among what specific segments. Thus, systematic deconstruction recommends itself as a profitable first step in the concept development process, albeit one that to date has not been fully utilized.

References

Moskowitz, H.R. (1984). Cosmetic Product Testing: A Modern Psychophysical Approach. New York: Marcel Dekker.

Moskowitz, H.R. (1996). Segmenting consumers worldwide: an application of multiple media conjoint methods. In: Proceedings of the 49th ESOMAR Congress, Istanbul. Amsterdam: ESOMAR [European Society of Marketing Research], pp. 535–552.

Moskowitz, H.R., Itty, B., Shand, A., and Katz, R. (2000). The mind of the customer: hot buttons and turn-offs regarding new methods of processing/preservation. Paper delivered at the Institute Of Food Technologists, Eastern Regional Meeting, Newark, Delaware, March 2000.

Moskowitz, H.R., Itty, B., Shand, A., and Krieger, B. (2002a). Understanding the consumer mind through a concept category appraisal: toothpaste. Canadian Journal of Market Research, 20: 3–15.

Moskowitz, H.R., Marketo, C., and Rabino, S. (2002b). Advertising energy: corporate communications, public policy issues, and the customer mind. In preparation.

Munoz, A.M., Chambers IV, E., and Hummer, S. (1996). A multifaceted category research study: how to understand a product category and its consumer responses. Journal of Sensory Studies, 11: 261–294.

Chapter 18

Bottom-up Innovation: Creating Product Concepts from First Principles

Introduction: Creating Ideas Through the Traditional Ideation Approach

For the most part, ideation is an exercise that depends on serendipity for its success. Each particular practitioner provides some basic form of ideation with an individuating twist. Quite often the approach is simply brainstorming, where the respondents sit together and generate ideas. There may be different ways to record these ideas and to facilitate the interaction between the participants, but the objective is to generate as many good ideas as possible. These ideas typically emerge as fully formed, albeit basic, concepts, not perhaps in the session, but edited later to be free-standing ideas. Success in this early ideation stage constitutes the creation of ideas that will later pass some screening criteria. Many of these ideas will then be passed off to an evaluation step, such as benefits screening or concept testing (see Chapter 2).

There are positives and negatives to the current methods. On the positive side, ideation leads to new ideas. The practitioners state, or at least imply, that it is the serendipitous emergence of new ideas that is key, and that the up-front discipline is necessary for that happy serendipity to emerge. It is left as a task for the professionals, who either facilitate the process or use the data to cull these elements down, polish them, and then advance into concept refinement and quantitative concept testing.

Many concepts spring from the minds of consumers who are faced with a situation and integrate the surrounding information around them to come up with an idea. This is often called the *ah-ha* experience. Experts in creativity are more capable of accomplishing that than are consumers, or at least the common wisdom holds that they are. Missing, however, is a systematized approach to concept development that neither needs creative consumers nor professional experts to work, but rather works in an algorithmic way, churning out the results in a testable format.

Given that serendipity is, by definition, unpredictable, and that "prediction is hard, especially when it is about the future" (a remark attributed Yogi Berra), perhaps there is a more disciplined way to create new concepts. This chapter introduces the notion of *first principles*, combining focused respondent attention *end use* with the metrics provided by conjoint analysis. The consumer integrates the ideas; the consumer organizes, rather than generates. First principles can be defined as the combination of concept ideas (the raw material; e.g., product features) by the integrator (e.g., the consumer) assessing this combination as appropriate for an end use. First principles thus create an idea using basic building blocks generated ahead of time.

Origin of First Principles in the Assessment of Appropriateness

A great deal of concept development is predicated on the use of a single key evaluative criterion, such as liking or purchase intent. The assumption is that the higher the liking score is, the more acceptable the concept will be. Whether the criterion attribute be liking or purchase intent often does not matter, because the objective is to maximize some hedonic rating. That may be the case for final success in the marketplace, but acceptance alone is not necessarily the only key evaluative criterion for a concept.

Over the past two decades, researchers in food science and technology have begun to explore the concept of *appropriateness* as a different type of dependent measure, in place of, or in addition to, the criterion measure of liking or purchase intent. Appropriateness requires the respondent to integrate knowledge of the food beyond simple likes or dislikes, in order to rate whether a food belongs in a specific situation (Schutz and Martens 2001). Appropriateness is a perfectly good rating attribute for foods and, according to Schutz (personal communication to H.R.M.), respondents seem to have no problem rating various foods on appropriateness for various day parts. There is every reason to believe that appropriateness as a dependent variable could therefore be used in conjoint measurement tasks. A respondent simply has to integrate the information in a product description and rate that vignette on a more complex, integrative attribute. In commercial work the senior author has used complex integrative scales for foods and for health and beauty aids with the independent variable being combinations of concept elements presented as small, easily understood concepts. There was nothing in the results to suggest respondent difficulty with appropriateness or fit to end use.

The conjoint analysis task generates two measures: the part-worth contributions of the different elements as they drive acceptance (liking or purchase intent), and appropriateness for one or several end uses. Appropriateness ratings thus provide another criterion by which to evaluate a product concept. That set of product features exhibiting both acceptance and appropriateness comprises the final consideration set for the new product, because the features have the property of driving liking and fitting an end use.

Creating the Product from First Principles: An Integrative Approach

Three tools come into play when we create concepts from first principles: elements, conjoint analysis studies, and the ability to have a respondent select a specific conjoint study in which to participate. The combination of tools enables developers to create product concepts from basic building blocks or statements of product features, with the property that these concepts are both acceptable and fit a specific end use or set of end uses.

The first-principles approach can be summarized by these eight steps:

1. *Raw materials.* Create a bank of raw materials or elements dealing with product features. This step uses conventional ideation procedures. The objective is to create a reservoir of elements for the actual conjoint analysis.

2. *Designed combinations.* Combine these elements into small, easily assimilated concepts about the product by using experimental design, following the conjoint measurement approach. This step follows the rules of conjoint analysis.

3. *Multiple studies with the same elements.* Create a set of these studies by using the exact same elements, but varying the end use or rating scale. Each end use constitutes a separate study. All studies have the same elements, but the rating scale focuses the respondent's mind on the specific end use. Examples of end uses are emotional states (viz., appropriate for happy times and appropriate for sad times) or day parts (appropriate for morning, appropriate for evening, etc.).

4. *Respondent choice of a relevant study.* Create a *wall* or a selection system, whereby the respondents can select a relevant or interesting end use. They do not know that the elements are identical across all of the studies on the wall.

5. *Specify the rating scale.* Present these to the respondents, instructing the respon-

dents to evaluate the combinations against a specific criterion (e.g., fit to a morning product, fit to a brand, or overall liking). For each respondent, use only one criterion for the evaluation of all concepts tested by that respondent in that particular evaluation. There is no reason to prevent a respondent from participating in multiple studies, as long as there is some time separating the different participations.

6. *Estimate utilities.* Obtain the ratings and, using statistical analysis, estimate the degree to which each product element *drives* the criterion rating (e.g., How much does each product element drive liking or drive fit to a morning beverage?).

7. *Compare the utilities against end uses.* Do the modeling for each respondent for each end use in order to determine how each concept element drives each end use (i.e., how each element drives interest, fit to a specific day part, and the like).

8. *Optimize.* Create new combinations of elements that satisfy one or several criteria (e.g., maximize acceptance, but ensure that the product is perceived as appropriate for morning).

Carbonated Beverage for a New Consumption Opportunity

The carbonated-soft-drink industry continually seeks new products and new drinking occasions. In the United States consumers drink carbonated beverages from morning until late evening, opening up the possibility of an early-morning carbonated beverage. Currently, the day part—early morning, breakfast—is not a popular occasion for carbonated beverages, although it is becoming more so over time. The research objective is to determine the features of a carbonated beverage appropriate for breakfast. The goal of the research is to identify the features of a product to fill a new opportunity.

This stepwise, systematic approach differentiates first-principles research from the more creative, intuitive, and generally serendipitous approaches that today characterize much of what we call innovation. First-principles research is systematic, not particularly innovative, and simply comprises the combinations of features that may either lie within one end use or cut across multiple, divergent end uses. The consumer's expectations of *fit* to the situation, as well as acceptance, are the driving factors here.

The actual specific study to create the concept for a morning-based carbonated beverage followed these six steps:

1. *Identify the different end uses* (see Table 18.1). The end uses comprised day parts fit to brand and to specific age groups of the end user. It is clear from this example that one can move beyond acceptance to appropriateness quite easily.

2. *Create a basic set of concept elements, providing principally product descriptions* (see Table 18.2). These elements can be created in many ways, such as looking at current communications, brainstorming, or having consumers provide their own ideas. The elements are both *close in* and *far out*, which is necessary for innovation, but not always recognized as such. Since the objective is to see how the different elements perform against a number of expectations, it is important to

Table 18.1. The different end uses for the carbonated-beverage study

End use	Base size	Additive constant
Overall interest	118	37
Fits 7-Up	111	45
Fits Coke	124	32
For breakfast	177	40
For midmorning	90	47
For lunch	114	43
For afternoon	108	45
For supper	141	39
For after supper	112	37
For after sport	89	42
For kids (ages 7–11)	62	32
For teens (ages 12–15)	52	44
For teens (ages 16–19)	53	41

Table 18.2. Concept elements and their performance when the criterion rating scale is interest (total panel and concept response segments): segment 1, conventional good citizens wanting a good tasting and good for their beverage; segment 2, health oriented; and segment 3, flavor explorers

A	B	C	D	E	F
				Segments	
		Total	1	2	3
	Base size	*118*	*32*	*63*	*23*
	Additive constant				
	Packaging	*37*	*47*	*35*	*29*
E01	The mini-drink 6-pack . . . the perfect size for children and people on the go	1	3	−2	5
E02	100% organic . . . healthy for you and the planet	3	12	2	−5
E03	A drink that appeals to your senses . . . with a unique aroma	−2	−5	−1	0
E04	Available in gallons to quench that giant thirst	−5	−5	−3	−9
E05	With a thermal barrier your drink will stay colder longer	1	1	2	−2
	Lightness and kick				
E06	A non-carbonated drink . . . that won't weigh you down	0	0	3	−10
E07	Enter a whole new universe with a blend of enticing aromas	−3	−6	−1	−5
E08	Slightly carbonated thirst quenching drink	−3	−10	1	−3
E09	Kick it up with a new highly carbonated drink	−5	−21	0	1
	Flavors				
E10	Invigorate your senses with shocking lemon-lime flavor	−1	−3	−3	8
E11	With a little splash of vanilla flavor . . . sure to delight	4	−1	5	10
E12	Comes in a variety of flavors and crazy opaque colors like punky purple, brilliant blue, and goofy green . . . you got to try them all	−6	−3	−11	6
E13	An eclectic mix of fruit and other intriguing flavors	1	3	−2	7
E14	Enjoy a smooth slightly translucent drink that's intriguing from the very first sip	5	3	4	7
E15	Introducing new and exciting flavors such as blueberry twist, and wacky pink watermelon	−2	8	−13	15
E16	Satisfy your thirst . . . with real plum juice, ginseng and honey	1	8	−11	22
E17	A thrilling burst of unique cherry flavor and a sweet, crisp taste that gives you "more to go wild for"	8	5	6	17
	Nutrition				
E18	Introducing new clear natural refreshments with a light hint of flavor	4	4	3	5
E19	Delivers at least 100% of the recommended daily intake of vitamin C, 15% of folate, and 14% of potassium per 8 oz. serving	7	8	7	5
E20	A healthful source of calcium	2	2	6	−9
E21	Provides you with the balanced nutrition you need to live a healthier life	1	4	3	−6
E22	Enjoy a delicious taste, but without the calories	4	5	5	−2
	Emotional benefits				
E23	A drink which eliminates stress	−1	2	0	−9
E24	Quenches your thirst and stimulates your mind	2	−1	3	6
E25	For the health conscious . . . a sweet drink with no sugar or aspartame	2	5	4	−6
E26	Created for today's naturally healthy lifestyle	3	−3	7	0
E27	Keep trim with a reduced calorie drink	1	2	1	−2
	Radical changes				
E28	Enjoy a daring, high-energy, high-intensity, active drink	2	5	2	1
E29	A bold, energetic, unstoppable drink in a glow-in-the-dark container	−2	0	−8	13
E30	A refreshing alternative to coffee . . . with a burst of caffeine	−4	−28	6	4

(continued)

Table 18.2. Concept elements and their performance when the criterion rating scale is interest (total panel and concept response segments): segment 1, conventional good citizens wanting a good tasting and good for their beverage; segment 2, health oriented; and segment 3, flavor explorers *(cont.)*

A	B	C	D	E	F
				Segments	
		Total	1	2	3
E31	Helps you to achieve peak performance when you need it most	3	3	4	−1
E32	Rich, and creamy with no caffeine . . . the perfect drink to satisfy the whole family	5	12	2	4
E33	A drink that kids thirst for and moms will love	1	−2	1	6
E34	An energizer that keeps you going . . . without the caffeine	6	13	4	0
E35	A jolt of caffeine to awaken your senses	−5	−25	2	6
E36	So light, so crisp, so refreshing	2	−5	7	−4

sample a variety of elements, not just a limited number that appear to fit a single end use.

3. *Create a wall or listing of studies available for participation.* A respondent who agrees to participate is taken to the wall and shown the available studies. In this case there were 13 studies. The respondent can then choose a study in which to participate. Studies that have a very high fill rate have more than their share of respondents and can then be temporarily suspended.

4. *Concept orientation and evaluation.* The respondent evaluates 60 combinations comprising the concept elements, taken 2–4 at a time. The respondent rates the combinations according to the end use he or she selected.

5. *Utility model created for an individual.* The results for each respondent generate a concept model, relating the presence or absence of the concept elements to the rating. As was discussed previously (see Chapters 2 and 5), the ratings were transformed into a binary scale (1–6 transformed to 0; and 7–9 transformed to 100). The model thus shows how each element drives appropriateness or inappropriateness for an end use. The data from the different respondents for a single end use are combined into one average model.

6. *Segment respondents, specifically those who have rated acceptance.* The respondents are also segmented on the basis of

the pattern of their responses. The segmentation uses k-means clustering and is based on the statistic ($1 − R$; R = Pearson correlation) discussed in Chapter 7. The segmentation, which reveals new subgroups of consumers with different mind-sets, can be carried on with the other studies, as well, and will be informative as long as the data do not get out of hand because of their potentially voluminous nature.

How Different Concept Elements for the Carbonated Beverage Interest Respondents

The conventional analysis looks at acceptance, so we will begin there. The first of the 13 studies asked the respondents to rate interest in a beverage, without any other positioning statement. A sense of which elements do well and which do poorly emerges from Table 18.2 (column C). Respondents want unusual flavors, not the same old flavors, yet they don't want flavors that are totally out of their everyday experience. Thus, a statement such as "A thrilling burst of unique cherry flavor and a sweet, crisp taste that gives you 'more to go wild for'" does well (utility = +9). In contrast, promising a funky taste "comes in a variety of flavors and crazy opaque colors like punky purple, brilliant blue, and goofy green . . . you got to try them all" performs poorly (utility = −6).

There are at least five interesting findings from this initial analysis:

1. *Total panel.* Only two elements perform well (cherry flavor and vitamins). Creating a winning beverage from winning elements would not be very easy were the developer and the marketer to consider winning elements alone.

2. *Segmentation.* The respondent population is not homogeneous. Three distinct segments emerge from the data. These segments are not of equal size.

3. *Segment 1 reflects the good citizen; respondents want good-tasting, good-for-you beverages.* This segment shows some strong winning elements. These respondents are interested in new health-related drinks of an exotic nature and in some unusual flavors. The flavors that do well deal with fruits (blueberry twist) or with truly different combinations (plum juice, ginseng, and honey). This segment shows a high additive constant (47), meaning that, even without the concept elements, these respondents are interested in the beverage. The elements do not have to do much work to drive acceptance.

4. *Segment 2 responds to messages about light and health.* These respondents are certainly not interested in novel flavors. The specific health messages attract them. They are moderately interested in the beverage, but the elements have to do a lot more work (additive constant = 35).

5. *Segment 3 responds strongly to novel flavors.* These respondents may look for sensory excitement. They are the *elaborates* who continue to reappear in study after study (see Chapters 20–22). They show the lowest constant (29) but strong-performing flavor elements. For these groups the elements drive acceptance.

Measuring the Opportunity

One of the most attractive aspects of Internet-based interviewing is the ability to attract many respondents. With a wall showing the different day parts or opportunities, researchers can measure the relative attractiveness or at least the curiosity/intrigue factor for each end use. Thus, Table 18.1 presents the frequency with which respondents completed each of the 13 studies. The most *intriguing time* for a carbonated beverage is breakfast, followed by supper, with lunch and after supper less popular. The breakfast day part is intriguing because the greatest number of respondents (177) participated in the study. The least popular are a midmorning and an after-sport drink. From a marketing perspective, therefore, the greatest interest and possibly the best opportunity is breakfast.

Beyond the frequency, however, one can learn about aspects of the new product from the nature of respondents participating in the different day periods. The numbers in Table 18.3 are proportions of respondents for each day-part study:

1. *Gender drives interest.* For example, women are most frequently interested in a lunch drink and least frequently interested in an after-supper drink.

2. Age makes a difference. Older respondents (51) are more frequently interested in a beverage for afternoon and for after supper, and as one might expect far less interested in after sports. Furthermore, older respondents are far less frequently interested in participating in a study that simply mentions a new drink.

3. Market makes a difference, as well. The East Coast respondents more frequently participate in, and thus are interested in, after supper, whereas the West Coast respondents are most interested in afternoon. The Midwest and Mountain respondents are most interest in after sport. Whether this difference represents true mind-set differences or is just an accident of one study remains for other research. However, the ability of the Internet to pull respondents from many locations makes the analysis of market differences more feasible.

Table 18.3. Who participated in the different studies for different day parts: numbers in the table body are percentages

	New drink	Breakfast	Midmorning	Lunch	Afternoon	Supper	After supper	After sport
Base size	*118*	*177*	*90*	*114*	*108*	*141*	*112*	*89*
Gender								
Female	80	75	76	82	71	74	61	57
Male	20	25	24	18	29	26	39	43
Age								
21–30	10	11	6	16	6	8	11	17
31–40	21	24	24	17	16	16	16	29
41–50	46	29	38	35	38	45	32	29
51–60	16	22	24	25	29	18	29	18
61+	6	13	8	6	10	12	12	6
Market								
Northeast	28	28	32	30	23	23	35	19
Southeast	21	21	12	22	19	26	21	19
Southwest	15	8	14	16	19	11	9	13
Northwest	12	9	4	6	12	10	7	4
Midwest	12	23	21	18	22	21	21	26
Mountain	5	3	3	4	0	3	3	7
East coast (total)	49	49	44	52	42	49	56	38
West coast (total)	27	17	18	22	31	21	16	17
Midwest + Mountain	17	26	24	22	22	24	24	33

Using the Results to Identify Consumer Requirements for Various Day Parts

The database provides sufficient material for many end analyses. A productive analysis looks at the elements relevant for different day parts. It is clear from Table 18.4 that the day parts have different needs:

1. *Breakfast.* Respondents want vitamins and a no-caffeine beverage here.

2. *Lunch.* Respondents want a healthful product that enables them to reach peak performance. This peak performance is not relevant for the breakfast product.

3. *Supper.* Respondents want a light product and stress reduction.

4. *Basic product acceptance positioned as a new product but without a specific end use.* This is a *check* on the basic acceptance of the other elements. Basic accept-

ance allows developers to understand whether an element chosen as appropriate for an end use is also acceptable to consumers. For example, the element "a drink which eliminates stress" is highly relevant to a supper beverage, but intrinsically is not an acceptable element. Its utility is −1.

Direction for Product Developers: Links Between Day Part and Flavor Statements

Flavor is a key aspect of beverage. *Flavor sorts* are a common task in consumer-marketing research because they show what flavor is desired. The typical flavor sort is done on a flavor-by-flavor basis, against a background of a specific end use. The conjoint measurement task generates a *fit of element to end use*, and the results can be used for flavor sorts. The nine concept elements dealing

Table 18.4. Key elements that drive perception of appropriateness for breakfast, lunch, and supper, and how these elements perform when positioned for a new product without any end use

	Breakfast	Lunch	Supper	New product
Base size	*177*	*114*	*141*	*118*
Additive constant	*40*	*43*	*39*	*37*
Breakfast day part				
Delivers at least 100% of the recommended daily intake of vitamin C, 15% of folate, and 14% of potassium per 8 oz. serving	7	7	5	7
An energizer that keeps you going . . . without the caffeine	6	3	1	6
A healthful source of calcium	6	5	5	2
Rich, and creamy with no caffeine . . . the perfect drink to satisfy the whole family	6	−3	3	5
Lunch day part				
Helps you to achieve peak performance when you need it most	2	9	−9	3
For the health conscious . . . a sweet drink with no sugar or aspartame	4	8	2	2
A drink which eliminates stress	3	7	6	−1
Delivers at least 100% of the recommended daily intake of vitamin C, 15% of folate, and 14% of potassium per 8 oz. serving	7	7	5	7
Supper day part				
A drink which eliminates stress	3	7	6	−1
So light, so crisp, so refreshing	3	3	6	2

with flavor (E10–E18) can be analyzed separately to determine whether a link exists between flavor description and either day part, brand, or age of respondent for whom the beverage is intended. This type of information is very useful to product developers.

The results from this flavor sort are presented in Table 18.5. Most of the flavors bear little relation to the end uses. Many of the utilities hover around 0 (+3 to −3). When there is a departure, it is usually negative, meaning that *most flavors do not fit most occasions.* Some flavors do adequately, but none dramatically, indicating the need for a segmentation strategy. No single flavor will be universally accepted for any day part, although some are promising and probably could be acceptable in the segments:

1. *Breakfast.* Avoid highly impactful flavors. However, the element promising "plum juice, ginseng and honey," while not doing well in acceptance, does best at breakfast.

2. *Midmorning.* This period is more forgiving for most flavors, but the only flavor element that does reasonably well (utility, +4) is the description "An eclectic mix of fruit and other intriguing flavors."

3. *Lunch.* This period is less forgiving. Avoid most unusual flavors. The only element that does reasonably well is "Introducing new clear natural refreshments with a light hint of flavor."

4. *Afternoon.* This period is also less forgiving. The only element that does well is "A thrilling burst of unique cherry flavor and a sweet, crisp taste that gives you 'more to go wild for.'"

5. *Supper.* This period is unforgiving. The only flavor element to do well is "An eclectic mix of fruit and other intriguing flavors."

6. *After supper.* Also unforgiving. Only "Enjoy a smooth slightly translucent

Table 18.5. Utilities of flavor descriptions for the six day parts: one end use

Elements	Breakfast	Midmorning	Lunch	Afternoon	Supper	After supper	After sport
E17 A thrilling burst of unique cherry flavor and a sweet, crisp taste that gives you "more to go wild for"	−4	3	3	4	1	−1	2
E14 Enjoy a smooth slightly translucent drink that's intriguing from the very first sip	−4	2	2	1	−1	4	−4
E11 With a little splash of vanilla flavor . . . sure to delight	1	−2	−1	−5	2	1	−4
E18 Introducing new clear natural refreshments with a light hint of flavor	−2	2	4	1	2	−1	2
E13 An eclectic mix of fruit and other intriguing flavors	2	4	0	3	4	2	0
E16 Satisfy your thirst . . . with real plum juice, ginseng and honey	2	−4	−9	−6	−6	−6	−2
E10 Invigorate your senses with shocking lemon-lime flavor	−11	−1	−1	−2	−4	−2	−1
E15 Introducing new and exciting flavors such as blueberry twist, and wacky pink watermelon	−8	−5	−7	−4	−6	0	−4
E12 Comes in a variety of flavors and crazy opaque colors like punky purple, brilliant blue, and goofy green . . . you got to try them all	−14	−11	−15	−7	−11	−5	−7

drink that's intriguing from the very first sip" does well.

7. *After sport.* This is very unforgiving. No flavor does well.

Getting Value from Appropriateness Ratings

End use calls into play more complex cognitive processes beyond just liking and disliking. These data suggest that respondents can, indeed, separate what they like from what is appropriate. It is clear from these data that the utility values for acceptance do not match the utility values for day part (or targeted user). For example, the element of a beverage that eliminates stress is disliked (utility = −1), although the element is appropriate for a lunch or a supper beverage. This element is less appropriate for a breakfast beverage. Day-part analysis can be thus

thought of as a way to divide the stimulus set into different subsets, each appropriate for a different use. This conceptualization is important for product development. Product developers may be able to fine-tune the focus early on by availing themselves of a database that shows both acceptance and appropriateness.

Creating the Concept for a Breakfast Beverage

One good way to tie the information together is by using the results to create a new product concept. Continuing with the idea of a breakfast carbonated beverage, let us see what the features would be of this product and how the researcher could use the data to identify these specific features. We follow the steps from identifying the opportunity to prescribing the components of the product concept. The outcome of these steps is presented in Table 18.6:

1. *What is the opportunity?* Looking at the number of individuals who actually participated in the study (Table 18.1), there is clearly a great deal of interest, or at least curiosity, in a carbonated breakfast beverage. A total of 177 respondents participated, which makes this the most intriguing day part. Thus, at first glance a breakfast beverage may be interesting to create. The judicious developer and marketer would also look at the potential competition that would be in place for breakfast beverages before embarking on the development. For the purposes of illustration here, it can be assumed that breakfast beverages are the most promising.

2. *What are the winning features?* The winning features for the concept are listed in Table 18.4, which shows the key elements. These need not be the only elements considered. One might wish, for other purposes, to substitute some less appropriate elements.

Given no additional knowledge nor additional imposed criterion, the concept might read as follows:

Delivers at least 100% of the recommended daily intake of vitamin C, 15% of folate, and 14% of potassium per 8 oz. serving.

An energizer that keeps you going . . . without the caffeine.

A healthful source of calcium.

3. *Are there any specific flavors for the product?* The results of the study revealed that no particular flavor was clearly very appropriate for breakfast. However, intuitively, one might consider the two elements that were marginally appropriate for breakfast (Table 18.5):

An eclectic mix of fruit and other intriguing flavors.

Satisfy your thirst . . . with real plum juice, ginseng, and honey.

Table 18.6. Features of the new breakfast beverage*

		New product	
	For breakfast	Total	Segment 3
Additive constant	*40*	*37*	*29*
Iteration 1: Without fine tuning			
Delivers at least 100% of the recommended daily intake of vitamin C, 15% of folate, and 14% of potassium per 8 oz. serving	7	7	5
An energizer that keeps you going . . . without the caffeine	6	6	0
A healthful source of calcium	6	2	−9
Rich, and creamy with no caffeine . . . the perfect drink to satisfy the whole family	6	5	4
Satisfy your thirst . . . with real plum juice, ginseng and honey	2	1	22
Total utility	*67*	*58*	*51*
Iteration 2: After fine tuning to appeal to segment 3			
Delivers at least 100% of the recommended daily intake of vitamin C, 15% of folate, and 14% of potassium per 8 oz. serving	7	7	5
An energizer that keeps you going . . . without the caffeine	6	6	0
Rich, and creamy with no caffeine . . . the perfect drink to satisfy the whole family	6	5	4
Satisfy your thirst . . . with real plum juice, ginseng and honey	2	1	22
Total utility	*62*	*56*	*60*

*Results are shown in terms of appropriateness for breakfast day part, interest by the entire panel, and interest by segment 3 (the target group for whom the beverage is being developed).

4. *Will this appeal to everyone or should a segmented strategy be used?* Table 18.6 can be used to estimate the likely appeal of the concept. The two iterations show how one can first use the information to lay out the possible elements of the beverage (iteration 1), identify what works and what doesn't (e.g., good source of calcium does not work), and then fine-tune the concept to increase the utility score (iteration 2). Iteration 2 has come up with a possible product blueprint that is highly acceptable to segment 3.

Creating a New Coffee Concept from First Principles by Using Emotion Data and Fit to End Use

Introduction

This case history looks at emotions, social occasions, and consumption venues rather than day parts. It deals with the creation of a coffee beverage. Coffee can be consumed as a fuel to start a busy day, during breakfast or even before breakfast, or as a break to recharge batteries and clear one's mind; it can be consumed quickly and alone, or be used for a chitchatting relaxing pause. Coffee is prepared at home for breakfast or as the best conclusion of a dinner with friends; or out of the home, as a quick refill or to enjoy the atmosphere of a coffee shop. All these situations have coffee as a common central element, but they differ profoundly one from the other. It is quite easy to imagine that the coffee or a coffee-based beverage itself will differ in these situations; that is, consumers' needs may differ depending on the end use. Depending on whether the coffee is intended as a kickoff or as a sensory experience to share with friends, the beverage features may be rather different.

Researchers all too often ignore the aspect of context and end use, or relegate it to a secondary position. Rather, researchers focus on the demographic or psychographic segmentation of the market. Failure to consider con-

text in product development may easily lead to a product that disappoints everybody. Mind-set segments are not intertwined with demographic groups, and the same people can have different needs and thus product requirements according to social occasion or emotional state.

To understand the role of end use, we move to the second study on first principles, which dealt with 28 aspects of coffee. The end uses dealt with general acceptance, as well as fit to different ages, brands, emotional states, physiological states, social occasions, times of day, and purchase venues. This second study expanded the notion of first principles to go beyond simple day parts and brands.

The test stimuli for the concepts comprised 36 elements relevant for a new coffee beverage and were identical for each study, as was the case for carbonated beverages. Table 18.7 shows the basic structure for the coffee elements. The concept elements were selected in order to be appropriate for all of the 28 studies; that is, there were no clear apparent contradictions between the concept elements and the end uses, although the results will show that some of the elements were clearly more appropriate for some end uses than they were for others. The approach paralleled that of the carbonated beverage study:

1. The same elements were used.
2. The study comprised both a conjoint portion and a self-profiling portion.
3. Respondents received an e-mail invitation and, when they accepted the invitation, were sent to a wall where they could choose a specific end use that they found interesting.
4. Respondents read an introductory screen that oriented them in the study.
5. Respondents evaluated 60 combinations comprising the 36 elements.
6. Respondents completed the self-profiling classification questionnaire.

Table 18.7. Concept elements for the coffee study

Category 1: Caffeine, organic, hot versus iced
E01   A lively decaffeinated coffee that won't weigh you down
E02   A coffee that's guaranteed to wake you up
E03   100% organic coffee . . . healthy for you and the planet
E04   Dark fancy house blend . . . an extremely rich cup of coffee
E05   A distinctive, well rounded cup of coffee . . . the ideal way to start a busy day
E06   A slightly caffeinated iced coffee drink . . . to help you get through your day
E07   A jolt of caffeine to awaken your senses
E08   Iced to the max . . . for those hot summer days
E09   The mini-drink six pack . . . iced coffee drinks for people on the go
Category 2: Flavors
E10   Invigorate your senses with Cinnamon Apple Spice & French Caramel
E11   A unique flavor, sure to delight . . . sweet and smooth rich cream compliment this delectable treat
E12   White chocolate mousse, and wild raspberry . . . a melt in your mouth dessert in a flavored coffee
E13   New classic combination . . . pistachio & maple walnut . . . unleash the nutty side in you
E14   Chocolate and cognac give this coffee a flair . . . try it once and you'll come back for more
E15   Enjoy the taste of toffee in a light cream . . . a new summer favorite
E16   Vanilla and chocolate fudge combined . . . a unique flavor that is sure to please
E17   Mocha and spicy Java create a one of a kind chocolate fantasy
E18   Thrilling burst of vanilla flavor and sweet, crisp taste . . . gives you "more to go wild for"
Category 3: Sensory promises (aroma, taste, body)
E19   The freshest cup of coffee possible
E20   A masterful combination of carefully chosen coffee from each year's harvest
E21   Highly aromatic, rich in taste with smoky overtones
E22   Wonderfully smooth with deep tones
E23   Its unique aroma will appeal to your senses
E24   Tangy taste, rich body and pleasing aroma
E25   Exceptional aroma and a deep mellow body
E26   Spicy aroma, medium body and clean flavor make this coffee stand out
E27   Aroma, body and flavor . . . perfectly balanced
Category 4: Brew, quality, origin
E28   Made from exotic Jamaican beans, experience the magic of another world
E29   Premier espresso made from the finest beans
E30   Made from a select combination of African and Central American beans
E31   A dynamic blend of washed Arabian coffee
E32   An Italian favorite . . . cappuccino with a flair
E33   What you always wanted . . . café Americano with all the works
E34   One of a kind coffee developed by top quality growers
E35   Dark exotic taste . . . a superb Turkish brew
E36   A robust strong coffee blend . . . made from dark roasted Brazilian beans

For the following discussion, three subsets have been chosen out of the 28 studies: moods (four studies), social occasions (four studies), and drinking venues (four studies). The objective is to identify the features of coffee beverages designed for these particular moods (tired, restless, happy, and sad), social occasions (alone, social, friends, and family) or with three different and popular venues in mind: Dunkin' Donuts, Starbucks, and Burger King. The underlying theme is whether a different occasion or different venue leads to different needs for the consumers and thus to different product features and messaging:

1. The mood can be a driver for different needs: Should the coffee desired by someone who is tired present the same character-

istics of the one desired by someone sad? This may sound like an unusual question, since it calls into play emotions as drivers of product features, an unusual topic for product development.

2. The three chosen venues are quite popular in the United States and represent out-of-home locations where coffee and coffee beverages are consumed. They are different types of venues, however, and may be associated with different types of new products that are appropriate.

Emotions and Social Occasions: Analysis of the Study-base Sizes

We first deal with the number of respondents participating in each of the studies. The issue is whether there are differences in the nature of the respondents and the possible reasons for those differences. Thus, this discussion deals with both the particular first-principles studies on coffee, a substantive topic, and on Internet research, a methodological topic:

1. *Respondent age is a driving factor.* There are almost no respondents in the teenager group. This was by design, because the e-mail invitations went to people over age 18 and stated that only those over age 18 should participate. Furthermore, there are very few respondents (less than 10% of the total) in both the older (over 61) and the younger (21–30) age groups.

2. *Alone and Sad as emotion end states show both the older and younger respondents participating.* The only study in which the older respondents reach 10% of the panel is the Alone study, whereas the only study comprising more than 10% for the young group is the Sad study. This age skew brings up the ever-present issue of interaction of respondent age with study participation. People participate in studies that they find interesting, so there may not be a true random sample. As far as the other age groups are concerned, major study-to-study differences in this dataset appear for the 31- to 40-year-

olds. However, both the 41–50 and 51–60 age groups appear in relatively higher proportions for the Alone study.

3. *Geographic distribution of respondents.* The majority of respondents (50%) live on the East Coast, and half as many (25%) live on the West Coast. The remaining 25% of the respondents distribute around the United States. As far as geographic skews for single studies relative to their participation in all of the studies, the Midwest and Southeast show unexpectedly high participation in the Alone study. The ability to reach many respondents on Internet-based research, and the study of emotion and situation end uses, make the respondent participation information for these studies worth deeper analysis.

4. *Coffee consumption.* Frequent coffee drinkers represent almost 50% of the respondents, followed by the very frequent consumers (heavy coffee drinkers) and general coffee consumers (~30%). The remaining respondents are those who say that they drink now and then (~15%), and on occasion or seldom (10%). The patterns are not clear:

   a. Those who "drink at every opportunity" and the "frequent drinkers" showed unusually high participation in the Alone study (relative to their other participation).

   b. Those who profiled themselves as "drink now and then" showed unusually high participation in the Social study.

   c. Those who drink coffee on occasion preferred the Friends study.

   d. The self-selection afforded by the wall design, with multiple studies available to any respondent, enables the researcher to identify the reach of interest in a specific study topic across many respondents.

5. *Time of day when respondents say that they drink coffee.* Early-morning drinkers account for more the 25% of the total panel, followed by late-morning drinkers (20%). Those who drink coffee during the other day parts (around lunchtime, midafternoon, early evening, or late evening) are about equally represented (10%–15%).

6. *Brand commonly used.* People who drink Folger's comprise almost 40% of the total panel, followed by those who drink Maxwell House (30%), Starbucks (25%), and Dunkin' Donuts (8%). Folger's and Maxwell House consumers comprise the overwhelming majority of respondents for the Alone study, but not the majority in the other studies.

7. *A working hypothesis.* The number of respondents to each study by itself provides information when cross tabulated against some of the segmenting variables from the questionnaire. The hypothesis is that the average daily amount and the pattern of consumption reflect different needs, and these differences generate the choice to participate in the different studies. A very frequent drinker may look for the energizing or stimulating properties of the coffee, which is seen almost as a *functional food* and has no reason for being consumed on a social occasion. On the contrary, someone who drinks coffee rarely considers it a rite or an occasion to be shared with someone else. What is expected is that the features of the coffee should be different.

Winning and Losing Elements: Total Panel

Since the results come from different studies we again better understand the differences by listing the elements that drive interest and that diminish interest. This comparison is even more powerful because the elements are identical across all studies. Moreover, it can be interesting to check whether there are correlations between the performance of the elements in different studies. If the correlations are high, then one would create the same product for different moods.

We begin this detailed analysis with the correlation matrix calculated on the utility values for the total panel of each study. Recall that the Pearson R shows the degree to which respondents in two studies show similar patterns of utilities. The Pearson R goes from a high of $+1.00$, denoting a perfect linear relation; to a middle value of 0, meaning no relation; to a low of -1.00, meaning a perfect inverse relation. We look for those pairs of end uses that show high correlations; these end uses thus exhibit similar patterns of consumer demands. We see from Table 18.8 that most of the end uses show similar patterns of utilities, so they have high Pearson correlations. Surprisingly, some particular combinations, specifically social, that we might feel are opposite (a coffee for when one is alone versus when one is with family) exhibit high correlations (Pearson $R = 0.92$). However, some end uses or emotional situations are not related at all (e.g., Alone vs Social, Pearson $R = 0.29$). These results mean that we will have to delve into the elements themselves and how they perform in end uses.

Let us now continue with a detailed analysis of the elements to identify patterns. Because of the voluminous nature of the data the results will not be shown but rather sim-

Table 18.8. Pairwise correlations of 36 utilities between pairs of end uses for the coffee study

	Alone	Family	Friends	Happy	Restless	Sad	Social	Tired
Alone	1.00							
Family	0.92	1.00						
Friends	0.78	0.81	1.00					
Happy	0.78	0.80	0.81	1.00				
Restless	0.79	0.81	0.92	0.84	1.00			
Sad	0.76	0.81	0.83	0.77	0.82	1.00		
Social	0.29	0.43	0.63	0.54	0.63	0.57	1.00	
Tired	0.82	0.88	0.84	0.94	0.87	0.84	0.61	1.00

ply described. Here are three observations from the many that can be made:

1. The winning elements are in some cases common across studies (e.g., E02, "A coffee that's guaranteed to wake you up"), but in quite a number of cases they differ from study to study.

2. For example, one of the winning elements for Friends and Sad is E28, "Made from exotic Jamaican beans, experience the magic of another world," which is substantially indifferent for the others.

3. Finally, in some cases, what drives interest in one study turns out to be a losing element in another study. For example, E20, "A masterful combination of carefully chosen coffee from each year's harvest," is a winning element for Family; strongly negative for Social, Happy, Restless, and Sad; and neutral for the rest.

A Closer Look at Two Studies with Opposite Mind-sets: Alone Compared with Social

Let's take a closer look at the results by choosing opposing consumption occasions, for example, Alone and Social. The Pearson correlation coefficient for these studies is rather low (0.29), meaning that what appears as appropriate for one emotional/situational state may be totally irrelevant for the other. Table 18.9 lists the data on which the observations are based:

1. *Additive constant.* This constant is similar for the two studies, meaning that the interest in the "naked" concept has the same success, even if Alone seems to be more popular (getting twice the number of participants compared with the other study).

2. *Winning element.* In both cases the element "wake-up coffee" is successful, whereas origin or brewing method is completely irrelevant.

3. *Differentiating elements.* Other elements, however, that really divide these two groups are listed in Table 18.9.

4. *Alone.* Elements fitting this situation deal with the notion of coffee as a fuel ("to wake you up" and "to start a busy day"). These respondents do not care about sensory promises and are substantially averse to iced coffee and to all flavors (mean value of -11, ranging from -3 to -23).

5. *Social.* Elements fitting this situation deal with sensory promises, such as rich and fine aroma and body. Respondents strongly reject some flavors, especially chocolate, whereas they are indifferent to the others.

Happy Compared with Sad

When taking into consideration particular moods—for example, happy and sad—we see that the base size is the same, but the additive constants are rather different (see Table 18.10). The additive constant for Happy (51) is much higher than the that for Sad (36). This means that, without any elements present in the concept, there is a lower affinity for elements to fit the emotion sad than to fit the emotion happy. The correlation is rather high between the elements for these two end uses (0.77, from Table 18.8), but this is due mostly to the respondents' aversion to some particular elements. In this case the element on which both groups agree is the decaffeinated coffee; they both feel that the element "decaffeinated" is inappropriate:

1. *Happy.* There is no real winner, even if the respondents tend to love some flavors. The developer should be careful, however, because when formulating for a happy feeling it is important to note that the respondents who chose to participate really hate some flavors (vanilla and chocolate). Another thing these respondents do not like is a decaffeinated coffee. Elements involving brewing methods, origin, or quality are irrelevant to this group.

2. *Sad.* These respondents are interested in one particular sensory characteristic, a spicy aroma, and one particular country of origin, Jamaica. Perhaps the association

Table 18.9. Winning and losing elements for Alone and Social studies based on the total panel

	Study	Social	Alone
	Base size	*123*	*222*
	Additive constant	*42*	*46*
	Winning element for both situations		
E02	A coffee that's guaranteed to wake you up	9	6
	Acceptable elements for Social		
E23	Its unique aroma will appeal to your senses	9	−2
E24	Tangy taste, rich body and pleasing aroma	8	1
E25	Exceptional aroma and a deep mellow body	8	1
E21	Highly aromatic, rich in taste with smoky overtones	4	−1
	Acceptable element for Alone		
E05	A distinctive, well rounded cup of coffee . . . the ideal way to start a busy day	2	5
	Losing elements		
E20	A masterful combination of carefully chosen coffee from each year's harvest	−11	2
E17	Mocha and spicy Java create a one of a kind chocolate fantasy	−11	−7
E14	Chocolate and cognac give this coffee a flair . . . try it once and you'll come back for more	−11	−11
E16	Vanilla and chocolate fudge combined . . . a unique flavor that is sure to please	−9	−11
E18	Thrilling burst of vanilla flavor and sweet, crisp taste . . . gives you "more to go wild for"	−6	−3
E15	Enjoy the taste of toffee in a light cream . . . a new summer favorite	−4	−8
E01	A lively decaffeinated coffee that won't weigh you down	−3	−12
E08	Iced to the max . . . for those hot summer days	−2	−11
E26	Spicy aroma, medium body and clean flavor make this coffee stand out	1	−5
E11	A unique flavor, sure to delight . . . sweet and smooth rich cream compliment this delectable treat	1	−6
E06	A slightly caffeinated iced coffee drink . . . to help you get through your day	1	−11
E09	The mini-drink six pack . . . iced coffee drinks for people on the go	1	−14
E12	White chocolate mousse, and wild raspberry . . . a melt in your mouth dessert in a flavored coffee	−1	−14
E10	Invigorate your senses with Cinnamon Apple Spice & French Caramel	2	−15
E13	New classic combination . . . pistachio & maple walnut . . . unleash the nutty side in you	−1	−23

with beaches, sea, sun, and reggae music helps keep away a blue mood. Not surprisingly, respondents who chose to participate in the Sad study are also slightly interested in the element of iced coffee (to be consumed on that beach?). They dislike vanilla and chocolate flavors, whereas they are indifferent to the others.

Mind-set Segmentation and Fit to Mood and Situation

As just shown, the results from the analysis of the total panel do not provide a detailed picture of the population. Often, rather different-minded segments hide behind average values. The low utilities of the winning elements hint at this possibility. Complementary segments that show opposite likes and dislikes cancel each other, generating average utilities close to zero for the total panel. A stronger approach is to use concept-response segmentation (see Chapter 7). These concept-response segments do not come from psychographic clustering of people who say they are similar in self-profiling classification, but rather from people who *respond* similarly to the concept elements.

Table 18.10. Winning and losing elements to fit the Happy and Sad end uses, based on the total panel

	Study	Happy	Sad
	Base size	*107*	*107*
	Additive constant	*51*	*36*
	Acceptable element: Happy		
E10	Invigorate your senses with Cinnamon Apple Spice & French Caramel	5	1
	Acceptable elements: Sad		
E09	The mini-drink six pack . . . iced coffee drinks for people on the go	2	5
E28	Made from exotic Jamaican beans, experience the magic of another world	−1	5
E26	Spicy aroma, medium body and clean flavor make this coffee stand out	0	4
	Losing elements		
E20	A masterful combination of carefully chosen coffee from each year's harvest	−17	−9
E01	A lively decaffeinated coffee that won't weigh you down	−14	−11
E14	Chocolate and cognac give this coffee a flair . . . try it once and you'll come back for more	−12	−10
E16	Vanilla and chocolate fudge combined . . . a unique flavor that is sure to please	−10	−7
E18	Thrilling burst of vanilla flavor and sweet, crisp taste . . . gives you "more to go wild for"	−8	−7
E17	Mocha and spicy Java create a one of a kind chocolate fantasy	−7	−13
E11	A unique flavor, sure to delight . . . sweet and smooth rich cream compliment this delectable treat	−5	−2
E02	A coffee that's guaranteed to wake you up	−2	−6

One of the issues with a very large dataset is exactly how to approach the segmentation analysis. It is important to recognize that in this large-scale study the respondents rated the concepts on different end uses. They did not rate the concepts on interest, except for the one study that requested that rating. Rather, the end uses are all different. Thus, it is inappropriate to put all of the respondents into one big group and segment them independently of rating question.

An alternative strategy is to segment the respondents from each study separately. Each study generates a specific number of segments that can be considered separately. Furthermore, across these separate segments (segment within end use) there may be some commonalities lurking. We can find these commonalities and create *supersegments* from the set of individual segments.

Within-study Segmentation

Let us begin with the dataset from social occasion. The four studies are Social, Friends, Family, and Alone, which are comparable because each deals with a social situation rather than with a mood. Let us segment the respondents from each of these four datasets separately, generating three segments for each dataset. The choice of three segments is arbitrary; there may be two, or four, or five, or even none. We are using three segments for illustrative purposes:

1. *The segmentation pulls out different consumer mind-sets from each study.* We know that at the level of statistical correctness the segmentation was done appropriately. Even though the respondents profiled appropriateness to a particular situation, which we might consider more *objective* than individual tastes, we still see differences among the segments. Correlations between pairs of segments are low (Table 18.11). This means that what drives interest for one respondent segment of a particular study is generally irrelevant to or even rejected by respondents in the other segments of the same study. This makes intuitive sense because by definition the segmentation generates clusters of individuals who are similar within a segment, but whose segments are unrelated to each other. What is important, however, is that even at the level of appropriateness there may be segments.

Table 18.11. Pairwise correlations of 36 utilities between segments of different studies

	Social			Friends			Family			Alone		
	1	2	3	1	2	3	1	2	3	1	2	3
Social												
1	1.0											
2	−0.2	1.0										
3	−0.1	−0.5	1.0									
Friends												
1	0.5	−0.2	0.3	1.0								
2	−0.2	0.9	−0.3	0.0	1.0							
3	−0.5	0.2	0.6	−0.1	0.2	1.0						
Family												
1	−0.3	0.3	0.4	−0.2	0.3	0.9	1.0					
2	0.1	−0.6	0.6	0.3	−0.6	0.1	0.0	1.0				
3	−0.2	0.9	−0.6	−0.2	0.8	0.0	0.2	−0.7	1.0			
Alone												
1	−0.4	0.4	0.4	−0.3	0.3	0.8	0.9	0.0	0.2	1.0		
2	−0.2	0.9	−0.6	−0.2	0.8	0.0	0.3	−0.7	1.0	0.2	1.0	
3	0.1	−0.6	0.7	0.3	−0.5	0.1	−0.1	0.6	−0.7	−0.1	−0.7	1.0

2. *Example from the Alone study* (Table 18.12). If we look, for example, at the three segments of the Alone study, we notice that what is rated as appropriate for Alone by segment 3 (flavors) is rated inappropriate by segments 1 and 2. Other elements show this segmentation, such as iced coffee, which is appropriate by segment 3, strongly inappropriate by segment 1, and slightly inappropriate by segment 2. Another example of this pervasive segmentation governing the appropriateness of sensory attributes to a social situation is "highly aromatic, rich in taste with smoky overtones." This element is appropriate according to segment 1 (+8), irrelevant to segment 2 (−2), and highly inappropriate by segment 3 (−8). Considering only the average value (−1 of the total panel) we would lose sight of these drivers of appropriateness for the respondents by homogenizing everyone into a single group.

3. *Example from the Family study* (Table 18.13). Flavors also segment the respondents in the Family study. They are generally rated appropriate by segment 2 but inappropriate by segment 3. Segment 1 feels that some specific origins like Brazil and Jamaica are very appropriate, whereas the other segments feel that

the elements are inappropriate. Again, averaging can be dangerous because it is the results of countervailing forces and the average results.

4. *Do these data simply restate liking.* One of the key issues that continues to emerge in concept research is the degree to which other attributes of a quasi-evaluative nature (e.g., appropriateness) really restated *liking*. Of course, the results here suggest that if only the respondents who like or dislike a coffee are really responding, then the pairwise correlations in Table 18.8 should be higher. The correlations are sufficiently lower than 1.00, denoting a perfectly linear relationship, which suggests that the respondents are adding some other criteria to their ratings. They may not be able to dissociate themselves from liking/disliking as much as we would like, however.

Segmenting the Segment Sets to Create Supersegments

When comparing the segments of Alone and Family studies (Tables 18.12 and 18.13), segments emerged in both studies that wanted "flavor and ice seekers." Moreover, a group also emerged that responded to "wake-up

Table 18.12. Most appropriate elements for Alone study: differences clearly emerge across the segments

Study: Alone		Alone			
		Total	1	2	3
Base size		*222*	*53*	*131*	*38*
Additive constant		*46*	*55*	*45*	*39*
	Most appropriate: Total panel				
E02	A coffee that's guaranteed to wake you up	6	5	7	4
E05	A distinctive, well rounded cup of coffee . . . the ideal way to start a busy day	5	4	4	8
	Appropriate: Segment 1 (hard to name)				
E17	Mocha and spicy Java create a one of a kind chocolate fantasy	−7	7	−20	16
E21	Highly aromatic, rich in taste with smoky overtones	−1	6	−2	−8
E02	A coffee that's guaranteed to wake you up	6	5	7	4
E18	Thrilling burst of vanilla flavor and sweet, crisp taste . . . gives you "more to go wild for"	−3	5	−13	18
	Appropriate: Segment 2 (wake-up)				
E02	A coffee that's guaranteed to wake you up	6	5	7	4
	Appropriate: Segment 3 (flavor and ice seekers)				
E16	Vanilla and chocolate fudge combined . . . a unique flavor that is sure to please	−11	0	−27	29
E12	White chocolate mousse, and wild raspberry . . . a melt in your mouth dessert in a flavored coffee	−14	−15	−25	25
E14	Chocolate and cognac give this coffee a flair . . . try it once and you'll come back for more	−11	−10	−19	18
E18	Thrilling burst of vanilla flavor and sweet, crisp taste . . . gives you "more to go wild for"	−3	5	−13	18
E17	Mocha and spicy Java create a one of a kind chocolate fantasy	−7	7	−20	16
E15	Enjoy the taste of toffee in a light cream . . . a new summer favorite	−8	−3	−16	12
E10	Invigorate your senses with Cinnamon Apple Spice & French Caramel	−15	2	−30	12
E11	A unique flavor, sure to delight . . . sweet and smooth rich cream compliment this delectable treat	−6	−1	−12	9
E06	A slightly caffeinated iced coffee drink . . . to help you get through your day	−11	−31	−8	9

coffee" and "coffee as a fuel." These commonalities may indicate *supersegments* emerging from this set of studies. The supersegments transcend the specific studies and constitute highly positively correlated segments in the different studies. For product development and marketing, these supersegments may provide the source of new opportunities. They emerge in different situations and may exhibit homogeneous preferences, leading to simplified consumer requirements for product features.

Let us create the supersegments. We first develop a database of *logical individuals*, one logical individual per segment. Each logical person comes from one end use (either social or emotional). Every end-use study, in turn, generates three different segments, or three different logical individuals. We can

create these logical individuals because we have the summary data for each end use and for the three segments within an end use. There are four mood studies—Tired, Restless, Happy, and Sad—so the three segments per mood study generate 12 logical individuals. There are four social situation studies—Alone, Social, Friends, and Family—so the three segments per social situation study generates 12 additional logical individuals.

The k-means clustering applied to this larger set of 24 logical individuals enables us to segment this new group of 24 respondents and to identify commonalities. The segmentation reveals three supersegments:

The Wake-up Supersegment (Table 18.14)

The first supersegment comprises the data from these segments: Alone2, Family3,

Table 18.13. Most appropriate elements for Family study: differences can be appreciated between different segments

Study: Family	Family Total	1	2	3
Base size	*109*	*46*	*9*	*54*
Additive constant	*46*	*49*	*29*	*45*
Appropriate: Total panel				
E20 A masterful combination of carefully chosen coffee from each year's harvest	6	13	−4	2
E02 A coffee that's guaranteed to wake you up	5	6	14	3
E22 Wonderfully smooth with deep tones	5	8	14	1
Appropriate: Segment 1 (caffeine, origins and sensory promises)				
E20 A masterful combination of carefully chosen coffee from each year's harvest	6	13	−4	2
E18 Thrilling burst of vanilla flavor and sweet, crisp taste . . . gives you "more to go wild for"	2	9	30	−8
E22 Wonderfully smooth with deep tones	5	8	14	1
E28 Made from exotic Jamaican beans, experience the magic of another world	1	7	0	−4
E36 A robust strong coffee blend . . . made from dark roasted Brazilian beans	0	7	−9	−3
E16 Vanilla and chocolate fudge combined . . . a unique flavor that is sure to please	−6	6	26	−21
E02 A coffee that's guaranteed to wake you up	5	6	14	3
E07 A jolt of caffeine to awaken your senses	2	5	−6	1
Appropriate: Segment 2 (flavor and ice seekers)				
E10 Invigorate your senses with Cinnamon Apple Spice & French Caramel	−12	−4	35	−26
E13 New classic combination . . . pistachio & maple walnut . . . unleash the nutty side in you	−18	−7	34	−36
E18 Thrilling burst of vanilla flavor and sweet, crisp taste . . . gives you "more to go wild for"	2	9	30	−8
E12 White chocolate mousse, and wild raspberry . . . a melt in your mouth dessert in a flavored coffee	−13	−3	27	−28
E16 Vanilla and chocolate fudge combined . . . a unique flavor that is sure to please	−6	6	26	−21
E15 Enjoy the taste of toffee in a light cream . . . a new summer favorite	−6	−4	24	−12
E02 A coffee that's guaranteed to wake you up	5	6	14	3
E22 Wonderfully smooth with deep tones	5	8	14	1
E24 Tangy taste, rich body and pleasing aroma	3	3	11	1
E17 Mocha and spicy Java create a one of a kind chocolate fantasy	−6	3	10	−16
E11 A unique flavor, sure to delight . . . sweet and smooth rich cream compliment this delectable treat	−4	−1	10	−9
E06 A slightly caffeinated iced coffee drink . . . to help you get through your day	−14	−30	7	−4
E08 Iced to the max . . . for those hot summer days	−15	−31	6	−6
E23 Its unique aroma will appeal to your senses	2	0	6	2
Appropriate: Segment 3 (fuel)				
E05 A distinctive, well rounded cup of coffee . . . the ideal way to start a busy day	3	2	1	5

Friends2, Happy2, Restless3, Sad2, Social3, Tired1, and Social1, reaching a total base size of more than 400 individuals. We know the base size because we know the number of respondents from each individual segmentation exercise. The key feature for the individuals of this segment is a wake-up coffee, but they also place extreme importance on the sensory characteristics of their coffee, namely, a pleasing, unique aroma, and a rich taste. This supersegment does not want flavored coffees.

The Uncomplicated Supersegment (Table 18.15)

This second group comprises the following segments: Happy1, Happy3, Restless1, Rest-

Table 18.14. Appropriate and inappropriate elements for supersegment 1: wake-up (base size = 411 respondents)

	Supersegment 1	Utility
	Appropriate	
E02	A coffee that's guaranteed to wake you up	20
E23	Its unique aroma will appeal to your senses	17
E25	Exceptional aroma and a deep mellow body	10
E21	Highly aromatic, rich in taste with smoky overtones	10
E24	Tangy taste, rich body and pleasing aroma	10
E22	Wonderfully smooth with deep tones	8
E20	A masterful combination of carefully chosen coffee from each year's harvest	6
E05	A distinctive, well rounded cup of coffee . . . the ideal way to start a busy day	5
E19	The freshest cup of coffee possible	5
	Inappropriate	
E16	Vanilla and chocolate fudge combined . . . a unique flavor that is sure to please	−11
E17	Mocha and spicy Java create a one of a kind chocolate fantasy	−11
E14	Chocolate and cognac give this coffee a flair . . . try it once and you'll come back for more	−12
E12	White chocolate mousse, and wild raspberry . . . a melt in your mouth dessert in a flavored coffee	−14
E10	Invigorate your senses with Cinnamon Apple Spice & French Caramel	−15
E13	New classic combination . . . pistachio & maple walnut . . . unleash the nutty side in you	−20

Table 18.15. Appropriate and inappropriate elements for supersegment 2: uncomplicated (base size = 292 respondents)

	Supersegment 2	Utility
	Appropriate	
E26	Spicy aroma, medium body and clean flavor make this coffee stand out	3
E10	Invigorate your senses with Cinnamon Apple Spice & French Caramel	3
	Inappropriate	
E23	Its unique aroma will appeal to your senses	−9
E16	Vanilla and chocolate fudge combined . . . a unique flavor that is sure to please	−10
E17	Mocha and spicy Java create a one of a kind chocolate fantasy	−10
E14	Chocolate and cognac give this coffee a flair . . . try it once and you'll come back for more	−14
E02	A coffee that's guaranteed to wake you up	−15
E01	A lively decaffeinated coffee that won't weigh you down	−15
E18	Thrilling burst of vanilla flavor and sweet, crisp taste . . . gives you "more to go wild for"	−21
E20	A masterful combination of carefully chosen coffee from each year's harvest	−25

less2, Sad1, Sad3, and Social2, with a total base size of almost 300 people. Building a winning product for this segment would not be easy. There is nothing that they like particularly, except for the element "spicy, medium body, clean flavor." These individuals strongly reject vanilla and chocolate flavors and are not interested in caffeine, but they do not want decaf coffee, either. They can best be described as the "give me my coffee and that's it" segment.

The Vanilla/Flavor-seeker Supersegment (Table 18.16)

This third supersegment comes from Alone1, Family1, Friends3, Tired2, Alone3, Family2, Friends1, and Tired3, with a total base size of

Table 18.16. Appropriate and inappropriate elements for supersegment 3: vanilla/flavor seeker (base size = 280 respondents)

Supersegment 3	Utility
Appropriate	
E16 Vanilla and chocolate fudge combined . . . a unique flavor that is sure to please	6
E02 A coffee that's guaranteed to wake you up	6
E17 Mocha and spicy Java create a one of a kind chocolate fantasy	5
E18 Thrilling burst of vanilla flavor and sweet, crisp taste . . . gives you "more to go wild for"	5
Inappropriate	
E03 100% organic coffee . . . healthy for you and the planet	−7
E13 New classic combination . . . pistachio & maple walnut . . . unleash the nutty side in you	−11
E01 A lively decaffeinated coffee that won't weigh you down	−26
E08 Iced to the max . . . for those hot summer days	−29
E09 The mini-drink six pack . . . iced coffee drinks for people on the go	−31
E06 A slightly caffeinated iced coffee drink . . . to help you get through your day	−31

280 people. These respondents love some flavors and dislike others. They hate iced coffee.

Coffee Venues

Beyond emotion the first-principles approach can work with appropriateness for different venues. Venues are important in coffee because they connote different types of expected experiences. This first-principles study with coffee explored three different venues from which the respondents could choose: Dunkin' Donuts, Starbucks, and Burger King. These three venues represent different types of situations in which a person drinks coffee. The data from the study can now reveal the features of coffee that most strongly correspond to each venue.

Interest in the Venue

Table 18.17 lists the additive constant, averaged across all of the respondents, for each venue. The constant ranges from a low of 42 (the Burger King study) to high of 65 (the Starbucks study). This additive constant means that the appropriateness of a coffee beverage, independent of elements, is highest at Starbucks. The respondent is immediately predisposed to fitting a coffee beverage with Starbucks. Without any element present in the

concept, 65% of the respondents would rate a coffee beverage as interesting if it were positioned at being from Starbucks.

The Additive Constant Is Stable for a Venue Across Subgroups

Although there are some differences in the value of the additive constants, for the most part the constants are quite similar within a particular study with a specific venue, but consistently (although not always) different for the same subgroup across different venues. Thus, it is the venue itself, rather than the subgroup, that drives basic interest.

Appropriate versus Inappropriate Concept Elements

The distribution of utility values is such that few elements are appropriate, but quite a number of elements are strongly inappropriate (see Table 18.17). This difference in the performance of the elements immediately signals that the respondents will not simply accept any product feature as appropriate to a venue, but discriminate among the features with regard to whether they fit a venue where coffee is consumed. Furthermore, what is appropriate in one venue may not always be acceptable in another. As an example, a coffee

Table 18.17. Classification distribution of respondents and additive constant for total panel of each of the three venues

	Venue					
	Starbucks		Dunkin' Donuts		Burger King	
	% Base	Additive constant	% Base	Additive constant	% Base	Additive constant
Total	*100*	*65*	*100*	*57*	*100*	*42*
Gender						
Male	30	62	27	51	30	37
Female	70	66	73	60	70	44
Age						
18–21	3	65	2	20	3	17
21–30	12	61	13	52	15	48
31–40	29	65	28	57	26	40
41–50	33	65	35	60	35	44
51–60	17	66	18	62	16	36
61+	5	68	4	51	6	50
Coffee consumption						
Drink at every opportunity	22	62	16	67	9	60
Frequently drink	47	70	36	64	33	47
Drink now and then	20	62	18	52	19	43
Drink on occasion	5	68	9	58	9	41
Hardly ever drink	5	45	8	60	16	38

beverage comprising "premier *espresso* made from the finest beans" is appropriate for Starbucks (utility = +3) but not for Burger King (utility = −4) and virtually irrelevant for Dunkin' Donuts (utility = −1).

Dunkin' Donuts and Burger King

When considering the appropriateness for Dunkin' Donuts and Burger King, the elements show a fairly narrow range of variation. There is no clear winner in terms of appropriateness, even if there are some losers. The appropriate elements for the Dunkin' Donuts venue are related to rather general categories and are connected to the general mind-set of coffee as a fuel. Inappropriate elements for both Dunkin' Donuts and Burger King talk about flavors.

Starbucks

Starbucks suggests a different type of venue with rather different elements that perform

well. Appropriate elements are related to quality ("A unique flavor, sure to delight . . . sweet and smooth rich cream compliment this delectable treat"), aroma ("Tangy taste, rich body and pleasing aroma") and, rather surprisingly, to a particular brewing method (Italian espresso or cappuccino).

Summing Up: Development Strategies from First Principles

The approach presented here assumes the development of product (or positioning) concepts *from the bottom up*. There is no assumption of prior knowledge, but rather a disciplined way to develop the concepts by using basic inputs (elements, raw materials); assessing performance and fit to emotion, situation, day part, or venue; and finally segmenting to particularize the product.

The real issue in development is the merger of systematized resource development with creativity. Instead of requiring

Table 18.18. Appropriate and inappropriate elements for the total panel in each coffee venue

		Dunkin' Donuts	Burger King	Starbucks
	Base size	*337*	*393*	*227*
	Additive constant	*57*	*42*	*65*
	Appropriate: Dunkin' Donuts			
E19	The freshest cup of coffee possible	5	1	0
E05	A distinctive, well rounded cup of coffee . . . the ideal way to start a busy day	3	5	1
E02	A coffee that's guaranteed to wake you up	3	3	0
	Appropriate: Burger King			
E05	A distinctive, well rounded cup of coffee . . . the ideal way to start a busy day	3	5	1
E02	A coffee that's guaranteed to wake you up	3	3	0
	Appropriate: Starbucks			
E11	A unique flavor, sure to delight . . . sweet and smooth rich cream compliment this delectable treat	0	−2	4
E24	Tangy taste, rich body and pleasing aroma	0	−1	3
E17	Mocha and spicy Java create a chocolate fantasy	−7	−9	3
E32	An Italian favorite . . . cappuccino with a flair	−3	−5	3
E29	Premier espresso made from the finest beans	−1	−4	3
	Inappropriate: Dunkin Donuts			
E13	New classic combination . . . pistachio & maple walnut . . . unleash the nutty side in you	−21	−16	−10
E14	Chocolate and Cognac give this coffee its flair . . . try it once and you'll come back for more	−16	−18	−8
E10	Invigorate your senses with Cinnamon Apple Spice & French Caramel	−13	−14	−7
E09	The mini-drink six pack . . . iced coffee drinks for people on the go	−11	−11	−10
	Inappropriate: Burger King			
E14	Chocolate and Cognac give this coffee its flair . . . try it once and you'll come back for more	−16	−18	−8
E13	New classic combination . . . pistachio & maple walnut . . . unleash the nutty side in you	−21	−16	−10
E10	Invigorate your senses with Cinnamon Apple Spice & French Caramel	−13	−14	−7
E09	The mini-drink six pack . . . iced coffee drinks for people on the go	−11	−11	−10
E12	White chocolate mousse, and wild raspberry . . . a melt in your mouth dessert in a flavored coffee	−9	−10	1
	Inappropriate: Starbucks			
E09	The mini-drink six pack . . . iced coffee drinks for people on the go	−11	−11	−10
E13	New classic combination . . . pistachio & maple walnut . . . unleash the nutty side in you	−21	−16	−10
E01	A lively decaffeinated coffee that won't weigh you down	−3	−1	−10

wholly new concepts created by talented individuals, perhaps serendipitously, the approach here provides an algorithm that may end up providing new nuggets of ideas by recombining features in new ways. *It is the set of features, not the combination, that requires the creativity.* Such features may be obtained from so-called creative consumers, from marketing insight, from market-research reports about the category and category trends, or simply by trolling the Internet to identify faint signals that hint at new op-

portunities. After the features are discovered the rest of the task becomes a disciplined exercise in combinatorics, which might be repeated several times until the patterns and specific answers begin to emerge clearly.

Reference

Schutz, H.G., and Martens, M. (2001). Appropriateness as a cognitive-contextual measure food attitudes. In: People, Food and Society: A European Perspective of Consumer Food Choices. Berlin: Springer-Verlag.

Chapter 19

Creating a Cyberspace Innovation Machine

Introduction: Early-stage Development

In previous years a great deal of attention in the consumer research community was focused on assessing products and concepts that were complete, polished, and ready to go. More recently, however, attention has shifted from this late stage of development to a much earlier stage, where the options for development are open and where changes in what is being developed do not impinge on existing systems of production and communication. The recognition that this stage is important has not only been through academic journals (e.g., *Journal of Product Innovation Management*) and in business, where the term *fuzzy front end* has been used to describe the stage. Perhaps the term itself stands as stark recognition of the importance of being right early in the process rather than waiting for the later stage, but the difficulty of doing so. The front end is where the opportunities exist, but the situation is fuzzy and not in clear relief (Moskowitz et al. 2002a).

Today's business environment has loudly proclaimed that it values innovation. Looking up the word *innovation* in a search engine like Google comes up with 9,800,00 hits. Refining the search to *business innovation* comes up with about 4,500,000 hits. Refining the search still further with the words *business innovation ideation* comes up with 5200 hits. Furthermore, one can scarcely read a business-related book without at some point coming up with a section or even a chapter on innovation. Innovation is thus held up to be one of the engines of growth for companies, whether the innovation be in new products, in business processes, or simply and more cynically in the buzz that is spread about the business.

A clear definition of the innovation was given almost 30 years ago by Zaltman and colleagues (1973). *Innovation*, using their definition, refers to any perceived deviation from existing practices or knowledge. Innovation may or may not be planned, deliberate, or involve action. It is easy to see, then, why the *fuzzy innovation stage* is often ignored by researchers, who prefer to concentrate their energy on the more defined, later stages of development where there are metrics, standard procedures, and recognized analytic approaches (Earle 1997). As a result, the important first stage of product development is poorly executed, leaving the product not optimized to survive in a Darwinian jungle (Fuller 1994; Peleg 1994).

In light of this interest in innovation, and recognizing that concept development can work at the so-called fuzzy front end, where all options are possible, how can we harness concept development to innovation? When all is said and done this fuzzy front end refers to the early stage of development [e.g., see Nuese (1995) and Stinson (1996)]. Certainly, concept development is closely linked to early-stage ideation sessions, where the business objective is to develop innovative ideas for new products. Through management *diktat*, innovation has been linked to concept development. Through practices of consultants who specialize in innovation, concepts have been the first tangible evidence presented for the fruits of the innovation process.

This chapter provides a structure to innovation, based on the identification and capture of trends, ideas, and finally concepts. The chapter comprises three different topics:

1. *Weak signals.* Identify unformed ideas in the environment.

2. *Innovative concepts.* Fit ideas to end uses.

3. *New tools.* Use state-of-the-art tools, many cyber based, to accelerate the development even more and place the innovation in its appropriate position in the world of consumer needs and the competitive marketplace.

Part I: Accelerating Development by Understanding Weak Signals and Strong Communications

There are a variety of both positives and negatives to the current methods for creating new ideas. On the positive side, ideation leads to new ideas. The tools that practitioners use are generally effective, although there do not seem to be any metrics about the quality of the ideas generated through the ideation process. Certainly, informal discussions and conference presentations on ideation would lead one to believe that ideas routinely emerge from these approaches. On the negative side, however, the ideation structure relies almost universally on the process and the practitioner, not on the outcome; that is, the process follows certain guidelines. The practitioners state, however, or at least imply that it is the serendipitous emergence of new ideas that is key, and that a defined up-front process is necessary for that happy serendipity to emerge. It is the practitioner's expertise that becomes the midwife for these new ideas.

The actual emergence of the ideas appears to fall into what James (1911) called a "blooming, buzzing confusion." As one idea rapidly follows its predecessor, cascading into a series of connected thoughts, the ideation process appears to deliver that which it promised. It is left as a task for the professional to facilitate the process. The ideas may be used immediately, polished and tested, or thrown back into the process to generate new ideas.

The Delphi Approach: Ideation by Experts

One way to come up with new ideas is to solicit the advice and input of experts. They are presumed to know the environment, and to keep attuned to it, although the precise way in which that attention and tuning takes place is never really explicated, nor for many users of the process need it be. An expert is assumed to integrate the information around him or her and then transmit impressions of that information. A good analogy to this is an integrative instrument that measures food quality by measuring a variety of factors in the food and coming up with a set of measures. The measures themselves may not have meaning, but they are consistent and relevant. An astute scientist in food research, like an astute social scientist in opinion and trend research, can take these measures by the integrative instrument and provide a transformation that converts the information into a relevant format.

The original Delphi method was developed as a technique to assess scenarios and future situations, based on the reactions of experts (Brown 1968). The process requires that information be presented by one expert to other experts. Eventually, through back-and-forth iterations the result is a coherent picture of the situation, digested by the consumer or expert, and during the process of interaction subsequently reshaped into a coherent format (Jolson and Rossow 1971).

The Delphi method can be applied to consumers. One can sense the future by using consumers as the integrative instruments in the same way that Delphi senses the future by using experts. One of the key goals of the Delphi method as applied to consumers is to identify so-called *weak signals* (Ansoff 1975). A weak signal can be defined as an idea or trend (which is not clearly in the consciousness of many people, nor easily evoked in an open-ended question) that is perceived as being relevant by a limited number of people and that can become much bigger with a larger audience. The weak signal represents an idea that has great, but as yet untapped, potential to change the way we do things. A weak signal is often ignored, but, when consumers see it, many immediately recognize it having potential. By their very nature weak signals are hard to define concretely because they are not popular or widely known. A weak signal may represent a new idea that the developer or the marketer needs in a category in order to obtain a momentary advantage. If consumers can be used to focus these ideas and improve them, then the marketer can anticipate the market with products and services based on the weak signal rather than follow the market like the competitors often do. It is vital, however, that the mechanism for weak-signal identification both recognize them and validate their potential through a rigorously evaluative stage implemented preferably at the same time as ideation.

Although weak signals are by necessity new and hard to recognize, they can be somewhat tamed by following a research algorithm to identify them, at least in the rough early stage. Flores and colleagues (2003) suggest four steps to incorporate the Delphi technique with consumers in the systematized creation of new products:

1. Open up the consumer's mind by an open-ended or *aufgabe question*, which provides an opportunity for the consumer to provide phrases that answer the problem.

2. Let subsequent respondents inspect some of the ideas provided by a consumer, choose the ones relevant to that person, and then offer up new ideas.

3. Let the same respondent rate some of the ideas offered before.

4. The process should pull in a large number of respondents, especially when the interview venue is on the Internet, where respondents are inexpensive and easy to find.

Evaluating the Strength of Ideas Through Conjoint Measurement

One needs an evaluative system *distinct and separated* from the ideation system in order to quantify the power of the ideas that are generated. Pure judgment might do. Judgment alone is commonly used because it is straightforward, incorporates the expertise of the judge, and generates a ranking of ideas on the degree of importance. It is also less expensive and more ego gratifying, especially to the judge. Judgment, however, may be biased, especially when applied to single ideas generated from the ideation. Judgment of single elements does not take into account how important the idea will be when incorporated into a concept or selling proposition and when it must fight for attention with the other ideas in the same concept. Furthermore, context is important. For example, price by itself has no context, and therefore one cannot judge the importance of a specific price to a concept. Brand plays an overwhelming role when alone, but often it is the message, not the brand, that does the work (Beckley and Moskowitz 2002) (see Chapter 23).

Conjoint analysis has an appropriate *follow-up role* to ideation generated through the Delphi method. Conjoint analysis provides three specific benefits in the early-stage process that can identify weak signals:

1. *The researcher must think through the concept elements, to make them simple, standalone ideas phrased in the active case.* All too often what emerges from ideation is a nugget of an idea, but the idea itself is poorly expressed. By forcing the researcher or ideation expert to edit the elements, one forces better thinking. The elements themselves mature from simple notions to better-expressed

statements. The exercise itself—homework after ideation—is as valuable as the ultimate discovery of weak signals. It makes a better research user by forcing consideration of the test stimuli.

2. *The performance of ideas is judged against different backgrounds, encouraging survival of the fittest.* Conjoint analysis uses experimentally constructed concepts comprising several different ideas. To the degree that a concept element performs well in multiple backgrounds, one can be sure that the idea is good. This rigorous test speeds up the development process because it creates an objective method to identify good ideas.

3. *Conjoint analysis leads immediately to segmentation, which may reveal the great promise of an idea, otherwise masked among the total panel.* Segmentation is key to achieving increasing acceptance because consumers do not all share the same mind-set. What appeals to one consumer may not appeal to another. Any hope of achieving a very big breakthrough in a new idea has to be tempered by the realization that only a fraction of the consumer population will be swayed by the new idea and perhaps an equal proportion will hate the idea.

Applying the Weak-signal Approach to a Trendy Beverage: Bottled Water

Bottled water exemplifies a growing category with new players. Bottled water had been touted for years as the smallest, yet fastest-growing, segment of the beverage industry. Bottled water offers portability, but also provides a healthy beverage alternative with their inclusion of, for example, fortified vitamins, minerals, and functional ingredients in the products. Despite the complexities of market maturation and ongoing segmentation, today's functional foods area offers beverage companies a solid but diverse hierarchy of product opportunities. However, there is a lack of information about consumer demands

for ongoing product and marketing development of bottled water. Both the marketing and the research and development (R&D) literatures lack the necessary structured information for new innovations.

Let us further specify the bottled-water product as being positioned either as a drinking-water or as an after-exercise product. We will use a combination of brand-Delphi followed by Internet-based conjoint measurement (IdeaMap.net). The two-phase approach should provide ideas and measure their performance.

Phase 1: Identify the features of, and communication about, the water product by using collaborative filtering. The objective of this phase is to identify features for a new bottled water through Internet-facilitated ideation with a large number of respondents. This method used the brandDelphi technique with consumers (Flores et al. 2003).

Phase 2: Identify the particularly strong elements that drive acceptance for total panel and segments by using conjoint analysis. The goal of phase 2 is to identify how strongly a subset of the consumer-provided elements drive interest as a regular table water and as an after-sports beverage, respectively.

Phase 1: Ideation

The ideation was done using a *collaborative filtering* system, in which many respondents offer ideas and evaluate one another's ideas, albeit sequentially. The respondents do not interact with one another, however. It is important to recognize that collaborative filtering must have as many respondents as participants, for it is the serial presentation, voting, and elimination of weak performers that enables the weak signals to emerge. This makes the Internet a particularly attractive venue for the method. The system, totally Internet driven, is schematized in Tables 19.1 and 19.2.

Table 19.1. Steps in the Internet-based ideation through collaborative filtering (brandDelphi)

Step	Action	Rationale
Step 1	Create *aufgabe* questions	Develop a set of open-ended questions
Step 2	Create classification questionnaire	Self-profiling
Step 3	Invite respondents to participate	Internet based e-mail, with explanation of study goals
Step 4	Obtain data	Use the respondents on the Internet both to obtain and to structure the data
Step 5	Identify elements	Follow the heuristic shown in Table 19.2

Aufgabe Questions

Respondents were given three open-ended, aufgabe questions to answer. These questions enable the respondents to provide new ideas. They set the scene for the end use. Figure 19.1 shows the questions as posed to the respondents.

Selecting Reasonable Elements

For each aufgabe question, the respondent first read the question. Afterward, the respondent was given a set of elements contributed by previous respondents (Figure 19.1, top). Only a partial set of elements could be shown to any single respondent. The respondent se-lected those elements from this set that he or she thought to be relevant. The respondent then offered his or her own elements (Figure 19.2, bottom).

The answers to these three questions generated over 400 elements, many of which presented the same core ideas in different ways, with slightly different phrasing. Since respondents did not see all of the previous ideas, they often offered the same idea as one previously given by someone else, but which they themselves did not actually see. It is clear from Figure 19.2 that the process does not edit the respondent data as it goes along, because editing would require human intervention and be contrary to the goal of speeding

Delphi Question 1: Please think about what kind of attributes and benefits you would like to see in bottled FORTIFIED WATER. What are the most important ingredients/nutrients (e.g., calcium, fiber, etc.) that you would like to see included in a new fortified water product?

Delphi Question 2: When looking at current FORTIFIED WATER products available at stores, you may not find a large variety of flavors and colors. What flavors and colors of FORTIFIED WATER would you love to see, which would drive you to buy this product?

Delphi Question 3: Think about bottles/containers and in store displays that are currently available for FORTIFIED WATER. What, if anything, do you dislike about them and what suggestions would you give for improvement

Figure 19.1 Delphi questions.

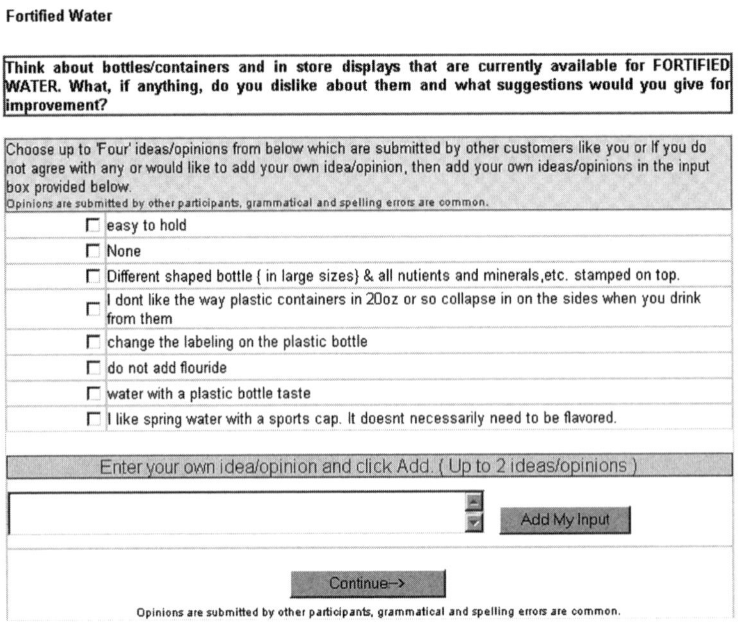

Fortified Water

Think about bottles/containers and in store displays that are currently available for FORTIFIED WATER. What, if anything, do you dislike about them and what suggestions would you give for improvement?

Choose up to 'Four' ideas/opinions from below which are submitted by other customers like you or If you do not agree with any or would like to add your own idea/opinion, then add your own ideas/opinions in the input box provided below.
Opinions are submitted by other participants, grammatical and spelling errors are common.

☐ easy to hold
☐ None
☐ Different shaped bottle (in large sizes} & all nutients and minerals,etc. stamped on top.
☐ I dont like the way plastic containers in 20oz or so collapse in on the sides when you drink from them
☐ change the labeling on the plastic bottle
☐ do not add flouride
☐ water with a plastic bottle taste
☐ I like spring water with a sports cap. It doesnt necessarily need to be flavored.

Enter your own idea/opinion and click Add. (Up to 2 ideas/opinions)

[   ] Add My Input

Continue→

Opinions are submitted by other participants, grammatical and spelling errors are common.

Figure 19.2. Screen shot for the *aufgabe* question dealing with bottles and displays.

up the process. Thus, when a respondent answers "None," this response is taken as a valid answer and may be sent down to other respondents as something to be voted on. Eventually, however, these low-information-content answers fall out because respondents do not choose them, and thus the algorithm treats them as infrequently selected answers. The algorithm is programmed to cease presenting these low-frequency selections.

Selecting the Promising Elements

Ideation develops many elements. Their abundance brings with it redundancy, nonsensical offerings, unclear ideas, and a host of other imperfections. One must select the meaningful elements from this blooming, buzzing confusion. Since one can generate hundreds or even thousands of elements, a selection system must be embedded. If there is no human editor, then the next best thing is a powerful selection device to sort through the material.

A heuristic was developed to identify new elements from the mass of text generated in the rapid ideation. The specifics of one such heuristic are listed in Table 19.2. The heuristic in Table 19.2, as any other heuristic dealing with this type of information, is a *work in progress*-a working system that will be modified, expanded, or trimmed based on ongoing experiences with the method. Fundamentally, the algorithm works as follows:

1. *Absolute selection frequency.* The elements that the algorithm identifies as weak signals must have been selected sufficiently frequently by respondents. No matter how good an element may be, the heuristic will not select it as being promising. There are simply not enough data to warrant that selection. This criterion, of course, works against those elements that emerge later in the ideation process, because there are fewer remaining respondents to select them. Call this number *Sel*.

2. *Relative selection frequency.* The element must be selected a certain proportion of times when it is presented. This criterion

Table 19.2. Heuristic to identify new elements from Internet-based ideation

Step	Action	Rationale
Step 1	Rank ideas by popularity	This identifies ideas that are selected often in an absolute sense (abbreviated as *Sel*).
Step 2	Eliminate redundant phrases and meaningless phrases	Sometimes the same idea may appear several times but stated differently without providing further value or *information*.
Step 3	Eliminate ideas with low frequency	Not gaining the support of other respondents in the process leads to the conclusion that the idea *value* is then low. The element does not really drive sufficient interest, but instead causes more noise than conveying information.
Step 4	Eliminate ideas that are relevant but not unique enough	Typically these ideas refer to *cost of entry* or *category drivers* for the category. Although relevant they are not differentiating enough to be true *weak signals*.
Step 5	Rank remaining ideas by level of popularity	The term *popularity* refers to the proportion an item was selected relative to the times it was seen. More popular means a more relevant signal (abbreviated as *Selp*).
Step 6	Focus on popular elements of high importance	These are ideas that are different—not the usual cost of entry, but when respondents see them they recognize them as relevant and important (abbreviated as *Rate*).

ensures that some items will be rejected because they are selected frequently, simply because they began to appear early in the ideation. Call this number *Selp*.

3. *Moderate to high importance when rated.* The element must be deemed sufficiently important. Ratings below a cutoff denote an element that is not deemed important. If the element always scores highly, this may be a warning signal that we are dealing with an element that is not really a weak signal, but simply a well-known factor whose presence is *cost of entry*. An element that scores too high may also be rejected. Since the goal is to discover ideas that are novel, it is assumed that these novel ideas would be important but not extremely important. For example, any element that is deemed to be extremely important (e.g., 8.5 or higher on a 9-point scale) is assumed to represent this cost of entry into the category and would not be considered to be a weak signal. Call this number *Rate*.

Ideation Output: Tentative Elements

Following the heuristic specified in Table 19.2, the ideation generated a list of ideas that can be expressed as elements, or simple, stand-alone phrases. Some of these ideas are listed in Table 19.3. We have seen lists of elements like these for conjoint analysis studies. The key here is that the elements were generated by consumers who did not see one another, but rather participated in isolated locations and interacted only by computer. Rather than separately presenting the emergent ideas separately from each of three questions (nutrition, flavor, and package design), Table 19.3 shows the intermixed set of ideas, which was then subject to the algorithm described in Table 19.2. The elements are presented in descending order of the proportion of times that an element was presented and chosen (i.e., the Selp statistic). It is clear that these different elements deal with a range of different topics, as one would hope.

An Overview of Phase 1

In this phase the ideation generated a lot of ideas. Rather than requiring an individual to come up with full concepts, or even to participate in an extended ideation session to come

Table 19.3. Examples of elements selected using the heuristic; elements are ranked by Selp

Question	Element text (after slight editing to create a simple, stand-alone phrase)	Sel	Selp	Rate
Flavor	Lemon	18	90	8.2
Health	Calcium and vitamin C	45	87	8.0
Health	All vitamins and minerals	47	73	8.3
Flavor	No aftertaste, please	33	73	8.5
Flavor	Water with natural fruit flavors, NOT artificial flavors	17	71	8.5
Flavor	No added colorings	36	64	8.6
Package	I prefer the sport top which allows you to drink the water without removing the cap—it minimizes spills.	23	55	8.6
Flavor	No taste and color differences. I prefer my water as it is supposed to be. I always buy just natural looking water.	30	53	8.6
Package	Lightweight containers—no glass	26	42	8.6
Package	Cap that seals well enough to contain spills if bottle is tipped over after it has been opened	22	41	8.6

up with snippets of ideas, the Internet-enabled ideation goes about the element generation in a different way. The organizing principle is that the ideas are resident in the mind of the consumers, but no consumer really has the entire idea in mind at the start of the session. Furthermore, at this initial stage the respondent's sole responsibility is to look at nuggets of ideas, identify those that are interesting, and provide a few more. The aufgabe questions serve to excite the respondent and open his or her mind, but not force new ideas. The result is a large array of raw material from consumers. The approach is cost and time effective because of the widespread availability of respondents by the Internet, and thus possibly tens of thousands of respondents can participate almost overnight, they can be scattered around the world, and human intervention is not necessary. All of the responses are recorded and immediately available for study by the developer. Finally, the heuristic allows the developer to identify promising ideas according to a structured set of quantitative rules.

Phase 2: Identifying Winning Elements by Using Conjoint Analysis

The foregoing ideation generated more than 400 elements. Even with the heuristic in place, however, the elements may or may not perform well subsequently. Just because an element appears in the list and is responded to by other individuals does not mean it will do well in the harsh competitive environment of a concept, where other elements compete with it. Conjoint analysis provides a good second stage to assess the power of these new elements. The typical evaluative criterion for conjoint analysis is purchase intent or liking, so the elements are judges in terms of their delivery against that criterion. Within the systematically varied concepts, all of the elements compete with one another to drive the ratings. Thus, the conjoint analysis becomes the acid test as to whether the ideas are good or poor.

This study used two evaluative criteria: water for refreshment and water as an after-sports drink. The two criteria necessitated two parallel conjoint studies, albeit with the same elements. These two end uses are not mutually exclusive, but the after-sports drink is more specific and was of particular interest. The after-sports drink may set up more concrete expectations of what the product should be.

This notion of changing the end use and identifying what elements fit that end use echoes the subsequent topic of developing products by using first principles (see Chapter 18) and the subsequent section of this chapter on innovating in cyberspace. By having respondents focus their mind on different end uses, different sets of new ideas can be crafted and winning weak signals identified for each end use.

The conjoint part of the bottled-water study investigated 36 elements: 29 elements from the output of the ideation session, as well as seven others that made sense for these end uses and had actually failed the heuristic, but nonetheless appeared interesting on the basis of judgment. It is important to keep in mind that elements emerging from ideation need not be the only stimuli to test. Other elements coming from deconstructing the competitive frame, or even from one's own intuition, are just as valid.

By e-mail invitation, the respondents were invited to participate in a study on bottled water. Those who responded were randomly taken either to the study positioning the product as regular bottled water or to the study positioning the product as a sports drink. The elements were the same in the two studies. The respondents knew they would be evaluating concepts for either bottled water or a sports drink.

From Table 19.4, which shows the utilities of the winning and losing elements, one can get a sense of the performance of these elements. It is clear from the table that the respondents liked the idea of fortified water, whether positioned for regular water or for a sports drink. The additive constant is 49 for concepts positioned as regular water and 44 for concepts positioned as a beverage for after sports. The 49 can be interpreted as meaning that 49% of the respondents would rate the concept as 7–9 (top-3 box) if no elements appeared. Recall that this additive constant is a purely theoretical one obtained from regression analysis, but it can serve as a benchmark. The slightly lower constant for the specific positioning (44 instead of 49) makes intuitive sense, since one might expect fewer respondents to be interested in water for sports than in water for regular use. Sports

drinks are more specific than general bottled water, appealing to a narrower population.

The elements disappoint, however, at least on the basis of the total panel. The best-performing element ("Fortified water . . . no color, or flavor added, just the crisp taste and smell of pure water") scores only a +8 for the total panel when positioned as a regular water and only +5 when positioned as a sports drink. The worst-performing elements, however, do far worse, with utility values in the −30s and lower. When working with weak-signal research and ideation we learn from Table 19.4 that it is important to measure the performance of the element in a subsequent study among consumers. Judgment alone will not work, because we see that the elements that passed the heuristic and those that were added did not do very well. At least five findings emerge from the results:

1. The consumers *are interested* in bottled water, as shown by the high additive constants of 49 and 44, respectively, for use as regular water and as a sports drink.

2. Few elements can add appreciably to that basic interest, at least for the total panel. The elements that do add fail to go beyond +10 and so, for the total panel, the elements can be said to perform only modestly as drivers of acceptance.

3. Many elements can detract from interest, especially if the elements are unusual. This is important. Weak signals are unusual, but a promising weak signal is not very unusual. Only by testing the concept elements in a subsequent and different analysis can we be sure that the unusual element has promise.

4. The poorest-performing elements present unusual features, suggesting that sensory novelty in this category will not appeal to the entire panel

5. There is little difference in the performance of the element when the concept is positioned as regular water versus when

Table 19.4. Winning and losing elements for the water concepts*

	Regular	Sports
Base size	*209*	*226*
Additive constant	*49*	*44*
Winning elements: Total panel		
Fortified water . . . no color, or flavor added, just the crisp taste and smell of pure water	8	5
Pure and clear . . . contains no color or heavy metals like lead and mercury	6	3
Pure satisfaction . . . no smell or aftertaste	6	3
Reliable labeling—Added ingredients are clearly labeled on the front . . . so you can see what you're getting or not getting	4	4
Losing elements: Total panel		
The refreshing taste of citrus or grapefruit to add a little tang	−11	−10
Watermelon or melon . . . a taste of summer in every sip	−15	−7
Enjoy the taste of winter and the holiday season . . . cinnamon, vanilla or almond flavor	−18	−19
Enjoy the light taste of peppermint or wintergreen flavors . . . oh, so refreshing	−29	−22
A touch of vegetable flavor like carrots, asparagus, tomato	−41	−34

*The average is the mean of the utilities for the concepts positioned as water and the concepts positioned as sport beverage, respectively.

the concept is positioned as bottled water. Where there are strong negative utilities, however, some differences appear between utilities of the end uses.

At this point we might ask, "Why proceed further?" Some elements perform adequately, whereas some elements perform poorly. There are no blockbusters. Shouldn't there be these blockbusters in ideation?

Concept-response Segmentation and the Identification of Stronger Ideas

Although one might wish to have strong-performing concept elements emerging from an ideation session, all too often the ideas that emerge do not perform particularly well for the total panel, unless the category is newly emerging. When it comes to product or service categories that are well explored and exploited commercially, more often than not the total panel data suggest few new ideas, whereas segmentation reveals greater possibilities. We saw this clearly with the case of toothpaste when the elements in the category were deconstructed (see Chapter 17 and Moskowitz et al. 2002b). This medioc-

rity of ideas is especially the case for food and beverage products, where new flavors can generate immediate dislike because of their unusual nature.

A clearer set of winning elements for water emerges after segmentation (Table 19.5). Although both studies have approximately the same number of respondents, the study with the positioning of concepts as *regular water* appears to comprise two interpretable clusters, whereas the study with the positioning of concepts as *sports beverage* appears to comprise three interpretable clusters. There are no hard-and-fast rules about the number of clusters. It makes more sense to opt for interpretability than to force the same number of clusters for each study.

The following four findings emerge after segmenting the respondents based on the pattern of their utilities:

1. The weak signals—that is, new and possibly unexpected ideas with high utility values—do not emerge from the total panel, but do emerge from the segmented results.

2. Each study has the same two clusters: a group seeking flavor and a group seeking the pure and natural.

Table 19.5. Winning elements for segments

Product positioned as regular water	Total	Regular	
		S1	S2
Base size	*209*	*113*	*96*
Additive constant	*49*	*46*	*53*
Segment 1: Flavor seekers			
The refreshing taste of lemon-lime, just quenches your thirst	−4	12	−21
Berry crazy flavors . . . choose from raspberry, strawberry, kiwi-strawberry	0	11	−13
Segment 2: Pure/natural			
Fortified water . . . no color, or flavor added, just the crisp taste and smell of pure water	8	2	15
Pure satisfaction . . . no smell or aftertaste	6	4	8

Product positioned as sports drink	Total	Sport		
		S1	S2	S3
Base size	*226*	*29*	*78*	*119*
Additive constant	*44*	*41*	*48*	*43*
Segment 1: Flavor seekers				
Apple, banana, peach or cherry flavors . . . just the right touch of fruit	−1	39	−26	6
Berry crazy flavors . . . choose from raspberry, strawberry, kiwi-strawberry	1	37	−23	9
Watermelon or melon . . . a taste of summer in every sip	−7	33	−32	1
Enjoy the taste of winter and the holiday season . . . cinnamon, vanilla or almond flavor	−19	21	−21	−27
Enjoy the light taste of peppermint or wintergreen flavors . . . oh, so refreshing	−22	15	−25	−30
The refreshing taste of lemon-lime, just quenches your thirst	−1	12	−12	3
Specially designed bottle with an easy to hold grip for people on the go	3	7	0	4
With a sports top that allows you to drink without removing the cap . . . minimizes spills	3	7	3	3
Add the flavor you like . . . drops/tablets attached on the bottle with different flavors such as strawberry (red), apple (green) etc.	−4	7	−11	−2
Segment 2: Pure/natural				
Fortified water . . . no color, or flavor added, just the crisp taste and smell of pure water	5	−11	12	4
Segment 3: Enhanced for performance				
Berry crazy flavors . . . choose from raspberry, strawberry, kiwi-strawberry	1	37	−23	9
More natural flavors without added sugar or other sweeteners	4	0	2	7
Enhanced with all the vitamins and minerals your body needs	4	−4	1	7
Enhanced with energy boosting vitamins and minerals or other functional ingredients for your health	3	−7	0	7

3. The flavor-seeking segment generates a lower constant, but the elements do well. The pure-and-natural segment generates a slightly higher constant, but the elements do poorly. This difference between the two segments means that, to achieve the same level of total acceptance from the sum of the constant and the element contributions, the elements will have to do more work among the flavor seekers than they will among the pure-and-natural group. This finding mirrors the author's previous observations on food. Segments wanting novel or elaborate flavors showed lower additive constants, but higher element utilities as compared with segments that were more classic and simple (see Chapter 21 on It!).

4. A third segment emerges—the functionality segment—but only for the set of

concepts positioned as a sports beverage. This is important because that segment did not emerge, even with the same elements, when the concepts were positioned as regular water.

Weak Signals for Concepts and the Role of Weak Signals in the Stage-gate Process

Food, beverage, and packaged goods companies in general embrace ideas such as the *stage-gate process* for systematizing development (Cooper 1993). A stage-gate system comprises a structured method to bring new products into the market by controlling the different stages of the innovation process. To maximize positive product performance, each product idea goes through several different types of tests before the product is launched. Today's process is the so-called *third-generation process*. The system is already a positive route toward more effective product innovation. It is missing a detailed up-front stage at the fuzzy front end, where there are many opportunities. The so-called *fourth-generation process* builds on the third-generation process by addressing three areas of concern: (a) focus on *weak-signal detection* to get the best idea out of consumers' minds, (b) decrease the time for product entry into market and (c) reduce risk associated with cross-functional teams and fuzzy gates.

The *fourth-generation process* for development might well employ the following six steps, of which steps 1 and 3 are relevant for this chapter:

1. *Stage 1: Opportunity identification.* This stage involves identifying unmet consumer needs and creating a bank of concepts using weak-signal detection. Weak signals as already defined fit here very well because they refer to simple ideas that consumers find intuitively interesting at an emotional or intuitive level. Marketers and developers may not be clearly aware of these signals in con-

sumers' rational minds. As already noted, though, the weak signals may possibly warn of a changing, evolving competitive environment. In the business world these weak signals represent *early alerts* or features that could be incorporated into new products. Weak-signal detection enables a company to use the collective mind of hundreds or even thousands of consumers to identify specific features of products corresponding to a defined state of need (e.g., time of day or defined health condition). Through the weak-signal metaphor and research approach, the corporation can spot trends emerging in the marketplace and embed those trends into products.

2. *Stage 2: General feasibility.* The objective of this stage is to discover whether the needs/ideas identified by consumers represent a feasible business opportunity for the company and whether the ideas are in line with the business strategies of the company. This stage is not particularly relevant for weak-signal detection.

3. *Stage 3: Systematic concept evaluation and optimization.* This stage involves testing systematically varied concepts with consumers, with the goal to create concept models, which reveal what particular elements in the concepts drive acceptance. At this stage the focus has gone beyond weak-signal discovery to concept creation. The conjoint analysis approach is particularly relevant here.

4. *Stage 4: Feasibility analysis.* This stage is again beyond concept work. These questions need to be answered: Can we make it? Can we sell it? Can we make money out of it?

5. *Stage 5: Development.* This stage takes the idea from concept into reality and has various substages, which include prototype development, sensory testing of prototypes, and determination of the final formula.

6. *Stage 6: Commercialization.* This involves two steps: production scale-up and product launch.

Weak Signals Plus Internet-based Collaboration Equals Opportunities for Ideation and Innovation

A key motif in weak-signal research is the use of consumers to collaborate via the Internet and to both identify and then filter new and winning ideas. Rather than relying on a single creative individual to come up with the totally new idea, weak-signal development suggests that the ideas can emerge from the sequenced, technologically enabled interaction of people with one another. The Internet facilitates this collaboration. The system creates a mechanism both to elicit ideas from consumers and to enable consumers to assess ideas previously elicited using both selection and rating.

The underlying organizing principle assumes that an idea sent by one person will be picked up by another person, spark a positive response, be selected, and perhaps generate a modification. The system allows for different ways to think about products (i.e., sensory based vs emotion based), since the respondents offer ideas in their answers to the open-ended questions. By using hundreds or even thousands of consumers rather than a handful, and by using the power of a program that is systematic yet encourages individual creativity, the approach presented here enables the creation of potentially newer and better concepts created by people who need not meet one another in a single session in order to provide their ideas. It is not the single idea or the single ideation venue, but the cascade of improved ideas over a short time from many individual respondents that lies at the ideation and innovation presented here.

Sustained Innovation versus Episodic Innovation

One of the covert messages in ideation is that the process is subject to a great deal of serendipity. Thus, various writers talk about the need to facilitate the ideation process through the use of play, changes in the local environment, and the selection of the correct types of respondents to participate such as the so-called *lead users* (Urban and von Hippel 1988). Some practitioners call these *creative consumers* or a similar term to denote that they are somehow special, such as being more articulate. Such approaches mean that, at best, the idea generation and innovation processes are episodic, and these measures to create special environments and select the right people simply increase the probability that the episodes will occur. In one sense the current thinking about innovation is akin to revered position of *sensory experts* in the food and fragrance industries. At one time it was thought that only these specialists could truly assess the sensory characteristics of a food or perfume, with the remaining individuals in the population able only to register whether they liked or disliked what they were sensing. The present study suggests three ways by which that sustained innovation can extend beyond the purview of experts devoted to the process:

1. *Enhanced sourcing of ideas.* By making the process Internet based, with thousands of potential respondents, it no longer resides in the mind either of the expert or among a limited number of special consumers. *Rather, the ideation and subsequent innovation process may reside somewhat in a process that is independent of any particular individual, no matter how talented.* Internet-based ideation can serve as a source of ideas. To the degree that one is unsuccessful with any round of ideation, one can rerun the ideation again and again, because the cost is very low and the reach is very high. If the ideation is still unsuccessful, then one might wish to change the aufgabe or mind-set of the respondent and rerun the study. If the ideation is still unsuccessful, then one may wish to further engage the unconscious of the respondent by providing out-of-the-box examples or otherwise change the way the consumer is stimulated. The enhancement to ideation comes

from the sheer operational ease of obtaining elements and the simplicity of modifying the work in midstream with different questions (aufgaben) or more respondents.

2. *Powerful metrics.* The second enhancement comes from the rapid, subsequent deployment of conjoint measurement with the elements newly discovered. Rather than waiting to complete new, fully formed concepts, which can take time, the researcher works with elements, which can be more easily polished than can full concepts, because the elements themselves represent simple declarative statements.

3. *Shifting mind-sets, shifting mind structures, and the increased potential value of the ideas.* The third enhancement to the process comes from the ability to use different end uses as the dependent variable in conjoint analysis. As the data suggest, the end use to which a respondent attends when assigning ratings to concepts (refreshment vs sports) can create different mind-sets, allowing some elements to emerge that would not have done so otherwise. These mind-sets also reveal themselves in the emergence of concept-response segments, representing more fundamental person-to-person differences. Thus, even should the ideation itself generate mediocre elements at the level of the total panel, the ability to direct an end use, coupled with the ability to subsequently segment the respondents based on their response profile, provides an additional opportunity for new ideas to emerge.

Part II: Syncretism—Further Accelerating Development by Working with Close-in and Far-out Stimuli in the Conjoint Analysis

The New Product Process: Retooling It for Innovation

In a variety of informal, unstructured conversations and interviews with clients, the senior author asked them to specify the steps used to create new products and to innovate. Most of the clients at both small and large companies responded that they had a formalized process, but the process they described was more of the standard sequence: from need recognition to ideation to concept development to testing. Although everyone talked about the early stages of need recognition and ideation, there were no formalized, systematic methods to identify needs and link them with concepts. There were some formalized ideation processes, primarily distinguished by the specific vendor who did the facilitation.

Concept development was another poorly described process. It was not clear from the discussions how the ideation stage led to concepts, other than that somehow the ideation stage produced the concepts. No one paid attention to the processes that led from the need to the concept, and no one talked about books on creativity and problem solving as being germane to this issue. Almost all of the discussions pointed to high satisfaction with the testing of these new concepts, and many emphasized that there were manuals and norms to guide them.

What became increasingly clear through these discussions was that the process of creating the complete ideas was nebulous, but somehow and inevitably got done. When it was time to test, specific procedures were in place with which everyone felt more or less comfortable. The clients said that they felt far less comfortable with the concepts that they were going to test; that is, the process was clear for quantifying, but the process was not clear for development of what would be quantified. The following three trends emerged from these discussions on innovation:

1. *The company closely followed competitors.* Competitive analysis, generally informal, was well established and acknowledged to be so. Indeed, it was probably in competitive analysis that most of the interviewees appeared to feel most comfortable. One possible reason is that one's own work

is not involved, so there is no personalization and no possibility of failure. The other reason is the innate joy that people have in learning about the strengths and weaknesses of the competitors. Learning about the competition constitutes, therefore, a risk-free exercise, because, as yet, it does not involve any creation of one's own ideas or products. The response to competitive analysis was similar to the response to the analysis of competitive products in the so-called *category appraisal*. Marketers and R&D love the category appraisal, because it provides information that is directional, does not involve effort in creating new products, can be easily commissioned, and can be talked about ad infinitum to all sorts of audiences. Category appraisal makes good copy in the business press and requires little development effort or resources. Indeed, it is in the area of competitive analysis that companies appeared to involve many individuals, perhaps because this initial stage could also serve as a bonding mechanism among the participants prior to the subsequent and potentially divisive efforts in the new product process.

2. *The client had fairly defined ideation processes in place.* Most of these processes were created by outside vendors. Very few of the processes were well documented or based on clearly enunciated, published, scientific principles. The majority of the processes were similar overall: people coming up with new ideas. What varied was the unique surround offered by the different ideation vendors. Some vendors had the participants go through exercises designed to increase their awareness and, in some cases, their creativity. Other vendors had respondents go shopping or perform other information-rich exercises prior to the session.

3. *There were some new product processes in place, but mostly* after *the prototypes had been created.* As already stated, the steps involved in the actual creation of concepts and products from the information were not clearly delineated. Occasionally, one got the sense of *deus ex machina*, some process whereby the idea suddenly appears from the unconscious of the participants. Whatever processes were going on tended to be more in the area of discussions of ideas or in quantitative evaluation of what had been in development. There was no sense of a dynamic back-and-forth process, of creation, testing, refinement, testing, and so on. These ideas may have been in place, but, if so, then they were quite well and quite universally disguised.

The Key Problem: Creating an Innovation Machine with Concrete, Testable Outputs

A recurrent issue that kept emerging in different forms was that clients recognized the need to innovate, but were very uncomfortable in the actual process. By recognizing this discomfort with the actual innovation process, it quickly became obvious that one had to work within the corporate *comfort zone*; that is, no matter how strongly the employees voiced the need to innovate as the lifesaver for the corporate future, most individuals when asked directly said that they were uncomfortable with the entire process. It became obvious that the fuzzy front end had to be systematized by a method that was at once creative, scientific, and public.

Strategy 1: Use Technology

The first solution to the problem brings technology into the picture. For some unknown reason, the clients agreeing to the discussion expressed no concerns (i.e., reservations or problems) with technology acting as a facilitator of innovation. There was no sense of machine-enhanced creativity as the result of a "thousand monkeys randomly typing, who eventually type out Shakespeare's plays." This complaint about technology was offered during the 1980s as a way to resist research methods in the creation of concepts.

Technology has, however, come a long way. It is no longer as threatening to the creatives as it once was. Creatives are accustomed to technology, whether in terms of composing text, or pictures, or even mining data for insights. Thus, technology comprises one part of the solution to innovation, especially insofar as technology provides answers *executed* at a higher quality level than ideas springing from a person's mind and written down. In some sense, technology appears to be able to smooth over the rough edges and present to clients a pretty picture that has face validity.

Strategy 2: Use Standard Research Procedures But Early in the Process

Most of the clients were familiar with, and accepted, consumer research. Many of the interviewees did research after the ideation session to assess the validity of the ideas. Many of the clients were familiar with conjoint measurement, not so much as a technique with which they are familiar but as a respected method that carries the imprimatur of scientific validity. The positive reaction, coupled with the lack of detailed knowledge, suggested that the innovation process could be helped along by technology, but the technology would have to be couched in familiar language. It would do no good to promise unusual things with the technology. It also appeared that client acceptance would be more easily gained by using familiar methods than by using novel methods; in a word, if the clients had heard something positive about the approach from several trusted sources, then they were prone to accept its use in innovation. For example, once the clients recognized the term *conjoint measurement* and said they were familiar with it, one major hurdle against using it in innovation had been overcome. Two lessons learned here were that clients would accept an approach that they had heard of even though not familiar with the approach, and that they would accept an approach when it was positioned for research as long as the positioning emphasized development, not final evaluation.

Strategy 3: Make the Approach into a System That Allows for the Individual But Also for the Machine

Innovation using concept development has to straddle two camps. On the one hand, it must produce new ideas in order to earn the label of innovation. On the other hand, it must be put into a system that is simple, routine and, in some ways, even rigid. One often hears about creative genius and the once-in-a-lifetime product. The reality appears to be a desire for new ideas slightly better than mediocre that would be amenable to systems and rote processes rather than rely on untamed brilliance.

Approaching the Problem in an Unthreatening Way

It was obvious from the discussions that, despite the need to innovate, there was an equal need not to stray from the corporate comfort zone. It appeared to make no sense to proclaim one's hope of changing the world. This comfort zone, however, poorly defined at the individual level, appeared to exist at least in an intuitive sense. Person after person mentioned the comfort zone. None of the interviewees was strongly positive about taking chances in innovating, despite the ongoing business wisdom and truism that one must innovate or die. Within this construct of the comfort zone, it appeared, rather surprisingly, that technology could actually help to create such a comfort zone. There were three clear reasons:

1. *Technology's wow factor mesmerizes.* The *wow factor* breaks down the resistance to innovating, perhaps because it sets the stage to accept things beyond a person's ability to explain. If the technology is clear, and if it is presented as technology and not as a proprietary black box, then the corporate folk

have a reason to believe. It is this simplicity, with the approach embedded *outside the individual*, that gives the corporate folk the feeling that this is safe. In some ways the innovation technology embedded in the personal computer (and now the Internet) makes it less threatening because the user does not feel intimidated by someone who may be superior. It is acceptable to say that the computer is a dumb genius. It is not acceptable to say that the person selling the innovation is either dumb or a genius. The first is socially unacceptable, and the second is threatening psychologically.

2. *Technology extends the capacity of individuals to perform ordinary tasks.* No longer is innovative thinking solely in the purview of the talented, unique individual who everyone really dislikes because of all-too-often capricious behavior. Technology becomes the servant of everyone and the superior of no one.

3. *Technology allows private feedback and feedback comforts.* It is important to note that shortening the time had little to do with the power of the system and all to do with the ability of the technology to supply anxiety-reducing feedback to the participants. Thus, in an innovation session, one does not have to wait weeks to measure the reactions to the test stimuli. Measurement is immediate, and the feedback comes virtually overnight. Furthermore, the feedback is private rather than in the marketplace, where one is subject to recriminations and humiliation for having made the wrong a decision.

4. *A developmental biology metaphor clarifies the idea.* A very cogent metaphor, which comes from developmental psychology, deals with the distinction between a very young child and a very old person, both of whom must walk across a rock-laden floor. The young child runs across, falls down and gets up, brushes himself off, and moves on. Very quickly, through many falls the child gets to the other side of the room. The older person, however, faced with the same situation, is

very reluctant to try anything but the flattest, least dangerous surface. Consequently, the older person spends a great deal of time scanning the environment to avoid problems, whereas the younger person goes for the goal, accepting temporary setbacks. In a sense this is what the innovation process should be: modeled after nature rather than modeled after excessive analysis. By having rapid hypotheses, tests, and feedback, one can suffer less painfully through numerous setbacks in the innovation process and yet end up with success.

Conceptualizing the System

The innovation system for concepts can be conceptualized in a simple way by using three metaphors: raw material, enabling technology, and information integrator. These features are combined into a low-risk, rapid, hypothesis-test-feedback-hypothesis system; that is, an iterative, low-risk system, based on trial and error, and biological type of relatively painless *feedback*. The goal is to institute a hypothesis-test-feedback-retry so that, even if the first iteration of this integrative system fails, the cost of failure is very low. The features can be described and elaborated upon as follows:

1. *The raw material.* Raw material comprises concept elements. As has been emphasized in so many chapters in this book, it is easier to create *snippets* of ideas than to create full concepts. Some of the ease in creating snippets is due to emotional factors. No one likes to be judged, and full concepts provoke judgment, for better or worse. The concept's completed structure invites the critical to measure it, comment on strengths and weaknesses, and otherwise proffer an opinion. By dealing with the raw materials of the concept in the form of snippets, everyone avoids judgment and embarrassment. Furthermore, these snippets could come from anyone, anywhere, including a deconstruction of the competition. To the degree that the snippets lie *close in*, the innovation exercise

will produce close-in ideas. To the degree that the snippets come from a variety of different fields, the innovation exercise will produce new, syncretistic ideas.

2. *Enabling technology.* The technology is a computer that mixes and matches these elements in some type of order to create new ideas. The computer is in its basic essence a dumb machine that simply combines inputs in preordained order without any knowledge whatsoever. There is no adaptive element in the system at all. Some of the professionals pointed that having an adaptive element might lead one down a path from which it would be impossible to recover. As soon as one hears the word *adaptive*, one assumes that the machine is intelligently programmed to come up with winning ideas. Thus, the notion of genetic algorithms for concept development leads one to assume that, with the *survival of the fittest*, this *fittest* is also the *best*.

3. *Information integrator.* This is the person. The person is shown a *setup* video, picture, or concept, setting forth the situation in which he might find himself. This setup stimulus could be presenting a current or future scenario. The setup helps the respondent's mind to evaluate all of the upcoming concepts against that framework, without dictating what those expectations should be. Then, the respondent looks at the concepts and rates the different concepts against the framework that was set up. Some of the concepts that the respondent evaluates will fit this end use or scenario, whereas others will not. At the end of the evaluation the researcher will have an idea of what features fit the scenario.

A Worked Example: New Tooth-cleaning Products

The example comes from the attempt to synthesize a new oral cleaning device. The approach created a set of elements, some close in to the topic of oral care, some from other disciplines and technologies dealing with

cleaning, and some benefits. The stimuli were multimedia. The respondents were told that they would be evaluating novel ideas about oral care, with some pictures and statements that "didn't seem to exactly fit oral care, but which might convey an idea of what was being sought." Each respondent rated different combinations of these features and benefits on three different rating scales: interest, uniqueness, and fit to end use. *End use* was defined by the phrase "A treatment that will allow me to use it once every 3 days, and not worry about plaque." The end use could have just as easily comprised a video, as well, to further elaborate and concretize it. The exercise generated three equations showing the part-worth contribution of every concept element to each attribute. Table 19.6 shows the coefficients of these three models.

A key benefit of knowing the utility values of consumers is the ability to create alternative concept scenarios. A scenario could be simply identifying the best combination of features from a set to maximize one or several objectives. Table 19.7 lists the synthesized concepts. In this table, the objective is to identify a new set of features that could generate a new product idea that both fits an end use and is acceptable.

The model for purchase shows the conditional percentage of respondents interested in the concept if no element is present (additive constant) and the additive conditional probabilities when individual elements are introduced into the concept. The same interpretation can be given to fitting the end use (conditional probability that the product fits the end use) and uniqueness (conditional probability that the product is unique).

With respect to innovation, the key thing to keep in mind is that the elements in Table 19.6 and the concepts in Table 19.7 either exist or can be readily realized. Thus, the innovation here is twofold:

1. The lateral thinking produced by using elements outside the current realm of oral

Table 19.6. How elements drive acceptance, perceived fit to end use, and uniqueness*

	Purchase	Fit end use	Unique
Additive constant	*42*	*36*	*22*
Picture			
Digital device showing the "amount of plaque"	9	−3	7
Picture of a laser gun	7	3	10
Cartoon picture—selective attack, by "little dots," of a tooth covered with plaque	6	2	3
Picture of foaming action	5	6	2
Picture of toothbrush	4	4	−6
Cartoon picture—ultrasonic device removing plaque deposit	4	9	4
Picture of sandblaster	−3	8	8
Mode of action			
Mechanical action against plaque	7	4	1
Builds a protective layer against plaque	5	−4	2
Combines with plaque to form inert compound	3	4	−1
Combines with plaque to loosen it	3	2	−3
Dissolves plaque	2	3	−3
Immunizes the body to prevent plaque	−5	2	8
Applicator device			
Chewing gum	6	−1	3
Long handle with reservoir	4	2	−3
Picture of a squeeze tube	3	2	−3
Liquid applied by brush	−2	5	−1
Picture of a toothpick	−5	−4	−5
Ingredient			
Effervescent materials that work on contact	6	4	4
Toothpaste-type surfactants	4	3	−4
Protective chemicals that coat the teeth	−1	3	5
Natural ingredients found in seaweed	−2	−4	6
Gritty material enrobed in a gelatin base	−4	−2	2

*The data are only partially shown and are sorted by purchase intent. *End use*: A treatment that will allow me to use once every 3 days and not worry about plaque.

care, but which have analogous properties and benefits.

2. The creation of new combinations (i.e., the new product concept). Even if the combination never before existed, as long as the user has an idea of what the product concept *could contain*, the data suffice to point the direction to the new product.

Platonic Ideas and Product Innovation

Beneath the development and research issues just discussed lies a philosophical underpinning. About 2500 years ago Plato discussed the notion of *form* as idea floating about in imperceptible ether. This Platonic notion of inchoate forms is both appealing and a compelling organizing principle. The basic idea is that innovative products exist in "ether." Consumers do not know how to describe these products, but will know them when they see them. Furthermore, what may be innovative today may have been unthinkable yesterday and may be trite tomorrow. Thus, innovation and creativity have the aspects and constraints of time, experience, and environment. Product ideas may be universal, but

Table 19.7. Optimal, innovative ideas created by selecting elements that drive fit to objectives

	Purchase	Fit end use*	Unique
Additive constant	*46*	*32*	*22*
Concept 1: Mechanical action (blasting/abrasive action)			
Picture of sandblaster	−3	8	8
Mechanical action against plaque	7	4	1
Long handle with reservoir	4	2	−3
Toothpaste-type surfactants	4	3	−4
Sum of the concept elements	*58*	*49*	*24*
Concept 2: Digital readout device coupled with a chemical protective action			
Digital device showing the "amount of plaque"	9	−3	7
Combines with plaque to form inert compound	3	4	−1
Liquid applied by brush	−2	5	−1
Effervescent materials that work on contact	6	4	4
Sum of the concept elements	*62*	*42*	*31*

*End use: A treatment that will allow me to use once every 3 days and not worry about plaque.

innovative products possess spatial and temporal limits.

The respondent's job is to react to combinations and judge the degree to which the concept presented (on a computer) is close versus far away from this unknown idea. The idea may be known to a respondent at an intuitive level, but not articulated, nor should it be. The idea functions like the Platonic ideal in the ether. In turn, the computer presentation of concepts that embody aspects of this Platonic ideal is akin to an individual hurling buckets of paint at a passing invisible object (viz., the idea). As the object (idea) traverses a path and as the individual hurls more and more buckets of paint, some of the paint will stick where the invisible idea *is* and will disappear where the invisible idea *isn't*. At the end of the path and with enough buckets of paint, the idea and all its lineaments should emerge more clearly delineated.

Let us return to the innovation machine for new ideas. The computer hurls test concepts toward the respondents, who simply react by saying "close" or "far away." Most likely the respondent reaction is intuitive rather than considered, because that is the way respondents respond when they are presented with many concepts. Eventually, with

enough concepts, some will be closer to the unformed, unexpressed idea, and some will be farther away. Perhaps no concept will exactly match the unexpressed idea, but that should not trouble us. Since the concept elements are combined by experimental design, it becomes straightforward to identify which particular elements push the concept closer to the unexpressed idea and which move it away. Furthermore, to aid innovation, one can put the respondents into a simulated environment (e.g., a future scenario) while conducting the study. To the degree that the idea changes in a respondent's mind (e.g., as a result of this simulated new environment or situation), the method can reveal changes in the features of the idea.

Observations and Comments on the Innovation Machine

At least five issues and aspects deserve comment, primarily with respect to the corporation:

1. *The process has low perceived risk and therefore is more likely to be adopted once the initial interest is ignited.* In corporations, risk aversion often becomes the professional's strongest motivation. As a person

becomes older there is a natural tendency to avoid doing things that could invite unwanted attention and perhaps lead ultimately to job loss. Despite what one reads in the business literature, the vast majority of researchers try to do things that are safe. Accountability is to be shunned by the risk averse. Often the researchers fight with the product developers and marketers, who, in contrast, are judged on performance and success rather than on adherence to a system and a knowledge-development process. The reduced risk and the process nature of the innovation machine described here increase the odds of the process being used or at least not immediately rejected.

2. *The process is positioned as being easy, not difficult and not feature rich.* In business nothing is as difficult as a task that is perceived to be difficult. Difficulty and complexity invite inspection, which causes anxiety among those who would rather follow the process than risk attention. By positioning the innovation system as a relatively easy task with clear guideposts, the innovation effort is less threatening. Such reduced threats also increase the probability of the innovation machine being adopted. A system that is relatively easy will almost always win over a system that may be more powerful but is relatively hard to implement. Fortunately, by making all the steps clear-cut the task is immediately perceived to be easier and within the scope of the corporate employees.

3. *Immediate, painless feedback.* Feedback about performance, especially constructive feedback, is very important. The psychology of learning and behavior is partially founded on the study of feedback and its impacts. Feedback becomes positive reinforcement to a business professional trying to introduce a new idea. Rather than having feedback play the role of judge (pass/fail), the approach incorporates feedback as simply another step, without emotional repercussions. Because the system is iterative, the feedback simply becomes a guide to what to do next rather than a report card of one's performance. This positioning of feedback as a guide also increases the odds that the innovation machine will be adopted. As the corporate participants begin to see that they are increasingly successful, with really no pain associated with that success, they are reinforced for their participation in the innovation exercise. The reinforcement, in turn, makes them feel good, and they become strong advocates.

4. *Iterativity.* Perhaps iterativity is the most important, yet unappreciated, aspect. No one really believes that one's first ideas are the best. When people know that they are free to make mistakes because the process itself is iterative, they can participate more freely. They need not worry about producing their absolutely best idea because there is always the promise of a follow-up low-cost, low-risk iteration. As a consequence, participants in the innovation process loosen up. They do not wait to polish their ideas, but in the best of circumstances they actively contribute more and more ideas. They become part of the iterative process. Like the low-risk, nonpunishing feedback, the iterative quality of the process is positive. There is no risk if one knows that the study will be repeated anyway, at low cost. Mistakes are welcome as opportunities to learn, not a reflections of one's inadequacies that up to now were hidden, even at the expense of suppressing one's own contributions.

5. *Sustainable, supportable, scalable.* The *sustainability* of a program over time without excess investment from the outside is a strong factor in the potential success of a program. Efforts that need constant investment of corporate time, money, and even management encouragement have a greater chance of failure. Sustainability is ensured because of available, cheap computer/Internet technology. Sustainability is further ensured by the relatively little effort it takes to create elements and have respondents participate. This simple process does not require much outside

help or encouragement. It can be implemented anywhere. *Supportability* means that the approach can be supported with the current resources. The system is set up to require minimum financial input. Finally, *scalability*, or the capability to multiply the effort in many locations with many different people, is ensured because the process is simple and inexpensive. The process simply requires a motivated task leader, a source of ideas, and a source of willing respondents. Since the system is simplistic, but uses the power of a computer to analyze the data through a fixed set of programs, there is no reliance on the talents of a single unique individual in a single location. Rather, the system is public. It is easy to set up and can be done on the spur of the moment if the computer program is available, and the results are easy to interpret. The process can be exported worldwide with little effort. The biggest enemies of the innovation machine are the ever-present inertia that plagues people and the hard-to-overcome smug satisfaction with the status quo. Both lead to rejection before adoption and are worse than the true failure that comes from doing things that don't work out

Part III: Integrating Tools: Combining Concept Development Methods at the Fuzzy Front End

Introduction

This final section demonstrates the integration of current research tools into an innovation system. We saw from the previous sections that the combination of Internet-facilitated ideation (brandDelphi) and Internet-based conjoint analysis (IdeaMap.net) can generate ideas quickly. We also saw that respondents can integrate ideas that are both close in and far out to help create innovative products, without the respondents themselves being particularly creative. With the plethora of new methods available to researchers, with the popularity and availability of conjoint meas-

urement approaches, and with the need to innovate continually, anything that provides a simple structure to this innovation may be welcome by companies. We finish this chapter with a larger-scale vision of the innovation approach that integrates the tools currently available into a *second-generation innovation process*.

The specific topic of the chapter is the development of specifications for a sandwich to be eaten in the car on the go. One might call this approach *boundary crossing*, because the approach goes beyond the current boundaries of cars as vehicles for transportation and looks at cars as venues for food consumption. It is such lateral thinking with new research methods that could provide a valuable tool for the consumer researcher.

Who Has the Tools?

Over the past decade the process of corporate innovation has generally been left to outside consultants and to internal/external teams specializing in the invention process (Wheelwright and Clark 1992; Griffin 1997). Researchers have shied away, preferring to act in the evaluative mode when ideas are created rather than in an active creative mode (Rosenau et al. 1995; Hoban 1998). A consequence of this situation is that consumer research and researchers are all too often neglected in this process, even though the newly emerging tools for knowledge creation can accelerate innovation. We saw this previously with conjoint measurement and ideation.

Given the ongoing development of consumer research as a field, and the evolutionary pattern that is pushing research into an evaluative role rather than a creative role, a new beginning is needed for consumer research in the development process (Moskowitz 1998). What is needed is a knowledge-development and knowledge-enhancing system operating efficiently and cost-effectively so that it can be adopted widely with low risk (Moskowitz 1997). The system should have these five properties to deliver knowledge in a way consistent with the worldview of consumer researchers:

1. Rapid and user friendly (days and weeks)
2. Consumer based (to ensure ongoing inputs from the end user)
3. Knowledge based (using data, not guesswork)
4. Reality based (using observation to identify actual behaviors leading to these new products)
4. Iterative (to allow for inexpensive changes in direction with new learning)

Creating the Marketing Research Toolboxes for Innovation

A key defining aspect of consumer research in the past ten years and, apparently for the next several, is the use of *tools*. Researchers graduating from universities today are awash in technological aids to creative thought. These aids span the range from computerized interviews either on personal computers or on the Internet to statistical analysis methods to high-level quantitative treatment of qualitative data. One of the key phrases often heard is the *toolbox*. A toolbox simply comprises a set of well-accepted research procedures that have been designated as appropriate to help solve a problem.

Although there is a method and analysis toolbox for many common research problems (e.g., product testing and tracking), there is no comparable *consumer research toolbox* for innovation. There are no algorithms, procedures, or analytic strategies to deal with the fuzzy front end of development, although numerous practitioners have provided one or another proprietary method. The business literature is replete with these methods, be they in the form of books or journal articles, in either the popular press or the archival academic press. A consumer research toolbox comprising high-level data acquisition and analysis

techniques for the early developmental stage would, therefore, be a very welcome addition to the researcher's armory. We saw the beginnings of this toolbox described previously, with the combination of Internet-enabled ideation and Internet-enabled conjoint analysis. The toolbox is feasible, but must be specified.

Intellectual Foundations: Models from Strategies of Adaptation in Evolutionary Biology

Although there are no well-accepted algorithms, procedures, or strategies to deal with the fuzzy front end of development, an emerging view suggests that organization improvisation and innovation may often arise from the recombination of previously successful subroutines (Borko and Livingston 1989). Models of adaptive systems as well as some evidence from evolutionary biology have shown that recombining routines provides one of the most fruitful sources of change. The recombination enables systems to prosper and to adapt to new circumstances, and yet use what has been developed so that the costs and demands are lower than would happen when the organism must start anew each time (Holland 1975; Levinthal 1991). In the same way, a firm using well-developed consumer research-oriented subroutines can use the principles of adaptive planning, recombining successful routines in rapid iterative learning cycles. Each subroutine should provide a clear piece of the puzzle so that practitioners can easily recognize *recombinatorial* possibilities as new learning occurs. The consequence would be the research-driven ability to produce and evaluate new ideas in response to unexpected learning or market changes (Flores and Briggs 2001).

Using adaptive subroutines provides a pragmatic answer to the general business conundrum of planning that tends to become an end, in and of itself. For business in general and for consumer researchers in particular, process definition and planning are necessary in order to ensure that action occurs along the defined critical path, as well as to prevent duplication of work. Yet, planning can take too much time, as many consumer researchers well know. Studies can be overengineered, research designs can be perfected, and meetings can multiply. As planning begins to take more time and costs more than the gains it promises, it actually detracts from the speed needed to create new products (Moorman and Miner 1995). Consumer researchers are especially prone to this overplanning because they do not operate with external sources of validation, such as profit and loss.

To round out this overview of intellectual foundations we can look at some of the critical new thinking around the concept of an *adaptive enterprise* that might embrace this new paradigm. Haeckel (1999) suggests that adaptive enterprises are *sense and respond* organizations rather than *make and sell* organizations. The barriers to innovation evaporate when the organization constantly searches for new ideas and recognizes publicly that it has an ongoing need to add new ideas to the conveyer belt: "Change is no longer a problem to be solved, but rather an indispensable source of energy growth and value" (Haeckel 1999).

An Organizing Principle for the Toolbox: Ideas as a Combination of Function and Form

One way to create a consumer researcher's toolbox involves adapting the proposition of Finke and colleagues (1995) that areas comprise both *functions* (viz., consumer needs) as well as their relation to *forms* (viz., solutions). They identified three types of cognitive search strategies that may be relevant to the creation of new product ideas:

1. Identify or define a function (viz., consumer need) and then perform an exploratory search for a suitable form (viz., solution).

2. Identify a form (viz., solution) followed by an exploratory search for a meaningfully related function (viz., consumer problem).

3. Generalize an already known function–form relation.

Goldenger and colleagues (1999) adopted this organizing principle to the context of ideation for new products, developing a classification that we can use here for the *subroutine*:

1. *Need spotting*, whereby need identification precedes product (viz., form) development.

2. *Solution spotting*, whereby either a form is identified and the inventor searches for a suitable need or both need and solution are concurrently identified, with the generalization following shortly afterward.

3. *Mental invention*, whereby there is a decision to innovate, and afterward both the need and the solution are developed interactively.

The foregoing organizing principle provides the theoretical groundwork for the paradigm presented here. The objective is to provide a system that identifies needs efficiently, identifies solutions equally efficiently, and provides an ongoing stream of alternative needs and solutions. *The goal is to remove serendipity as a key factor when creating new products.* Thus, the paradigm presented here fits with the other parts of this book, which advocates a systematic, knowledge-based system for concept development.

The Paradigm

The paradigm presented comprises a set of discrete steps, each of which has been used in a variety of applications. The sequence and combination could provide the necessary insight for innovation and continual development. The paradigm is grounded in the notion that innovation should not be at the whim of chance and happenstance, but rather should be harnessed in a systematized, public fashion. The paradigm comprises steps that focus on each of the cognitive search processes relevant to ideation: specifically need spotting, solution spotting, and interactive mental invention. The process encompasses two dynamically interacting opposites: innovation (new, different, unique, and perhaps radical) and systematized/public fashion (current, conventional, and the same).

The paradigm comprises five specific stages: The first two stages are informational. They provide context (It! studies as foundational) and, within that context, identify behavior that may lead to a new product opportunity (ethnography). In the aforementioned scheme, these are subsumed under need spotting:

1. *Foundation study.* The foundation study (e.g., the It! study; see Chapter 20) comprises a body of knowledge about a product or service category that is available prior to the need for the innovation. The foundation study presents the landscape currently known to the developer and the marketer, albeit in a more complex way than conventional means. It uses well-defined methods, such as conjoint measurement, to identify those specific features of a current product or service category that are attractive to consumers, versus those that repel the consumers. It also provides indications about the existence and nature of segments in the consumer population (Beckley and Moskowitz 2002). One might consider the foundation study to provide a corpus of publicly available knowledge that can be used as a background within which to interpret observations, problems, and so forth. The importance of a foundation study cannot be overemphasized, for it comprises systematized learning (Veganti 1997).

2. *Observation (e.g., ethnography).* Observation means looking at actual behavior in the environment rather than considering reported behavior. Ethnography is becoming very popular today as a way to understand the customer in-depth. Ethnography by itself, however, simply provides snapshots of behavior. When merged with a foundation study, ethnographic observation puts the behavior into a context (e.g., typifies a problem to be solved, thus revealing an opportunity; or shows the way a person solves an everyday problem). All too often it is easy to obtain ethnographic data by video camera and other recording devices, but hard to locate this behavior within a matrix that reveals the business opportunity. Detail often overwhelms, hindering insight. Foundation studies enhance the potential usefulness of ethnographic information (Ericson and Stuff 1998; Stewart 1998; Abrams 2000).

The remaining three stages deal with the use of consumers to provide the innovation, that is, solution spotting. The assumption here is that the consumers are not particularly verbal, may not express themselves, are not lead users, cannot easily think out of the box (Cleveland 1997) and, in general, are not highly motivated to provide award-winning insights. Rather, the consumers who will provide the innovation are conventional, but can respond to specific stimuli and provide simplistic, albeit occasionally inspired, answers. These last three stages thus work with the modal type of consumer rather than with the articulate individual or lead user. The innovation is designed to come from the *masses*, not from the elite.

3. *In-depth computer-based interviews, with artificial intelligence to analyze the language and identify key ideas.* The objective here is to create an interview by using the computer as the interviewer, similar to that done by an intelligent interviewer who probes. The in-depth probing uses rule-based software methods rather than an interviewer (Cleveland 2001). The software is set up to help respondents elaborate on key themes by selectively repeating certain key phrases in the *probe* mode. The verbatims from dozens, or perhaps

even hundreds or thousands, of interviewees are then analyzed to identify linkages of ideas. This area of artificial intelligence has been rapidly developing over the years and has reached the point where the computer analysis of the verbatims can isolate key ideas (Cleveland 1986). Of course, critics might argue that machine probing could never replace human interviewers—a criticism that is still valid. On the other hand, with dozens, hundreds, or even thousands of interviews available, there is a plethora of raw material from which to extract key ideas, and sufficient data from which one or a few key, new ideas might well emerge. What the computer lacks in truly profound intelligence may be partly compensated for by its ability to access the minds of hundreds of people in a cost-effective and time-effective fashion. The output of this third step is a particular problem or situation with which the respondents can identify and that forms the frame of reference for concept development and solution. The availability of relatively inexpensive computer interview administration and analysis of the interview makes the in-depth interview affordable. Furthermore, the ease of administration makes the potential for iterative interviewing quite real, enabling researchers to change the focus of interviews as more knowledge is obtained.

4. *Create raw materials for concepts by framing a situation in the respondent's mind (the aufgabe).* The objective here is to obtain elements for concept creation from consumers, with these consumers placed in a specific mind-set. We saw this explicated previously for bottled water in the explication of weak-signal research. Step 3 (in-depth interview by computer) provides the necessary information from which to formulate a frame of reference. Step 4 presents that frame of reference to consumers and then instructs them to provide elements that could either expand the problem or provide part of the solution or the benefit. The problem statement serves as a springboard, catalyzing a respondent's creativity and focusing a respondent's output into a specific and relevant

direction. Through a Delphi-like procedure (Jolson and Rossow 1971; Flores 2002), the respondents in a variety of locations provide the elements and also react to elements provided by others. Through this process each individual respondent is not required to be particularly creative but rather to provide some few elements and to judge a few other elements provided by respondents previously participating in the process. The outcome, however, comprises a rich set of raw materials that serves as the basis for concept development and optimization. One of the key aspects here is that hundreds of respondents participate, removing the onus of creativity from any single respondent. Another aspect is that the elements are polished and voted on as they move through the system. The output provides a ranking of the relevance of the most important to least important elements, as well as a complete list of the elements offered by the participants. Another feature of the approach is the iterativity. With relative ease, researchers can return again and again, with new questions or mind-sets, to create more information with the consumer's help.

5. *Evaluation and optimization by online conjoint analysis.* The final stage in the development comprises the creation and evaluation of test concepts through conjoint analysis, an approach dealt with extensively in this book (see Chapter 5). By selecting winning concept elements from the conjoint exercise and recombining them, even better concepts can be engineered than those originally tested. The conjoint interview is run on the Internet, with hundreds or even thousands of respondents, as was already shown for the study on bottled water. The data are analyzed to identify the utility value of each element. It is worth noting again that the Internet makes the conjoint approach quite cost effective.

A Case History: Car as Restaurant— New Food Product, New Car Product

An easy way to understand the paradigm comes from a simple case history that deals with the common problem of time pressure and its effect on eating and driving. As the demands on consumer time increase, available time decreases. This shows up in an increasing number of individuals eating in cars. Indeed, it is becoming obvious that more and more fast-food restaurants are featuring drive-up windows where customers can order food to eat in the car. The frequency of "eating on the go" continues to increase, as do the problems associated with this behavior, such as messy cars. The case history deals with the creation of a product to fit with this new lifestyle trend.

Step 1: Foundation Study as the Baseline for Available Knowledge

Crave It!, the foundation study on food preferences, dealt with 30 different food products, ranging from hamburgers to French fries to ice cream, etc. The goal of the foundation study was to identify what particular features of foods drove craving (see Chapter 21, and Beckley and Moskowitz 2002). The foundation study comprised 30 different conjoint studies, one per food. Each conjoint study explored 36 different aspects of a single food, such as physical characteristics, brand, benefit, and emotional aspects. Through segmentation based on the pattern of the individual utility values, the foundation study revealed three radically different mind-sets of consumers, transcending the 30 different foods. These were defined as *classics* (who want food the traditional way), *elaborates* (who want lots of variation of their food, including toppings, flavors, etc.), and *imaginers* (who are responsive to emotion and promise, and less so to descriptions of food). These three segments appeared in all the foods, with a great many respondents falling into the elaborates. Table 19.8 shows the results of segmentation for the segments with respect to hamburger.

Table 19.8. Utility values for hamburger arranged by three segments

		Segment		
	Total	1	2	3
Base size	*150*	*54*	*72*	*24*
Additive constant	*30*	*52*	*9*	*47*
Segment S1: Classic				
Lots of crispy bacon and cheese on a juicy grilled hamburger on a lightly toasted bun	17	11	34	−22
Segment S2: Elaborate				
Lots of crispy bacon and cheese on a juicy grilled hamburger on a lightly toasted bun	17	11	34	−22
With all the toppings and sides you want . . . pickles, relish, jalapenos . . . lettuce, tomato, chips . . . whatever	10	2	19	3
Burger smothered in onions and cheese	5	−7	18	−5
A grilled aroma that surrounds a thick burger on a toasted bun	10	3	17	4
Layers of burger, sauce, pickles, and lettuce on a moist sourdough sesame seed bun	7	6	17	−21
So tasty & juicy you practically have to lick your lips twice after each bite	8	4	14	1
A char-grilled hamburger with a taste you can't duplicate	7	3	14	−2
Juicy burger with the crunch of lettuce and tomato	5	−3	13	2
Premium quality . . . that great classic taste, like it used to be	7	5	12	−3
Segment S3: Imaginer				
Fresh from the grill, especially for you . . . by you	5	−1	7	13
You can imagine the taste as you walk in the door	7	6	6	12

Step 2: Ethnographic Observation

By itself the foundation study provides a corpus of information for marketers and developers, but does not seek out problems. Ethnographic observation of people in their daily lives does, however, reveal these problems, even when the problems are not easy to articulate in a questionnaire. In a later project that studied fast-food consumption, Jacqueline Beckley and Hollis Ashman of the Understanding and Insight Group studied consumers in cars (J. Beckley and H. Ashman personal communication). These consumers were fast-food customers who purchased and ate some of the food in cars. During some of the interviews it became obvious that a key issue was *cleanliness*. Although this issue did not surface directly, observation of behavior revealed that customers were having problems eating food neatly in cars.

Ethnographic observation is not survey research. The observation records behavior, but it is up to the analyst to put a structure around that behavior. The video records of the consumer observations suggested that a possible link between the category of elaborates discovered in the foundation study and the messy situation in which customers found themselves when trying to eat fast food in cars. The respondents did not make the connection at a conscious level, and even if they had an inkling of what might be the cause of their car behavior, they were not privy to the foundation study. The connection did occur to the researchers that perhaps those individuals eating messily in cars might belong to the same class of respondents classified as elaborates. With a large number of elaborates in the foundation study (exceeding 40% across all 30 categories) this connection of messiness and elaborates took on additional meaning. The connection suggested an opportunity for new products, designed for fast-food restaurants, geared toward in-car food consumption.

It is noteworthy that this type of connection comes from the availability of the foundation study, which sets a framework for understanding the behavior that is observed. Were the foundation study unavailable, and thus the categorization of people into classics, elaborates, and imaginers unrecognized, the link with eating in the car might never have been made.

Step 3: In-depth Interviews Using Computerized Methods on the Internet

The recognition that eating in cars was a messy situation (ethnographic output) that might afflict the group of consumers in the elaborate segment (foundation study output) led to the need to define the problem more thoroughly. By means of in-depth interviews powered with artificial intelligence and analyzed quantitatively (Cleveland 2002) it soon became apparent that a number of issues were involved in eating fast food, especially with children and in cars. Through evaluations with several dozen respondents and analysis of key issues, the findings showed that the foods purchased were too large and too messy, and that there was an opportunity to downsize the product to make it easier to eat in cars (Table 19.9).

Step 4: Creating the Elements for the Concept by Using the Delphi-like Method.

The open-ended questioning in step 4 provides a sense of the problem and, at some level, one or two solutions. These solutions are provided by respondents, almost in a serendipitous way, during the depth interview. What is needed, however, is a concentrated attack on the problem, once it has been formally identified. This attack is launched in step 4 by using a Delphi-like process (see both the beginning of this chapter and Chapter 3).

The setup question emerging from the in-depth interview was phrased to the respon-

Table 19.9. Key topics and conclusions from an in-depth Internet survey and automatic analysis of interviews

Core text list of key vocabulary phrases used in interview
  Food that works in a car
  Bite or bites (size food)
  Plain or hamburgers
  Barbecue (sauce container)
  French or fries
  Wrappers (that hold food in)
  Finger or food (not fall part)
  Candy and Wrapper (type)
  Chicken and nuggets (type food)

Key conclusions
  Eating food in cars is messy for mothers. The causes of messy food are foods that are uncontrolled, too much ketchup, etc., or foods that fall apart or food that cannot be handled by the hands they are given to, such as "little" hands.
  The fact that mothers have strict criteria of what they will allow in their cars and what they will not allow yet there is still a mess says that the criteria, no matter how strict, are not working for all mothers:

1. The food they receive does not conform to what they ordered, too much ketchup.
2. The food they order comes in portions too large to easily control.
3. The food they order does not come in wrappers that keep them contained, such as drinks, shakes and burgers. There are leaks.

  Small portions, controllable wrappers and getting what they ordered are the keys to controllable and non-messy eating in the car.
  The number one idea for a nonmessy food in the car was the burger bite, a single toddler size bite.

dents as follows: "We are interested in making a new type of bread/bun for breakfast. We are looking to develop new buns that will make 'eating on the go' easier and less messy (e.g., in cars, trains, walking). We welcome your ideas that will help us make better designed bread/buns." This paragraph, presented at the start of each Delphi-like exercise, to each of the 480 respondents on the Internet, sets the stage for creativity. The paragraph makes the creativity somewhat more constrained in the right direction, but does not provide any additional ideas.

The respondents were invited to participate by e-mail for a 10-minute session. During this session,

1. They read the orientation paragraph.
2. They looked at eight ideas previously provided by other respondents and checked off the ideas that they thought were relevant.
3. They provided two new ideas to complement those that they had seen.
4. They rated the ideas that they had selected on an importance scale.
5. They completed a classification questionnaire.

With several hundred respondents participating and rating different ideas as well as providing their own, the exercise generated a matrix of ideas that can be ranked in terms of relevance and importance. Table 19.10 lists some of the results of this exercise after 6 hours on the Internet. What is important here is that the creativity is directed by the setup paragraph or by an image, even perhaps a video. Furthermore, the creativity is done on a large scale by virtue of hundreds of respondents. The task is iterative and collaborative, because later respondents cycle through and evaluate the contribution of earlier respondents. Finally, the task is quantifiable because

Table 19.10. Ranked elements from a Delphi-like exercise on the Internet*

	Rank
Top-tier elements	
Something that holds all of the contents (breakfast sandwich) without the possibility of overspill when you bite into it.	286
Whatever the product, it should not be flaky or too pastry-like as that is likely to create crumb problems all over the front seat of a vehicle. Wrappers that could catch crumbs are ideal. As is a container that one could put on their lap so it would catch the food.	284
Be able to handle with one hand.	274
Can not have too many wet ingredients, like sauce that can come shooting out the bottom.	241
Easy open packages (able to open one handed if possible).	238
Don't make sandwich so greasy the bread falls apart in your lap.	233
Something closed like a pocket so you won't make a mess. Then you can fill this with egg, bacon or whatever you like.	231
Needs to be self-contained. If there are too many individual layers they can slip and cause a mess.	212
Middle-tier elements	
Bite-sized pieces (less mess)	202
Do it with less flaky bread and less gooey filling for less mess.	199
Less oil on sandwich	160
Finger foods in bite size pieces	151
Pocket bread (pita)	150
Something small, like cereal bar-sized.	142
Using a bread that will not fall apart with the first bite and no runny mustard, Mayo or catsup. Something like a Calzone without the grease.	142
Elongated so that they can be held by a paper towel or aluminum foil more effectively.	141
A smaller sealed sandwich. Like bite size hot pockets or small sealed finger sandwiches. Something you can buy a half dozen of on the way to work and pop in as you go.	133
Something more pocket like, i.e.: pita style	124
Make non-greasy	122
Packaging that keeps product warm and can be easily set down in car without falling over.	118
No sesame seeds because I hate those seeds everywhere in the car.	115
Use a pocket bread	109
Burrito Style	108
Pocket bread like the ones sold in a bakery	106
A good tasting low calorie breakfast sandwich	100
Bottom-tier elements	
Ingredients baked right into a biscuit, such as egg, bacon and cheese.	97
Minimal packaging	96
A half-pita style bagel bread would hold nicely, outside crust would prevent leaks, taste would work with any filling.	84
Wrapped in another layer of wrapping to prevent messes before they happen.	80
Picture a 3-4 year old trying to eat in the backseat then you will have a pretty good idea of what a mess is.	79
Pita Pockets	77
Toasted Hamburger buns not just warmed.	77
More of a pocket-type sandwich. If the insides aren't falling out, then the toppings are oozing out.	74
Similar to a subway sandwich bun, very soft. I eat these often when I'm driving. And the filling (whatever that may be) does not fall out.	66
Cutting a sandwich in half when it is wrapped in paper makes it much easier to eat.	63
Sealed bagel with spread, for putting in the toaster.	63
Cut bread or sandwich in small hand size, fits in one hand.	61
A sandwich which is sealed on all sides to hold the contents inside.	60

(continued)

Table 19.10. Ranked elements from a Delphi-like exercise on the Internet* *(cont.)*

	Rank
A better tasting toaster muffin (like the Kellogg's ones) that taste like real muffins.	56
Bite size with filling, several in one Velcro type package.	56
Make something that you would combine all the ingredients together.	55
Something sweet and soft really hits the spot with me in the mornings, or some kind of packaging that you tear off a strip and it heats product so it's warm. but not hot enough to burn you or a child.	55
Healthy—no preservatives, low sugar, no hydrogenated or partially hydrogenated oils; + whole grains, dried fruit.	52
Wraps perform well to keep wet or moist product contained.	49
Somehow have bread completely encase the sandwich so the inside can't slide out.	48
Use puff pastry dough. French call it pate chou. Shape into a ball about one inch in diameter (Bite size). Stuffed with sausage & country gravy, ham & scrambled egg, bacon & cheddar cheese, etc. Nuke to warm or eat at room temp.	47
Something like a burrito.	46
Make package easy to open—one of the problems with those at gas stations is getting them open.	44

*Results were obtained from 480 respondents during one evening. Ranking was done by looking at the Slep (high-proportion selections when the element was shown to other respondents).

there are measures of both importance and frequency of selection.

It is important to remember that this type of creative exercise can be repeated with different setup paragraphs, images, or videos or a combination of both, time after time, and with relatively little difficulty, until a very large number of elements have been created. Of course, there will be many redundancies with a lot of respondents participating, but this plethora can be narrowed down at the developer's leisure, after the material has been collected. It is worth noting that this approach is equivalent to a large-scale bioassay of the respondents' minds; that is, if the Internet acquisition of the elements is continued for a week or more, with tens of thousands of elements, one can begin to count the frequency of appearance of each type of idea. The Delphi system, conducted on a regular basis, also measures the customer mind for solutions in a tracking-like fashion.

Step 5: Concept Optimization

The key business issue underlying most development research is to craft a concept for a new product (or service) that is acceptable, unique, and answers the particular problem. All the steps leading up to the concept optimization are designed to address this issue. The problem is made clear in the in-depth interviews and in the nature of the ethnographic observation (e.g., messy cars resulting from messy eating). The business opportunity is clear from the foundation study, which shows the large number of individuals who eat messily in their cars, perhaps because they belong to the segment known as elaborates. The potential solutions to the problem come from the elements proffered by the respondents in the Delphi-like exercise when presented with the problem situation. These are all informational. The final step is to synthesize the solution from raw materials, guided by the insights developed in the first four steps. The outcome should be a winning concept.

One alternative is not to create the concept by using consumers, but rather to take the information about messy eating and craft one's own best ideas. A different alternative identifies the impact of the different features for this new concept by using conjoint measurement, in which case optimal concepts can be created. There is the perennial dynamic between wanting to run with ideas to craft a solution and feeling the necessity to move more slowly and deliberately with data and with research in order to optimize the solution.

Two studies were run on the Internet, each with 35 respondents. Both studies were run with the same set of 24 elements combined into 40 different combinations. Each respondent evaluated the 24 elements in a unique set of combinations. The respondents used anchored 9-point scales. Half of the respondents rated the concepts on the following 1–9 easiness scale: *1 = very hard to eat . . . 9 = very easy to eat.* Half of the respondents rated the concept on the following 1–9 liking scale: *1 = hate . . . 9 = love.* Table 19.11 shows a collection of promising elements se-

lected from the Delphi-like exercise and their utility values in the conjoint study.

It is clear from Table 19.11 that elements driving the perception of "ease of use" may not be the same as those that are liked. The conjoint measurement results suggest different dynamics for these elements in the concept, depending on the mind-set of the consumer. Indeed, there may be no elements that are very highly liked and very easy to use. The optimal concept, therefore, is some combination of these elements, as shown in Table 19.11. Depending on the relative importance of "ease" and "liking," different concepts can be created through optimization.

Table 19.11. Utility values for 24 elements evaluated by conjoint analysis on two scales (hard vs. easy to use, and hate vs. love)

EL*	Text	Easy	Like
	Additive constant	*54*	*56*
C2	Special container designed for use in the car . . . no more crumb problems all over the front seat.	9	1
D2	It's sealed on all sides . . . so you don't mind eating in the car.	8	−1
D5	Longer edges make it easy to hold the sandwich.	7	−1
B3	Bite-sized sandwiches . . . more convenient to eat while driving.	7	−3
B1	Bite-size sandwich . . . easy to eat on-the-go.	7	1
D6	Easy to eat while driving.	6	2
D4	All sides covered, so nothing can slide out.	5	3
A4	Bread that won't fall apart after the first bite . . . no runs, no grease, no mess!	5	0
A1	A new bread that holds everything together . . . no more spills.	4	−6
A5	A bread that keeps wet ingredients like sauce from shooting out.	3	−1
B6	Bite-size . . . just like finger sandwiches	2	−4
B4	With a spread already inside . . . easy to toast, easy to eat.	2	−5
C4	Comes in a double layered wrapper.	2	−1
A3	Bread that keeps your sandwich intact . . . won't slip and cause a mess.	1	−2
D1	A smaller portion . . . easily fits in palm of your hand.	1	−3
C1	Comes in a package you can open single-handedly!	1	5
C3	Packaged to stay warm and can easily be placed in the car without falling over.	0	4
A6	A pita pocket that keeps toppings where they belong . . . inside the sandwich.	0	−2
C6	An easy to open package, so you don't have to hassle when you're on the road.	0	1
D3	No preservatives, low in sugar, no saturated oils . . . made from whole grain and dried fruit.	−1	6
C5	Sweet and soft to get you going in the morning . . . in a self heating package, just tear off the strip and it warms up.	−1	−7
A2	A new pita pocket bread . . . so you can fill it up with whatever you like and not worry about making a mess.	−5	−4
B2	Egg, bacon and cheese baked inside the bread.	−6	−13
B5	Made with French 'pate chou' pastry dough . . . stuffed with sausage, country gravy, ham and scrambled egg, bacon or cheddar cheese.	−18	−19

*EL, elements.

An idea of what might emerge from this process is presented in Table 19.12, which shows three concepts created from the same dataset in Table 19.11 and thus which emerge from the development process. The concepts range from 100% utilitarian (maximize "ease of use") to 100% hedonic (maximize "liking" without any attention to ease of use). The intermediate concept shows one of the many possible combinations of concept elements that constitute a compromise.

Paradigm Foundations: The Necessity and Desirability of Collaboration and Syncretism

As the technology today becomes increasingly sophisticated, no "one-stop shop" can provide all of the details. Even knowledge of the available tools can only be superficial if an individual researcher is expected to span the range from ethnography to databases to depth interviews to ideation to conjoint analysis. The range is simply too large and the task demands too great. Most researchers today, overwhelmed by concrete business problems to solve, simply cannot stay abreast of the large number of available technologies. Furthermore, even if a person could know the nature of these techniques, it is almost impossible today to be able to weave together a system by combining the techniques. Different groups, across companies and across disciplines, must work together in a collaborative and syncretistic mode to combine their expertise into a coherent whole. The individual parties in this combination do not, how-

Table 19.12. Three optimum combinations for the new easy-to-eat sandwich created to maximize perception of *easy*, perceptions both of *easy* and *liking* (a compromise), and perceptions of *liking*, respectively*

EL	Text	Easy	Like
	Additive constant	*54*	*56*
	Maximize *easy* alone		
A4	Bread that won't fall apart after the first bite . . . no runs, no grease, no mess!	5	0
B3	Bite-sized sandwiches . . . more convenient to eat while driving.	7	−3
C2	Special container designed for use in the car . . . no more crumb problems all over the front seat.	9	1
D2	It's sealed on all sides . . . so you don't mind eating in the car.	8	−1
	Total (additive constant + elements)	*82*	*54*
	Maximize both *easy* and *like* (compromise optimum)		
A4	Bread that won't fall apart after the first bite . . . no runs, no grease, no mess!	5	0
B1	Bite-size sandwich . . . easy to eat on-the-go.	7	1
C1	Comes in a package you can open single-handedly!	1	5
D4	All sides covered, so nothing can slide out.	5	3
	Total (additive constant + elements)	*72*	*67*
	Maximize *like* alone		
A4	Bread that won't fall apart after the first bite . . . no runs, no grease, no mess!	5	0
B1	Bite-size sandwich . . . easy to eat on-the-go.	7	1
C1	Comes in a package you can open single-handedly!	1	5
D3	No preservatives, low in sugar, no saturated oils . . . made from whole grain and dried fruit.	−1	6
	Total (additive constant + elements)	*66*	*69*

*These combinations come from the conjoint measurement exercise and from optimization of utilities to satisfy one or two objectives, respectively. EL, elements.

ever, lose their identity, but rather effectively combine to create a more powerful business organism.

The nature of the paradigm just presented here spans the often incommensurate, incompatible range from observation to discussion, from qualitative to quantitative, from the so-called touchy-feely to database numbers. Individuals expert in one of these areas are unlikely to be working side by side with individuals in another of these areas. It is organizations that must cooperate, no longer simply individuals in a large organization. Perhaps this is why the system represents an ecological chain of different organizations competing for some businesses, but also cooperating to achieve the objective that any one of the companies could not achieve on its own (Katz 1998).

Tools and Technology Are Necessary But Not Everything, Yet Neither Is the Inspired Analyst Working Alone in Splendid Isolation

One of the hall conversations often heard at professional meetings in consumer research is the growing popularity of technology tools for creative data acquisition, analysis, and knowledge development (e.g., see Ciborra and Patriotta 1998, and Pawle and Cooper 2001). To some individuals, tools become the panacea that promises to bring consumer research to a new level of sophistication. At the other end of the spectrum stand the traditionalists, deeply suspicious of consumer-research technologies, longing for the good old days, and staunchly refusing to abandon the insight of research to the mindless, heartless, soulless machines that often acquire the data and perform the analysis. To be sure, tools and technologies are necessary for the new paradigm sketched out earlier. The Internet above all provides the means to reach and engage the hundreds or thousands of respondents in a parallel model. The computer, the server, and the Internet explorer

are tools that allow researchers to reach consumers. The specific computer applications replace the interviewer with automated presentation, data acquisition, and structured analysis. The cost, the time, the scope, and the power could not be duplicated without enormous expense in a world lacking machines (Moskowitz and Ewald 2001).

Yet, machines cannot work alone. Without the humans framing the problem, identifying an issue, formulating a mind-set or aufgabe question, and selecting the correct elements, all that is done in the paradigm is to create a high-tech monkey typing Shakespeare; that is, without soul, without knowledge and intuition, and without the experience to recognize a business problem, the paradigm will simply lead to combinations of features that have neither reality nor reason to exist (Cooper 1999).

Perhaps the best that can be said about the paradigm is its ability to engage consumers in the development of innovative ideas in a structured manner. The ingoing assumption is that respondents need not be particularly innovative, nor articulate, but they should have some sense of the problem and an ability to intuit whether a solution is meaningful. The joint effort of dozens or hundreds of respondents, interacting on the computer in real time, provides the necessary process to create solutions and fine-tune and then optimize them. The existence of a foundation study beforehand puts things into context and provides an organizing principle into which these solutions can be located. Finally, ethnographic observation enables the developer to intuit how the solution might fit a problem that has become real through the actual observation of people experiencing the problem.

Importance of the Paradigm for Different Constituencies

The different groups involved in the paradigm can extract separate benefits relevant

for their needs. Here are the potential benefits as the authors see them right now, in
2004:

1. *For the corporation.* Enhanced discovery of new opportunities by using
knowledge-based methods rather than reliance on pure insights. A great deal of
new opportunity and product development
comes from the insight of professionals. To
the degree that this insight, and the knowledge it brings, can be formalized in a system
that can be accessed by everyone, the corporation will flourish.

2. *For the research profession.* A new role
and opportunity for consumer research. Traditionally, and in the eyes of high management in corporations, consumer research is
assigned the role of reporting what consumers want (consumer insight) or what consumers have done (tracking and ad hoc studies). The role of invention is assigned to
consultants or to other groups inside the corporation. Consumer researchers are relegated
to the level of lower-order knowledge workers constrained to operate in a far less dynamic role than consultants or marketers. A
paradigm with actionable results might
change that situation.

3. *For the science of consumer-research
methods.* The ability to use high-level research tools in an intelligent, time-efficient,
and cost-efficient manner. Researchers can
now enter fully into the innovation process,
using an integrated array of tools. The combination of powerful tools identifies opportunities (database analysis and ethnographic
observation) and creates new products that
fit these opportunities (using conjoint measurement, empowered by consumer-provided
inputs). Additionally, these high-level consumer-research tools go well beyond the traditional role of research, enabling the research process to evolve from data delivery
to an insight delivery. This evolution is possible through the availability of research
methods and products that package intelligence in a user-friendly manner (see von

Bertalanffy 1969; Terrano et al. 1995; Vosges
2001).

References

Abrams, B. (2000). The Observational Research Handbook. Lincolnwood, IL: NTC Business Books
(AMA).

Ansoff, H.I. (1975). Managing strategic surprise by response to weak signals. California Management Review, 8: 21–33.

Beckley, J., and Moskowitz, H.R. (2002). Databasing
the consumer mind: The Crave It!, Drink It!, Buy It!
& Healthy You! Databases. Anaheim, CA: Institute of
Food Technologists.

Borko, H., and Livingston, C. (1989). Cognition and improvisation: Differences in mathematics instruction
by expert and novice teachers. American Educational
Research Journal, 26: 473–498.

Brown, B.B. (1968). Delphi Process: A Methodology
Used for the Elicitation of Opinions of Experts. Santa
Monica, CA: Rand.

Ciborra, C.U., and Patriotta, G. (1998). Groupware and
teamwork in R&D: limits to learning and innovation.
R&D Management, 28: 43–52.

Cleveland, C.E. (1986). Defining customer expectations
using computer content analysis. In: Handbook on Research: Techniques to Solve Common Marketing Problems. Washington, DC: Bank Marketing Association.

Cleveland, C.E. (1997). Developing ideas when consumers are not very creative. Paper delivered at the
annual Genesis symposium, Genesis Institute,
Spokane, WA.

Cleveland, C.E. (2001). Quali-quant techniques with
Socrates and Aristotle. In: Quester Text Processing.
Spokane, WA: Genesis Institute, chapter 3.

Cleveland, C.E. (2002). Research project to determine
what the key ideas are when eating fast food in the car.
Quester Research Project 02845. Spokane, WA:
Quester.

Cooper, R.G. (1993). Winning at New Products: Accelerating the Process from Idea to Launch, 2nd edition.
Reading, MA: Addison Wesley.

Cooper, R.G. (1999). The invisible success factors in
product innovation. Journal of Product Innovation
Management, 16: 116–133.

Earle, M.D. (1997). Changes in the food product development process. Trends in Food Science and Technology, 8: 19–24.

Ericson, K., and Stuff, D. (1998). Doing team ethnography: warnings & advice. In: Qualitative Research
Methods Series, 42. Thousand Oaks, CA: Sage.

Finke, R.A., World, T.B., and Smith, S.M. (1995). Creative Cognition Approach. Cambridge: MIT Press.

Flores, L. (2002). Making idea generation and innovation available on the decision-maker desktop. Work-

ing Paper. Amiens, France: Amiens Graduate School of Business.

Flores, L., and Briggs, R. (2001). Beyond data gathering, implications of CRM systems to market research. In: Proceedings of the 54th ESOMAR Congress, Rome. Amsterdam: ESOMAR [European Society of Marketing Research].

Flores, L., Moskowitz, H.R., and Maier, A. (2003). From "weak signals" to successful product development: using advanced research technology for consumer driven innovation. In: Proceedings of ESOMAR Conference: Technovate2, Cannes. Amsterdam: ESOMAR [European Society of Marketing Research].

Fuller, G.W. (1994). New Food Product Development: From Concept to Marketplace. Boca Raton, FL: CRC.

Goldenger, J., Lehmann, D.R., and Mazursky, D. (1999). The primacy of the idea itself as a predictor of new product success. Working Paper. Cambridge, MA: Marketing Science Institute.

Griffin, A. (1997). PDMA research on new product development practices: updating trends and benchmarking best practices. Journal of Product Innovation Management, 14: 429–458.

Haeckel, S. (1999). Adaptive Enterprise, Creating and Leading Sense and Respond Organizations. Boston: Harvard Business School Press.

Hoban, T.J. (1998). Improving the success of new product development. Food Technology, 52: 46–49.

Holland, J.H. (1975). Adaptation in Natural and Artificial Systems: An Introductory Analysis with Applications to Biology, Control and Artificial Intelligence. Ann Arbor: University of Michigan Press.

James, W. (1911). Percept and concept: the import of concepts. In: Some Problems of Philosophy. Reprinted 1948; London: Longmans, Green.

Jolson, M.A., and Rossow L. (1971). The Delphi process in marketing decision making. Journal of Marketing Research, 8: 443–448.

Katz, F. (1998). How major core competencies affect development of hot new products. Food Technology 52: 46–52.

Levinthal, D.A. (1991). Organizational adaptation and environmental selection: interrelated processes of change. Organization Science, 2: 307–333.

Moorman, C., and Miner, A. (1995). Walking the tightrope: improvisation and information use in new product development. Working paper. Cambridge, MA: Marketing Science Institute.

Moskowitz, H.R. (1997). From a process to a transaction: implications for research and the research community. In: The 1997 CASRO Journal. Port Jefferson, NY: Council of American Survey Research Organizations, pp. 81–85.

Moskowitz, H.R. (1998). Designing new products in cyberspace: research driven innovation. In: CASRO Annual Journal. Port Jefferson, NY: Council of American Survey Research Organizations, pp. 69–78.

Moskowitz, H.R., and Ewald, J. (2001). Always on: bringing consumer research down to the development engineer, closer to the customer, and into the vortex of product development. In: Proceedings of the 54th ESOMAR Congress, Rome. Amsterdam: ESOMAR [European Society of Marketing Research].

Moskowitz, H., Flores, L., Beckley J., Mascuch, T., Cleveland C., and Ewald J. (2002a). Crossing the knowledge and corporate to systematize invention and innovation. In: Proceedings of the ESOMAR Congress, Barcelona. Amsterdam: ESOMAR [European Society of Marketing Research].

Moskowitz, H.R., Itty, B., Shand, A., and Krieger, B. (2002b). Understanding the consumer mind through a concept category appraisal: toothpaste. Canadian Journal of Market Research, 20: 3–15.

Nuese, C.J. (1995). Building the Right Thing Right: A New Model for Product Technology Development. New York: Quality Resources.

Pawle, J.S., and Cooper, P. (2001). Using Internet research technology to accelerate innovation. In: Proceedings of Net Effects[4]. Barcelona: European Society of Market Research, pp. 11–30.

Peleg, M. (1994). Darwinian evolution patterns in food product. Critical Review in Food Science and Nutrition, 34: 95–108.

Rosenau, M.D., Jr., Griffin, A., Castellion, G.A., and Anschuetz, N.F. (1995). The PDMA Handbook of New Product Development. New York: John Wiley and Sons.

Stewart, A. (1998). The ethnographer's method. In: Qualitative Research Methods Series, 46. Thousand Oaks, CA: Sage.

Stinson, W.S., Jr. (1996). Consumer packaged goods (branded food goods). In: Rosenau, M.D., Jr., Griffin, A., Castellion, G.A., and Anschuetz, N.F., editors. Product; New Product Development; Product Development. New York: John Wiley and Sons, pp. 297–312.

Terrano, T. Ishino, Y., and Yoshinaga, K. (1995). Integrating machine learning and simulated breeding techniques to analyze the characteristics of consumer goods. In: Biethahn, J., and Nissen, V., editors. Evolutionary Algorithms in Management Applications. Berlin: Springer, pp. 211–224.

Urban, G.L., and von Hippel, E. (1988). Lead user analyses for the development of new industrial products. Management Science, 34, 5: 569–82.

Veganti, R. (1997). Leveraging on systemic learning to manage the early phases of product innovation projects. R&D Management, 27: 377–392.

Von Bertalanffy, L. (1969). Complexity: The Emerging Science at the Edge of Order and Chaos. London: Viking.

Vosges, K. (2001). Using evolutionary algorithm techniques for the analysis of data in marketing. Cyber-Journal of Sport Marketing, 1; http://www.ausport.gov.au/fulltext/1997/cjsm/v1n2/VogesNo21.htm.

Wheelwright, S.C., and Clark, K.B. (1992). Revolutionizing Product Development: Quantum Leaps in Speed, Efficiency and Quality. New York: Free Press.

Zaltman, G., Duncan, R., and Holbeck, J. (1973). Innovations and Organizations. New York: John Wiley and Sons.

Part V

Databasing

Chapter 20

Creating an Integrated Database from Concept Research: The It! Studies

Introduction

One of today's pressing needs is to better understand the mind of the consumer in order to spot newly emerging trends in the market and to capitalize on them. Data themselves are no longer the choke point in the market, for researchers are awash in data. Data that cannot be easily obtained by subscribing to information services can be obtained by commissioning a custom study. In one way or another marketers, product developers, and researchers can answer most of the questions about consumer attitudes, new and current products, and advertisements.

Given this abundance of information one might naturally ask the following: Why bother with more data? What is missing? Why would a rational businessperson invest in new data when there is so much data from which to choose? The answer, as provided in this chapter, is quite simple. There are lots of sources of data, but unfortunately there is no systematically developed database about the mind of the consumer that can be interrogated to identify patterns and trends. The situation in marketing today is much like the amalgam of differently sourced computer programs that perform a gamut of tasks from spreadsheets to presentations to document control and preparation. Until Microsoft came out with its Office Suite there were many relatively unconnected alternatives for each task. Going from one task to another—for example, from document preparation to computation to presentation—meant learning all sorts of new tasks, finding how to do things, and then performing the task. In the meantime the effort was spent on learning the steps to move from one system to another, rather than on the information to be communicated.

What Is Available Today

Most of the knowledge resides in unrelated sources, such as corporate offices, trade and academic journals, and the experience of development and marketing professionals. A lot of information resides in disparate documents available to the public and accessed either by some intelligent search engine such as Google or by some "pay as you go" system such as Lexis/Nexis.

Creating a cross-sectional *and* longitudinal database to understand the algebra of the consumer mind is a major contribution to academic and business-oriented product development, marketing, and consumer sciences, respectively. If such a database can be developed easily, at low cost, with data that excites the user in terms of scientific and commercial applicability, then the notion of this database provides a unique business opportunity.

Much of the knowledge today about consumers as customers comes from one of three main types of standard research that have business value:

1. *Qualitative*, such as focus groups, probing in depth the motivations of consumers for a particular product or service (knowledge building and insight development), run with a few individuals (e.g., 10–60 over several sessions), without numbers that can be "sliced and diced" for new insights. Qualitative research gets into the inside of the

consumer mind, but requires trained researchers to pull out the insights.

    a. Hard to do worldwide without true experts participating in the research and analysis phase

    b. Provides strong insights

    c. Thus not scalable

2. *Primary quantitative*, such as surveys, which deal with reasons underlying certain behaviors or responses to concepts, run on a one-off basis with several hundred respondents and addressing a single issue.

    a. Can be done worldwide

    b. Does not need experts

    c. Provides weak insights into the consumer mind

    d. Data not valuable enough for general database use by the general buyer

    e. The single-issue focus prevents it from having strong long-term economic value

    f. Can be scaled, but data not valuable enough for general database use

3. *Systematized databases* arising from tracking studies, either sponsored by one company for its own use or sold on a syndicated basis by a research/data supplier. These systematized databases provide either answers to a limited number of questions over time or observations of behaviors.

    a. Can be done worldwide

    b. Does not need experts, except to run it, and provides answers from "queries"

    c. Provides very weak insights into the consumer "mind"

    d. Lends itself to database and can be sold

    e. Not typically actionable, except in term of showing "what's happening in general"

Creating an Integrated Concept Database

This chapter presents a different approach, albeit one that has been hinted about in this book. The approach makes knowledge more manageable by creating a database that shows the impacts or utilities of different messages in a general area, or *supercategory*, spanning several related topics. This organizing approach was followed in a series of large-scale studies called the It! mega-studies. The studies deal with such related issues for food and drink as:

1. Food acceptance (Crave It! study)—for adults, teens, and US compared with European consumers

2. Healthful foods (Healthy You!)

3. Beverages (Drink It!)

4. Social issues (Deal With It!)

5. Fast-food and quick-serve restaurants (Grab It! And Go)

The organizing idea behind the It! studies is to learn about what specific concept elements make people respond strongly, yet ensure that the learning transcends a particular product. The patterns that emerge across many products may be far more instructive than the patterns obtained by a deep investigation of only one particular category. Thus, the approach focuses simultaneously on many related categories. The It! approach uses the now-familiar method of self-authoring, Internet-enabled conjoint measurement. The approach is now affordable, works with many respondents, and covers topics in a category. Further information about the respondent is developed by having the same respondent who participates in a conjoint study also profile himself or herself on a conventional, large-scale attitude and usage questionnaire (classification). The approach is similar to the first-principles studies (see Chapter 18).

Four key structural features about the It! studies differentiate them the conventional segmentation studies:

1. *Each mega-study comprises 20–30 smaller studies.* As a consequence, the Crave It! study discussed below really comprises 30

different studies. Each deals with related concepts about a specific food or beverage, for example, hamburger, potato chips, and coffee. In contrast, most segmentation studies, even using conjoint analysis, are of much smaller scope.

2. *Elements in one study are comparable to elements in another.* Each It! study has a conjoint analysis study comprising 36 elements, divided into four *silos*. The structure of the elements was set up ahead of time so that each of them has a raison d'être or rationale for being. Once this rationale is developed/determined (e.g., brand, simple versus complex product description, or emotion), the actual text of the element can be easily created separately for each of the 30 studies. The text of the specific element was appropriate for the particular study, but conformed as closely as possible to the basic overarching design. Furthermore, in quite a number of cases the same text can be used across the 30 studies. This common structure across the studies allows for meta-analyses, showing patterns transcending a particular study. Table 20.1 shows an example of the elements for hamburger and chocolate candy, respectively, presenting the rationale for the element and the way the element is phrased for a specific study. Emphasis in the study is always on the balance between functional elements and emotional elements (Lautman and Percy 1983).

3. *Each study uses the identical self-profiling classification questionnaire.* The same set of questions is used for self-profiling for each of the studies in a particular It! study, allowing for comparison across studies.

4. *Respondents select the study that interests them.* Rather than allocating respondents to studies in a forced manner, the respondents go to a wall that presents the available studies. Respondent are free to choose any study that interests them. In this way one can measure the incidence of interested respondents.

Field Execution

The Internet-based execution is straightforward, following these steps:

Step 1: Invitation Letter

The respondents are invited to participate by using an e-mail "field house" (Open Venue, Toronto, Canada). Figure 20.1 shows the invitation letter. Sending respondents an interesting invitation letter generates a large number of participants. Furthermore, offering a chance to win a prize also increases interest.

Step 2: The Crave It! Wall

The respondents are guided to a *wall*, where they can choose to participate in any of a set of different but related conjoint studies. The wall is set up so that the least-popular study (fewest respondents) lies at the top left and the most-popular study (most respondents) lies at the bottom right. The location of the different studies on the wall is redone every 15 minutes. This strategy ensures that the studies are not biased by location. When the base size reaches a specific cutoff, the study option disappears and the button for the particular project disappears from the wall. Figure 20.2 presents an example of the wall.

Step 3: Conjoint Study

The actual study comprises 36 elements, combined in short, two- to four-element combinations and presented on a screen. Respondents rate the particular concept on the rating scale appropriate for that study. For example, in the Crave It! study, the respondents rated the craveability of the product as it was described (see Figure 20.3)

For any particular study in a single megastudy, 80 different experimental designs were created, embodying the 36 elements in 60 combinations. A respondent was randomly

Table 20.1. Example of elements (E) and their rationale for Crave It!

	Category	Rationale	Hamburgers	Chocolate candy
E01	Primary attributes	Basic physical attributes	Fresh-grilled hamburger	A smooth, dense piece of chocolate
E02	Primary attributes	Continuum: basic to complex/detailed physical attributes (in some cases . . . "healthy")	A char-grilled hamburger with a taste you can't duplicate	Smooth appearance with a light chocolate flavor and a creamy texture
E03	Primary attributes	Continuum: basic to complex/detailed physical attributes	A grilled aroma that surrounds a thick burger on a toasted bun	Crispy wafers coated in thin layers of milk chocolate
E04	Primary attributes	Continuum: basic to complex/detailed physical attributes (in some cases . . . "real")	Moist bites of bun, burger, and onion	Real chocolate made with ingredients like chocolate, cocoa butter, vanilla, and sugar
E05	Primary attributes	Continuum: basic to complex/detailed physical attributes	Juicy burger with the crunch of lettuce and tomato	White chocolate with crunchy cookie pieces throughout
E06	Primary attributes	Continuum: basic to complex/detailed physical attributes	Gooey grilled burger with rich sauce and fresh lettuce and tomato	Heavy dense chunk of chocolate with complex flavors, velvet appearance . . . enticing aroma
E07	Primary attributes	Continuum: basic to complex/detailed physical attributes	Layers of burger, sauce, pickles, and lettuce on a moist sourdough sesame seed bun	Dense chocolate with swirls of dark chocolate and chocolate sprinkles on the surface
E08	Primary attributes	Continuum: basic to complex/detailed physical attributes	Lots of crispy bacon and cheese on a juicy grilled hamburger on a lightly toasted bun	Clusters of chocolate and nuts, with caramel and marshmallow throughout
E09	Primary attributes	Complex physical attributes; details	Burger smothered in onions and cheese	Golden milk nougat with whole almond pieces on top, caramel drizzled over them and enrobed with semi-sweet chocolate
E10	Secondary attributes/mood	Party pleaser/inviting	Burgers are a party pleaser	When it's cold outside, chocolate is cozy and inviting
E11	Secondary attributes/mood	Beverages	With a chilled glass of water . . . or carbonated beverage	With a hot cup of coffee, tea, hot cocoa . . . or carbonated beverage

(continued)

Table 20.1. Example of elements (E) and their rationale for Crave It! *(cont.)*

	Category	Rationale	Hamburgers	Chocolate candy
E12	Secondary attributes/mood	With . . .	With great-tasting french fries . . . and that special sauce	Bite-sized pieces; ready for a fast taste . . . with a chocolate truffle filling
E13	Secondary attributes/mood	Premium quality/classic taste	Premium quality . . . that great classic taste, like it used to be	Premium quality . . . that great classic taste, like it used to be
E14	Secondary attributes/mood	Savor it . . .	You can just savor it when you think about it during work and school	You can just savor it when you think about it during work and school
E15	Secondary attributes/mood	All natural/changing flavors	100% natural . . . a real beef burger!	100% natural . . . and new choices every month to keep you tantalized
E16	Secondary attributes/mood	With all the extras you want . . .	With all the toppings and sides you want . . . pickles, relish, jalapenos . . . lettuce, tomato, chips . . . whatever	With fruit fillings in any flavor you want
E17	Secondary attributes/mood	Imagine the taste . . .	You can imagine the taste as you walk in the door	You can imagine the taste as you walk in the door
E18	Secondary attributes/mood	Lick your lips twice . . .	So tasty and juicy you practically have to lick your lips twice after each bite	So good . . . you practically have to lick your lips twice after each bite
E19	Emotional	Quick/fun/alone	Quick and fun . . . eating alone doesn't have to be ordinary	Quick and fun . . . eating alone doesn't have to be ordinary
E20	Emotional	Have to have it . . . can't stop	When you think about it, you have to have it . . . and after you have it, you can't stop eating it	When you think about it, you have to have it . . . and after you have it, you can't stop eating it
E21	Emotional	Fills that empty spot . . .	Fills that empty spot in you . . . just when you want it	Fills that empty spot in you . . . just when you want it
E22	Emotional	Cheers you up . . .	When you're sad, it makes you glad	When you're sad, it makes you glad
E23	Emotional	Escape routine/ celebrations	Now you can escape the routine . . . a way to celebrate special occasions	Now you can escape the routine . . . a way to celebrate special occasions
E24	Emotional	Multidimensional sensory experience	A joy for your senses . . . seeing, smelling, tasting	A joy for your senses . . . seeing, smelling, tasting

(continued)

Table 20.1. Example of elements (E) and their rationale for Crave It! *(cont.)*

	Category	Rationale	Hamburgers	Chocolate candy
E25	Emotional	With family and friends	An outrageous experience . . . shared with family and friends	An outrageous experience . . . shared with family and friends
E26	Emotional	Ecstasy . . .	Pure ecstasy	Pure ecstasy
E27	Emotional	Satisfies hunger . . .	It feeds THE HUNGER	It feeds THE HUNGER
E28	Brand or benefit	Basic brands/experiences	At QSR* A	From brand Q
E29	Brand or benefit	Continuum: basic to premium brands	At QSR B	From brand R
E30	Brand or benefit	Continuum: basic to premium brands	At QSR C	From brand S
E31	Brand or benefit	Continuum: basic to premium brands	At QSR D	From brand T
E32	Brand or benefit	Continuum: basic to premium brands	At QSR E	From brand U
E33	Brand or benefit	Premium brands/experiences	At QSR F	From brand V
E34	Brand or benefit	Fresh . . . for you . . . by you	Fresh from the grill, especially for you . . . by you	Made fresh . . . especially for you
E35	Brand or benefit	Best in world . . .	Simply the best burger in the whole wide world	Simply the best chocolate in the whole wide world
E36	Brand or benefit	Safety . . .	With the safety, care and cleanliness that makes you trust it and love it all the more	With the safety, care, and cleanliness that makes you trust it and love it all the more

*QSR, a restaurant name.
Data courtesy of It! Ventures, Inc.

Take our survey and win CASH!! The more surveys you take, the more chances to win!!

Are you the kind of person who loves FOOD and DRINK? Whatever you eat or drink, here is a survey for you! Tell us what kinds of foods and beverages you CRAVE ... choose from up to 30 different kinds ... then qualify to receive one of four CASH PRIZES and join one of The Understanding & Insight Group's exclusive consumer panels.

Simply click on the link below (if your email does not support hotlinks, cut and paste the link into your browser) and choose one of the easy-to-answer surveys.

http://12.109.160.54/uics2y4/craveit2002.asp

You will only have until NOON on XXX to complete this survey, so be sure to respond as quickly as possible. Your opinions are important to us!

Depending on your connection speed, each survey should take between 15 and 20 minutes to complete.

Forward this survey to your friends and family so that we know what kinds of food and beverages they CRAVE, too! Everyone's opinion is important!

Please be assure that any information you provide will be held in the strictest confidence. You will not be contacted by any sales or other research organization as a result of your participation in this survey.

If you have any difficulty accessing the survey, you can contact Tom Farrar at testmaster@theuandigroup.com

Thanks in advance for your input, and have fun !

The Crave IT! Team

Figure 20.1. Invitation letter to participate in the Crave It! study. Courtesy of It! Ventures, Inc.

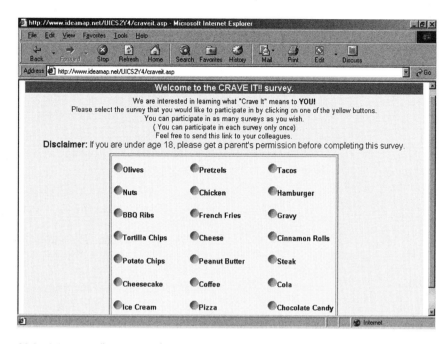

Figure 20.2. Crave It! wall. Courtesy of It! Ventures, Inc.

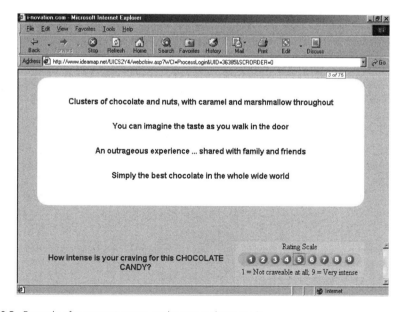

Figure 20.3. Example of a concept screen as the respondents see it.

allocated to a specific experimental design. No more than six respondents were ever allocated to the same design. This strategy of working with multiple experimental designs ensured that there would be minimal bias due to a particular combination.

Step 4: Self-profiling Questionnaire

At the end of the concept ratings the respondents completed the self-profiling questionnaire. For example, in the Crave It! study, each respondent completed an extensive classification, dealing with demographics (gender, age, and market), attitudes toward the product [acceptance using the FACT (food action) scale (Schutz 1965)], self-rated hunger, and the importance of both situations and product features as drivers of "craveability." The term *craveability* is used in the colloquial sense of high degree of liking, not in the medical sense of an addiction.

Step 5: Respondent Feedback

The respondents were then given an option to see the concept that they liked best versus the concept that everyone else liked best (Figure 20.4). The optimal concept for the total panel was created anew every 3 minutes.

How to Look at the Results of a Mega-study

With a large-scale study, comprising 20–40 smaller studies, and with a combination of self-profiling and conjoint analyses in each study, researchers have a plethora of information organized in ways that allow for detailed analyses. Five different approaches to looking at the data appear below, but they just scratch the surface of the analytic potential:

1. *Participation.* How many people participated, and who participated? This statis-

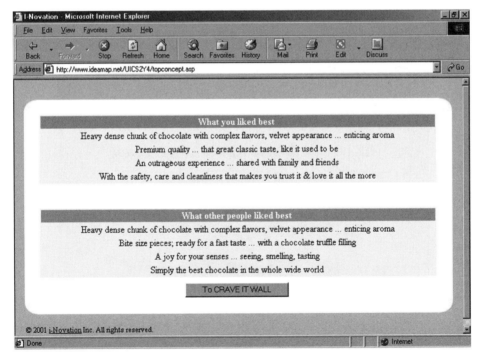

Figure 20.4. Example of feedback given to a respondent.

tic reveals a great deal about the popularity of the issue.

2. *Check-off batteries.* What are the self-reported factors that drive the "acceptance" of foods? This information comes from the classification questionnaire and is similar to the type of information obtained from conventional attitude and usage studies.

3. *Conjoint measurement.* What phrases drive "acceptance" in particular food categories? Rather than asking respondents to list these or check them off, the conjoint measurement uses a stimulus-response approach. Respondents simply rate the concept (stimulus). The pattern of responses is analyzed to determine the key drivers.

4. *Meta-analyses.* How do the same concept elements fare across different product categories?

5. *Segmentation.* Are there fundamental segments in the population that repeat from product category to product category?

Participation

Participation in the e-mail interview showed the majority of respondents to be women, but not always in the same proportion. Table 20.2 shows the participation for the first 20 studies

run for the Crave It! project. The remaining studies were run later. It is clear from these results that, although women participate more, gender-linked product preferences drive more men to participate in a study about steak than in one about a product such as cheesecake.

What Is Important: From the Self-defined Profile (Classification)

Self-profiling allows the respondents to describe themselves and, by so doing, provide in-depth information about what is important. Self-profiling is the conventional type of research collected by consumer researchers. The self-profiling questionnaire can be short or long, depending on the depth of information that a researcher wishes to collect.

In the It! studies the self-profiling classification generates a snapshot of how a respondent sees himself or herself. Furthermore, the same questionnaire is used for many foods so that the data can be analyzed from the perspective of each particular food. For example, in the Crave It! study the respondents were instructed to check reasons why they craved a specific food. One of the reasons was "mood." As Table 20.3 shows, there are "mood foods," such as chocolate, and then ice cream, cola, nuts, pretzels and coffee, respectively. From these data one can create a profile of foods and occasions/situations when they are most craved. It is important to note,

Table 20.2. Participation of men compared with women in the 2001 Crave It! study

Food	Total	% Men	% Women	Food	Total	% Men	% Women
Chocolate candy	478	14	86	French fries	151	19	81
Pizza	324	33	67	Taco	151	21	79
Ice cream	321	26	74	Pretzels	151	25	75
Cola	239	26	74	Nuts	151	33	67
Coffee	208	31	69	BBQ ribs	151	38	62
Cheesecake	173	16	84	Hamburger	151	40	60
Steak	168	44	56	Tortilla chips	150	20	80
Potato chips	153	24	76	Olives	150	24	76
Chicken	153	27	73	Cheese	150	27	73
Cinnamon rolls	152	20	80	Peanut butter	150	31	69

Data courtesy of It! Ventures, Inc.

Table 20.3. Proportion of respondents saying that mood is key fact in craveability for 20 of the foods in the 2001 Crave It! study

Food	Total panel	Mood is important	Mood/total (%)
Chocolate	472	226	48
Ice cream	316	115	36
Cola	237	86	36
Nuts	149	52	35
Pretzel	148	50	34
Coffee	206	69	33
Olives	147	42	29
Tacos	148	42	28
Tortilla chips	148	41	28
Potato chips	151	41	27
Cheesecake	172	45	26
French fries	148	36	24
Hamburger	150	31	21
Cinnamon rolls	149	29	19
Cheese	149	28	19
Chicken	148	27	18
Peanut butter	149	27	18
Pizza	318	55	17
Steak	168	22	13
BBQ ribs	149	17	11

Data courtesy of It! Ventures, Inc.

however, that these results are strictly from the self-profiling done in the classification questionnaire.

Expanding the Picture: Combining Conjoint Analysis and a Self-profiling Questionnaire

The heart of the It! studies is the conjoint measurement part, where respondents rated interest in descriptions about food or beverage. These descriptions comprise statements about product features, emotion, brand, situation, etc., encompassing an array of different types of elements. All the elements are text. From the responses the researcher creates a model at the individual respondent level relating the presence/absence of each of the concept elements to the respondent's ratings. We have seen this approach explicated in Chapter 5 on conjoint measurement and Chapter 16 on self-authoring systems.

The key to the data is the structured array of concept elements combined with a self-profiling questionnaire. From this array it becomes possible to create a snapshot of the respondents and how they react to the different food statements. We can gain a sense of the depth of this information by looking at one particular study from the 2001 Crave It! study dealing with hamburger. The principles developed from that analysis can then be applied to analyzing the other studies in the same data sets and also analyzing the other data sets in the It! series.

We begin with the self-profiling questionnaire that tells us about the respondents as they see themselves. Such questionnaires are the stock and trade of market researchers. Questionnaires can be short or long and go into agonizing detail. One of the perennial and thorny issues is to create such a questionnaire that is both reasonably detailed, yet can be completed in the Internet interview in no more than 5 minutes. Otherwise, in a 15- to 20-minute interview there will be no time for a respondent to do the conjoint portion of the study.

We get a sense of how the respondents profile themselves by looking at the results of the questionnaire for hamburgers (Table 20.4). The *analytic subgroups* for the data are presented in the rows:

1. The first five questions require the respondents to tell who they are, where they live, their age, and what time of day they are participating in the study.

2. The second set of questions asks the respondents to profile their acceptance of the particular food, here hamburger, using the FACT scale for acceptance, and to select up to three reasons for craving the food, as well as three locations where the food is bought and when the food is craved.

3. The third set of numbers asks about the respondent's state of "health" for the oral cavity and the respondent's willingness to participate in upcoming panel studies.

4. The fourth part of the table presents five derived profile values. The first comprises membership in the "hamburger" acceptor group, defined as those respondents who checked the top-3 rating scale values for hamburger on the FACT scale. These are individuals who say that they like hamburger.

5. The remaining analytic subgroups are derived from the individual utility values. Since the utility values are available for each respondent and for each element, researchers can interrogate the database to identify acceptors or rejectors of specific elements. The analytic subgroups are created by incorporating individuals who showed positive utility values for the four elements.

The importance here is that one can combine the self-profiling questionnaire with the conjoint analysis. The conjoint utilities can be analyzed in at least two ways:

1.   Create different groups based on the self-profiling questionnaire (e.g., self-profiled males versus females)

2.   Create different groups based on the individual's utility values (e.g., acceptors versus rejectors of McDonald's), which puts the respondent into a group

We continue with the summarized elements for hamburger, which are presented in Table 20.5. The table presents the average utility ratings for the 171 respondents and for each of the 36 elements, as well as the average utility values for gender and for three concept-response segments that emerged from the data. The gender data are obtained directly from the self-profiling questionnaire. The segments were developed from the correlation between pairs of respondents on their concept utilities (see Chapter 7). The actual results are not important here. Rather, the table shows the depth of information to be obtained from this type of database.

We see the same depth of information in Table 20.6, which shows the utility values for subgroups derived from either the self-profiling questionnaire or from the utility values themselves. In the interest of space the table shows only the winning elements for each subgroup. The results in the table differ from those of most consumer research studies because it shows subgroups created from utility values (those positive versus negative to McDonald's; and those positive versus negative to element E22, "When you're sad, it makes you glad"). Responses to McDonald's can be obtained from either self-profiling or the conjoint study. Responses to an element, such as "When you are sad it makes you glad" are not easy to obtain from a self-profiling questionnaire, and certainly with such a questionnaire it is hard to know what a respondent means by the rating. In contrast, by creating a subgroup of those who respond positively to this element we create a subgroup that behaviorally we know to respond positively to concepts with this element; that is, we already know the behavior toward the element

Table 20.4. Summary of the respondent panel, based on the self-profiling questionnaire for hamburger and based on derived subgroups

Question	Response	Base	%
Part I: Geo/demographics			
1 Gender	Female	107	63
	Male	64	37
2 Self-profiled level of hunger	1	26	15
	2	56	33
	3	60	35
	4	29	17
3 Midnight to 3 A.M.	3	12	7
3 to 6 A.M.	6	5	3
6 to 9 A.M.	9	10	6
9 A.M. to 12 noon	12	20	12
12 noon to 3 P.M.	15	25	15
3 to 6 P.M.	18	42	25
6 to 9 P.M.	21	33	19
9 P.M. to midnight	24	24	14
4 Age	Younger than 18	1	1
	18–30	51	30
	31–40	46	27
	41–50	35	21
	51–60	30	18
	61–75	8	5
5 Market	Midwest	34	20
	Southeast	31	18
	Northeast	29	17
	Southwest	29	17
	Northwest	17	10
	Canada/Mexico	14	8
	Europe	10	6
	Mountain	5	3
	Alaska/Hawaii	2	1
Part II: Response to food			
6 FACT scale	4 Hardly eat	4	2
	5 If available	6	4
	6 Now and Again	36	21
	7 Frequently	60	35
	8 Often	46	27
	9 Every opportunity	19	11
7 Where I buy	Food stores	42	25
	Convenience stores	4	2
	Warehouse stores	5	3
	Supermarkets	7	4
	QSR restaurants	131	77
	Local chain rest	56	33
	Local eatery	45	26
	Specialty stores	7	4
	Make from scratch	1	1
	Do not buy	4	2

(continued)

Table 20.4. Summary of the respondent panel, based on the self-profiling questionnaire for hamburger and based on derived subgroups *(cont.)*

	Question	Response	Base	%
8	Reasons for craving	Appearance	83	49
		Aroma	97	57
		Texture	29	17
		Taste	161	94
		Memories	12	7
		Associations	3	2
		Brand	27	16
		Advertisement	13	8
		Package	2	1
		Portion size	44	26
		Social occasion	3	2
		Mood	34	20
9	Time when craved	Breakfast	3	2
		Midmorning	4	2
		Lunch	97	57
		Midafternoon	41	24
		Dinner	61	36
		After dinner	8	5
		While Shopping	12	7
		Going to work	21	12
		After school	4	2
		Watching TV	26	15
		While alone	8	5
		With friends	23	14
		To celebrate	12	7
		When bored	4	2
		When hormones acting up	12	7
7		When kids bug me	2	1
		When I want to escape	10	6
	Part III: Oral health and panel membership			
10	State of mouth	All teeth	92	54
		Some teeth missing, teeth match	33	19
		Some teeth missing, teeth don't match	22	13
		Braces	1	1
		Cavities	72	42
		Retainer	0	0
		Crown	27	16
		Bridge	12	7
		Denture	14	8
		Caps	10	6
11	Join the panel		150	88
	Part IV: Derived profile values from the questionnaire			
12	Top–3 box Fact scale		125	73
13	Element–22 acceptor: "When you're sad, it makes you glad"		64	37
14	Element–24 acceptor: "A joy for your senses . . . seeing, smelling, tasting"		84	49
15	Element–30 acceptor: At McDonald's		68	40
16	Element–31 acceptor: At Wendy's		76	44

FACT scale, food action scale.
Data courtesy of It! Ventures, Inc.

Table 20.5. Utility values for the 36 elements (E) and the additive constant, for total panel, gender, and the concept-response segments

			Gender		Segment		
		Total	Male	Female	Classics	Elaborates	Imaginers
	Additive constant	29	34	27	44	12	56
	Base size	171	64	107	68	85	18
E01	Fresh-grilled hamburger	5	5	6	5	5	7
E02	A char-grilled hamburger with a taste you can't duplicate	7	9	5	6	8	2
E03	A grilled aroma that surrounds a thick burger on a toasted bun	11	9	12	5	17	8
E04	Moist bites of bun, burger, and onion	−2	−1	−2	−8	3	0
E05	Juicy burger with the crunch of lettuce and tomato	7	0	11	1	14	−1
E06	Gooey grilled burger with rich sauce and fresh lettuce and tomato	1	−1	2	−8	7	8
E07	Layers of burger, sauce, pickles, and lettuce on a moist sourdough sesame seed bun	7	10	6	3	18	−28
E08	Lots of crispy bacon and cheese on a juicy grilled hamburger on a lightly toasted bun	17	12	21	7	33	−20
E09	Burger smothered in onions and cheese	6	6	6	−9	17	7
E10	Burgers are a party pleaser	−2	1	−3	2	−5	−2
E11	With a chilled glass of water . . . or carbonated beverage	0	−1	1	1	−2	4
E12	With great-tasting french fries . . . and that special sauce	6	5	7	3	9	3
E13	Premium quality . . . that great classic taste, like it used to be	7	5	8	8	7	1
E14	You can just savor it when you think about it during work and school	0	1	0	3	−1	−3
E15	100% natural . . . a real beef burger!	9	9	9	9	9	4
E16	With all the toppings and sides you want . . . pickles, relish, jalapenos . . . lettuce, tomato, chips . . . whatever	10	13	8	2	19	1
E17	You can imagine the taste as you walk in the door	8	8	8	9	5	12
E18	So tasty and juicy you practically have to lick your lips twice after each bite	8	7	9	5	13	1
E19	Quick and fun . . . eating alone doesn't have to be ordinary	0	−1	1	2	0	−3
E20	When you think about it, you have to have it . . . and after you have it, you can't stop eating it	7	7	7	6	7	8
E21	Fills that empty spot in you . . . just when you want it	6	3	7	5	6	7
E22	When you're sad, it makes you glad	−4	−6	−3	−2	−6	1
E23	Now you can escape the routine . . . a way to celebrate special occasions	0	−2	2	3	−2	1
E24	A joy for your senses . . . seeing, smelling, tasting	3	3	3	7	1	0
E25	An outrageous experience . . . shared with family and friends	0	−3	1	3	−1	−6
E26	Pure ecstasy	0	−1	1	2	−1	−2

(continued)

Table 20.5. Utility values for the 36 elements (E) and the additive constant, for total panel, gender, and the concept-response segments *(cont.)*

	Total	Gender		Segment		
		Male	Female	Classics	Elaborates	Imaginers
E27 It feeds THE HUNGER	2	2	3	6	1	−3
E28 At White Castle	−5	−6	−4	−29	8	24
E29 At Jack-in-the-Box	−17	−22	−14	−29	−5	−25
E30 At McDonald's	−8	−10	−6	−15	5	−39
E31 At Wendy's	2	2	2	0	6	−8
E32 At Burger King	−1	3	−4	−5	3	−5
E33 At Fuddruckers	−4	−5	−4	−3	0	−29
E34 Fresh from the grill, especially for you . . . by you	5	3	5	1	5	15
E35 Simply the best burger in the whole wide world	5	5	5	3	6	8
E36 With the safety, care, and cleanliness that makes you trust it and love it all the more	−1	−4	0	−2	−2	2

Data courtesy of It! Ventures, Inc.

from the conjoint utilities, and therefore we know whether the element is behaviorally a driver of acceptance or rejection.

Meta-analyses: How the Same Element Performs When Presented in the Context of Different Foods

A key aspect of the database is the use of *meta-analyses*, which are analyses that transcend one study and go across all the studies. Continuing with the Crave It! data as an example, let us look at the performance of one element across the different foods. The element is a claim "Simply the best [food name] in the whole wide world." For cinnamon rolls this statement is a persuasive communication (utility $= +7$), whereas for steak this is not a persuasive communication (utility $= -1$). Table 20.7 shows these results for total panel and for gender. Men, in turn, tend to be more swayed by this type of claim than are women, but the pattern is somewhat ambiguous. The key here is that the same element can take on different utility values depending on the context in which it is presented. The value of the integrated database is its ability to show this performance of the element when the element is incorporated in the body of a concept.

Using It! Database to Drive Product Development

Concepts provide the blueprints from products and specifically for product features. We saw in previous chapters that one could fit a product to a concept or vice versa (Chapter 7). In those cases the concept research was specific to a particular product in a particular place and time. This specificity is a benefit to developers but also a problem. Conventional concept studies provide specific direction, solving the momentary problem. There is no systematized learning, except from the insight of the consumer researcher and the product developer. This systematized learning quickly fades as both individuals move on to their next tasks.

With the It! studies, one can get a more global picture of winning elements for product concepts. The direction is not as specific, but the integrated learning across many categories is much greater and of significant potential to the company. Unlike the solution of a momentary problem, the It! studies provide a more general database from which one can obtain new ideas for years afterward.

We get a sense of this information from the specific data from hamburger, shown in

Table 20.6. Utility values for the 36 elements (E) and the additive constant, for total panel, from those who say that they like or dislike hamburgers [FACT (food action) scale], and from those who behaviorally respond positively or negatively to the McDonald's name in a concept or to an emotional phrase in the concept (E22)

		FACT scale	
		High	Low
	Additive constant	*34*	*16*
	Winning elements: high FACT scale (acceptors)		
E08	Lots of crispy bacon and cheese on a juicy grilled hamburger on a lightly toasted bun	17	18
E16	With all the toppings and sides you want . . . pickles, relish, jalapenos . . . lettuce, tomato, chips . . . whatever	10	10
E03	A grilled aroma that surrounds a thick burger on a toasted bun	9	16
	Winning elements: low FACT scale (rejectors)		
E08	Lots of crispy bacon and cheese on a juicy grilled hamburger on a lightly toasted bun	17	18
E03	A grilled aroma that surrounds a thick burger on a toasted bun	9	16
E35	Simply the best burger in the whole wide world	3	12
E18	So tasty and juicy you practically have to lick your lips twice after each bite	7	11
E17	You can imagine the taste as you walk in the door	6	11
E02	A char-grilled hamburger with a taste you can't duplicate	5	11
E06	Gooey grilled burger with rich sauce and fresh lettuce and tomato	−3	11
E16	With all the toppings and sides you want . . . pickles, relish, jalapenos . . . lettuce, tomato, chips . . . whatever	10	10
E15	100% natural . . . a real beef burger!	8	10
E32	At Burger King	−5	9

		Response to McDonald's element	
		Negative	Positive
	Additive constant	*37*	*17*
	Winning elements: negative to McDonald's		
E15	100% natural . . . a real beef burger!	8	11
E16	With all the toppings and sides you want . . . pickles, relish, jalapenos . . . lettuce, tomato, chips . . . whatever	8	14
	Winning elements: positive to McDonald's		
E08	Lots of crispy bacon and cheese on a juicy grilled hamburger on a lightly toasted bun	6	35
E30	At McDonald's	−27	22
E03	A grilled aroma that surrounds a thick burger on a toasted bun	7	17
E13	Premium quality . . . that great classic taste, like it used to be	2	15
E16	With all the toppings and sides you want . . . pickles, relish, jalapenos . . . lettuce, tomato, chips . . . whatever	8	14
E05	Juicy burger with the crunch of lettuce and tomato	3	14
E02	A char-grilled hamburger with a taste you can't duplicate	2	14
E18	So tasty and juicy you practically have to lick your lips twice after each bite	5	13
E07	Layers of burger, sauce, pickles, and lettuce on a moist sourdough sesame seed bun	3	13
E20	When you think about it, you have to have it . . . and after you have it, you can't stop eating it	3	13
E12	With great tasting french fries . . . and that special sauce	1	13

(continued)

Table 20.6. Utility values for the 36 elements (E) and the additive constant, for total panel, from those who say that they like or dislike hamburgers [FACT (food action) scale], and from those who behaviorally respond positively or negatively to the McDonald's name in a concept or to an emotional phrase in the concept (E22) *(cont.)*

		Response to McDonald's element	
		Negative	Positive
E15	100% natural . . . a real beef burger!	8	11
E09	Burger smothered in onions and cheese	3	11
E01	Fresh-grilled hamburger	1	11
E17	You can imagine the taste as you walk in the door	6	10
E34	Fresh from the grill, especially for you . . . by you	2	9
E35	Simply the best burger in the whole wide world	2	9

		Element 22	
		Negative	Positive
	Additive constant	*36*	*18*
	Negative to element 22		
	"When you're sad, it makes you glad"		
E08	Lots of crispy bacon and cheese on a juicy grilled hamburger on a lightly toasted bun	13	24
E03	A grilled aroma that surrounds a thick burger on a toasted bun	8	15
E16	With all the toppings and sides you want . . . pickles, relish, jalapenos . . . lettuce, tomato, chips . . . whatever	8	13
	Positive to element 22		
E08	Lots of crispy bacon and cheese on a juicy grilled hamburger on a lightly toasted bun	13	24
	"When you're sad, it makes you glad"		
E22	When you're sad, it makes you glad	−17	18
E07	Layers of burger, sauce, pickles, and lettuce on a moist sourdough sesame seed bun	2	16
E03	A grilled aroma that surrounds a thick burger on a toasted bun	8	15
E21	Fills that empty spot in you . . . just when you want it	0	15
E18	So tasty and juicy you practically have to lick your lips twice after each bite	5	14
E16	With all the toppings and sides you want . . . pickles, relish, jalapenos . . . lettuce, tomato, chips . . . whatever	8	13
E20	When you think about it, you have to have it . . . and after you have it, you can't stop eating it	3	13
E12	With great tasting french fries . . . and that special sauce	2	13
E15	100% natural . . . a real beef burger!	7	12
E02	A char-grilled hamburger with a taste you can't duplicate	3	12
E05	Juicy burger with the crunch of lettuce and tomato	5	11
E17	You can imagine the taste as you walk in the door	6	10
E13	Premium quality . . . that great classic taste, like it used to be	5	10
E24	A joy for your senses . . . seeing, smelling, tasting	−1	10
E27	It feeds THE HUNGER	−2	9
E19	Quick and fun . . . eating alone doesn't have to be ordinary	−5	9
E25	An outrageous experience . . . shared with family and friends	−5	9
E26	Pure ecstasy	−5	9

Data courtesy of It! Ventures, Inc.

Table 20.7. Utility values for the element "Simply the best [food name] in the whole wide world"

	Total	Men	Women
Cinnamon rolls	7	3	8
Ice cream	5	3	6
Hamburger	5	5	5
Tacos	5	7	4
BBQ ribs	4	6	3
Chocolate candy	4	5	4
Pizza	4	6	3
Olives	3	6	3
French fries	3	4	3
Cheesecake	3	7	2
Peanut butter	3	0	4
Tortilla chips	3	−4	4
Coffee	2	−1	4
Chicken	2	−1	3
Nuts	2	1	2
Pretzels	1	4	0
Cheese	1	0	1
Potato chips	0	0	0
Cola	0	3	−1
Steak	−1	2	−2

Data courtesy of It! Ventures, Inc.

Table 20.8. The table presents the winning elements for the total panel and for the three segments. The segments become one organizing principle for product development. Beyond that information, however, it becomes possible to select pictures that demonstrate the product concepts, based on the segments. Figures 20.5–20.8 present examples for the three segments of what some foods and beverages might look like. Although these pictures do not provide concrete direction, they set the stage for more focused, principles-based direction—a direction that has been sadly lacking in the food and beverage industries.

Using the It! Studies

There is a vast amount of information to be obtained about consumers that is simply

Classic Hamburger

Elaborate Hamburger

Imaginer Hamburger

Figure 20.5. Pictures that represent hamburgers for the three segments.

Table 20.8. Winning elements for total and three concept response segments for hamburgers: the winning elements comprise statements about product features that the developer can use

		Total	Segment 1	Segment 2	Segment 3
	Additive constant	*29*	*44*	*12*	*56*
	Base size	*171*	*68*	*85*	*18*
	Total panel				
E08	Lots of crispy bacon and cheese on a juicy grilled hamburger on a lightly toasted bun	17	7	33	−20
E03	A grilled aroma that surrounds a thick burger on a toasted bun	11	5	17	8
E16	With all the toppings and sides you want . . . pickles, relish, jalapenos . . . lettuce, tomato, chips . . . whatever	10	2	19	1
	Segment 1: Classics				
E15	100% natural . . . a real beef burger!	9	9	9	4
E17	You can imagine the taste as you walk in the door	8	9	5	12
E13	Premium quality . . . that great classic taste, like it used to be	7	8	7	1
	Segment 2: Elaborates				
E08	Lots of crispy bacon and cheese on a juicy grilled hamburger on a lightly toasted bun	17	7	33	−20
E16	With all the toppings and sides you want . . . pickles, relish, jalapenos . . . lettuce, tomato, chips . . . whatever	10	2	19	1
E07	Layers of burger, sauce, pickles, and lettuce on a moist sourdough sesame seed bun	7	3	18	−28
E09	Burger smothered in onions and cheese	6	−9	17	7
E03	A grilled aroma that surrounds a thick burger on a toasted bun	11	5	17	8
E05	Juicy burger with the crunch of lettuce and tomato	7	1	14	−1
E18	So tasty and juicy you practically have to lick your lips twice after each bite	8	5	13	1
	Segment 3: Imaginers				
E34	Fresh from the grill, especially for you . . . by you	5	1	5	15
E17	You can imagine the taste as you walk in the door	8	9	5	12

Data courtesy of It! Ventures, Inc.

lacking in other studies. The information ranges from self-profiling on a variety of features, to understanding the product and communication *hot buttons* for specific products. The key benefits and conclusions from the It! databases are the following:

1. *Discipline invoked that is not found in corporations.* Most corporations do not have the discipline to invest in studies that create databases without having a specific goal in mind, such as the solution of an immediate problem. By creating the database separately from the corporate aegis, it becomes possible to contribute to the field of product and con-

cept development by using a more disciplined, less knee-jerk approach.

2. *Simplicity of design and execution.* An integrated database can be created in a short time period (weeks rather than months/years)

3. *Ease of navigation to find results and insights.* Setting up the structure ahead of time makes the database easier to use. By having a coherent structure it becomes possible to find the right data and, where necessary, step back and look at patterns across foods or other categories (e.g., insurance).

4. *Deep learning from the data that transcends a momentary problem.* There are

Classic Cheesecake   Elaborate Cheesecake

Imaginer Cheesecake

Figure 20.6. Pictures that represent cheesecake for the three segments.

Classic Coffee   Elaborate Coffee

Imaginers Coffee

Figure 20.7. Pictures that represent coffee for the three segments.

Classic Ice Cream   Elaborate Ice Cream

Imaginers Ice Cream

Figure 20.8. Pictures that represent ice cream for the three segments.

segments, but these segments are far more profound than one might have thought, based on previous data. Much of the traditional segmentation is attitudinal, but there is difficulty in bringing this segmentation down to the realm of the actionable (Wells 1975). By segmenting the respondents on the basis of the conjoint results, one can create strong segments that are homogeneous with respect to the types of messages that they find interesting.

Acknowledgments

The authors thank the Insight and Understanding Group for ongoing collaboration in the It! Ventures, Inc. We also thank Mc-Cormick and Company for supporting the 2001 and 2002 Crave It! studies and for providing ongoing assistance in analysis and in materials support for this and other papers dealing with Crave It! Photographs courtesy of McCormick and Company.

References

Lautman, M.R., and Percy, L. (1983). Cognitive and affective responses in attribute-based versus end-benefit oriented advertising. Proceedings of the Association for Consumer Research, 11: 11–17.

Schutz, H.G. (1965). A food action rating scale for measuring food acceptance. Journal of Food Science, 30: 365–374.

Wells, W.D. (1975). Psychographics: a critical review. Journal of Marketing Research, 12: 196–213.

Chapter 21

Highlights and Insights from The It! Studies: Crave It! and Eurocrave

Introduction

What makes a food so good that it evolves to become the gold standard representative of a given food category? What makes it so good that it becomes a food you constantly think about and are delighted to eat? When offered a wide variety of foods, what makes a person want one type of food more than another type?

Food acceptance and even cravings for foods have been studied for decades. Key studies have focused on chocolate and carbohydrates (Hill and Heaton-Brown 1994; Pelchat 1997). No one has looked across many, many food categories to begin to understand how the drivers of craving are similar or different for these foods. The Crave It! study and the ensuing database, introduced in Chapter 20, enable researchers to identify key factors across foods that drive craving or at least very strong degrees of acceptance. Moskowitz and colleagues (2003) have suggested that these cross-product studies are analogous to the genomic studies that are now very popular. Rather than studying the effect of one gene, biologists study the effects of many genes and, by so doing, understand how their combination generates an individual. Here, the study of responses to many products, with either one person or many people, shows the genomics of the *consumer mind*. The assumption was that, if one were to look broadly, one would learn something new and possibly discover some common behavior that occurs across all these eating situations.

What Are the Components of the Heightened Desired Food Experience?

The Crave It! study looked at trade-offs that consumers make when faced with the ultimate food experience or foods they crave. The trade-offs work with aspects of the food, the eating situation, and emotion. What makes the Crave It! study important for a book on food concepts is that the entire study was an integrated concept study with different product categories. The Crave It! study used the approaches described in Chapter 20. The systematic mining of the information contained therein enables researchers to understand the type of information that can be obtained when one goes from a single concept to a large-scale concept study.

The aspects of the foods, the eating situation, and the resulting emotions were captured by four sets of attributes:

1. *Primary descriptors* of the food itself, ranging from simple descriptions to more complex descriptions
2. *Situational attributes*, such as beverages with which the food is eaten
3. *Emotions*, which comprise the feelings one has about the food and about oneself
4. *Brands and benefits*, including the ubiquitous food safety element

There was a variety of different foods and beverages in the Crave It! study:

1. *Beverage:* cola, coffee, and iced tea

2. *Foods eaten in a restaurant:* BBQ ribs, steak, chicken, tacos, french fries, gravy, hamburgers, pizza, shellfish, meatloaf, mashed potatoes, hot dogs, bread, and salad

3. *Snacks:* nuts, pretzels, potato chips, tortilla chips, cheese, bacon, popcorn, and snack mix

4. *Sweets:* cheesecake, cinnamon buns, ice cream, chocolate, chocolate chip cookies, fresh fruit, and donuts

A key goal was to discover what was most highly craved and what was less craved. There are a number of ways to look at this ordering of craving:

Frequency of Selection

The *frequency* corresponds to the studies selected while the Internet study was live. This is a measure of what foods would be most highly linked with cravings in the consumers' minds. After they agreed to the invitation, as the respondents began the study they entered the study wall, which comprised a computer screen that rotated the study order of the food categories and their corresponding buttons so there would be no first-order or placement effects. A visual way to depict this wall is to imagine that a consumer walks over to a cupboard, opens it, and sees these 30 different foods there. The key information we are looking at is the food that the consumer respondent would select first. The studies that respondents chose first were chocolate candy, ice cream, and pizza. These three foods had the highest number of completions (at least twice the level of other categories) in the first day. Oddly, cinnamon rolls, cola, and meatloaf had a higher selection rate, as well (similar to the selection rate for chocolate candy, ice cream, and pizza), but a much lower completion rate (similar to the level shown by the other, less craveable foods). Chocolate candy, ice cream, and pizza correspond to foods that we typically consider *craveable*. Foods such as cinnamon rolls, cola, and meatloaf could be considered highly craveable, but would not be the ones that people might choose first.

Magnitude or Intensity of Craveability

The goal here was to understand the intensity of craveability, not its frequency. The conjoint analysis portion of the study enables us to create an index of craveability called *total craveability*. This index, presented in Table 21.1, was *operationally defined* as the sum of two numbers:

1. The *additive constant*, which shows the craveability level without the benefit of the elements. The additive constant itself reveals the level of interest consumers bring to a category before any of the descriptive elements are added in; that is, how strong the craveability rating is before the respondents are confronted with concepts that have food descriptors, emotions, brands, and so on.

2. The *average utility of the first nine elements*, which shows the average part-worth craveability of the elements that describe the food's physical characteristics

This definition of craveability is expressed by this simple equation:

$$\text{craveability} = \text{additive constant} + \text{average of food descriptor elements}$$

Based on this definition of magnitude of craveability, cheesecake turned up as the most highly craved food for the entire population of respondents. This was followed by fresh fruit, steak, BBQ ribs, and shellfish. Only after these highly craved foods did the food studies that were chosen first from the study wall show up: ice cream, and chocolate candy. This is an important observation because it tells us that there is a difference between self-reported behavior and the more closely measured actual behavior. Whereas one might say that one craves chocolate candy and be very interested in telling the interviewer about that craving of chocolate, the actual depth of craving for

Table 21.1. Total craveability, defined as the sum of columns A and B

	A	B	C
	Additive constant	Average utility of the nine product descriptors	Total craveability
Cheesecake	53	10	63
Fresh fruit	54	4	58
Steak	50	7	57
BBQ ribs	49	8	57
Shellfish	51	5	56
Ice cream	49	6	55
Chocolate candy	46	8	54
Bacon	46	7	53
Popcorn	48	4	52
Donuts	49	2	51
French fries	48	3	51
Tacos	47	4	51
Cinnamon rolls	45	5	50
Chocolate chip cookies	40	9	49
Coffee	48	0	48
Salad	43	5	48
Cola	47	1	48
Chicken	47	1	48
Hamburger	43	5	48
Pizza	45	3	48
Nuts	45	2	47
Mashed potatoes	47	0	47
Cheese	42	4	46
Iced tea	48	−2	46
Pretzels	42	3	45
Bread	40	5	45
Gravy	38	5	43
Meatloaf	38	4	42
Hot dogs	40	1	41
Potato chips	36	3	39
Tortilla chips	29	6	35
Snack mix	31	3	34

cheesecake is much higher. Furthermore, there were clear gender differences. Men appeared to be driven more by protein-based foods (e.g., BBQ Ribs and steak), whereas women were more driven by carbohydrate-based and fat-based foods (e.g., chocolate candy and ice cream).

What Drives the Craving Experience Overall for the Entire Population?

Looking at the impact of the four silos or categories in the design tested (food descriptors, situational attributes, emotions, and brand and benefits) allows the researcher to assess the impact on the overall craving experience for the entire respondent group. The average of each category shows the impact of the category:

1. *The primary food descriptor elements had the largest impact on overall cravings for most of the categories.* For categories like cheesecake, chocolate chip cookies, BBQ ribs, and chocolate candy, the product descriptors generally have the largest impact on overall cravings. For categories like mashed potatoes and iced tea, the product descriptors on average have a negative impact on overall cravings.

2. *Other types of elements.* These show lower impacts. Across all food categories the situation, emotions, and brands and benefits across all the food categories have a smaller impact.

3. *Situational elements.* These vary by product. Situational attributes exert the largest positive impact for products such as tacos and tortilla and the strongest negative effect for beverages such as coffee and cola.

4. *Emotion-relevant elements.* These show the largest positive impact for foods such as BBQ ribs and bread and have the least impact for foods such as ice cream and salad.

5. *Brands and benefits.* These elements exert their impact on foods such as fresh fruit, ice cream, and tortilla chips. Brands and benefits have their great negative impact on fast-food products such as pizza, chicken, and hamburger.

The key to craveability for most people is the *product description*. Describing the product itself will have the largest impact on the craving experience. Elements describing situations, emotions, brands, and benefits can have an impact on craving overall, but they are *attributes of the craving experience, not the food*. These rules of thumb apply for the craving experience overall, but when looking within a specific category these generalities can shift. For example, in the category of iced tea, describing the iced tea generally does not enhance craveability. Describing the emotional experience of iced tea can enhance craveability. The key to building an ultimate food experience is using the elements that have the greatest impact on craving for the segment one is trying to persuade for the specific product.

How Often Do Consumers Crave Food and What Impact Does Frequency Have on Craveability?

One of the key questions is whether foods are craved because they are not eaten frequently. In general, foods that are not eaten as often are craved more, although individual food categories can vary:

1. Foods and beverages typically consumed once a day or more frequently (e.g., coffee, cola, bread, iced tea, fresh fruit, or cheese) on average show a moderate level of craveability. The only exception is fresh fruit, which has high craveability

2. Foods that are eaten several times a week, like chicken, hamburgers, and french fries, also show moderate craveability, although some foods, such as steak and ice cream, can have a high craveability.

3. Foods that are typically eaten a couple of times a month have a higher average craveability (Table 21.2).

How frequently a specific food is consumed affects the types of attributes the consumer looks for in the food experience. For example, for potato chips the overall rated craveability decreases among respondents who say that they consume potato chips less frequently. Respondents who consume potato chips more frequently crave them more. Respondents who consume potato chips only 2–3 times a month have a very low craving for the chips (see Figure 21.1).

Consumption patterns also drive responses to product features. Consumers who eat potato chips once a day or more focus on the classic taste and an emotional experience. The brand is very important. Consumers who eat potato chips once a week or more focus

Table 21.2. Average craveability (the constant plus average of first nine elements) as a function of how often the food is eaten

On average, how often do you drink/eat this food?	
Once a day or more	49
Multiple times a week	47
Multiple times a month	54
Once a month or less often	53

Data courtesy of It! Ventures, Inc.

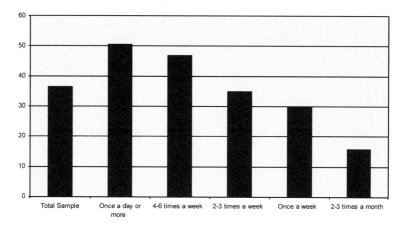

Figure 21.1. Craving for potato chips compared with frequency of consumption. Data courtesy of It! Ventures, Inc.

on the texture of the potato chips. A few brands can satisfy these consumers. Consumers who eat potato chips 2–3 times a month focus on the flavor and texture experience of the potato chip. Brand is less important to this group (Table 21.3).

Using the Crave It! Database to Understand the Key Attributes of a Food

What are the key attributes that consumers look for in a given food? What is most important to consumers? Looking across all the categories and using the self-profiling classification, one quickly sees that taste is chosen as the most important attribute. Table 21.4 shows this ranking across all of the products from the classification of respondents.

Individual foods are driven by different attributes, however. Taste is not the major driver for some foods. We can get a feeling about the importance of different sensory attributes from Table 21.5, which shows the selection of products based on taste, aroma, appearance, and texture. For example,

1. With coffee, popcorn, bacon, cinnamon rolls and bread, aroma is the most important attribute and taste second.

2. With iced tea and cola, thirst is the most important attribute. Taste is the next important.

3. Chocolate candy, ice cream, and mashed potatoes are driven by a respondent's general mood, which may explain how chocolate got its reputation as a mood food. Cola, potato chips, and hot dogs are driven by brand.

4. Fresh fruit, salad, and cheesecake are driven by product appearance. Cheesecake, mashed potatoes, and bread are driven by texture.

Day Part and Craveability

Just as with consumption rates, the day part can have an impact on the types of food and the attributes of the food that are desired:

1. The self-profiling questionnaire shows that food is craved all day long.

2. Not surprisingly, coffee is the most highly craved food the first thing in the morning.

3. The foods craved in the morning are bacon, donuts, and cinnamon rolls.

4. The foods craved at lunchtime are french fries, hamburgers, hot dogs, cheese, and colas.

Table 21.3. How concept elements drive craveability for potato chips as a function of the frequency of a respondent's potato chip consumption

Once a day or more frequently	
Base size	*33*
Additive constant	*51*
When you think about them, you have to have them . . . and once you have them, you can't stop eating them	10
Classic taste . . . the way you remember it	10
So delicious, just thinking about them makes your mouth water	9
A special treat . . . you will savor every bite	9
From Lay's	8
Once a week	
Base size	*40*
Additive constant	*30*
From Lay's	13
Thin sliced and lightly salted potato chips . . . a light taste and a crispy crunch	13
Potato chips with ridges . . . perfect for holding dips or spices	11
Kettle style . . . for a crispier crunch that is so satisfying!	11
Potato chips with the unique texture that only comes from special potatoes . . . seasoned for a one of a kind experience	11
So delicious, just thinking about them makes your mouth water	11
With your favorite beverage	10
From Ruffle's	9
Available at a value price	9
2-3 times a month	
Base size	*20*
Additive constant	*16*
Covered with savory spices like onion & garlic, sour cream & chive, cheddar, BBQ, vinegar, whatever	27
Brushed with olive oil and topped with parmesan cheese, garlic, and basil	16
So delicious, just thinking about them makes your mouth water	15
Classic taste . . . the way you remember it	14
Marinated before cooking for a unique flavor, golden color, and extra crunch	14
A joy for your senses . . . seeing, smelling, tasting	12
When you think about them, you have to have them . . . and once you have them, you can't stop eating them	12
Potato chips . . . baked golden instead of fried	11
Potato chips with the unique texture that only comes from special potatoes . . . seasoned for a one of a kind experience	10
Kettle style . . . for a crispier crunch that is so satisfying!	9
Thin sliced and lightly salted potato chips . . . a light taste and a crispy crunch	9
Available at a value price	9
Everything you want . . . all in one place . . . a mixture of tastes and textures	9

Data courtesy of It! Ventures, Inc.

5. The foods craved in the midafternoon are snack mix, chocolate candy, nuts, pretzels, chocolate chip cookies, iced tea, fresh fruits, and tortilla chips.

6. The foods craved at dinner are gravy, mashed potatoes, steak, meatloaf, shellfish, BBQ ribs, salad, chicken, pizza, tacos, and bread.

7. The only food craved after dinner is cheesecake.

8. The food craved just before bedtime is ice cream.

Table 21.4. Attributes selected as those that most influence cravings

Which 3 attributes MOST influence your craving for this food? [check three]	% Choosing
Taste	64
Thirst	44
Aroma	42
General mood or how you feel	30
Brand	23
Product appearance	19
Texture	15
Portion size	12
State of being, e.g., relaxed or on the go	7
Environment and SOCIAL SITUATION	7
Associations with family or friends	6
Advertising	4
Season	4
Stress level	3
Weather	3
Packaging	2
None of the above	2

Data courtesy of It! Ventures, Inc.

Table 21.5. Percentage of respondents selecting different sensory attributes as key to craveability

	Percentage			
	Taste	Aroma	Appearance	Texture
Average across foods	*90*	*45*	*43*	*33*
Taste is key				
Ice cream	95	7	46	50
French fries	95	50	46	39
Potato chips	93	24	24	40
Cheese	93	41	43	40
Hamburger	93	57	50	18
Cola	92	19	10	7
Pizza	91	53	61	21
Coffee	91	92	15	11
Taco	91	54	46	17
Peanut butter	91	49	21	61
Nuts	90	47	39	19
Aroma is key				
Coffee	91	92	15	11
Chicken	89	72	54	26
BBQ ribs	89	72	54	26
Cinnamon rolls	84	69	62	17
Appearance is key				
Cinnamon rolls	84	69	62	17
Pizza	91	53	61	21
Cheesecake	87	17	61	58
Steak	83	52	59	38
Texture is key				
Peanut butter	91	49	21	61
Cheesecake	87	17	61	58
Ice cream	95	7	46	50

Data courtesy of It! Ventures, Inc.

9. The foods craved while watching TV are popcorn and potato chips.

The conjoint portion of the Crave It! study goes beyond listing the specific foods and reveals what specific attributes become psychologically important as the day part changes. *Day part within the context of the Crave It! study is now defined as the time of day when respondents say that they crave the food.* The desired attributes of the food can change as a function of the day part. For example, coffee overall is about premium quality and freshness across all the day parts. When the respondent says that he craves coffee the first thing in the morning, though, then the elements denoting quality, fresh, and made the way you want it become more important. Coffee at breakfast time is basic Colombian, fresh, and premium. Coffee at midmorning begins to take the basic coffee and add emotional elements to it like refreshing, special treat, and making the stress just go away. By midafternoon the fresh coffee is a cappuccino, with milk and branded. By after dinner the coffee is fresh, a special treat, and premium (see Table 21.6).

Mind-set Segments and Experiences

In *The Dream Society* by Rolf Jensen (1999), emotional branding is called the "market for feelings." He says, "The consumer buys feelings, experiences, and stories. This is the post-materialistic consumer demanding a story to go with the product. Food that is good quality, tasty, and nutritious will no longer be sufficient. It must appeal to the emotions with a built-in story of status, belonging, adventure, and lifestyle." The key question whether this idea of the experience applies equally to everyone.

Chapter 7 on segmentation talked about dividing respondents into similar groups that share similar utility patterns. One can look across all the foods to determine whether there exist specific, overarching segments that share common responses to concept stimuli. Consumers who fall into different mind-set segments are often similar in their ages, demographics, and more conventional lifestyle patterns, yet they differ radically from one another in coherent patterns best revealed by their responses to concept elements. This *mind-set segmentation* reflects actual behavior in terms of responses to stimuli rather than behavior that is self-profiled on a set of scales. The segmentation is hard to fake by using politically correct answers.

The Crave It! database suggests that the same three segments emerge for the different products, albeit in different proportions and with different degrees of clarity. The naming of the segments is left to the researcher, but the identification of the segmentation and the assignment of respondents to the segments is a purely statistical exercise based on objective, quantitative criteria that are based in mathematics. The three emergent segments can be defined as follows:

1. *Classics* are interested in the product category and focus on the basic, classic nature of the product.

2. *Elaborates* are not as interested in the product category and focus on the product as it is varied to lend sensory excitement.

3. *Imaginers* are interested in the product category and focus on all the aspects around the product (i.e., brand, emotion, and specific product attributes).

The creation of overarching mind-set segments helps developers understand how to create products and sensory/product experiences for different groups showing different preference patterns. Many marketers like Rolf Jensen talk about creating a story experience that goes with the product when building an experience-based brand. The mind-set segmentation suggests that the optimal experience may differ across the segments, and therefore successful product developers will have to create different concepts and different experiences for these segments.

Table 21.6. How coffee elements perform in the conjoint study among respondents who crave coffee at different times of the day

	Time of day when respondent says he craves coffee					
	Total panel	As soon as I wake up in the morning	At breakfast time	Mid-morning	Mid-afternoon	Just after dinner
Base size	*274*	*210*	*76*	*67*	*71*	*54*
Additive constant	*48*	*49*	*45*	*53*	*47*	*44*
As soon as I wake up in the morning						
Premium quality . . . the best Coffee in the whole world	8	9	10	11	4	5
Fresh ground and brewed coffee	7	7	8	9	6	3
Prepared just to your liking . . . add whatever your heart desires	6	7	4	9	3	7
At breakfast time						
Premium quality . . . the best Coffee in the whole world	8	9	10	11	4	5
Fresh ground and brewed coffee	7	7	8	9	6	3
Made with 100% Columbian coffee beans	6	7	7	9	0	3
Midmorning						
Premium quality . . . the best Coffee in the whole world	8	9	10	11	4	5
Fresh ground and brewed coffee	7	7	8	9	6	3
Prepared just to your liking . . . add whatever your heart desires	6	7	4	9	3	7
Made with 100% Columbian coffee beans	6	7	7	9	0	3
A quick refresher for when you're on the run	2	2	−1	7	−1	3
A special treat . . . you will savor every sip	4	4	4	6	3	5
Drinking it makes all the stress just melt away	2	2	3	5	2	−3

Data courtesy of It! Ventures, Inc.

Classics

The optimal experience for classics focuses on the food itself. Classics seek foods prepared in the classic or traditional manner. They will notice whether the flavors, texture, and other sensory attributes differ from what they have come to expect or what they expect from reading a description. Classics are quite interested in the overall category itself, but will not be moved by specific brands unless the brands deliver consistently. They will also not be moved by emotional positioning or by situations. They want to know that the experience offered is all about the food and will accept the food when it *does not* present unusual flavors or textures. For example, in the food category of chicken the classics (the largest segment) would enjoy "a branded, ready to eat, premium quality chicken, coated in their favorite batter and spices and deep fried for a crunchy good taste." Classics do not want another branded chicken, "cut into strips and served with onions, pepper, cheese, tomatoes, sour cream" or "with a wine sauce" or "with a touch of red pepper spice."

Elaborates

The optimal experience for the elaborates focuses on what can be added to the food in terms of flavor or texture and other sensory attributes. Elaborates are not particularly interested in the food itself, but rather respond strongly to added or overlaid flavor and texture experiences. The food becomes a substrate to get more flavors and stuff into their mouth. What becomes clear after reading the following list is that elaborates are all about having as much choice as possible. For example, with chicken elaborates would enjoy

1. "chicken cut into strips with onions, peppers, cheese, tomatoes, sour cream"
2. "baked into a pot pie"
3. "with french fries or baked/mashed potatos"

4. "coated in batter and spices and deep fried"
5. "fresh and slow roasted"
6. "as a chicken salad"
7. "char grilled"
8. "marinated"
9. "with a wine sauce"
10. "with red pepper spice"

Imaginers

The optimal experience for imaginers focuses on the brand, the company in which the food is eaten, the emotions, the setting, or all of these. Imaginers are interested in the food itself and everything around it. This is the group that will appreciate a story to accompany the food. Imaginers want more from the experience than the food itself. The food is only a small part of what will make the imaginers happy. For example, in the food category of chicken an imaginer would enjoy fresh chicken, premium chicken, eaten with family and friends, as a special treat. Imaginers look for the chicken to be good tasting, but having people around them, feeling like it's a special treat, makes the experience even better.

Creating Innovation Using the Crave It! Study

Innovation typically happens at the outer edges where systems overlap. Using this idea, one can take a category like cheesecake, which is highly craved, and use it to understand how to improve the craving for another unrelated category like mashed potatoes. This type of innovative combination of two product categories by using the It! studies was pioneered by Christensen and Foley at Frito Lay and has been called the *Christensen-Foley method* for innovative recombination (Maier et al. 2003).

As an example, when looking at the top-rated elements craved for cheesecake and for mashed potatoes (Table 21.7), we find these key drivers:

1. Texture: dual texture for cheesecake and single texture for mashed potatoes
2. Melt in your mouth
3. Homemade
4. Real ingredients
5. Flavor varieties
6. Premium quality
7. Classic taste

By working jointly with the attributes of potatoes and cheesecake, the developer begins to realize that innovative mashed potato products may include an ability to deliver multiple textures and or additions of flavors. The best mashed potato products would also have "melt in your mouth" textures, be close to homemade (this is difficult to execute), and have real ingredients, premium quality indicators, and classic tastes. Knowing this kind of information takes a good product developer to a different level of insight in terms of what could or should be delivered to the market place from this food product.

Porting the It! Approach Beyond the United States: The Eurocrave! Database

Eurocrave! represents the extension of the Crave It! study to three major European countries: the United Kingdom, France, and Germany. The objective was to develop a database similar to that developed in the United States, but with local elements. Since the smells and tastes of foods are introduced at childbirth, the selection and acceptance of flavors, ingredients, and preparation methods are strongly linked to the culture and tradition of families, population groups, and geographic origins (Prescott and Bell 1995). As a result, food preference and acceptance are culturally determined and vary significantly from region to region. The results found for the United States may not hold up in Europe or elsewhere.

Country-to-country differences become increasingly important when we recognize the increase in the number of global food and beverage companies (Bech-Larsen et al. 2001; Steenkamp and Hofstede 2002). In today's market environment, food-science research tends to be performed in one central

Table 21.7. Integrated table for cheesecake and mashed potatoes that shows winning elements of each

Cheesecake that melts slowly to release delicate, *intense flavor* and has a *rich, silky texture* that *just melts in your mouth* . . . so sinful!	19
Smooth, fluffy cheesecake . . . with a *light, creamy flavor and texture* and a graham cracker crust	16
A *smooth, dense* slice of New York Style cheesecake	14
Real cheesecake *made with ingredients like eggs, cream cheese*, sugar, vanilla, and lemon juice	13
Creamy mashed potatoes that *melt in your mouth* as you eat them	11
With whipped butter melting in the center	10
Rich, creamy cheesecake . . . *swirled together* with ribbons of chocolate chips in a chocolate *crust*	10
Homemade	9
Prepared just to your liking . . . add whatever your heart desires	8
Blended with butter, garlic, salt and pepper . . . then whipped	8
Premium quality . . . the best Cheesecake in the whole world	8
Made from real Idaho potatoes	7
Premium quality . . . the best Mashed Potatoes in the whole world	7
Classic taste . . . the way you remember it	7
Reminds you of great home cooking	6

Italicized words are appropriate for either cheesecake or mashed potatoes. Data courtesy of It! Ventures, Inc.

research and development (R&D) site for implementation on a global scale. Therefore, researchers in one geographic location are expected to have a profound understanding of the needs, wants, and preferences of consumers from across the world in order to aid in the product development and marketing of successful new products. Furthermore, sensory scientists are expected to track global changes in brand awareness, product usage, and product quality that might influence the demand and profitability of existing products.

Specifics of Eurocrave! and the Database

Eurocrave! was designed on the pattern used for Crave It! previously discussed. The European studies were designed both to be appropriate for three European countries (France, Germany, and the United Kingdom) and for comparison with the data from the United States. Thus, the data were both local and global. Furthermore, as in the Eurocrave study, segments or respondents were identified based on their pattern of utilities.

In total, 6768 consumers participated in the study across France ($n = 2762$), Germany ($n = 2174$), and the United Kingdom ($n = 1832$). More women than men participated. This trend follows the phenomenon identified in other Internet-based conjoint analysis tests (Cappuccio et al. 2002) and has three possible explanations:

1. Women might be more likely to have the time to respond to these surveys.

2. Women might be more interested in voicing their opinions about food items because they comprise the population who are most frequently doing the grocery shopping.

3. Women most frequently visit the Internet sites that provide links to the Internet-based studies (Luckow et al. 2003b).

Common Elements Translated into the Local Language

Where possible the same products were tested across the different countries. The products had to be familiar ones to the respondents. The elements were translated locally by a native speaker of the language. This is important because in concept work sometimes simple nuances can dramatically change the meaning (see Chapter 8 on transnational research). Table 21.8 gives a sense of the nature of the elements for French-speaking and English-speaking respondents.

Single Country-based Segments: Tomato Soup in the United Kingdom

Just as the Crave It! study suggested three segments in the United States, the Eurocrave! study revealed several segments, although for a single product these segments were not necessarily the same from country to country. To make the data more concrete, consider soup in the United Kingdom. First, let us start with the total panel, which reveals the following pattern of utilities for the different elements (Table 21.9):

Table 21.8. Comparison of four elements about tortilla chip flavor given to the English-speaking (United Kingdom) respondents compared with French-speaking respondents

Tortilla chips with cheese flavor	Des chips Mexicaines piquantes et épicées
Tortilla chips with different flavors, e.g., vegetable, onion, cheese	Des chips Mexicaines aux différentes saveurs, au légume, aux oignons ou au fromage
Tortilla chips seasoned with a mild spice blend	Des chips Mexicaines assaisonnées avec un mélange d'épices doux
Tortilla chips, hot and spicy	Des chips Mexicaines aromatisées au fromage

Table 21.9. Strong-performing concept elements for soup tested within the United Kingdom

	Utility
Total panel: Additive constant = 21	
Hot chicken soup with juicy chunks of chicken, noodles or vegetables	15
With a crispy roll or a slice of fresh bread	14
Tomato soup with a dash of real cream	11
When it's cold outside, a cup of hot soup is warm and inviting	10
When you think about it, you have to have it . . ., and once you have it you want more	8
Classics (54%): Additive constant = 39	
Hot chicken soup with juicy chunks of chicken, noodles or vegetables	33
Tomato soup with a dash of real cream	18
With a crispy roll or a slice of fresh bread	15
Minestrone, original Italian soup with vegetables	14
Progressives (27%): Additive constant = 20	
Creamy mushroom soup	22
Soup from sun ripened tomatoes	17
Tomato soup with a dash of real cream	17
With a crispy roll or a slice of fresh bread	14
Loyalists (19%): Additive constant = 15	
Creamy hot chicken soup	13
Hot chicken soup with juicy chunks of chicken, noodles or vegetables	12
Premium quality . . .. that great traditional taste	12

1. *The best element.* "Hot chicken soup with juicy chunks of chicken, noodles or vegetables" is a strong-performing element. Using this element as a part of the product and market concepts for chicken soup and ancillary products would be a strong message.

2. *Other winning elements.* "Tomato soup with a dash of real cream," "when it's cold outside, a cup of hot soup is warm and inviting," and "with a crispy roll or a slice of fresh bread" are also strongly performing concept elements that would attract British consumers.

3. *The three segments.* Within the data for the United Kingdom exist three distinct and different consumer segments. The three segments could be labeled *classics* (who follow trend of population average, love all flavors of soups, like images of soup as "hot" and "original," and have brand favorites), *progressives* (who are willing to try new, creative flavors of soup, but dislike conventional chicken soups and broths), and *loyalists* (who are enthusiastic about any variety of

chicken soup, but dislike flavors and varieties of soup that do not contain chicken).

4. *Product development opportunities in light of the segmentation.* Companies that wish to enter the UK soup market may choose several options. They can compete for the average consumer population by catering to the needs of the classics. This would be difficult because this group responds strongly to current brands. However, by using the strongest, most significant concept elements, "hot chicken soup with juicy chunks of chicken, noodles, or vegetables," "tomato soup with a dash of real cream," and "with a crispy roll or a slice of fresh bread," companies can develop a variety of products that capture a segment of the UK classic consumers.

Marketing opportunities

Existing companies that would like to develop new soup products, or new companies that are hoping to capture a small, loyal niche of the UK soup market, may wish to cater to the progressives or the loyalists. Based on their self-profiling classification questionnaire these segments describe themselves as not brand aware and say that they are more willing to try new things. Progressives say that they are attracted to interesting and novel flavor developments that draw their attention when "flagged" on the grocery shelf. In contrast, loyalists say that they would try new soup brands that heavily advertised a variety of high-quality chicken soup options.

Using the Eurocrave Database for a Cross-country Comparison of Cola

In some cases the same product was tested in Eurocrave! across three countries and tested separately with some overlapping elements in the United States with Crave It! The data enable researchers to compare the utility values of common elements across four countries. One particularly interesting dataset enabling this multinational comparison is cola,

an exceptionally popular carbonated beverage worldwide (Luckow et al. 2003a). Table 21.10 presents an example of this comparison across countries and provides a sense of the distribution of utilities by country (e.g., range of utilities and the number of positively performing versus negatively performing elements). Many of the elements describing physical, flavor variety, brand, and emotional characteristics were tested in each of the four countries. Elements describing usage occasion and product benefits had to be adjusted to cater to the different needs of each participating country and therefore are not included in this analysis.

Respondents from each country connected very strongly to the cola category. Cola has become such a staple in the diet and is consumed at such a high frequency that consumers have strong opinions about the elements describing colas. Furthermore, both European and American respondents are very capable of providing negative feedback about concepts that do not fit in with their definition of a good, refreshing, cola drink.

Table 21.10. Cola data from Eurocrave and Crave It! databases: utilities for elements dealing with product feature, brand, and emotional connection

	France	Germany	UK	USA	Average	Range
Additive constant	*55*	*25*	*42*	*42*	*41*	*30*
Number of negative elements	21	20	19	20	20	2
Number of positive elements	15	16	17	16	16	2
Product features						
Cola . . . carbonated and sparkling, just the						
right amount of taste and bubbles	6	12	4	4	7	8
Diet cola . . . refreshment without the calories	−6	−5	−18	−15	−11	12
With a slice of lemon	5	12	−2	−16	0	16
Brands						
Coca-Cola brand	15	16	9	7	12	9
Pepsi Cola brand	−15	2	3	−5	−4	18
Private label/store brand	−23	−11	−16	−13	−16	−12
Emotional elements						
When you think about it, you just have to have it,						
and once you have it, you can't stop eating it.	−1	−2	3	0	0	4
A joy for your senses . . . Seeing, smelling, tasting.	1	2	0	4	2	4
To be enjoyed while surrounded by family and friends.	−2	2	−2	−1	−1	4

In general, brand elements were the most influential in determining consumer preference, which stands in contrast to the weak performance of brands in general (see Chapter 23). Overall, all consumers preferred globally recognized cola brands from large cola companies to the regional private label and store brands. The Coca-Cola brand generated more rated craveability than did any other cola brand. However, other categories of elements contributed to craving, giving researchers a sense of the country-to-country differences, even for the same product:

1. German consumers were positively affected by physical cues describing "carbonated," "sparkling" colas with "bubbles."

2. French and German consumers were interested in colas containing lemon flavors.

3. Finally, US and UK consumers responded modestly to the emotional elements.

4. UK participants agreed that the thought of cola led them to craving colas.

5. Americans felt that cola was a "joy for the senses . . . seeing, smelling, tasting."

Issues and Opportunities in Large-scale Databases Such as Crave It! and Eurocrave

Although researchers recognize the value of cross-national research, there is little in the way of formalized databases that compare respondents in different countries on their preferences and their reaction to concepts across many product categories. Many multinational companies commission studies that cover different countries in an attempt to understand generalities and uncover differences. These studies are done with commercial goals and thus focus on specific issues. Companies are affected by internal politics and monetary considerations, however. Despite the desirability of large-scale databases to understand consumers, the internal environment of a company militates against these types of databases, for the following five reasons:

1. *There is no immediate payout.* Companies need payouts to justify research. The brutal competitive environment requires that any investment pay out quickly. Occasionally, there are long-term investments, but usually in technology rather than in consumer knowledge.

2. *The history of consumer insights is a history of focus groups and research reports rather than databases about consumer mindsets.* The corporate culture of multinationals is accustomed to large-scale databases only for market-level data, which are too expensive to collect for a single company. Market-level data, such as store sales, can be sold to many companies without impacting the strategic decisions of a company. Market-level data is raw information, and companies are accustomed to paying for raw information.

3. *The mind-set of consumers is considered to be a strategic asset.* Companies want to know about these mind-sets. The strategic asset is typically considered on an ad hoc basis. The research community is accustomed to providing this information on a need-to-obtain basis rather than on a database.

4. *Companies have not thought about the consumer habits and responses to concepts in terms of databases.* The idea is novel. It takes 3–10 years before novel ideas become real to buyers. Early adopters may try out the idea, but most multinationals need time for the database idea to achieve a level of reality. Typically, once the idea becomes real and a purchasing code is assigned to the multinational database, the data become real and are used

5. *In today's action-oriented world, clients demand immediate* actionability *from data that they buy.* With multinational databases on the consumer mind, one has a hard

time convincing buyers to purchase a product and then changing the internal culture so that the buyers actually know what to do with the data.

The Crave It! and Eurocrave! studies provide product developers and marketers with a new type of information that may answer some of the issues just raised. Up to now, studies about food concepts and profiling one's behavior typically comprised *one-off* studies, usually commissioned by a corporation and left to reside in the corporate knowledge vaults. University studies have also been commissioned and published, but usually of smaller scale than corporate studies and usually with convenience samples of students and others in the university environment. The data presented here for the Eurocrave and Crave It! databases provide a different type of resource, for the following reasons:

1. The scope is far larger than the typical one-off study

2. The data comprises both attitudinal information (i.e., self-profiling as respondents describe themselves) and responses to communications from the conjoint analysis portion. These two types of information can be studied and mined separately or in conjunction with each other.

3. The self-profiling goes into a great deal of depth, including geodemographics, attitudes, and behaviors.

4. The concept elements also go into a great deal of depth and breadth, covering a wide range of topics, from product descriptions to emotions to situations to benefits.

5. The comparability of data and trends across products, countries, and time provides a powerful additional benefit, because the classifications are the same and the structures of the concept elements are the same.

References

Bech-Larsen, T., Grunert, K.G., and Poulson, J.B. (2001). The acceptance of functional foods in Denmark, Finland, and the United States: a study of consumer's conjoint evaluations of the qualities of functional foods and perceptions of general health factors and cultural values. MAPP Working Paper 73. Aarhus: Department of Marketing, Aarhus School of Business.

Cappuccio, R., Krieger, B., Katz, R., Itty, B., and Moskowitz, H. (2002). Coffee: development of product concepts for drinking venues from first principles. Foodservice Technology (submitted).

Hill, A.J., and Heaton-Brown, L. (1994). The experience of food craving: a prospective investigation in healthy women. Journal of Psychosomatic Research, 38: 801–814.

Jensen, R. (1999). The Dream Society: How the Coming Shift from Information to Imagination Will Transform Your Business. New York: McGraw-Hill.

Luckow, T., Aarts, P., and Moskowitz, H.R. (2003a). A comparison of purchasing habits and sensory preferences for cola consumers across France, Germany, the United Kingdom, and the United States. Journal of Food Technology, 1: 84–96.

Luckow, T., Moskowitz, H.R., Beckley, J., Hirsch, J., and Genchi, S. (2003b). The four segments of yogurt consumers: preferences and mind-sets. Journal of Food Products Marketing (in press).

Maier, A.S., Christensen, K., Foley, M., and Beckley, J. (2003). Innovation by fusing multiple winning elements across product categories. Paper presented at the Fifth Pangborn Conference, Boston.

Moskowitz, H.R., German, J.B., and Saguy, I.S. (2003). Unveiling health attitudes and creating good for you foods via informatics and innovative web-based technologies. (In review.)

Pelchat, M.L. (1997). Food cravings in young and elderly adults. Appetite, 28: 103–113.

Prescott, J., and Bell, G. (1995). Cross-cultural determinants of food acceptability: recent research on sensory perceptions and preferences. Trends in Food Science and Technology, 6: 201–205.

Steenkamp, J.B.E.M., and Hofstede, F.T. (2002). International market segmentation: issues and outlook. International Journal of Research in Marketing, 19: 185–213.

Chapter 22

Highlights and Insights from the Drink It! Study

Introduction

The drink industries have come a long way in the last 100 years. Before then, beverage choices were relatively limited. However, the rapid growth in food technology has changed this and created a new world of choice. Today, new beverage products are continually being developed with each aiming to capture the imagination of consumers and grab part of their daily consumption. The goal to increase one's business by changing people's choices and even drinking habits is called by marketers increasing *share of stomach*. The consequence of this objective and related events is that consumers throughout the world now have a wider range of drinks to choose from than ever before. The recent growth in high-energy drinks is a prime example of how a new product can quickly capture significant proportions of the beverage market.

Whereas this ever-increasing, ever-more-competitive market is great news for consumers, it makes the life of the producers more difficult. No longer can any company in the beverage industry simply assume that its market share will remain unchanged. Consumer tastes can change, and competition may also erode a company's share of the market. It is with this in mind that producers require a deep understanding of consumers so as to avoid either of these outcomes.

The Drink It! database, part of the It! databases, was created in the same way as the other data from the It! studies (Beckley and Moskowitz 2002). Chapters 20 and 21 present the It! approach, so the method need not be repeated here. The emphasis in the Drink It! database was on beverages rather than on foods. Specifically, the studies investigated the types of product elements and the emotional outcomes of drinking beverages. Given this background, let us immediately deal with the data from the Drink It! database across 30 different beverages. The analysis focuses on segments rather than on the total panel, because there is a great deal to be learned from the segmentation results about the consumer mind-sets and how they vary by beverage.

Nonalcoholic Drinks

Coffee

Coffee is one of the most popular beverages in the world. In fact, world coffee sales top $6.8 billion per year (Gobi International 2001). Whereas coffee consumption has long been popular, it still continues to grow, especially in the American market (Cappuccio et al. 2002). This is typified by the success of the Starbucks chain of stores: there are now more than 6000, with 5000 of them in the United States (Starbucks 2002).

Before we interpret the available data regarding what is important to the consumers of coffee in terms of their decisions to buy or not buy a certain product, it is instructive to understand the variety of situations in which coffee is consumed. Since coffee contains caffeine, it's frequently used as a stimulant; that is, coffee is consumed in the morning to kick-start the day or to prevent tiredness during the day or night. It is often a social drink that consumers share with friends or colleagues and may be served in a variety of forms. While at home or the workplace, most consumers simply drink a straight serving of coffee with milk or cream added. However, smaller coffeehouses offer consumers a much wider range of options, including cappuccino, café latte, and mocha to name a few. These options are further differentiated by variations in the types of additions to coffee, such as milks or flavors. One question of interest that the Drink It! data answers is the proportion of the population that is interested in these specialty brews. Another is the type of features in coffee beverages that excite consumers.

In Table 22.1, representative elements and their utility values are presented for four coffee market segments that emerged from segmenting the respondents on the basis of their utility values. These elements were selected to give a sense of the range of acceptance and rejection generated by the coffee elements. If we first look at the additive constants, it is clear that all coffee consumers show high basic interest in coffee; that is, they are highly motivated to want to purchase the product. Segment 4 shows significantly lower basic interest than do respondents in the other three segments. Although their utility values are higher for many elements, this segment, labeled the *variety seekers* (similar to the elaborates in the Crave It! study), need more convincing to be interested in buying coffee products. However, if we look at the base size, it is clear that this group comprises less that 8% of the population.

Now we must try to understand the mindsets of the different types of consumers. These four segments of consumers have be labeled as *no frills* (segment 1), *traditionalists* (segment 2), *impressionables* (segment 3), and *variety seekers* (segment 4). These names have been chosen to parallel similar mindsets identified previously for a whole range of products (e.g., see Hughson et al. 2002):

1. *No frills.* The first segment is the largest and makes up almost 40% of the total sample. These consumers are called no frills because they show high basic interest in coffee but, other than that, no elements significantly add

to their interest. In fact, concept elements were much more likely to detract from their product acceptance. Any complex description of the product or even the Starbucks brand strongly reduces the liking of the coffee for these consumers. Thus, these respondents seem to enjoy coffee, but only basic coffee. They don't like their coffee to be fancy. It seems likely that the no-frills consumers probably consume coffee primarily for its stimulant value rather than based on trends or enjoyment.

2. *Traditionalists.* These consumers constitute the second largest group and comprise almost 35% of the sample. They also show very high basic interest in the product. However, as opposed to the no-frills consumers, a number of features were able to add to acceptance, albeit only modestly. The theme of these elements is the traditional forms and types of coffee. Traditionalists want their coffee to be natural and fresh. What they do not like is images that seem to be new or untraditional types of coffee. For example, iced coffee, decaffeinated coffee, coffee flavored with caramel, and even branded coffees, such as from Starbucks and Nescafe, reduced consumer acceptance. They want a simple cup of coffee served in the traditional manner. It seems likely that their apparent dislike of Nescafe and Starbucks suggests that traditionalists like to make their own coffee or buy it at specialist coffee shops rather than at a supermarket or coffee chain.

3. *Impressionables.* These consumers are "sensitive to and responsive to other emotional and brand benefits" (Moskowitz et al. 2004). They definitely show a response to emotive messages, as is illustrated by the positive coefficients for elements such as "refreshing," "perfect," "hearty," and "fresh." However, they fail to respond to brand benefits. In fact, it appears that the response of impressionables to brands across all the segments is more negative than positive. This is quite unexpected.

4. *Variety seekers.* These consumers like products with sophisticated and variety-seeker product descriptions. Variety seekers

Table 22.1. The four concept-response segments for coffee

		Segment			
		1	2	3	4
	Base size	*90*	*78*	*40*	*18*
	Additive constant	*69*	*69*	*66*	*43*
	Segment 1: No frills				
E28	From Starbucks	−6	−7	3	−3
E16	With all the milk, cream, and toppers you want . . . cinnamon, nutmeg, chocolate sprinkles, sugar cubes, whipped cream . . . whatever	−18	−4	5	20
E08	Brewed coffee blended with cream and caramel then topped with heavy whipped cream	−35	−6	−1	25
	Segment 2 (Traditionalists)				
E02	Fresh coffee made from 100% Columbian coffee beans	2	9	6	7
E01	Fresh ground and brewed coffee	4	9	5	5
E15	100% natural coffee beans	2	5	2	4
E08	Brewed coffee blended with cream and caramel then topped with heavy whipped cream	−35	−6	−1	25
E28	From Starbucks	−6	−7	3	−3
E31	From Nescafe	−2	−8	−6	2
E04	Decaffeinated whole bean coffee for those who want all the taste and none of the caffeine	−23	−24	−22	19
E03	Cool and refreshing Iced Coffee	−24	−36	11	2
	Segment 3 (Impressionables)				
E09	A hearty cup of Cappuccino, frothy with foam and with the rich taste of espresso	−20	−1	12	21
E03	Cool and refreshing Iced Coffee	−24	−36	11	2
E07	Coffee and milk . . . blended just right for the perfect latte	−16	−4	7	14
E02	Fresh coffee made from 100% Columbian coffee beans	2	9	6	7
E31	From Nescafe	−2	−8	−6	2
E29	From Folgers	2	0	−6	−3
E30	From Maxwell House	0	−2	−7	5
	Segment 4 (Variety Seekers)				
E08	Brewed coffee blended with cream and caramel then topped with heavy whipped cream	−35	−6	−1	25
E09	A hearty cup of Cappuccino, frothy with foam and with the rich taste of espresso	−20	−1	12	21
E16	With all the milk, cream, and toppers you want . . . cinnamon, nutmeg, chocolate sprinkles, sugar cubes, whipped cream . . . whatever	−18	−4	5	20

Data courtesy of It! Ventures, Inc.

show a slightly lower basic interest in coffee than do the other consumer segments, although it is not low in comparison with other products. No elements strongly detract from the acceptance of coffee. Variety seekers like a range of flavors and sensations in their coffees even if those sensations are far from traditional, such as the addition of caramel.

We can get a visual sense of the three major segments by looking at the stock visuals for coffee presented in Figure 22.1. Often such stock pictures drive home the nature of the segments, as they did for ice cream in Crave It! (see Chapter 20).

It is interesting to note that the proportion of variety seekers in the coffee sample is much lower than for the majority of other beverages in the Drink It! study. This low proportion and the fact that the no-frills consumers and traditionalists are the most popular segments, as

Traditionalists

Variety Seekers

Impressionables

Figure 22.1. Visual examples of coffees for the three major concept-response segments: traditionalists, variety seekers, and impressionables.

well as the minimal effect of brand on consumer acceptance, suggest that coffee drinkers are generally quite conservative. They generally like their coffees to be traditional and simple. Anything that differs from that conservatism is likely to alienate a large proportion of consumers. Also, coffee drinkers are generally more likely to lose interest rather than gain interest in a product when some new element is added, as illustrated by the overall element values for all the segments combined (Table 22.2). This would suggest that, to market to the largest proportion of coffee consumers, a coffee product should be presented as nothing more than a simple coffee made in the traditional way.

Tea

Tea is in many ways very similar to coffee. Tea contains caffeine and, thus, often is used as a stimulant. It also has long-established traditions similar to those for coffee, suggesting that consumer preferences may be largely conservative, as they are for coffee drinkers. Tea is also similar to coffee in that it is often consumed in social situations. Thus, it seems likely that the segments found in the coffee and tea markets are quite similar. To some extent this is true, but there are also important differences. Table 22.3 lists the results. The concept-response segmentation revealed three clear segments:

1. *The variety-seeker segment.* This segment is very similar to that found in the coffee study and constitutes a relatively small proportion of the population. Their basic interest in tea, although high, is not as high as the basic interest shown by the other two segments, meaning that the variety seekers must be encouraged to a greater extent than the other consumers to be interested in a product. As is commonly found for variety-seeker segments, they are interested in products with complex flavor and texture descriptions. Such elements increase consumer interest substantially. Importantly, none of the elements reduced the consumers' interest, showing that any product

Table 22.2. Representative element utilities for the whole sample

E01	Fresh ground and brewed coffee	6
E02	Fresh coffee made from 100% Columbian coffee beans	6
E33	Multiserve containers . . . so you always have enough!	−5
E09	*A hearty cup of Cappuccino, frothy with foam and with the rich taste of espresso	−5
E28	From Starbucks	−5
E31	From Nescafe	−5
E07	*Coffee and milk . . . blended just right for the perfect latte	−6
E16	*With all the milk, cream, and toppers you want . . . cinnamon, nutmeg, chocolate sprinkles, sugar cubes, whipped cream . . . whatever	−6
E34	*Now also available in a self-heating can	−8
E08	*Brewed coffee blended with cream and caramel then topped with heavy whipped cream	−14
E04	Decaffeinated whole bean coffee for those who want all the taste and none of the caffeine	−20
E03	*Cool and refreshing Iced Coffee	−20

Novel or nontraditionalist features of the coffee are prefaced by an asterisk. Data courtesy of It! Ventures, Inc.

details are unlikely to increase the interest of the variety-seeker segment.

2. *The no-frills segment.* These respondents, who represent 27% of the sample, display high basic interest but, beyond that, respond only to the simplest concept elements for tea. Any elements that describe any kind of fancy or complex tea generally reduces acceptance.

3. *The health-conscious segment.* This segment, which represents over 60% of the respondents, does not appear among coffee drinkers, at least noticeably. It seems likely that the appearance of this segment is based on the recent evidence showing the health benefits of tea consumption (e.g., see Trevisanato and Kim 2000). The segment shows very high basic interest in tea. The elements that interest the respondents in this segment deal with the tea being natural and healthy. They are not interested in a tea with healthy effects, but rather those teas that are healthful, without chemical alteration or addition. This is illustrated by the fact that elements talking about the addition of Saint-John's-Wort, *Echinacea*, and decaffeinated teas decrease interest, whereas elements such as chamomile tea and peppermint tea increase consumer interest. These respondents also find statements mentioning milk also negative. The results suggest that this segment responds to tea with milk as being unhealthy.

The fact that this segment is found among tea consumers but not coffee consumers, and that it makes up such a large proportion of the tea market, suggests that, in at least one way, consumers of coffee and tea differ from each other. More specifically, product choices for many tea consumers are driven by how healthful and natural the tea appears, as opposed to the choices made by coffee consumers. Coffee buyers appear to place less value on health issues when choosing a coffee product. One related finding of interest from a similar study showed that the number of health problems that consumers faced did not appear to covary with the respondent being a health-conscious consumer or not; that is, both self-profiled "health conscious" and self-profiled "not health conscious" respondents choose products based on their perception of health-related consequences (Hughson et al. 2003). The inverse of this finding suggests that the health-conscious consumers are probably not less healthy than are the other respondents; rather, they merely try harder to attend to the health-related outcomes of the drinks they consume.

Cola Drinks

Whereas cola beverages appear quite dissimilar to coffee and tea, they do share one important component: most cola drinks contain

Table 22.3. Segments identified among tea consumers

		Segment		
		1	2	3
	Base size	*17*	*74*	*143*
	Additive constant	*50*	*64*	*67*
	Segment 1: Variety seekers			
E06	Wellness teas with ginger, ginseng, Saint-John's-Wort or *Echinacea* made with pure spring water	21	−28	−6
E08	Teas with frothy milk, sugar, and spices for that feeling of the exotic	21	−22	−28
E09	Teas infused with exceptional fragrance, superb flavors and a hint of spice	17	−23	2
E01	Blended black teas . . . strong, potent and dark	16	2	−3
E16	Mixed berry, honey, lemon ginseng, cinnamon apple, chamomile, peppermint, lemon, orange . . . whatever you're looking for	15	−17	5
E04	Teas with a milky flavor to counter balance the flavor strength of the leaves	14	−16	−26
E03	Black teas with bold fruit flavors	14	−20	−2
E11	With a spritz of lemon and a cookie	13	−11	−7
E05	Teas . . . high in antioxidants, vitamins and minerals	11	−6	5
E02	Green teas with a delicate, clean taste and pale green color	10	−14	3
E07	Herb teas with naturally occurring flavonoids and antioxidants	9	−17	3
E27	It quenches THE THIRST	9	1	0
E13	Premium quality	7	2	3
E19	Quick and fun . . . ready to drink	6	0	−1
E32	Made from exotic teas . . . like Roobois, Mate, Pu Erh, Honeybush	5	−13	−8
E24	Looks great, smells great, tastes delicious	5	3	3
E23	Now you can escape the routine . . . a way to celebrate special occasions	5	0	0
	Segment 2: No frills			
E28	From Lipton	4	5	1
E10	Drinking hot tea is so inviting	0	5	3
E05	Teas . . . high in antioxidants, vitamins and minerals	11	−6	5
E12	Decaffeinated or diet . . . whatever you need	−2	−8	−5
E11	With a spritz of lemon and a cookie	13	−11	−7
E32	Made from exotic teas . . . like Roobois, Mate, Pu Erh, Honeybush	5	−13	−8
E02	Green teas with a delicate, clean taste and pale green color	10	−14	3
E04	Teas with a milky flavor to counter balance the flavor strength of the leaves	14	−16	−26
E16	Mixed berry, honey, lemon ginseng, cinnamon apple, chamomile, peppermint, lemon, orange . . . whatever you're looking for	15	−17	5
E07	Herb teas with naturally occurring flavonoids and antioxidants	9	−17	3
E03	Black teas with bold fruit flavors	14	−20	−2
E08	Teas with frothy milk, sugar, and spices for that feeling of the exotic	21	−22	−28
E09	Teas infused with exceptional fragrance, superb flavors and a hint of spice	17	−23	2
E06	Wellness teas with ginger, ginseng, Saint-John's-Wort or *Echinacea* made with pure spring water	21	−28	−6
	Segment 3: Health conscious			
E05	Teas . . . high in antioxidants, vitamins and minerals	11	−6	5
E16	Mixed berry, honey, lemon ginseng, cinnamon apple, chamomile, peppermint, lemon, orange . . . whatever you're looking for	15	−17	5
E12	Decaffeinated or diet . . . whatever you need	−2	−8	−5
E34	Now available in a self-heating can	−3	−3	−5
E06	Wellness teas with ginger, ginseng, Saint-John's-Wort or *Echinacea* made with pure spring water	21	−28	−6
E11	With a spritz of lemon and a cookie	13	−11	−7
E32	Made from exotic teas . . . like Roobois, Mate, Pu Erh, Honeybush	5	−13	−8
E04	Teas with a milky flavor to counter balance the flavor strength of the leaves	14	−16	−26
E08	Teas with frothy milk, sugar, and spices for that feeling of the exotic	21	−22	−28

Courtesy of It! Ventures, Inc.

caffeine unless specifically formulated not to and, as a consequence, are sometimes used as stimulants. However, colas are also quite different. Cola drinks have little of the tradition and perceived sophistication of coffee and tea. Another important difference is that coffee and tea drinkers are generally past their teenage years, whereas the cola drinkers are found in a variety of age groups.

From the concept-response segments illustrated in Table 22.4, it is clear that there is one very important difference between those identified for colas and those identified for coffee and tea, the two other beverages examined so far. For all but one small segment, brand was the most important positive element for consumer acceptance. Before we examine this further, let's take a closer look at the five market segments:

Segment 1: The classic Coca-Cola drinkers. For them, only two elements are significantly positive: the brand "Coca-Cola" and an "ice cream float." These respondents report from their conjoint results that any other branded cola drink is unacceptable. Segment 1 finds diet colas to be unacceptable, suggesting that these consumers are interested only in the standard Coca-Cola. This group also shows very high basic interest in the product. Segment 1 constitutes more than 23% of the sample.

Segment 2: The Pepsi drinkers. This group made up by far the highest proportion of respondents and included over half the sample. They, like the Coca-Cola drinkers, show high basic interest in consuming cola drinks. They, too, show great brand loyalty in that almost any brand that is not Pepsi significantly decreases their liking for the product. Diet drinks and low-calorie drinks are again unpopular elements.

For elements 3–5, we return to the mix of segments that emerged for coffee and tea. However, brand is still an important element for two of these three segments, and they constitute only small proportions of the market.

Segment 3: The traditionalists. This segment is relatively small, constituting 14% of the respondents. They show positive reaction to traditionalist styles of cola from Coca-Cola and Dr Pepper and also respond positively to a cola drink described as traditionalist or as being served in a traditional manner, such as "A thick slushy of cola and ice." On the other hand, this segment dislikes other brands (e.g., Pepsi). It seems that these brands are seen as less traditional. Perhaps this may be due to companies such as Pepsi purposely marketing their product as a newer style of cola drink. Again, diet drinks fail to impress this segment.

The fourth and fifth segments are both statistically very small and together make up only around 10% of the sample.

Segment 4: The impressionables. They show the lowest basic interest in the product, illustrating the need for product elements to entice these consumers. This segment shows not only high interest in the Pepsi and Dr Pepper brands, but also other nonfood, emotional elements that improve the experience, such as the beverage being "refreshing" and "relaxing."

Segment 5: The health conscious. This segment shows the strongest positive response to diet colas and a one-calorie drink.

The data from the cola-drink study are quite unique in that utilities are largely dominated by brands and the way those brands are marketed. It is clear that Pepsi and Coca-Cola are the most important brands, illustrated by the proportions of respondents in the Coca-Cola-drinking and the Pepsi-drinking segments.

From the segmentation it is clear how different mind-sets are attracted to various types of cola beverages. For example, drinkers who are traditionalists are more likely to respond to elements of what we would call a traditionalist-style beverage presented in the traditional manner. They

Table 22.4. Segments identified among cola consumers

		Segment				
		1	2	3	4	5
	Base size	*61*	*137*	*37*	*12*	*15*
	Additive constant	*67*	*68*	*59*	*49*	*65*
	Segment 1: Classic, Coca-Cola drinkers					
E32	From Coca-Cola	10	0	10	2	−1
E09	An ice cream float-cola, ice cream . . . chilled and tasty	5	−8	7	−5	−7
E06	Diet cola with a slice of lemon . . . the world's most perfect drink!	−14	−34	−17	12	17
E31	From Pepsi	−18	5	−14	22	7
E30	From Dr Pepper	−18	−7	14	13	−5
E29	From Tab	−24	−19	−20	−5	−1
E28	From RC Cola	−24	−6	−9	−9	3
	Segment 2: Pepsi drinkers					
E31	From Pepsi	−18	5	−14	22	7
E11	Fortified with important vitamins and minerals for your body	0	−6	−3	−12	−9
E28	From RC Cola	−24	−6	−9	−9	3
E12	With twice the jolt from caffeine . . . gives you just the added energy you need	−3	−7	−6	−15	−28
E30	From Dr Pepper	−18	−7	14	13	−5
E09	An ice cream float—cola, ice cream . . . chilled and tasty	5	−8	7	−5	−7
E16	Flavored colas . . . cherry, vanilla, lemon . . . whatever you're looking for	−1	−10	5	2	−2
E04	Cola . . . The dark brown color, faint smell of vanilla, and bubbles tell you, you have real cola	1	−12	2	−10	−19
E05	Cola . . . all the taste but only one calorie	3	−13	−4	−13	8
E29	From Tab	−24	−19	−20	−5	−1
E06	Diet cola with a slice of lemon . . . the world's most perfect drink!	−14	−34	−17	12	17
	Segment 3: Traditionalists					
E30	From Dr Pepper	−18	−7	14	13	−5
E32	From Coca-Cola	10	0	10	2	−1
E09	An ice cream float—cola, ice cream . . . chilled and tasty	5	−8	7	−5	−7
E01	A Traditionalist cola . . . just the way you like it	3	4	6	9	6
E07	A thick slushy of cola and ice	4	−1	6	5	1
E16	Flavored colas . . . cherry, vanilla, lemon . . . whatever you're looking for	−1	−10	5	2	−2
E02	Carbonated, sparkling Cola . . . just the right amount of taste and bubbles	3	3	5	6	−4
E10	Drinking Cola is so inviting	2	1	5	5	−6
E12	With twice the jolt from caffeine . . . gives you just the added energy you need	−3	−7	−6	−15	−28
E03	A perfect beverage . . . with breakfast, lunch, a break, or dinner	0	1	−7	−5	−5
E28	From RC Cola	−24	−6	−9	−9	3
E31	From Pepsi	−18	5	−14	22	7
E06	Diet cola with a slice of lemon . . . the world's most perfect drink!	−14	−34	−17	12	17
E29	From Tab	−24	−19	−20	−5	−1
	Segment 4: Impressionables					
E31	From Pepsi	−18	5	−14	22	7
E30	From Dr Pepper	−18	−7	14	13	−5
E18	So refreshing . . . you have to drink some more	1	2	2	12	−6
E06	Diet cola with a slice of lemon . . . the world's most perfect drink!	−14	−34	−17	12	17

(continued)

Table 22.4. Segments identified among cola consumers *(cont.)*

		Segment				
		1	2	3	4	5
E01	A Traditionalist cola . . . just the way you like it	3	4	6	9	6
E08	Cola . . . the perfect mixer for everything you drink	1	−2	1	7	−7
E22	Relaxes you after a busy day	1	2	−2	6	−4
E02	Carbonated, sparkling Cola . . . just the right amount of taste and bubbles	3	3	5	6	−4
E07	A thick slushy of cola and ice	4	−1	6	5	1
E10	Drinking Cola is so inviting	2	1	5	5	−6
E20	When you think about it, you have to have it . . . and after you have it, you can't stop drinking it	2	1	−1	−5	1
E23	A great way to celebrate special occasions	0	2	1	−5	−1
E09	An ice cream float—cola, ice cream . . . chilled and tasty	5	−8	7	−5	−7
E29	From Tab	−24	−19	−20	−5	−1
E03	A perfect beverage . . . with breakfast, lunch, a break, or dinner	0	1	−7	−5	−5
E35	Resealable single serve container . . . to take with you on the go	1	2	2	−6	2
E28	From RC Cola	−24	−6	−9	−9	3
E15	100% natural	0	0	2	−9	1
E04	Cola . . . The dark brown color, faint smell of vanilla, and bubbles tell you, you have real cola	1	−12	2	−10	−19
E36	With the safety, care and quality that makes you trust it all the more	1	0	1	−10	−3
E11	Fortified with important vitamins and minerals for your body	0	−6	−3	−12	−9
E05	Cola . . . all the taste but only one calorie	3	−13	−4	−13	8
E12	With twice the jolt from caffeine . . . gives you just the added energy you need	−3	−7	−6	−15	−28
	Segment 5: Health conscious					
E06	Diet cola with a slice of lemon . . . the world's most perfect drink!	−14	−34	−17	12	17
E05	Cola . . . all the taste but only one calorie	3	−13	−4	−13	8
E31	From Pepsi	−18	5	−14	22	7
E01	A Traditionalist cola . . . just the way you like it	3	4	6	9	6
E03	A perfect beverage . . . with breakfast, lunch, a break, or dinner	0	1	−7	−5	−5
E30	From Dr Pepper	−18	−7	14	13	−5
E10	Drinking Cola is so inviting	2	1	5	5	−6
E24	Looks great, smells great, tastes delicious	3	1	3	0	−6
E18	So refreshing . . . you have to drink some more	1	2	2	12	−6
E08	Cola . . . the perfect mixer for everything you drink	1	−2	1	7	−7
E09	An ice cream float—cola, ice cream . . . chilled and tasty	5	−8	7	−5	−7
E11	Fortified with important vitamins and minerals for your body	0	−6	−3	−12	−9
E04	Cola . . . The dark brown color, faint smell of vanilla, and bubbles tell you, you have real cola	1	−12	2	−10	−19
E12	With twice the jolt from caffeine . . . gives you just the added energy you need	−3	−7	−6	−15	−28

like Coca-Cola. Impressionables, who respond to elements that are non-food-related emotional benefits, also like Pepsi. Health-conscious consumers also find Pepsi to be the most attractive brand. Both traditionalists and impressionables like Dr Pepper. As is discussed later, this knowledge about the type of mind-set of consumers enables developers to target beverage creation and marketers to identify persuasive messages, as well as to recognize when brand name is a detriment.

Sports Drinks

Although widely consumed, sports drinks are relatively new to the beverage market. Generally sports drinks are formulated for energy:

they contain sugars, as well as vitamins and minerals. They were originally designed to aid athletic performance and the postcompetition recovery of athletes. Now the market for these drinks has grown. Today, they are often consumed as soft drinks. The segment results are presented in Table 22.5:

Segment 1: The health conscious. When we view the concept-response segments closely it

Table 22.5. Segments identified among consumers of sports drinks

		Segment			
		1	2	3	4
	Base size	*79*	*41*	*68*	*44*
	Additive constant	*59*	*60*	*72*	*53*
	Segment 1: Health conscious				
E08	Vitamins, minerals, and herbal energy to replace the essential body elements that are depleted during daily activities	13	4	−2	1
E04	A sports drink with all the vitamins, minerals, and energizing ingredients you need	11	4	−1	1
E03	A sports drink that's engineered to refuel your working muscles	11	3	−1	4
E07	Liquid hydration and a mix of energy releasing B vitamins	10	3	−1	−3
E06	An optimal combination of fluids, carbohydrates, and minerals designed to quench your body's deepest thirst	10	4	−1	1
E05	A sports drink . . . a light taste with complete balanced nutrition to help you stay healthy, active, and energetic	8	2	0	6
E09	A system of high performance beverages that fuse science with an understanding of human physiology	7	−2	−3	−1
E02	Liquid hydration and energy drink	6	2	−1	−1
E16	Orange Ice, Passion fruit, Citrus cooler, Fruit punch and all the ices or freezes you could ever need	5	5	1	21
E29	From Capri Sun	−6	6	−4	3
	Segment 2: Impressionables				
E30	From Gatorade	−1	17	9	2
E28	From Minute Maid	−4	9	3	4
E35	Resealable single serve container . . . to take with you on the go	1	7	2	3
E29	From Capri Sun	−6	6	−4	3
E31	From SoBe	−1	5	−13	1
E32	Gives you all the mental and physical stamina you need	4	5	1	−2
E16	Orange Ice, Passion fruit, Citrus cooler, Fruit punch and all the ices or freezes you could ever need	5	5	1	21
E24	Looks great, smells great, tastes delicious	3	−5	1	2
E22	Relaxes you after a busy day	1	−5	1	1
	Segment 3: Gatorade drinkers				
E30	From Gatorade	−1	17	9	2
E31	From SoBe	−1	5	−13	1
	Segment 4: Variety seekers				
E16	Orange Ice, Passion fruit, Citrus cooler, Fruit punch and all the ices or freezes you could ever need	5	5	1	21
E15	100% natural	3	4	3	6
E05	A sports drink . . . a light taste with complete balanced nutrition to help you stay healthy, active, and energetic	8	2	0	6

Data courtesy of It! Ventures, Inc.

is not surprising that the largest proportion of respondents in the sports drink study fall into the health-conscious segment. One might expect this high proportion because of the positioning of the sports beverage. The health-conscious segment shows high basic interest in sports drinks. It also responds positively to elements presenting health messages and elements talking about high-quality athletic performance. This strength of the two types of messages seems reasonable based on these beverages having been seen historically as healthy and used by athletes. Thus, many people who fit this image, or would like to, are interested in health-related and performance-related elements.

Brand is important for segments 2 and 3. For them, as was the case for cola drinks, there is clearly an important role played by brand, but in a slightly dissimilar way.

Segment 2: The impressionables. This segment, which constitute 18% of the sample, shows high basic interest in sports drink products. The interesting thing about the respondents in this segment is that, whereas they strongly react to the Gatorade brand, they also react positively to other brands. Actually, just about any brand is positive. Whereas performance-based elements such as stamina and a range of flavors are of some interest, the brand of the product is of primary interest.

Segment 3: The Gatorade drinkers. These respondents, who constitute around 30% of the sample, show high product interest. The difference between these respondents and those in segment 2 (the impressionables) is that the Gatorade segment is interested only in the Gatorade brand. No other elements significantly increase the interest of these respondents. The only other element that matters is SoBe, which is a strong negative. So while brand is important as for segment 2, segment 3 respondents are interested only in

Gatorade rather than in a broad cross section of brands.

Segment 4: The variety seekers. These respondents, who constitute about 20% of the sample, show fairly high basic interest in Sports drinks. This segment shows no reaction to brand, but reacts strongly to a wide range of sensations and flavors.

Juice

Juice is produced by the extraction of liquid from fruit, such as oranges and apples. Whereas juice is commonly thought of, or at least idealized as, being freshly squeezed, commercial drinks are often reconstituted from concentrate, usually with the addition of various elements, such as sugar, to improve the taste. One of the most important reasons that consumers would choose to drink juice is its perceived health qualities. For example, citrus juice is generally high in vitamin C. Thus, it seems likely that health-related factors play some role in the interest of consumers in juice products.

The first issue that stands out when we examine the concept-response segments for juice is that, in relation to the previous beverages (coffee, tea, and cola), brand plays very little, if any, role for juice consumers (see Table 22.6). As suggested previously, however, health factors are very important:

Segment 1: The health conscious. This segment comprises about half of the total number of respondents. The respondents show high basic interest in the product as a whole. They want healthy, *real* juice, with an array of vitamins and minerals, or even *veggie juice*, as long as it is natural and healthy. A product that is only 30% juice or mixed with milk detracts from their interest, perhaps because it is perceived as being less healthful.

Segment 2: The traditionalists. This segment, which comprises 45% of the sample,

Table 22.6. Segments identified among juice consumers

		Segment		
		1	2	3
	Base size	*110*	*105*	*20*
	Additive constant	*68*	*68*	*43*
	Segment 1: Health Conscious			
E01	A bold fruit flavored drink made with real juice, not from concentrate	5	5	17
E05	Juice with a full days supply of vitamins and minerals	5	2	6
E06	Veggie juice . . . the all natural way to get a full serving of vegetables in every glass	5	−30	11
E09	A juice spritzer . . . lightly carbonated . . . made with 30% real fruit juice	−8	−10	11
E08	Juice mixed with real milk to give you a creamy taste	−30	−29	3
	Segment 2: Traditionalists			
E01	A bold fruit flavored drink made with real juice, not from concentrate	5	5	17
E03	All nectar juice with antioxidants	3	−5	9
E09	A juice spritzer . . . lightly carbonated . . . made with 30% real fruit juice	−8	−10	11
E08	Juice mixed with real milk to give you a creamy taste	−30	−29	3
E06	Veggie juice . . . the all natural way to get a full serving of vegetables in every glass	5	−30	11
	Segment 3: Variety seekers			
E16	Orange and white cranberry, apple, fruit punch, pear, raspberry, pineapple or tomato, carrot, or veggie blends . . . whatever you're looking for	3	−1	25
E01	A bold fruit flavored drink made with real juice, not from concentrate	5	5	17
E12	Exotic blends naturally sweet with real pulp	3	−4	17
E06	Veggie juice . . . the all natural way to get a full serving of vegetables in every glass	5	−30	11
E09	A juice spritzer . . . lightly carbonated . . . made with 30% real fruit juice	−8	−10	11
E03	All nectar juice with antioxidants	3	−5	9
E05	Juice with a full days supply of vitamins and minerals	5	2	6
E14	So refreshing you want to savor how it makes you feel	0	0	6
E11	With Calcium, Vitamins A and the energy releasing B vitamins . . . or what ever you need	1	2	6

Data courtesy of It! Ventures, Inc.

also shows high basic interest in juice. These respondents are interested only in a traditional glass of orange juice, a beverage that has been produced naturally. Unlike the health-conscious consumers in segment 1, the traditionalists are interested neither in veggie juice nor in a healthful juice with antioxidants. Thus, their main focus is tradition rather than health.

Segment 3: The variety seekers. They comprise less than 10% of the market. Variety seekers show their greatest interest in beverages that have a range of flavors, sensations, and qualities. As the basic interest of these respondents is lower than the basic in-terest for the other segments, it is more important to entice these respondents with something they like. Acceptance does not come automatically with them.

Alcoholic Beverages

So far, this chapter has considered nonalcoholic drinks, but alcoholic drinks constitute a large proportion of the beverages consumed throughout the world. Before we consider individual alcoholic drinks, it is important to recognize how these beverages differ from those considered previously. Obviously, alcoholic beverages contain alcohol, a mild intoxicant. What this means is that people drink alcoholic drinks for reasons that differ from those given for nonalcoholic drinks. Whereas nonalcoholic drinks may be consumed to refresh, alcohol is generally consumed to relax. Not only does the motivation differ, but so does the nature of the consumer. For example, the law in most countries prohibits the consumption of alcohol until at least age 16. In some countries such as the United States, this age limit rises to 21. Also, in some religions, alcohol consumption is discouraged or even forbidden. Thus, the consumption of alcohol is quite distinct from that of nonalcoholic drinks.

The profound differences between nonalcoholic and alcoholic beverages—in occasion, nature of respondent, motivation, etc.—raise the distinct possibility that the mind-set of alcohol consumers differs from that of nonalcohol consumers, even if the same person is involved. Since the majority of alcohol drinks are often perceived as unhealthy, with the possible exception of red wine and white wine, we might expect that health-conscious segments would be largely absent for alcohol beverages.

Beer

We begin with beer. Its market around the world is massive. Recent estimates suggest that it is worth over $200 billion annually (Mindbranch 1997). Before we look at the market segments in the beer market, it is important to understand the consumer view of beer. Beer is a relatively low-alcohol beverage, with the percentage of alcohol ranging, on average, from 2% to 5%. For this reason it is often seen as drink for mild relaxation or refreshment. Beer is also often consumed at a variety of occasions. Much is consumed in bars, clubs, or homes as a social lubricant. Beer can also be used to accompany a meal, although wine is probably more often seen in this role, especially in Europe. There are two clear, very large concept-response segments (Table 22.7):

Segment 1: The traditionalists. These respondents are interested in traditional types and styles of beer. They want a beer that is cold and refreshing and from a large, traditional brewer. What they do not want is a beer that does not fit their traditional vision of what a beer should be. For example, beer made by a more boutique brewer is of little interest to traditionalists. Furthermore, they are not interested in a beer that has a complex description of flavors. They want a beer that tastes of beer and beer alone.

Segment 2: The variety seekers. They are the opposite of the traditionalists. Variety seekers respond strongly to concepts about complex flavor descriptions, including the addition of very nontraditional flavors, such as those of cherries and tropical fruits. They also show low basic interest, so they require interesting product elements and features to entice them to choose the beverage. It is the description, not the basic interest, that drives them. Variety seekers are also highly accepting: novel product elements do not seriously reduce their interest in the product.

There is one particularly interesting implication from these market segments and their relative sizes identified here: whereas the majority of consumers want a simple traditional styled beer, there are also a great number who are interested in beers that have a range of strange and wonderful flavors. As the majority of beers are fairly traditional in their style, these results suggest that there may be commercial opportunities for new products that have a wide range of fairly novel flavors.

Low-alcohol Flavored Drinks

Low-alcohol flavored drinks are the newest style of alcoholic beverage. Generally they comprise premixed liquor, often combined with wine or coolers, as well as some kind of flavoring. Generally the alcohol level is similar to that found in beer. As a conse-

Table 22.7. Segments identified among beer consumers

		Segment 1	Segment 2
	Base size	*118*	*100*
	Additive constant	*48*	*24*
	Segment 1: Traditionalists		
E30	From Anheuser Bush	6	3
E34	Icy cold	5	4
E18	So refreshing . . . you have to drink some more	5	−5
E29	From Labatts	−6	−1
E16	Raspberry, Berry, Lemon and Lime, Cinnamon, Honey flavors . . . whatever you're looking for	−8	10
E09	Brewed with 5 specially roasted malts for a deep red color and hearty robust taste with the crisp finish of mixed berries	−19	18
E07	Amber ale with blue agave nectar and a natural flavor of Mexican tequila and lime	−20	16
E03	A beer blended with exotic tropical flavors	−21	12
E02	A beer mixed with real fruit juices and lightly carbonated	−21	15
E08	Michigan cherries with a generous portion of wheat malt for a bright lively ale with a crisp finish	−22	19
E04	A lager with a citrus hop flavor throughout for a fruity assertive flavor	−24	16
E06	Smooth rich cream pilsner with roasted chocolate flavor	−27	0
	Segment 2: Variety seekers		
E08	Michigan cherries with a generous portion of wheat malt for a bright lively ale with a crisp finish	−22	19
E09	Brewed with 5 specially roasted malts for a deep red color and hearty robust taste with the crisp finish of mixed berries	−19	18
E07	Amber ale with blue agave nectar and a natural flavor of Mexican tequila and lime	−20	16
E04	A lager with a citrus hop flavor throughout for a fruity assertive flavor	−24	16
E02	A beer mixed with real fruit juices and lightly carbonated	−21	15
E03	A beer blended with exotic tropical flavors	−21	12
E16	Raspberry, Berry, Lemon and Lime, Cinnamon, Honey flavors . . . whatever you're looking for	−8	10
E05	Amber ale with Nutmeg and Raspberry flavors	−29	7

Data courtesy of It! Ventures, Inc.

quence, these drinks, also known as RTDs (ready to drink), are consumed in situations where one might drink beer and for similar reasons. However, they are also quite different in that they are new and often perceived to be fashionable. Thus, we would expect to see differences between the segments found for beer and those found for RTDs. From the segments illustrated in Table 22.8, it is clear that there are disparities in the different markets to which beer and RTDs appeal:

Segment 1: The traditionalists. This group, which comprises over 65% of the sample, shows interest in the traditional types of RTDs: flavored coolers and breezers made from juices and white liquors from companies (e.g., Smirnoff and Bacardi) associated with alcoholic drinks. However, other types of drinks that could serve as the basis of RTDs—beer or whisky—strongly diminish the traditionalists' interest in the RTD. It is interesting to note that, despite the fairly new arrival of such beverages on the market, it appears that a traditionalist image of the product quickly develops and a large number of consumers show interest in it. It seems quite possible that the newness of these RTD products is such that the traditionalists do not have a fixed idea of what they are and thus

Table 22.8. Segments identified among consumers of low-alcohol-flavored drinks

		Segment		
		1	2	3
	Base size	*145*	*59*	*16*
	Additive constant	*45*	*31*	*40*
	Segment 1: Traditionalist			
E16	Raspberry, Berry, Lemon and Lime flavors . . . whatever you're looking for	10	25	1
E09	A mix of the freshest juice, the best spirits, and natural flavors	9	24	19
E02	Wine based coolers or breezers in so many flavors	8	33	8
E29	From Bacardi	8	3	24
E28	From Smirnoff	7	−2	18
E01	Hard cider, alcoholic lemonade, and all types of coolers and breezers	−7	27	−8
E05	A malt beverage with a light crisp flavor	−16	5	0
E07	Lemonade with a lite hit of malt whiskey . . . cool and refreshing	−20	16	−13
E04	A splash of beer, a sweet tingly flavor, fizz and some juice . . . it goes together to create a sensation for your senses	−24	18	12
E03	Lemon and a touch of beer, just the right touch	−25	13	−11
	Segment 2: Variety seekers			
E02	Wine based coolers or breezers in so many flavors	8	33	8
E06	Vodka and juice . . . not too sour, not too sweet . . . just a refreshing sensation!	1	27	16
E01	Hard cider, alcoholic lemonade, and all types of coolers and breezers	−7	27	−8
E16	Raspberry, Berry, Lemon and Lime flavors . . . whatever you're looking for	10	25	1
E09	A mix of the freshest juice, the best spirits, and natural flavors	9	24	19
E04	A splash of beer, a sweet tingly flavor, fizz and some juice . . . it goes together to create a sensation for your senses	−24	18	12
E07	Lemonade with a lite hit of malt whiskey . . . cool and refreshing	−20	16	−13
E03	Lemon and a touch of beer, just the right touch	−25	13	−11
E10	Hits the spot on a hot summer day	2	6	11
E20	When you think about it, you have to have it . . . and after you have it, you can't stop drinking it	1	−6	−36
E21	Simply the best	0	−7	4
	Segment 3: Impressionables			
E31	From Gallo	−3	−1	25
E29	From Bacardi	8	3	24
E09	A mix of the freshest juice, the best spirits, and natural flavors	9	24	19
E28	From Smirnoff	7	−2	18
E06	Vodka and juice . . . not too sour, not too sweet . . . just a refreshing sensation!	1	27	16
E35	Easy to drink right from the bottle	2	1	14
E23	A great way to celebrate special occasions	−1	3	14
E04	A splash of beer, a sweet tingly flavor, fizz and some juice . . . it goes together to create a sensation for your senses	−24	18	12
E27	It quenches THE THIRST	−1	−4	12
E15	100% natural	2	0	11
E10	Hits the spot on a hot summer day	2	6	11
E02	Wine based coolers or breezers in so many flavors	8	33	8
E25	A wonderful experience . . . shared with family and friends	4	−2	8
E30	From Anheuser Bush	−5	0	8
E19	Quick and fun . . . ready to drink, no bartender required	5	1	7
E11	Goes down smooth and easy	−1	0	6
E12	Made in the Traditionalist way	0	−2	−6
E32	Doesn't even taste like alcohol	1	−2	−6

(continued)

Table 22.8. Segments identified among consumers of low-alcohol-flavored drinks *(cont.)*

		Segment		
		1	2	3
E22	Relaxes you after a busy day	1	−3	−8
E01	Hard cider, alcoholic lemonade, and all types of coolers and breezers	−7	27	−8
E36	With the safety, care and quality that makes you trust it all the more	0	2	−9
E03	Lemon and a touch of beer, just the right touch	−25	13	−11
E08	Flavored alcoholic energy drinks with a little buzz	−4	3	−11
E07	Lemonade with a lite hit of malt whiskey . . . cool and refreshing	−20	16	−13
E14	So refreshing you want to savor how it makes you feel	1	3	−19
E18	So refreshing . . . you have to drink some more	2	2	−23
E20	When you think about it, you have to have it . . . and after you have it, you can't stop drinking it	1	−6	−36

Data courtesy of It! Ventures, Inc.

can find alternative styles of the same beverage to be highly acceptable. In contrast, for well-established beverages, the traditionalists usually have a fixed, rigid idea of what the product should be and find departures from that image somewhat disquieting.

Segment 2: The variety seekers. This group, which comprises almost 27% of the sample, shows the most interest in elements that present flavor descriptions. The respondents are only moderately interested in RTDs without the flavor description, meaning that the elements are relatively important in their beverage choice. In contrast to the traditionalists, the variety seekers show interest in all types of flavor aspects, including nontraditional RTDs (e.g., with beer or whisky). Interestingly, the most negative-performing element for this group was "Simply the best," illustrating that quality is not necessarily what they are looking for. Rather, as with all other variety seekers, they seek a drink that provides a wide range of sensations and flavors.

Segment 3: The Impressionables. This group, which constitutes the smallest segment (7%), shows the most interest in emotional and brand-related messages; that is, the respondents like the nonflavor benefits that tie in with their emotions. Examples of such messages that are strongly positive for the impressionables is that the beverage is

"refreshing" and "a great way to celebrate special occasions," as well as the Gallo and Bacardi brands.

Some elements convey possible addiction. A point to note here is that the most negative element was "When you think about it, you have to have it." This element is supposed to tie into a craving for a particular beverage. For the variety seekers and the impressionables, though, this element reduces consumer interest. It seems likely that this response is based on respondents drawing a link between alcohol craving and the highly negative outcome of it: alcoholism.

Wine

Whereas wine is an alcoholic beverage, its profile is quite different from those of the other alcoholic beverages. Wine has traditionally, and to this day remains, largely consumed as an accompaniment to food. There is also a longer, richer tradition associated with wine than with any of the other alcoholic beverages. Wines are respected beyond all other beverages. For example, many of the wines from France and, in particular, Burgundy and Bordeaux are revered throughout the world thanks to their long history of fine quality. There is also a somewhat elitist attitude about wine that does not exist for beer and RTDs. These factors suggest that one segment that may be important in the wine market is the traditionalists; that is, individuals who want their wines to be traditional in style. There are four segments for red wine (Table 22.9):

Segment 1: Red wine variety seekers. This segment, which comprises 22% of the sample, shows very low basic interest in red wine as a whole. This comparatively low interest, especially in relation to other beverages, suggests that these red wine consumers need product elements to entice them. They show greatest interest in elements that provide descriptions of flavors, tastes, and sensations related to the products. Of interest is that these descriptions need not fit the traditionalist image of red wine. Elements such as "chilled" or "fizzy" red wine, which were unpopular with the remaining segments, were highly attractive to red wine variety seekers.

Segment 2: Red wine traditionalists. This segment, which comprises 45% of the respondents, shows high basic interest in red

Table 22.9. Segments identified among consumers of red wine

		Segment			
		1	2	3	4
	Base size	*48*	*99*	*49*	*24*
	Additive constant	*19*	*39*	*18*	*54*
	Segment 1: Variety seekers				
E06	A pale red wine with a light flavor of raspberries . . . so soft and sweet	29	−19	2	−14
E05	A heavy red wine with a warm plumy fruit flavor and a little bit of dryness	24	8	28	−27
E03	Deep red colored wine with a sweet complex fruity flavor and delicate flowery fruit aroma	20	0	11	−4
E09	Bubbly fresh and sweet with a hint of raspberry tart flavor	18	−19	3	−18
E04	Ruby red color with a black currant flavor and a hint of oak	16	4	26	−44
E07	A chilled light bodied red wine with a fresh fruity flavor	15	−6	5	12
E02	Pale pink colored wine with fizz and the flavor of fruit	11	−22	−10	−9
E01	Wine with a pale ruby color and a fruity taste	10	−5	3	9
E23	A great way to celebrate special occasions	10	3	10	0
E22	Relaxes you after a busy day	9	0	1	3
E08	Red wine with added antioxidants	6	−6	−3	−5
E29	From Gallo	−6	3	−4	−9
E32	Tastes great . . . without the taste of alcohol	−10	−4	−17	1
	Segment 2: Traditionalists				
E16	Dry, Sweet, Semi Sweet . . . whatever you're looking for	2	7	5	−7
E12	Made in the tradition of the greatest wine producers all over the world	1	6	8	2
E33	Multiserve containers . . . so you always have enough!	−1	−6	−13	6
E08	Red wine with added antioxidants	6	−6	−3	−5
E07	A chilled light bodied red wine with a fresh fruity flavor	15	−6	5	12
E35	Resealable single serve container . . . to take with you on the go	1	−7	−10	4
E34	Icy cold	−4	−9	−9	−7
E09	Bubbly fresh and sweet with a hint of raspberry tart flavor	18	−19	3	−18
E06	A pale red wine with a light flavor of raspberries . . . so soft and sweet	29	−19	2	−14
E02	Pale pink colored wine with fizz and the flavor of fruit	11	−22	−10	−9

(continued)

Table 22.9. Segments identified among consumers of red wine *(cont.)*

		Segment			
		1	2	3	4
	Base size	*48*	*99*	*49*	*24*
	Additive constant	*19*	*39*	*18*	*54*
	Segment 3: Impressionables				
E05	A heavy red wine with a warm plumy fruit flavor and a little bit of dryness	24	8	28	−27
E04	Ruby red color with a black currant flavor and a hint of oak	16	4	26	−44
E23	A great way to celebrate special occasions	10	3	10	0
E12	Made in the tradition of the greatest wine producers all over the world	1	6	8	2
E15	100% natural	2	2	6	−1
E30	From Kendal Jackson	−2	4	6	−4
E27	It quenches THE THIRST	0	−2	−7	0
E34	Icy cold	−4	−9	−9	−7
E02	Pale pink colored wine with fizz and the flavor of fruit	11	−22	−10	−9
E35	Resealable single serve container . . . to take with you on the go	1	−7	−10	4
E33	Multiserve containers . . . so you always have enough!	−1	−6	−13	6
E32	Tastes great . . . without the taste of alcohol	−10	−4	−17	1
	Segment 4: No frills				
E11	Goes down smooth and easy	2	1	3	10
E21	Simply the best	4	3	2	8
E36	With the safety, care and quality that makes you trust it all the more	−5	2	−3	6
E33	Multiserve containers . . . so you always have enough!	−1	−6	−13	6
E34	Icy cold	−4	−9	−9	−7
E16	Dry, Sweet, Semi Sweet . . . whatever you're looking for	2	7	5	−7
E13	Premium quality	−1	5	5	−8
E02	Pale pink colored wine with fizz and the flavor of fruit	11	−22	−10	−9
E29	From Gallo	−6	3	−4	−9
E06	A pale red wine with a light flavor of raspberries . . . so soft and sweet	29	−19	2	−14
E09	Bubbly fresh and sweet with a hint of raspberry tart flavor	18	−19	3	−18
E05	A heavy red wine with a warm plumy fruit flavor and a little bit of dryness	24	8	28	−27
E04	Ruby red color with a black currant flavor and a hint of oak	16	4	26	−44

Data courtesy of It! Ventures, Inc.

wine, and only a few elements significantly improve their product interest. All of these winning elements focus on the traditional idea of red wine; that is, the traditional choice of wine styles, and wines made in the traditional way. Possibly the most telling factor is the elements that the traditionalists strongly reject. It appears that all elements that differ from the standard style of red wine (e.g., cold, fizzy, sweet, single-serve containers, and added antioxidants) decrease the interest of these traditionalists.

Segment 3: Red wine impressionables. This segment, which comprises 22% of the

sample, shows positive reactions to traditional flavor descriptions. These respondents also respond positively to emotive messages, such as the wine being "natural" and "to celebrate special occasions," as well as brand. One interesting thing to note about this segment is that, much like the traditionalists, it responds negatively to less traditional styles of red wine. This distaste for the nontraditional in red wine illustrates the importance of tradition in the wine market.

Segment 4: Red wine, no frills. This segment, which comprises 11% of the respondents, shows very high basic interest in red wines, but beyond that high level these individuals do not respond to any kind of complex product description or the history of the wine. All they find interesting are messages about the fact that the wine is good, convenient, easy to drink, and will not cause them any problems. This segment also wants the wine to conform to the traditional product stereotype. This is further shown by this segment showing a similar dislike for nontraditional styles of red wine.

There are three segments for white wine (Table 22.10):

Segment 1: White wine variety seekers. This segment, which comprises 31% of the respondents, shows relatively low basic interest in the white wine itself, again suggesting that it is the specific elements and the descriptions that drive interest.

Segment 2: White wine traditionalists. This segment, which comprises 57% of the

Table 22.10. Segments identified among consumers of white wine

		Segment		
		1	2	3
	Base size	66	120	26
	Additive constant	26	39	23
	Segment 1: Variety seekers			
E03	A lightly carbonated pale white wine with a hint of raspberries . . . so soft and sweet	33	−19	11
E06	Golden colored wine with a sweet complex fruity flavor and delicate flowery fruit aroma	22	−3	9
E04	A light bodied white wine with a fresh, fruity flavor	20	1	10
E05	Pale yellow colored wine with a smooth creamy texture and the flavor of green apples, melon and hazelnut	18	−22	13
E08	Bubbly fresh and very sweet white wine, with a hint of citrus lemon, lime, and melon flavors	17	−15	18
E01	Wine with a pale yellow color and a fruity flavor	14	−3	10
E07	Golden yellow colored wine with a dry flavor of grape and berry with a hint of oak	11	−12	−4
E09	Bubbly fresh and sweet with a hint of lemony tart flavor	9	−18	5
E12	Made in the tradition of the greatest wine producers all over the world	7	8	11
E11	Goes down smooth and easy	6	3	−5
E20	When you think about it, you have to have it . . . and after you have it, you can't stop drinking it	6	−2	−12
E32	Tastes great . . . without the taste of alcohol	−9	−4	−7
E29	From Gallo	−9	−2	9
	Segment 2: Traditionalists			
E16	Dry, Sweet, Semi Sweet . . . whatever you're looking for	0	11	10
E12	Made in the tradition of the greatest wine producers all over the world	7	8	11
E10	Hits the spot on a hot summer day	−2	6	−6
E15	100% natural	0	6	4
E25	A wonderful experience . . . shared with family and friends	2	6	10
E07	Golden yellow colored wine with a dry flavor of grape and berry with a hint of oak	11	−12	−4
E08	Bubbly fresh and very sweet white wine, with a hint of citrus lemon, lime, and melon flavors	17	−15	18
E09	Bubbly fresh and sweet with a hint of lemony tart flavor	9	−18	5
E03	A lightly carbonated pale white wine with a hint of raspberries . . . so soft and sweet	33	−19	11
E05	Pale yellow colored wine with a smooth creamy texture and the flavor of green apples, melon and hazelnut	18	−22	13

(continued)

Table 22.10. Segments identified among consumers of white wine *(cont.)*

		Segment		
		1	2	3
	Segment 3: Impressionables			
E31	Imported from France	5	3	23
E08	Bubbly fresh and very sweet white wine, with a hint of citrus lemon, lime, and melon flavors	17	−15	18
E30	From Kendal Jackson	−4	2	17
E05	Pale yellow colored wine with a smooth creamy texture and the flavor of green apples, melon and hazelnut	18	−22	13
E23	A great way to celebrate special occasions	3	5	12
E03	A lightly carbonated pale white wine with a hint of raspberries . . . so soft and sweet	33	−19	11
E12	Made in the tradition of the greatest wine producers all over the world	7	8	11
E24	Looks great, smells great, tastes delicious	3	1	11
E16	Dry, Sweet, Semi Sweet . . . whatever you're looking for	0	11	10
E01	Wine with a pale yellow color and a fruity flavor	14	−3	10
E25	A wonderful experience . . . shared with family and friends	2	6	10
E04	A light bodied white wine with a fresh, fruity flavor	20	1	10
E06	Golden colored wine with a sweet complex fruity flavor and delicate flowery fruit aroma	22	−3	9
E29	From Gallo	−9	−2	9
E28	From Northern California	3	5	7
E36	With the safety, care and quality that makes you trust it all the more	−4	1	6
E10	Hits the spot on a hot summer day	−2	6	−6
E32	Tastes great . . . without the taste of alcohol	−9	−4	−7
E02	White wine with added antioxidants	−3	−3	−7
E18	So refreshing . . . you have to drink some more	4	2	−10
E20	When you think about it, you have to have it . . . and after you have it, you can't stop drinking it	6	−2	−12

Data courtesy of It! Ventures, Inc.

respondents, shows interest in a traditional style of white wine. These traditionalists also show strong disinterest in variety-seeker wine descriptions, especially for sweet and carbonated wines. Again, here we see how strongly traditional types of wine drive acceptance of different products.

Segment 3: White wine impressionables. This segment, which comprises 12% of the respondents, strongly responds to messages related to brands, where the wine is made, and any emotional benefits of the product (e.g., "A great way to celebrate special occasions").

One important disparity exists between the segments found for red wine and white wine respondents. More specifically, that a no-frills market segment was found for red but not white wine drinkers. One possible explanation for this relates to the sophistication often associated with wine consumption and with red wine consumption in particular; that is, some consumers do not want to be tagged with the elitist attitudes sometimes found associated with wine. Their response may be to become no-frills consumers who are only interested in a simple product. It seems likely that the appearance of the no-frills consumers in the red wine study, but not white wine study, may reflect that such attitudes are more associated with red wine than with white wine.

Tequila

So far, in our analysis of the consumer mind for alcoholic drinks, we have considered only those that have low to medium alcohol levels. However, a large part of the alcoholic beverage market is made up of *hard liquor*: drinks containing at least 25% alcohol. It seems reasonable that this type of higher-alcohol beverage should generate very different mind-sets than do beer or wine. From the segments illustrated in Table 22.11, it seems clear that this is not the case. A strong similarity exists between the segments found in the tequila market and those for other beverages:

Segment 1: The variety seekers. Unlike the other alcoholic drinks, a very high proportion of tequila respondents (77%) can be classified as variety seekers. In fact, over 77% of subjects were in this market segment. These respondents again showed low basic interest, typical of variety seekers, illustrating the need for products to attract these consumers.

Segment 2: The no frills. This group, which comprises 8% of the respondents, simply wants clean, safe, and easy-to-drink tequila. Complex flavor descriptions strongly reduce interest.

Segment 3: The traditionalists. This group, which comprises 15% of the respondents, wants a natural, cold, and refreshing tequila without the addition of fancy mixers and flavors.

Summing Up: What Do Consumers Want?

In this chapter, we have outlined the results from ten beverage studies. Each study from Drink It! investigated the mind-sets underlying market segments for a particular beverage. These studies illustrate five clear segments of consumers that show differing patterns of what interests them in a beverage. Not surprisingly, the mix of segments differs between beverages. Much of these disparities appear to be based on the nature of the bever-

age in question and the consumers they attract. For example, wine consumers are mainly interested in traditionally styled drinks, whereas tequila drinkers want a greater range of flavors and sensations. However, it is also important to note that the general types of consumers found across a range of beverages are quite similar, and these similarities parallel the basic types of people that continue to appear in the It! studies [e.g., traditionalist or classic, variety seeker or elaborate, impressionable or imaginer (Moskowitz et al. 2004)]. The other two segments, health conscious and no frills, probably exist in the Crave It! studies as well, but simply have not been pulled out (see Chapters 20 and 21).

Although there are five broad types of segments, there are important differences across the same named segments for different beverages. For example, if we look at the segments of impressionables for wine and cola drinks, the cola-drink segment is dominated by the role of brand, whereas brand plays very little role for the wine drinkers. Thus, the segments just discussed are a general description of the characteristics of each of these types of consumers. The descriptions should be used as pointers to the types of consumers that may be found for any beverage. For a full understanding of a particular beverage's consumers, the pattern of responses in the segment must be interpreted in relation to the specifics of a beverage and the market in question:

Traditionalists. They generally show high basic interest in a product. For these individuals the most important elements are those that reflect the traditional forms and flavors of a particular beverage, as well as that the beverage is served in the traditional way. They are not interested in product innovation. New styles of product and complex drink descriptions are unattractive to these consumers.

Variety seekers. They show relatively low basic interest in a beverage, so they must be wooed to buy a particular product. What

Table 22.11. Segments identified among tequila consumers

			Segment		
			1	2	3
	Base size		*148*	*16*	*28*
	Additive constant		*31*	*54*	*68*
	Segment 1: Variety seekers				
E04	A mix of tropical exotic flavors and tequila . . . clear and clean		21	13	−14
E06	Tequila from blue agave nectar mixed with the natural flavor of Mexican tequila and lime		20	−17	−3
E09	A mix of the freshest fruits from real juice, the best spirits, and natural nectars		19	6	−14
E16	Raspberry, Berry, Lemon and Lime, Cinnamon, Honey flavors . . . whatever you're looking for		17	−3	−34
E03	Tequila already mixed with salt, natural lime flavors, and triple sec		16	−18	−3
E05	Tequila and juice that's not too sour, not too sweet . . . nice refreshing sensation!		14	−8	−2
E02	Lemon/ Lime and Tequila . . . a bit of sweet, tangy flavor		12	−7	−5
E07	Mandarin and Lime flavored tequila . . . so tart and refreshing		12	−18	−8
E08	Real tequila blended with natural flavor and a hint of carbonation		11	1	−7
E01	Naturally flavored tequila		10	1	5
E19	Quick and fun . . . ready to drink, no bartender required		8	−4	−5
E11	Goes down smooth and easy		7	0	3
E10	Hits the spot on a hot summer day		7	2	−4
E18	So refreshing . . . you have to drink some more		7	1	7
E21	Simply the best		6	−2	−1
E35	Easy to drink right from the bottle		6	10	3
	Segment 2: No frills				
E04	A mix of tropical exotic flavors and tequila . . . clear and clean		21	13	−14
E36	With the safety, care and quality that makes you trust it all the more		0	11	−4
E35	Easy to drink right from the bottle		6	10	3
E34	Icy cold		2	8	7
E09	A mix of the freshest fruits from real juice, the best spirits, and natural nectars		19	6	−14
E02	Lemon/ Lime and Tequila . . . a bit of sweet, tangy flavor		12	−7	−5
E05	Tequila and juice that's not too sour, not too sweet . . . nice refreshing sensation!		14	−8	−2
E06	Tequila from blue agave nectar mixed with the natural flavor of Mexican tequila and lime		20	−17	−3
E03	Tequila already mixed with salt, natural lime flavors, and triple sec		16	−18	−3
E07	Mandarin and Lime flavored tequila . . . so tart and refreshing		12	−18	−8
	Segment 3: Traditionalists				
E13	Premium quality		3	4	11
E15	100% natural		5	−2	9
E34	Icy cold		2	8	7
E18	So refreshing . . . you have to drink some more		7	1	7
E32	Tastes great . . . without the taste of alcohol		3	2	−6
E33	Multiserve containers . . . so you always have enough!		3	1	−6
E08	Real tequila blended with natural flavor and a hint of carbonation		11	1	−7
E07	Mandarin and Lime flavored tequila . . . so tart and refreshing		12	−18	−8
E20	When you think about it, you have to have it . . . and after you have it, you can't stop drinking it		1	−5	−9
E09	A mix of the freshest fruits from real juice, the best spirits, and natural nectars		19	6	−14
E04	A mix of tropical exotic flavors and tequila . . . clear and clean		21	13	−14
E16	Raspberry, Berry, Lemon and Lime, Cinnamon, Honey flavors . . . whatever you're looking for		17	−3	−34

Data courtesy of It! Ventures, Inc.

most interests them are the wide and fancy descriptions of how a product will feel and taste. They have no preexisting conception of how a beverage should taste or, if they do, the conception is not judgmental. To a variety seeker, there is no right answer. Rather, anything—no matter how strange—will interest these consumers. In fact, very few things can reduce the interest of these consumers.

Impressionables. They want to get into the experience of a beverage, perhaps revel in it, but not just drink it. As a consequence, they show high interest in elements that relate to emotions, such as a drink being great for special occasions or imported from France, and brand-related benefits. With regard to brand, there appears to be some variation in the nature and mind-set of the impressionables regarding different beverages. For most beverages, brand adds to the interest for impressionables. However, this is not true for coffee drinkers. Furthermore, some brands detract from the interest of cola drinkers. It appears that this reaction by impressionables is beverage and brand specific. An explanation of this could be that brands that arouse some kind of emotional interest (e.g., Coke and Pepsi) are of interest to impressionables, whereas those that fail to arouse this emotional interest or bond (e.g., Maxwell House and RC Cola) are rejected. For some beverages, there were consumers highly attracted to one particular brand (e.g., Pepsi drinkers). These respondents fall under the broad classification of impressionables because the brand benefit is what interests them most. However, this may be an artifact, because their interest lies in only one brand, not in the experience.

No-frills consumers. They look upon beverages as nothing more than refreshments. They are not interested in a product having specific flavors or tastes, nor do they respond strongly to any particular brand. They want a simple, safe drink that will cause them no problems. No-frills consumers strongly dislike complex product descriptions.

Health-conscious consumers. They are interested primarily in the healthy outcomes to be obtained by consuming a beverage; that is, elements that are natural and healthy strongly increase their interest. Any element that appears unhealthy or not natural strongly reduces their interest.

One factor that has not as yet been covered is the demographics of the different types of segments. Demographics have often been seen as important factors for product marketing. As a consequence, it is important to outline how such factors relate to the concept-response segment that exists over a range of ages and income brackets; *that is, these broad segments are largely independent of demographic factors.*

Using the Drink It! Data for Marketing

The Drink It! studies illustrate that the consumer mind-set is a fundamental factor in determining the types of beverages that an individual will find interesting. As a consequence it also is probably the most important factor in the decision to purchase a product. For these reasons the marketing must tie in with the major mind-sets.

For example, imagine that we are trying to devise a marketing campaign for a new red wine. First, the wine would need to be portrayed in a traditional manner, because tradition is important to almost all of the wine consumers. Such a wine product would also appeal specifically to the traditionalist segment, which is the largest in this study. If we wanted to create a product that appeals to a small boutique market, then other messaging would be necessary. For example, the interest of the variety seekers would be aroused by a campaign that highlights the wide range of tastes and sensations found within a particular product, although such a strategy would alienate traditionalist consumers. For the impressionables, descriptions or pictures relating to how the wine is made or where the grapes are grown could be useful. For the no-frills consumers, just a simple label without much detail might be best. Obviously, the best strategy would be a product that appeals to all consumers. However, this would be extremely difficult in the red wine market because of the wide disparity in the elements that appeal to the traditionalists and the variety seekers.

The Role of Brand

Why does brand play such an important role for some beverages, but not for others? One key factor is that the markets for some products, such as juice, are not dominated by large brands, as they are for cola drinks and sports drinks. As a consequence, no one brand is likely to dominate such surveys of individuals from a cross section of regions and socioeconomic groups. A second key factor is that some products do not lend themselves to be marketed as a mass-produced beverage. For example, the traditional, and perhaps even romantic, image of wine means that massive multinational brand names, such as Pepsi or Coca-Cola, are unlikely to dominate the wine industry in the same way that the two brands dominate cola drinks.

The Final Word

In the beverage industries, it seems clear that a powerful new tool in understanding the behavior of consumers involves the consumer mind-set as revealed by a conjoint analysis task. More specifically, understanding the different mind-sets and how they affect a range of beverage markets can provide producers with a greater ability to predict purchasing behaviors. Such knowledge can also enable producers to develop new drinks more effectively and efficiently that appeal to consumers. However, although we now may have greater access to the consumer mind, this access does not guarantee greater sales and, in turn, profits. To achieve sales growth and profits, effective marketing must be developed that ties in with the mind-sets identified here, so that consumers are truly drawn to a specific product.

Acknowledgment

The authors and It! Ventures, Inc., gratefully acknowledge the sponsorship of the 2002 and the 2003 Drink It! by WILD Flavors, Erlanger, Kentucky.

References

Beckley, J., and Moskowitz, H.R. (2002). Databasing the Consumer Mind: The Crave It!, Drink It!, Buy It! & Healthy You! Databases. Anaheim, CA: Institute Of Food Technologists.

Cappuccio, R., Krieger, B., Katz, R., Itty, B., and Moskowitz, H.R. (2002). Coffee: development of product concepts for drinking venues from first principles. Foodservice Technology (submitted).

Gobi International (2001). Gobi International world business information: soft commodities market. Retrieved 1 December 2002 from www.gobi.co.uk/softcomm.htm.

Hughson, A., Ashman, H., De la Huerga, C., and Moskowitz, H.R. (2002). Mind-sets of the wine consumer. Journal of Sensory Studies (in press).

Hughson, A., Hirsch, J., and Moskowitz, H.R. (2003). The role of health-related factors for consumers of coffee and tea. (Submitted.)

Mindbranch (1997). World beer report 1997–2005. www.mindbranch.com.

Moskowitz, H.R., Silcher, M., Beckley, D.M., Minkus-McKenna, D., and Mascuch, T. 2004. Sensory benefits, emotions and usage patterns for olives: using Internet-based conjoint analysis and segmentation to understand patterns of response. Food Quality and Preference (in press).

Starbucks (2002). Starbucks reports strong November revenues [Press release]. www.starbucks.com/aboutus.

Trevisanato, S., and Kim, Y. (2000). Tea and health. Nutrition Reviews, 58: 1–10.

Chapter 23

Understanding Brand Names in Concepts

Introduction: About brands

The concept of brand has a long, venerable history. A lot of this history has to do with the development of organization structures that recognize the importance of brands. According to Aaker and Joachimsthaler (2000), the birth of Brand Management took place when, in 1931, Neil McElroy, the Promotion Department Manager of Procter and Gamble, wrote a three-pager that created a marketing organization based on competing brands managed by dedicated groups of people.

According to McEnally and De Chernatony, and in light of this book being about concepts, this chapter differentiates the understanding of *brand* into two constructs. On the one hand we have brand concepts developed by management, that is, a vision and brand values. On the other hand, we have brand images developed by consumers, for example, forming associations and images, and perceiving usage situations (McEnally and De Chernatony 1999). The impressions that consumers have for brands are extensive and include such things as communication of particular aspects of benefit and lifestyle through the use of specific brands (Goffman 1959; Gordon 1991). Managers of companies or departments of advertising embark on a strategy knowing that their visions of brands must mesh with those of their target consumers. This chapter shows how to assess the success of these brand-building efforts by measuring the utility values of brand names.

Creating, developing, implementing, and maintaining a successful brand require a strategic perspective whereby strong brand concepts are communicated and presented to well-targeted segments. The target should result in favorable brand images that reflect a brand's identity (Gardner and Levy 1955; Reynolds and Gutman 1984; De Chernatony 1998). Thus, concepts play a dual role. First, on the "giving end," concepts define the brand: these are so-called brand concepts. Second, on the "taking end," concepts use the brand to add power and persuasiveness.

The literature about brand typologies includes many various approaches of strategists who have planned actual brand concepts for the market (e.g., see Gardner and Levy 1955, Leahy 1994, and De Chernatony and Riley 1997). An extensive review about brand elements by Kappor (2003) presents the 18 abilities that a brand must have to be successful: reliability, availability, acceptability, stability, credibility, responsibility, serviceability, desirability, visibility, respectability, adaptability, profitability, accountability, unpredictability, affordability, durability, capability, and usability. Each of the 18 abilities of a concept can be quantified separately through concept research, although in concept development they are all integrated into the acceptance measure.

In the early stage of its life when a product is new, the goal of brand management is to explain product-related *functional information*—what the product is and does, and how it can benefit consumers. People remember those brands that are personally self-relevant and are more likely to remember information they consider useful (Callaghan and Wilson 1998). The brand name plays a very important role at this stage. The *brand concept* identifies the brand's functional benefits with a distinctive name thereby differentiating it from other brands (Jones 1986; Brown 1992). Studies by Jacoby and colleagues (1971, 1977) indicate that consumers find the brand name to be the most useful piece of information when making a purchase decision. They offer consumers a choice of impacts like brand name, price, or other product attribute information. Respondents often choose brand name first and price second. Thus, the *label name* represents the central component of brand and becomes the most important distinguishing feature. Through the name the product receives its individuality and the strength to stand out in relation to other products. When we use the brand name of a product we summon up its image and a surrounding set of feelings and impressions. However, a brand is much more than the name. Due to the positioning of a name the consumers can now improve product/brand selection because they can identify brands and distinguish between them (Hoyer and Brown 1990).

After asking what the product is, consumers want to know "What can it do for me?" and they draw the brand closer by creating a relationship between the brand and themselves (Fournier and Yao 1997). At this stage the challenge of the management's brand concept is to ensure new product brand features (e.g. emotional values or enjoyment), because otherwise the distinction between the product and brand becomes blurred (Prentice 1987; Southgate 1994).

To differentiate their brand, marketers try to enhance its functional characteristics by incorporating emotional values into the brand, portraying this through the metaphor of brand personality. Brand personalities are strengthened by advertising through the use of advertising characters, slogans, packaging, user imagery, and other elements of the marketing mix (Plummer 1985; Batra et al. 1993). At this stage the management's brand concept must connect the brand personalities consonant with the emotional values of the brand and the lifestyle of the target consumers so that the consumers and brand personalities are brought into alignment (Schiffman and Kanuk 1996).

Another step in brand concept development is the brand's extendibility. If a brand becomes a symbol, then quite often the brand can be used to stand for something beyond itself. Consumers feel so much closer to brands that can be said to reside with them (Gordon 1991). These brand images are strong enough to stand on their own: Nike is winning, Rolls Royce is the epitome of luxury and status, and Coca-Cola has been the brand that refreshes (McEnally and De Chernatony 1999).

In postmodern marketing, the brand may be the company or align itself with social causes. Company and brands often align with social and political issues, such as corporate social responsibility. Hilton (2003) describes three specific ways in which brands can move from responsibility to leadership:

1. By harnessing the cultural power of brands to inspire positive social change

2. By using the innovation process to develop new products and services that turn social and environmental needs into market opportunities

3. By using corporate resources to tackle the source of most of the world's social and environmental problems

Brand Equity

Brand-equity research identifies the rational and the emotional values a brand possesses, but also can be used to identify the monetary value associated with the brand. The various approaches of brand strategists have spawned the growth of literature on brands and brand equity. A study cites some 30 or more conceptualizations of different brand-equity measurement techniques (Mackay et al. 1997). Chapters about branding, brand value, and ancillary issues can be found in many marketing textbooks and provide interesting reading for marketing students (e.g., see Levitt 1983, Boone and Kurtz 1992, Aaker 1996, Kotler and Armstrong 1996, and Weitz and Wensley 2002).

According to Farquhar (1989), brand equity is about understanding the amount of *value added* to a product with the use of branding. In a similar vein, Wells (1990) believes brand equity is the value added to a product by its brand name. According to Aaker and Joachimsthaler (2000), *brand equity* is the loss of some or all assets or liabilities if the brand's name or symbol changes. They also make some provision for shifts to a new name and symbol.

Recent research on brands and brand equity suggests that the aim of consumer brand equity is to maintain and reinforce a brand's long-term market position with respect to the views, memory associations, attitudes, and the existing relationship the brand holds with con-

sumers (Kamakura and Russell 1993; Keller 1993; Lassar et al. 1995; Krishnan 1996).

Brands are a continuing area of study among market researchers, advertising agencies and, of course, corporations. Often, in financial buy/sell transactions, the sales price is far more than the actual current profit of the company. The intangibles that increase the sales price often can be traced to the brand. Table 23.1, which lists the values of some major global brands, reveals that the intangibles quantified by brand value can lead to a substantial dollar value for a well-known brand.

The Decline of Brand Loyalty

In the last few years an increasing number of studies have reported that brand loyalty is declining (Dekimpe et al. 1997). The decline calls for changing brand strategies by the companies, as well as increasing consumer

Table 23.1. Value of global brands as measured by Interbrand*

	Brand	Book value	Market value	Difference (brand value)	Brand as proportion of market value
1	Coca-Cola	58	142	84	59
2	Microsoft	215	272	57	21
3	IBM	115	158	44	28
4	GE	295	328	34	10
5	Ford	25	57	32	58
6	Disney	20	53	32	58
7	Intel	114	144	30	21
8	McDonald's	15	41	26	64
9	AT&T	78	103	24	24
10	Marlboro	91	112	21	19
11	Nokia	26	47	21	44
12	Mercedes	31	48	18	37
13	Nescafe	60	78	18	23
14	Hewlett-Packard	38	55	17	31
15	Gillette	27	43	16	37
16	Kodak	10	25	15	60
22	BMW	5	17	11	77
28	Nike	2	11	8	77
36	Apple	1	6	4	77
43	Ikea	1	5	4	75
54	Ralph Lauren	1	3	2	66

*All numbers in billions, except brand proportion.
Adapted from Raymond Perrier, "Interbrand's World's Most Valuable Brands," report of a June 1999 study sponsored by Interbrand and Citigroup, 1999.

sophistication. With these two trends in full force, brand building generates two contradictory demands (Blumenthal 2002):

1. Constancy to build awareness and credibility
2. Change to remain relevant in an ever-evolving market

McEnally and De Chernatony (1999) offer an excellent overview about the interplay between how management communicates its brand strategies and how consumers respond to management's moves. According to Goodyear's (1996) branding evolution model, there is a clear relationship and mutual interdependence of brand concept with level of development of the product and with the sophistication of the consumer. In concept research, this should show up as a change in the utility value of the brand over time as management varies its strategy. Indeed, by measuring the utility value of the brand name, one might be able to understand the net effect of these changes by management, and the consumer response to those changes.

The Role of Conjoint Measurement to Understand Aspects of the Brand

This chapter provides a quick tour through brand value, from the aspect of concept testing and specifically from the perspective afforded through conjoint measurement. The data show clearly that respondents can integrate brand information into their evaluations of product concepts.

Branding plays out in the world of food products. Researchers have long known that branding the stimulus by giving it a name can change the ratings. In product research, for instance, one can increase or decrease product acceptance and even change the perceived or at least the rated sensory characteristics of a product by labeling it with a well-known brand name. Even identifying the product by a well-known name, without the label, often

suffices to change the rating assigned to the product. Not all brand names add to the acceptance rating or make the sensory ratings more positive. Sometimes a brand name diminishes acceptance because the brand itself stands for poor product quality.

When dealing with concepts, branding comes into play in at least two ways:

1. *Brand names may be a repository for the different aspects of the product.* The different features of a message or a communication either fit or do not fit the brand. To operationalize this aspect of branding, the researcher makes the brand into an attribute. We saw this approach in Chapter 18 on first principles. Each respondent is given a test stimulus, such as a single promise statement or even a full concept or package, and instructed to rate the degree to which the test stimulus is congruent with the brand. To the degree that the brand sets up specific expectations the respondent will find some stimuli *close* to the brand and other stimuli *far away* from the brand. Researchers, marketers, product developers, and package designers all should be aware of the biases inherent in this type of research. Quite often respondents who like a test stimulus will say that it fits a brand even if it does not. One way to counter this bias is to give the respondents several such brand names and ask how the names fit the different brands. A second way is to instruct respondents to use a bipolar scale, with one end corresponding to Brand X and the other end corresponding to Brand Y. The second strategy ensures there are no right versus wrong, or good versus bad, scale regions and prevents the exercise from degenerating into a rating of acceptability.

2. *Brand names can become elements in a concept, on a par with other elements.* When brand becomes an element, the data show how powerfully the brand draws or repels respondents. One can measure the impact of a brand name in the same way that one measures the impact of a price or a benefit. To the respondents the brand name is simply another part of the communication. The

benefit of this approach is that it enables researchers to put all brand names, and indeed all elements, on a common utility scale. If branding is extremely important, then one should see that the utility value of the brand name is very high. If that is not the case, then the failure of the brand name to add utility means that branding as a name is not an acceptance driver. Branding may lead to expectations, but it is the expectations, and not the branding itself, that are critical.

Measuring the Performance of a Brand as an End Use

We will look at brand as an attribute by using two case histories, both for beverages and both of which deal with well-known brand names. Each of the studies follows the strategy outlined in the development of concepts from first principles that is discussed at some length in Chapter 18. This analysis looks at the fit of the concept elements to brand name.

Case History 1: Revisiting Carbonated Soft Drinks

Two of the end uses were fit to two major, well-known brands: 7-Up and Coca-Cola. A third end use was really an evaluation of interest in the beverage without positioning that beverage. It is important to keep in mind that some of the elements were conventional phrases that might apply to soft drinks, whereas others were novel elements that might be used for other beverages, but could be applied to carbonated soft drinks.

We can understand the impact of brand by looking at the data in Table 23.2. The elements are sorted so that those fitting 7-Up most strongly are in the first group. This includes the expected element "Invigorate your senses with shocking lemon-lime flavor" and three others that are associated with 7-Up. Only one nonspecific element, however, fits Coca-Cola: "Kick it up with a new highly carbonated drink." The other elements fit

moderately. The only element fitting both Coca-Cola and 7-Up simultaneously is the less specific statement "So light, so crisp, so refreshing." Elements that fit Coca-Cola do not fit 7-Up. Many of the elements do not fit either Coca-Cola or 7-Up.

A key aspect of studies investigating *fit to brand* is the richness of the data. Thus, from Table 23.2, an analyst gains a variety of insights about the brand Coca-Cola compared with the brand 7-Up. Some of the insights will be critical, others will be "nice to know," and still others will be irrelevant. For example, here are five additional insights that an analyst might glean from Table 23.2. There might be several more, if the analyst is privy to other market-level information:

1. *The concepts of Coca-Cola and 7-Up are reasonably well defined.* Coca-Cola is an exciting beverage. 7-Up is a lemon-lime beverage. Both brands allow for some, but not extreme, flavor departure from current products in the market.

2. *Flavor elements influence fit to brand.* They can increase or decrease fit to concept. For example, there is a clear separation of brands with the term *vanilla flavor*: "With a little splash of vanilla flavor . . . sure to delight." This vanilla element does not fit 7-Up, but it does fit Coca-Cola. Cherry fits 7-Up, not Coca-Cola. There is a history of Cherry Coke, but at the concept level it is not a big idea.

3. *Unusual flavors tend not to fit brands.* Other flavors like ginseng (E16), blueberry twist (E15), and "lurid colors" like punky purples (E12) do not fit these beverages, meaning that there is a strong congruence of flavor and brand.

4. *The term* caffeine *diminishes the fit of a concept to a product.* The element *caffeine* is rejected strongly with the element 7-Up (-10), but this does not mean that it is associated with Coca-Cola (-3).

5. *Although health is a popular topic, it does not promote fit to either of the two carbonated soft-drink brands.* The features or

Table 23.2. Fit of beverage ideas to 7-Up versus Coca-Cola, as well as fit to a new beverage that is simply defined as a carbonated beverage

		Acceptance of new beverage	Fit to 7-Up	Fit to Coke	7-Up minus Coke
	Base size	*118*	*111*	*124*	
	Additive constant	*37*	*45*	*32*	*13*
	Elements that fit 7-Up far more				
E10	Invigorate your senses with shocking lemon-lime flavor	−1	12	−2	14
E36	So light, so crisp, so refreshing	2	9	4	6
E17	A thrilling burst of unique cherry flavor and a sweet, crisp taste that gives you "more to go wild for"	8	7	3	4
E18	Introducing new clear natural refreshments with a light hint of flavor	4	6	−2	8
	Elements that fit Coke far more				
E09	Kick it up with a new highly carbonated drink	−5	1	11	−10
E05	With a thermal barrier your drink will stay colder longer	1	0	7	−7
E11	With a little splash of vanilla flavor . . . sure to delight	4	−7	6	−13
E26	Created for today's naturally healthy lifestyle	3	3	5	−2
E33	A drink that kids thirst for and moms will love	1	1	4	−4
E28	Enjoy a daring, high-energy, high-intensity, active drink	2	0	4	−4
E34	An energizer that keeps you going . . . without the caffeine	6	3	4	−1
	Elements that fit neither (very far away from one brand or both)				
E16	Satisfy your thirst . . . with real plum juice, ginseng and honey	1	−12	−14	2
E35	A jolt of caffeine to awaken your senses	−5	−10	3	−13
E04	Available in gallons to quench that giant thirst	−5	−10	0	−10
E30	A refreshing alternative to coffee . . . with a burst of caffeine	−4	−10	−3	−7
E12	Comes in a variety of flavors and crazy opaque colors like punky purple, brilliant blue, and goofy green . . . you got to try them all	−6	−7	−7	0
E19	Delivers at least 100% of the recommended daily intake of vitamin C, 15 % of folate, and 14% of potassium per 8 oz. serving	7	−6	1	−7
E15	Introducing new and exciting flavors such as blueberry twist, and wacky pink watermelon	−2	−5	−9	4

messages *health* (one exception, E26) and *energizer* (one exception, E34) do not fit the either brand.

Case History 2: Coffee

The goal of this study was to better understand the features of a *new coffee-based beverage* that could be sold in different venues (e.g., Starbucks, Burger King, and Dunkin' Donuts), as well as the features of the beverage that best fit certain emotional states. The data of interest here are the fit to the three major coffee brands: Folger's, Maxwell House, and Taster's Choice.

Table 23.3. Fit of coffee elements to three coffee brands: Folger's, Maxwell House, and Taster's Choice

		Taster's Choice	Maxwell House	Folger's
	Base size	*108*	*239*	*282*
	Additive constant	*57*	*56*	*57*
	Elements that fit Taster's Choice			
E02	A coffee that's guaranteed to wake you up	7	4	7
	Elements that fit Maxwell House			
E04	Dark fancy house blend . . . an extremely rich cup of coffee	1	6	3
E05	A distinctive, well rounded cup of coffee . . . the ideal way to start a busy day	3	6	6
	Elements that fit Folger's			
E02	A coffee that's guaranteed to wake you up	7	4	7
E05	A distinctive, well rounded cup of coffee . . . the ideal way to start a busy day	3	6	6
E13	New classic combination . . . pistachio & maple walnut . . . unleash the nutty side in you	−24	−30	−27
E12	White chocolate mousse, and wild raspberry . . . a melt in your mouth dessert in a flavored coffee	−17	−26	−25
E10	Invigorate your senses with Cinnamon Apple Spice & French Caramel	−22	−24	−27
E14	Chocolate and cognac give this coffee a flair . . . try it once and you'll come back for more	−15	−22	−22
E16	Vanilla and chocolate fudge combined . . . a unique flavor that is sure to please	−16	−21	−24
E18	Thrilling burst of vanilla flavor and sweet, crisp taste . . . gives you "more to go wild for"	−11	−16	−14
E15	Enjoy the taste of toffee in a light cream . . . a new summer favorite	−9	−14	−9
E09	The mini-drink six pack . . . iced coffee drinks for people on the go	−12	−13	−14
E17	Mocha and spicy Java create a one of a kind chocolate fantasy	−12	−13	−17
E06	A slightly caffeinated iced coffee drink . . . to help you get through your day	−9	−12	−10
E08	Iced to the max . . . for those hot summer days	−8	−10	−10
E35	Dark exotic taste . . . a superb Turkish brew	−2	−7	−7
E32	An Italian favorite . . . cappuccino with a flair	−2	−6	−5
E11	A unique flavor, sure to delight . . . sweet and smooth rich cream compliment this delectable treat	−4	−5	−7

Table 23.3 lists the results, sorted by fit to brand, for the three coffees. Taster's Choice shows only one strong fitting element: "A coffee that's guaranteed to wake you up." Maxwell House shows two elements that fit well, and Folger's shows three elements. What is key is that the degree of fit of the elements to the brand is fairly low for most elements, meaning that the brands are quite specific. What is also interesting for the three coffees is that a number of the best-fitting elements overlap, meaning that these elements are probably generic to coffee.

The data for coffee suggest that there are relatively few specific expectations about coffee brand. What fits each of these three coffees is pretty similar to what fits the other brands. It is also clear that many possible elements for a coffee beverage do not automatically fit the three brands. It is quite possible that with sufficient advertising some of the elements could eventually fit the three brands. Right now, however, the image of the coffee brand is quite defined, and most unusual elements do not fit in.

Again, by using brand as an attribute to be fit, the analyst can obtain some insights about what fits different coffee brands, and what does not, such as these four observations:

1. *No element possesses a clear relation to a single coffee brand.* In other words, among the elements tested here, no coffee brand *owns* an element, such that the element lies very close to that coffee brand. There may be other elements that a specific coffee owns, but these remain to be discovered.

2. *Some elements fit all three brands.* Those elements that communicate "wake you up" and "start a busy day" (e.g., E02, "A coffee that's guaranteed to wake you up") fit all three coffee brands. Thus, respondents generally find the notions of "wake you up" and "start a busy day" lie close to each brand. In other words, to create a unique brand it is advisable to avoid the conventional messages.

3. *The branded coffee is a stand-alone, pure product with a defined identity, although the identity may reside in the product, not in the brand.* A suggestion of unusual attributes, like "vanilla flavor," "chocolate fantasy," "pistachio," and "walnut maple" (E13, E10, E12, and E16), shows that respondents cannot link these attributes to a common brand, at least for coffee.

4. *Country of origin does not drive brand fit.* The stimuli dealing with country or continent of origin, such as African (E30), Jamaican (E28), and Turkish (E35), does not set up any expectations about fit to any of the brands of Taster Choice, Maxwell House, or Folger's. Folger's advertising and packaging emphasize the term *mountain grown*, but this does not translate into a country of origin.

Measuring the Performance of a Brand as an Element

In conjoint analysis the elements can be brand names as well as message. A number of the messages perform well, but other messages perform poorly. The same cannot be said for brand names, though, as we will see in the following. Brand names, for the most part, do not do particularly well, nor do they do particularly poorly, although overall there is a wide range. This means that the brand name as an element in the concept is not a particularly important factor, despite the extraordinarily high value placed on the brand, as Table 23.1 shows. It could be that the brand name itself, as part of a *white board* concept, does not play a particularly important role, despite its true importance.

We can see the relatively small contribution of brand name quite clearly from the results of the Crave It! studies run in 2001, 2002, and 2003. The mega-study database has been described in length in Chapters 20–22. The data listed in Table 23.4 come from the 2001 database, both for adults and for teens. Since the brand names are treated as elements just as other elements such as product features or emotions, one quickly discovers how well the brand names perform on a product-by-product basis. Table 23.4 also presents phrases about in-store branding, such as "From your local butcher shop" for bacon. This phrase can also be construed as a brand. At least nine observations emerge from inspecting Table 23.4:

1. *Wide range of utility values.* There is a wide absolute range in brand values. The highest brand value is $+19$ (cheesecake: "From the Cheesecake Factory"). The lowest brand value is -19 (cola: "From Tab").

2. *Range of utilities within a product category.* Even within a product category there can be a wide range of utility values for names, meaning that the brand name rather than the product drives brand value. If brand names were irrelevant, then they would all be equal, and all lie around 0.

3. *Difference between teens and adults.* Teens show a greater tendency to react to brands as compared with adults.

4. *No-name products (those without a brand).* They are often highly acceptable.

Table 23.4. Brand values for six brands, 29 food and beverage categories, from the 2001 Crave It! and Teen Crave It! databases (adults compared with teens)

	Average	Adult	Teen
Bacon			
From Oscar Mayer	4	2	6
From your local butcher shop	−1	2	−3
From Armour	−1	1	−3
From Smithfield Farms	−3	1	−6
From Bob Evans	−3	−2	−4
From Eckrich	−3	−1	−6
Chicken			
At Kentucky Fried Chicken	4	0	8
At Popeye's	2	−6	10
At Wendy's	0	−7	7
At Boston Market	−1	−3	0
At TGI Friday's	−2	1	−4
At McDonald's	−8	−12	−3
Hot dogs			
At Nathan's	8	4	12
From Ball Park Franks	7	3	12
A Kosher Frank	7	0	15
From Oscar Mayer	2	5	−1
From the local street vendor	−1	−2	1
Available at your favorite sporting event	−2	−2	−3
Tacos			
At Taco Bell	14	3	25
Kits available at your store from Taco Bell . . . make them your way	13	−3	29
Made fresh . . . especially for you . . . by you	6	2	10
From your favorite local Mexican eatery	3	3	2
At Chi-Chi's	−1	−7	5
At Chevy's	−2	−6	2

	Average	Adult	Teen
BBQ ribs			
With Jack Daniel's	9	−7	26
At TGI Friday's	8	−3	18
At Tony Roma's	3	−4	10
With Open Pit	3	1	5
With K.C. Masterpiece	−3	0	−6
At Lloyd's	−7	−11	−2
Hamburger			
At Wendy's	6	2	9
At McDonald's	3	−8	13
At White Castle	0	−5	5
At Burger King	−2	−1	−2
At Fuddruckers	−3	−4	−2
At Jack-in-the-Box	−5	−17	7
Steak			
At The Outback Steakhouse	8	3	14
At Applebee's	1	−7	9
At Morton's Steak House	−3	−2	−3
At Ruth Chris' Steakhouse	−3	−2	−4
At Steak Escapes	−7	−6	−8
At Sizzler	−12	−9	−14
Fries			
At Wendy's	10	−2	21
At McDonald's	9	3	14
At Morton's Steak House	7	0	15
At a Diner	2	0	3
At Burger King	−2	−4	−1
From Ore-Ida	−4	−3	−5

(continued)

Table 23.4. Brand values for six brands, 29 food and beverage categories, from the 2001 Crave It! and Teen Crave It! databases (adults compared with teens) *(cont.)*

	Average	Adult	Teen		Average	Adult	Teen
Brownies				Chocolate chip cookies			
At Mrs. Fields	7	2	12	From Nestlé	4	2	6
From Entenmann's	7	-3	17	From Chips Deluxe	3	-4	9
From Pillsbury	6	2	10	From your local bakery or pastry shop	2	1	2
From Sara Lee	5	1	9	From Pepperidge Farms	1	-2	3
From a shop specializing in making chocolate baked goods and fudge	3	3	4	At Mrs. Fields	0	-2	3
At Starbucks	2	-2	7	From Chips Ahoy!	-5	-5	-6
Cinnamon rolls				Cheesecake			
At Bob Evans	-6	-5	-8	Available by mail order	-5	-10	1
At Burger King	-10	-11	-9	From a fine local pastry shop	5	0	9
At Dunkin' Donuts	2	-2	6	From Philadelphia Brand	11	2	20
From Entenmann's	3	-1	7	From Sara Lee	7	2	13
From Pillsbury	2	0	5	From the Cheesecake Factory	19	4	34
From your favorite local bakery or pastry shop	-1	3	-6	Made by Eli's	1	-4	7
Donuts				Bread			
At Krispy Kreme	18	10	26	From the local bakery	9	2	16
At Dunkin' Donuts	10	1	19	From the special artisan baker	3	5	0
From Entenmann's	7	-5	18	From Pepperidge Farm	1	4	-2
Freshly made, especially for you	6	3	10	From your favorite store brand	1	-2	3
At your favorite bakery or pastry shop	3	2	4	From Au Bon Pain	-2	2	-5
From your favorite local diner or restaurant	-9	-8	-10	From Wonder Bread	-7	-5	-9
Ice cream				Chocolate candy			
From Haagen Dazs	11	5	17	From Hershey's	10	4	16
From Ben & Jerry's	11	2	19	From Godiva	8	4	12
At Baskin Robbins	11	2	19	From M&M/Mars	3	2	5
From Breyers	5	2	9	From Nestlé	1	1	0
From Edy's/Dreyers	4	-2	10	From Lindt	-2	0	-3
From Good Humor	1	-4	7	From Cadbury	-2	2	-5

(continued)

Table 23.4. Brand values for six brands, 29 food and beverage categories, from the 2001 Crave It! and Teen Crave It! databases (adults compared with teens) *(cont.)*

	Average	Adult	Teen
Fresh fruit			
Fresh fruit . . . just simple and pure	10	10	11
From Dole	7	-1	14
Grown on your own land and hand picked for freshness	6	3	10
From your local grocer . . . known for farm fresh produce every time	3	-3	8
From Sunkist	2	-1	5
From your local farm stand	-2	0	-4
Nuts			
From Blue Diamond	10	1	18
From Planter's	9	3	15
From Fisher's	4	1	7
From Mauna Loa	3	-1	6
From Frito Lay	2	-3	7
From Harry & David	1	-3	4
Peanut butter			
From Reese's	7	2	13
From Jif	6	4	9
From Skippy	3	3	3
From Smuckers	2	-2	6
From Peter Pan	0	1	-1
From Health Valley	-4	-3	-4
Popcorn			
From Cracker Jack	-1	-3	2
From Orville Reddenbacher	5	3	7
From Pop Secret	5	1	8
From Blockbuster	-1	-6	3
From the vendor at the fair, picnic, amusement park . . . wherever	-6	-1	-12
At the movie theater amusement park . . . wherever	7	2	12
From Herr's	-6	-1	-12

	Average	Adult	Teen
Cheese			
From Land O' Lakes	4	1	6
From Kraft	0	-3	2
From Velveeta	-1	-4	2
From Sargento	-2	2	-7
From Treasure Cave	-3	-1	-6
From Healthy Choice	-5	-6	-4
Olives			
From California	5	0	11
From Greece	3	-1	7
From Italy	2	3	2
Freshly prepared . . . especially for you	2	3	1
From Dean & Deluca	0	-1	2
Select from the Deli	-2	-2	-2
Pizza			
From Domino's	6	-2	14
At Pizza Hut	6	2	10
From DiGiorno	4	-3	11
At Papa John's	2	-2	7
From Tony's	2	-4	9
At Donato's	-2	-5	0
Pretzels			
From Combo's	6	0	12
From Rold Gold	5	2	8
Hot—from your favorite local street vendor . . . or warmed in your own oven	-1	7	-9
At Aunt Annie's	-2	2	-5
From Old Tyme	-3	1	-7
From Herr's	-5	0	-10

(continued)

Table 23.4. Brand values for six brands, 29 food and beverage categories, from the 2001 Crave It! and Teen Crave It! databases (adults compared with teens) *(cont.)*

	Average	Adult	Teen		Average	Adult	Teen
Potato chips				Tortilla chips			
From Lay's/Ruffles	16	3	29	From Doritos	0	5	−5
From Pringles	10	1	20	From Taco Bell	0	−5	4
From Herr's	5	−8	17	From Tostitos	0	4	−5
From Bistro Gourmet	4	−5	12	From Utz	−1	−4	2
From Utz	−4	−7	−1	From your favorite local Mexican eatery	−2	−2	−2
From Cape Cod	−5	−6	−5	From Guiltless Gourmet	−6	−2	−9
Water				Cola			
From Poland Spring	15	2	29	From Coca-Cola	14	13	14
From Evian	15	4	25	From Pepsi	12	0	25
From Aquafina	14	3	24	Dispensed fresh from the fountain,	7	3	11
By Brita	12	5	19	especially for you			
From Dasani	6	2	11	Classic cola	4	6	1
From your local bottled water	−3	−1	−4	From Dr Pepper	2	−5	8
company				From Shasta	−8	−20	4
				From Tab	−19	−23	−15
Coffee							
Freshly ground at your favorite food	10	1	19				
store from Starbucks							
Fresh ground and brewed coffee	7	10	4				
From Maxwell House	1	1	1				
From Folger's	0	−2	2				
From Nescafe	−5	−6	−5				
From Chock Full o'Nuts	−7	−11	−2				

Data courtesy of It! Ventures, Inc.

Table 23.5. Brand range versus total range (brand sensitivity): amount of the total range of element utilities spanned by the brands in the 2001 Crave It! study*

	Average	Adult	Teen
Average	*41*	*35*	*62*
Standard deviation	*14*	*17*	*19*
Potato chips	71	49	89
Donuts	65	49	78
Cheesecake	60	45	70
Water	58	24	94
Cola	56	91	60
Fries	55	33	90
Steak	54	34	72
Tacos	47	40	68
BBQ ribs	47	42	80
Coffee	47	67	62
Peanut butter	47	30	71
Popcorn	44	26	86
Chocolate candy	42	14	78
Bread	41	23	76
Cinnamon rolls	38	47	43
Hamburger	38	55	75
Chicken	37	40	47
Fresh fruit	36	43	45
Pretzels	36	38	50
Chocolate chip cookies	34	24	54
Cheese	32	28	54
Hot dogs	30	18	67
Ice cream	28	27	35
Olives	28	16	57
Pizza	27	23	42
Nuts	22	20	30
Bacon	20	7	48
Tortilla chips	19	33	41
Brownies	13	21	30

*Percentage is defined as [(range of brand utilities)/(range of all utilities)]. Numbers in the table body are percentages. Data courtesy of It! Ventures, Inc

Thus, the products from local grocery stores (+8 for fresh fruit), pastry shops (+9 for cheesecake), or specialty chocolate shops (+4 for brownies) are accepted, suggesting high implicit brand value.

5. *Brand sensitivity of teens.* Teens are exceptionally brand sensitive and differentiate between the individual products that a company makes. Thus, a company can have different brand values. For example, TGI Friday's has a good/high brand (+18) for

BBQ ribs, but a bad brand for chicken (−4). McDonald's has a good/high brand for hamburger (+13) and for fries (+14), but a bad brand for chicken (−3).

6. *Age differences in brand value for pizza.* Adults and teens differ most in their reactions to the brand values for pizza. Adults are all negative to pizza brand, whereas teens are nearly all positive.

7. *Universal positive brand values.* Some products, such as water and ice cream, show positive brand value for just about every brand tested.

8. *Biggest differences between adults and teens.* The biggest differences between adults and teens are for water, ice cream, and potato chips brands.

9. *Importance of brand name in general for a specific category.* A sense of the importance of brand name can be obtained by looking at the range of utilities (maximum to minimum) for brand names in the database, compared with the total range of utilities for all 36 elements. This ratio generates a value that can be compared across foods and across teens and adults. Table 23.5 shows the differences by food and by age groups. Teens are more brand sensitive than are adults, but the degree of brand sensitivity (compared with degree of sensitivity to all elements) varies dramatically by food. For example, potato chips are most brand sensitive, and brownies are the least brand sensitive. Brand sensitivity may have most to do with advertising budgets rather than with the nature of the food itself. It would be an interesting exercise to correlate advertising expenditures in a category with brand sensitivity to determine how strongly such expenditures drive response to brand.

European Brand Values: The Eurocrave! Database

Modeled on the Crave It! database, the Eurocrave! database comprised 36 elements for 20 foods and beverages for three countries

Table 23.6. Top three brands and bottom three brands in each of the three countries by total panel

Country	Food category	Food type	Brand	Total sample	Women	Men
Best-performing brands across three European countries						
France	Candy	Fruity sweets	De la marque Haribo	15	13	20
	Beverage	Cola	De la marque de Coca-Cola	15	16	13
	Candy	Mild chewing gum	De la marque Freedent	7	6	10
UK	Beverage	Cola	From Coca-Cola	9	9	10
	Candy	Mild chewing gum	From Wrigley's—Orbit	9	7	15
	Fast food	Hamburger	At Burger King	9	10	6
Germany	Beverage	Cola	Von Coca-Cola	16	19	12
	Snack	Chocolate	Von Lindt	11	10	12
	Candy	Fruity sweets	Von Haribo	10	8	15
Worst-performing brands across three European countries						
France	Beverage	Cola	De la marque Look Cola	−34	−33	−35
	Beverage	Cola	De la marque American Cola	−26	−31	−23
	Beverage	Cola	De la marque Virgin	−23	−25	−22
UK	Fast food	Chicken	At McDonald's	−17	−13	−25
	Fast food	Chicken	At Burger King	−16	−12	−25
	Beverage	Cola	From Tesco	−16	−15	−17
Germany	Fast food	Chicken	Bei Burger King	−16	−14	−20
	Fast food	Hamburger	Bei Heisse Hexe	−15	−11	−22
	Fast food	Chicken	Bei McDonald's	−11	−10	−15

Data from the Eurocrave! database, courtesy of It! Ventures, Inc.

(Germany, France, and the United Kingdom; see Chapter 21). Each country had its own set of products, with a fair number of overlapping products. Each country had its own concept elements, with a fair number of common elements. Finally, for the purposes of this chapter, each country had its own relevant brand set. Eurocrave! thus gives a sense of the brand values for brand names in each country, as rated by the inhabitants of each country. Furthermore, the Eurocrave! dataset allows for a comparison of men and women because the sample was more balanced in terms of gender than was Crave It! and other It! studies.

The first analysis simply looks at the top three brands in each country to get a sense of what does well. The supporting data are presented in Table 23.6. These four results emerge:

1. Brand Coca-Cola (a beverage) places among the top three elements in all three countries when the entire set of brand names is ranked independently of particular products.

2. In Germany and in the United Kingdom, the Haribo brand (a candy) lies in the top three.

3. There are no clear differences between men and women for these brands in the countries. Whatever differences exist seem to be random.

4. The worst-performing brands are competitors of Coca-Cola, especially those in France.

The Effect of Brand in General

The large number of brands and products across three countries allows us to look at the effect of brands, in general, by country. The analysis is presented in Table 23.7, showing brand sensitivity, and Table 23.8, showing the actual highest and lowest utility values for the brand by product and by country. We first need to specify the ranges of utilities

Table 23.7. Influence of brand name for 18 food products by European country*

Food item	Brand influence		
	France	Germany	United Kingdom
Fast food			
Hamburger	Strong	Strong	NS
Chicken	Strong	Strong	Strong
Pizza	Strong	Strong	NS
Snacks			
French fries	Strong	NS	NS
Tortilla chips	NS	NS	NS
Potato crisps	Mild	NS	NS
Healthy foods			
Cheese	Strong	NS	NS
Yogurt	Strong	NS	NS
Cereal bar	NS	NS	NS
Soup	NS	NS	NS
Candy			
Chewing gum	NS	Mild	Mild
Gummies	Strong	Mild	NS
Mint	Mild	Mild	Mild
Chocolate	Mild	Mild	NS
Beverage			
Cola	Strong	Strong	Strong
Coffee	NS	NS	NS
Iced tea	Mild	NS	NS
Orange juice	Mild	NS	NS

*Results based on utility values of brand name. NS, not significant.
Data courtesy of It! Ventures, Inc.

for strong-performing compared to weak-performing elements. The specification below comes from observations made in thousands of studies. The specification could be made in terms of statistical differences, but it is easier to develop specifications that relate to performance in the marketplace, such as utilities corresponding to winning versus losing products:

Level of performance	Utility value
1. Strong performer	>10
2. Modest performer	5–10
3. Weak performer	0–5
4. Poor performer	<0

From Tables 23.7–23.9, we see the following seven trends:

1. *Where brand name has its greatest impact.* In general, brand has the most influence on consumers of cola and fast-food items, which items tend to be heavily advertised by global companies, and therefore consumers might be more both brand aware and brand loyal.

2. *Where brand name has its smallest impact.* In general, brand has the least influence for snacks and for select health foods (cereal bars and soups). Cereal bars are a new product category, and therefore consumers may not have had sufficient exposure to the category to become brand loyal. Soups, french fries, and chips are often purchased in restaurants and other food-service locations. In these locations, brand name is not generally available as a label on the product at the time of consumption. Therefore, consumers may be less aware of brand name for these food categories.

3. *Difference by country to brand name.* French consumers are the most influenced by

Table 23.8. Highest versus lowest utilities achieved by brand names, in the Eurocrave! database, for comparable product categories across countries

Food item	Highest utility achieved by a brand name			Lowest utility achieved by a brand name			Utility range for brand names		
	France	Germany	UK	France	Germany	UK	France	Germany	UK
Fast food									
Hamburger	5	9	9	−16	−15	−9	21	24	18
Chicken	9	−5	2	−21	−16	−17	30	11	19
Pizza	15	15	8	−7	−10	−8	22	25	16
Snacks									
French fries	13	5	5	−18	−5	−7	31	10	12
Tortilla chips	2	5	6	−7	−3	−8	9	8	14
Potato crisps	2	7	4	−11	−6	−8	13	13	12
Healthy foods									
Cheese	4	6	2	−14	−7	−9	18	13	11
Yogurt	6	7	4	−16	−2	−4	22	9	8
Cereal bar	4	3	7	−9	0	−2	13	3	9
Soup	1	2	3	−6	−3	−5	7	5	8
Candy									
Chewing gum	7	10	9	−6	−6	−10	13	16	19
Gummies	15	10	3	−8	−8	−9	23	18	12
Mint	6	4	5	−13	−10	−10	19	14	15
Chocolate	10	11	8	−6	−4	−3	16	15	11
Beverages									
Cola	23	16	9	−34	−11	−16	57	27	25
Coffee	7	4	1	−7	−2	−6	14	6	7
Iced tea	2	6	1	−12	−8	−3	14	14	4
Orange juice	2	6	−1	−12	−8	−6	14	14	7

Data courtesy of It! Ventures, Inc.

brand name, followed by German consumers and then by UK consumers.

4. *Impact of brand name compared with all elements.* We can look at the relative impact of brand versus all elements for each country, for each food, for the total panel, and for men compared with women (Table 23.9). The data show clear differences in the sensitivity to brands. The statistic for the effect of brand is the range in element utilities attributed to brand divided by the total range. That percentage shows how much of the total range can be traced to range in brand. To the degree that the ratio approaches 1.00 or the percentage value approaches 100, we can conclude that brand is increasingly important. The utility values for brands are just as

different as the utility values for the other elements. When the ratio approaches 0, we must conclude that the brand is far less important.

5. *Food brand sensitivity.* Foods and beverages such as cola and traditional mints show the highest brand sensitivity. Surprisingly, tea and coffee show the lowest brand sensitivity.

6. *Gender difference in brand sensitivity.* Women show somewhat higher sensitivity to brands than do men, but not unusually higher.

7. *Country difference in brand sensitivity.* Brand sensitivity varies somewhat by country, with France highest, followed by the United Kingdom and then Germany.

Table 23.9. Brand sensitivity for foods in Europe: percent of the total range utility range across elements spanned by the utilities of the brands*

	All Europe			UK			France			Germany		
		Gender			Gender			Gender			Gender	
	Total	F	M	Total	F	M	Total	F	M	Total	F	M
Mean	*49*	*54*	*44*	*47*	*47*	*56*	*51*	*57*	*46*	*44*	*52*	*51*
Standard deviation	*18*	*17*	*20*	*18*	*21*	*26*	*28*	*26*	*25*	*25*	*26*	*18*
Cola	87	83	94	78	72	93	100	100	100	83	77	90
Traditional mint	77	82	70	57	64	42	100	100	88	75	83	79
Chewing gum	76	71	68	100	88	83	67	57	65	63	68	55
Fruity sweets	73	65	75	50	52	53	87	77	100	83	65	72
Potato crisps	61	68	38	61	69	18	65	59	65	56	76	31
Hamburger	61	66	69	54	66	51	69	71	71	59	63	86
Chocolate	58	57	46	42	40	36	57	61	52	77	70	50
Pizza	57	51	66	40	34	67	74	69	66	0	0	0
Yogurt	56	57	57	37	39	39	89	85	89	41	47	42
Ice cream	51	51	47	47	48	39	0	0	0	55	53	54
Lemonade	50	28	71	35	26	72	61	17	86	55	42	54
Cheese	46	38	56	36	46	30	49	32	60	53	35	78
Chicken	42	38	52	53	44	67	41	45	39	33	25	50
Iced tea	41	35	49	19	16	42	43	50	36	60	40	68
Fries	39	46	47	42	56	48	44	48	36	33	34	57
Crispy cereal bar	36	38	32	35	47	17	58	43	52	15	24	27
Orange juice	34	49	42	21	48	56	38	43	38	43	56	32
Soup	34	26	60	42	41	49	43	19	69	18	16	63
Tortilla chips	33	26	44	24	29	17	40	20	37	37	30	77
Tea	26	30	32	35	47	32	17	13	31	0	0	0
Coffee	23	27	27	16	14	29	36	52	26	18	14	27

*Numbers in the body of the table are percents. F, female; and M, male. Data from the 2002 Eurocrave! study.

A Tour Through Some Foods in Europe

The Eurocrave! database shows a number of interesting patterns for brand value. We look at a few of these in the next sections on a food-by-food basis. There are no clear overriding patterns for brands, nor should we expect there to be. The brand effect is a function of product, country, competitive frame, gender, and so on. However, the results suggest that respondents have no trouble integrating brand into their evaluations, because if the respondents had trouble we would expect to see utilities that are strongly positive, strongly negative, or hovering around 0. They do not. The respondents discriminate among the brands. The brand names vary in their utilities just as do the other aspects of the food and beverage concepts. The one

clear outcome is that branding value can be assessed with conjoint analysis worldwide and not just in the United States.

Chicken in a Fast-food Restaurant

Fast-food restaurants do poorly in brand name when their brand name is associated with chicken (see Table 23.10). In the chicken studies, respondents knew that they were evaluating concepts dealing with chicken. The respondents down-rated all of the restaurants when the context was a chicken product, even down-rating specialty restaurants such as Kentucky Fried Chicken. The low rating is due to the brand, and not to the negative response to chicken. It could be that the chicken served at restaurants is not of the same quality as what consumers would like, and therefore the respondents down-rate the restaurant name.

Table 23.10. Utilities for fast-food restaurants in Europe when the restaurant names are embedded in a study of chicken as a product

Country	Brand-chicken	Total sample	Women	Men
	Kentucky Fried Chicken			
France	Chez Kentucky Fried Chicken	−21	−21	−22
UK	At Kentucky Fried Chicken	2	2	1
Germany	Bei Kentucky Fried Chicken	−5	−7	−1
	McDonald's			
France	Au Mc Donald's	−20	−21	−19
UK	At McDonald's	−17	−13	−25
Germany	Bei McDonald's	−11	−10	−15
	Burger King			
UK	At Burger King	−16	−12	−25
Germany	Bei Burger King	−16	−14	−20

Data courtesy of It! Ventures, Inc.

Cola

The performance of the colas clearly varies tremendously, as shown in Table 23.11. What is so remarkable is the enormous brand advantage held by Coca-Cola across the three European countries, and the equally but negative utilities shown by competitors.

Candy (Haribo)

Haribo candy shows a mixed performance (Table 23.12). It is strong in France and weak in the United Kingdom. Haribo is stronger among men than among women in all three countries.

Cheese and the Country as a Brand

In Europe, cheeses are identified by their country of origin. Such country-specific branding has an effect on brand value, most clearly for France (Table 23.13). French respondents evaluated the brand "French cheese" more positively than they rated cheese of other countries. Likewise the English respondents assigned their own cheese and the French cheese good brand images. Only in Germany were respondents positive to all country-level branding.

Country names as brand names for cheese and potentially for other products is not new. A number of papers over the years have ad-

dressed the use of country as a brand name (Kotler and Gertner 2002; Papadopoulos and Heslop 2002). For over three decades, researchers have investigated the effects of country image on consumers' product evaluations (Srikatanyoo and Gnoth 2003). Country-based marketing is often underused or misdirected because of inadequate understanding of the meaning of *country branding*. Indeed, a previous study on the magnitude of brand loyalty toward country, state, and service provider showed that brand loyalty toward the country is strongest, followed by that toward the state and service provider, and even more stable than loyalty toward the service brand (Paswan et al. 2003).

Cross-product Brand (McCain)

McCain occupies an interesting position. McCain crosses a number of categories, from fries to hamburger to pizza. In all areas it performs either neutral or poorly (Table 23.14).

Brand Name versus No Brand (Artisan Chocolate)

A good brand need not be bound to a name. The image of individual chocolate shops and the preparation of handmade chocolate got a better evaluation than did the chocolate of some international companies (Table 23.15).

Table 23.11. Utilities for cola brands in Europe

Country	Brand—cola	Total	Women	Men
France	De la marque de Coca-Cola	15	16	13
	De la marque de Pepsi Cola	−15	−21	−9
	De la marque Virgin	−23	−25	−22
	De la marque American Cola	−26	−31	−23
	De la marque Look Cola	−34	−33	−35
Germany	Von Coca-Cola	16	19	12
	Von Pepsi Cola	2	1	3
	Von Africola	0	−2	2
	Von Jolt	−9	−10	−9
	Von River Cola	−10	−10	−10
	Von Sinalco Cola	−11	−9	−12
UK	From Coca-Cola	9	9	10
	From Pepsi Cola	3	5	−3
	From Virgin	−12	−11	−13
	From Sainsbury's Classic	−13	−11	−17
	From Tesco	−16	−15	−17

Data courtesy of It! Ventures, Inc.

Table 23.12. Performance of Haribo candy brand in three countries

Country	Brand—Haribo	Total	Women	Men
France	De la marque Haribo	15	13	20
Germany	Von Haribo	10	8	15
UK	From Haribo	0	−1	5

Data courtesy of It! Ventures.

Table 23.13. Utility of "country as brand" for cheese

Country	Brand—cheese	Total	Women	Men
France	De France	4	4	5
	De Hollande	−14	−7	−19
	De Grèce	−10	−8	−13
	De Suisse	−7	−1	−11
	D'Italie	−3	3	−9
Germany	Aus Deutschland	2	3	−2
	Aus Holland	4	6	−2
	Aus Griechenland	3	5	−1
	Aus der Schweiz	2	2	3
	Aus Frankreich	6	4	9
	Aus Italien	4	5	2
UK	From Britain	2	2	−2
	From Holland	−7	−4	−18
	From Greece	−7	−6	−10
	From Switzerland	−3	−2	−9
	From France	1	3	−6
	From Italy	−4	−4	−4

Brand Value Within the Context of "Good for You" Foods: The Healthy You! Database

The Healthy You! database comprises 29 product categories, each having 36 elements per category, following the format of the other It! studies. Among these 36 elements are four in each study representing brand names relevant to the particular product category. Respondents in the Healthy You! studies

Table 23.14. Performance of the brand name
McCain across categories

Country	Brand	Total
	Fries	
France	De McCain	−4
Germany	Von McCain	−2
UK	From McCain	−1
	Fast food	
France	Hamburger de la marque McCain	−16
UK	Pizza from McCain	−6

were invited to participate in an Internet-based study dealing with *good for you* products. Thus, all of the individual studies, one per product, began with the same basic positioning, and all of the brand values must be considered in the context of a good-for-you product.

Let us look at the brand sensitivity for these good-for-you foods in the same way that we looked at brand sensitivity for other products (Tables 23.5 and 23.9). Five trends emerge from Table 23.16:

1. *General brand sensitivity for good-for-you foods.* This sensitivity is lower for such foods. On average the brands cover approximately 33% of the full range of the concept elements. This range is lower than the 41% range achieved by brands in the Crave It! database and the 49% range achieved by brand in the Eurocrave! database. Thus, the brands appear to cover about a third of the full utility range.

2. *Gender and age effects on general brand sensitivity.* Neither gender nor age has a clear differential effect on relative sensitivity to brand versus the other, nonbrand elements

3. *Wide range of utilities.* There is a very wide range in sensitivity to brands across food and beverage categories.

4. *Large food-to-food differences.* Some products, such as frozen fish and canned fruit, are enormously brand sensitive. These brand-sensitive foods tend more to be commodity items, where brand does its differentiating. We don't think of sensory characteristics for frozen fish or canned fruit. In contrast, salsa and milk shakes are not brand sensitive, or at least there is little differentiation across brands. This may be the case because people don't pay attention to brands or because the brands are all poor.

5. *Gender and age effects on the utility of brands, whereas age does not.* Gender affects the utility of brands, whereas age does not. For example, brand differences play a greater role for men than for women when it comes to frozen fish, canned fruit, rice mix, nuts, salsa, and chocolate candy, but play a greater role for women than for men when it comes to bread, pasta, water, and salad dressing. Age does not seem to show any clear pattern except for yogurt and coffee.

Table 23.15. Performance of chocolate brands, including branding as an "artisan" product

Country	Brand-chocolate	Total	Women	Men
France	Fait à la main par un artisan chocolatier	10	7	13
UK	Handmade, from a chocolate shop	5	4	6
France	De la marque Nestlé	−2	−2	−3
UK	From Nestlé	−3	−3	3
France	De la marque Mars	−6	−8	−3
UK	From Mars	−1	−1	−1
France	De la marque Suchard	0	−1	1
UK	From Suchard	−1	−2	1

Table 23.16. Brand sensitivity for the Healthy You! database*

	Total	Gender		Age				
		Male	Female	20–29	30–39	40–49	50–59	60–69
Average	*33*	*37*	*32*	*33*	*32*	*40*	*37*	*40*
Standard deviation	*13*	*19*	*12*	*17*	*18*	*14*	*17*	*19*
Frozen fish	66	78	58	61	56	68	76	63
Canned fruit	52	89	43	14	53	43	63	66
Pretzels	50	51	49	26	55	63	48	30
Margarine	47	49	47	35	35	36	68	74
Vegetable burger	47	41	49	36	43	35	58	49
Peanut butter	45	48	33	35	69	34	37	35
Soup	42	46	36	42	40	38	27	47
Milk	41	44	40	49	62	29	31	59
Bread	40	16	42	66	9	60	25	34
Rice mix	39	58	37	21	45	55	13	85
Water	38	23	42	50	35	36	21	48
Juice: citrus	38	33	33	55	44	23	20	33
Coffee	37	32	36	16	33	27	42	53
Pasta sauce	36	34	37	22	40	43	36	35
Tea	33	34	35	35	16	48	19	18
Cracker	32	36	31	47	19	17	40	41
Yogurt	30	44	30	24	16	44	39	52
Frozen meal	29	15	30	14	11	18	58	33
Pasta	26	7	35	18	13	53	9	10
Cold breakfast cereal	26	29	25	12	29	27	23	27
Nuts	24	52	18	52	50	67	40	28
Chews	22	28	13	14	21	41	42	11
Salad dressing	21	11	25	49	15	26	39	32
Chocolate candy	21	51	19	10	37	48	46	31
Cheese	20	21	19	56	1	34	39	42
Energy bar	19	26	23	31	15	51	31	7
Juice: noncitrus	18	13	21	18	22	34	45	18
Salsa	14	30	8	28	13	36	34	56
Milk shake	12	27	14	21	37	15	15	33

Sensitivity is defined as the percentage of the total range of element utilities spanned by the brand names within a study. All numbers in the body of the table are percentages.
Data courtesy of It! Ventures, Inc.

Learning from an In-depth Analysis of Specific Brand Names

A more granular analysis of the brand values comes from assessment of the individual brand names from the Healthy You! database. The data for individual product categories and brands are listed in Table 23.17. Though one misses the grand overview, some insights can be gained from inspection of cross-sectional data across different products. Some of the insights are the following:

Most brand names perform only modestly. A few brands do quite well, such as the Godiva brand for chocolate, the Lean Cuisine brand for frozen meals and vegetable burgers, and chocolate from Hershey's. Brands can be quite negative, though. For example, two brands of milk do quite poorly: Lactaid and Organic Valley. The remaining brands do

Table 23.17. Utility values for brands from the Healthy You! study

Category	Positive-performing brands	Utility	Category	Negative-performing brands	Utility
Chocolate	From Godiva	8	Nuts	From Planters	0
Frozen meal	From Stouffers/Lean Cuisine	7	Chews	From Quaker	0
Vegetable burger	From Morningstar Farms	7	Chews	From GNC	0
Chocolate	From Hershey's	7	Bread	From Wonder Bread	0
Tea	From Lipton	6	Juice: noncitrus	From V8	0
Chocolate	From Nestlé	6	Peanut butter	From Peter Pan	0
Cheese	From Kraft . . . Cracker Barrel brand	6	Water	From Poland Spring	0
Juice: noncitrus	From Ocean Spray	5	Pretzels	From Auntie Anne's	0
Pretzels	From Rold Gold	5	Coffee	From your favorite grocery store brand	0
Water	From Aquafina	5			
Frozen fish	From Van de Kamps	5	Margarine	From Parkay	0
Margarine	From I Can't Believe It's Not Butter	5	Pasta	From Buitoni	0
Peanut butter	From Jif	5	Pasta sauce	From Five Brothers	0
Salsa	From Old El Paso	5	Tea	From Twinings	0
Cracker	From Nabisco	5	Salad dressing	From Newman's Own	−1
Soup	From Campbell's	5	Energy bar	From Nutri-Grain	−1
Rice mix	From Uncle Ben's	5	Soup	From Lipton	−1
Pretzels	From Snyder's of Hanover	5	Yogurt	From Stonyfield Farm	−1
Rice mix	From Rice A Roni	5	Chews	From Calci-Wise	−1
Cold breakfast cereal	From Kellogg's	5	Energy bar	From Luna/Clif	−1
Yogurt	From Yoplait	5	Peanut butter	From Smuckers	−1
Canned fruit	From Del Monte	5	Margarine	From Smart Balance	−1
Frozen fish	From Mrs. Pauls	4	Nuts	From Mauna Loa	−1
Vegetable burger	From Boca	4	Pasta	From Ronzoni	−2
Juice: citrus	From Tropicana	4	Canned fruit	From your favorite grocery store brand	−2
Cold breakfast cereal	From Post	4	Margarine	From Take Control	−2
Pasta sauce	From Prego	4	Salad dressing	From Ken's	−2
Cheese	From Sargento	4	Cracker	From Sunshine	−2
Juice: noncitrus	From Minute Maid	4	Milk shake	From Kashi Go Lean	−2
Canned fruit	From Dole	3	Peanut butter	From Skippy	−2
Cracker	From Pepperidge Farm	3	Juice: citrus	From Veryfine	−2
Frozen meal	From Swanson's	3	Energy bar	From Balance	−2
Pasta sauce	From Ragu	3	Tea	From Tazo	−2
Salsa	From Ortega	3	Milk	From Dean's	−2
Salad dressing	From Kraft	3	Canned fruit	From Wild Oats	−3
Chocolate	From M&M's	3	Soup	From Healthy Choice	−3
Coffee	From Maxwell House	3	Vegetable burger	From Quorn	−3
Water	From Dasani	3	Bread	From Arnold	−3
Vegetable burger	From Gardenburger	3	Milk shake	From EAS	−3
Cheese	From Land O'Lakes	3	Pretzels	From Bachman	−4
Yogurt	From Breyers	3	Milk shake	From Slim-Fast	−4
Pasta	From Barilla	3	Milk shake	From Ensure	−4
			Energy bar	From PowerBar	−4

(continued)

Table 23.17. Utility values for brands from the Healthy You! study *(cont.)*

Category	Positive-performing brands	Utility	Category	Negative-performing brands	Utility
Salsa	From Pace	3	Pasta sauce	From Healthy Choice	−5
Yogurt	From Dannon	3	Frozen fish	From Fisher Boy	−5
Frozen fish	From Gorton's	3	Rice mix	From Near East	−5
Nuts	From Blue Diamond	2	Frozen meal	From Weight	
Rice mix	From Lipton	2		Watchers	−5
Milk	From your favorite		Coffee	From Chock Full	
	local dairy	2		o'Nuts	−5
Cold breakfast			Juice: citrus	From Fresh Samantha	−6
cereal	From General Mills	2	Milk	From Organic Valley	−7
Juice: noncitrus	From Welch's	2	Milk	From Lactaid	−8
Soup	From Progresso	2	Bread	From Pepperidge Farm	2
Nuts	From Diamond of		Bread	From Earth Grains	2
	California	2			
Juice: citrus	From Minute Maid	2			
Salad dressing	From Wishbone	2			
Water	From VitaminWater	2			
Cheese	From Borden	1			
Pasta	From Mueller's	1			
Frozen meal	From Uncle Ben's	1			
Tea	From Celestial				
	Seasonings	1			
Chews	From Viactiv	1			
Cracker	From Keebler	1			
Salsa	From Tostitos	1			
Chews	From WalMart	1			
Coffee	From Starbucks	1			

modestly. Utilities for most of the other brands lie in region between +5 and −5.

A comparison of utilities across databases shows reasonable but not perfect reproducibility. The utility values are lower for the Healthy You! database than for the Crave It! database. The utility values will be lower than the utility values for the brands in the Crave It! database (Table 23.4). However, the same element may perform either better or worse in Healthy You! compared to Crave It! For example, Land O' Lakes cheese was tested as a brand in both. It scores +4 when in Crave It! and +3 in Healthy You! In contrast, Sargento cheese scored −2 in Crave It! but +4 in Healthy You! Although there are flip-flops, there are no enormous reversals that would lead one to think that the utility of

the brand is a function of the in-going positioning of the study.

Brands are not necessarily all positive or all negative within a single product category. There are categories in which the brands of these companies are generally rated positively. On the other hand, one finds categories where the brands of these companies are generally rated negatively (energy bar, milk, and milk shake). The exception for milk is the nonbrand "From your favorite local dairy." Likewise, there are categories of products where brand is almost irrelevant, such as chews (from −1 to +1).

The same brand can do well in one product and not well in another. An example is Lipton, which does well for tea, but not for soup.

Sometimes a product does exceptionally well in a product category, suggesting that it has bonded with that category. For example, in frozen meals, Stouffer's does exceptionally well. For frozen fish, the Van de Kamp brand does well. In contrast, other brands, such as Weight Watchers and Fisher Boy, do not do well.

Overview

This chapter provides a quick tour through brand value, from the aspect of concept testing and specifically from the aspect of conjoint measurement. The data show clearly that respondents can integrate brand information into their evaluations of product concepts and that the brand information plays a role. What emerges consistently, however, is that the role can be either positive or negative but in general is usually less than that of the winning element. Brands cannot do too much harm, but they also do not do much good in the concept. Despite what one might say, the actual messages themselves appear to outweigh the brand name, although one would not wish to begin the marketing process with a name that has a substantial negative utility. The good news is that almost no elements in the United States have such large negative utilities that it is impossible to overcome them. The bad news is that perhaps in Europe these very low, strongly negative utilities do exist and can drag down any concept in which they appear.

References

Aaker, D. (1996). Building strong brands. New York: Free Press.

Aaker, D., and Joachimsthaler, E. (2000). Brand Leadership, 1st edition. London: Free Press Business/Simon and Schuster.

Batra, R., Lehmann, D.R., and Dipinder, S. (1993). The brand personality component of brand goodwill: some antecedents and consequences. In: Aaker, D.A., and Biel, A.L., editors. Brand Equity and Advertising. Hillsdale, NJ: Lawrence Erlbaum.

Blumenthal, D. (2002). Beyond "form versus content" Simmelian theory as a framework for adaptive brand strategy. Journal of Brand Management, 10: 9–18.

Boone, L.E., and Kurtz, D.L. (1992). Contemporary Marketing. Hinsdale, IL: Dryden.

Brown, M. (1992). How It All Began. Otley, UK: Smith Settle.

Callaghan, W.M., and Wilson, B.J. (1998). The role of the category in brand equity studies: a brand attitudinal segmentation perspective. Working paper. Melbourne: RMIT University.

De Chernatony, L. (1998). Developing on effective brand strategy. In: Egan, C., and Thomas, M., editors. The Chartered Institute of Marketing Handbook of Strategic Marketing. Oxford: Butterworth-Heinemann.

De Chernatony, L., and Riley, F.D.O. (1997). The chasm between managers' and consumers' views of brands: the experts perspectives. Journal of Strategic Marketing, 5: 89–104.

Dekimpe, M.G., Steenkamp, J.B.E.M., Mellens, M., and Abeele, P.V. (1997). Decline and variability in brand loyalty. International Journal of Research in Marketing 14: 405–420.

Farquhar, P.H. (1989). Managing Brand Equity. Marketing Research, 1 (September): 24–33.

Fournier, S., and Yao, J. (1997). Reviving brand loyalty: a reconceptualization within the framework of consumer brand relationships. International Journal of Research in Marketing, 14: 451–472.

Gardner, B.B., and Levy, S.J. (1955). The product and the brand. Harvard Business Review, 33: 33–39.

Goffman, E. (1959). The Presentation of Self in Everyday Life. Garden City, NY: Doubleday.

Goodyear, M. (1996). Divided by a common language: diversity and deception in the world of global marketing. Journal of the Market Research Society, 38: 105–122.

Gordon, W. (1991). Assessing the brand through research. In: Cowley, D., editor. Understanding Brands. London: Kogan Page.

Hilton, S. (2003). How brands can change the world. Journal of Brand Management, 10: 370–378.

Hoyer, W., and Brown, S. (1990). Effects of brand awareness on choice for a common repeat purchase product. Journal of Consumer Research, 17: 141–148.

Jacoby, J., Olsen, J., and Haddock, R. (1971). Price, brand name and product composition characteristics as determinants of perceived quality. Journal of Applied Psychology, 55: 570–579.

Jacoby, J., Szybillo, G., and Busato-Schach, J. (1977). Information acquisition behavior in brand choice situations. Journal of Consumer Research, 3: 209–216.

Jones, J. (1986). What's in a name? Lexington, MA: Lexington Books.

Kamakura, W.A., and Russell, G.J. (1993). Measuring brand value with scanner data. International Journal of Research in Marketing, 10: 9–22.

Kappor, J. (2003). 18 brand astras: using brand abilities as weapons for CRISP brand building. Mumbai, India: Samsika Marketing Consultants.

Keller, K.L. (1993). Conceptualizing, measuring, and managing customer-based brand equity. Journal of Marketing, 57: 1–22.

Kotler, P., and Armstrong G. (1996). Principles of Marketing. Upper Saddle River, NJ: Prentice Hall.

Kotler, P., and Gertner, D. (2002). Country as brand, product, and beyond: a place marketing and brand management perspective. Journal of Brand Management, 9: 249–261.

Krishnan, H. S. (1996). Characteristics of memory associations: A consumer-based brand equity perspective. International Journal of Research in Marketing, 13: 389–405.

Lassar, W., Mittal, B., and Sharma, A. (1995). Measuring customer-based brand equity. Journal of Consumer Marketing, 12: 11–19.

Leahy, T. (1994). The emergence of retail brand power. In: Stobart, P., editor. Brand Power. Basingstoke, UK: Macmillan.

Levitt, T. (1983). The Marketing Imagination. New York: Free Press.

Mackay, M.M., Romaniuk, J., and Sharp, B. (1997). A typology of brand equity research. In: Australia and New Zealand Marketing Educators Conference, Melbourne, 1–3 December, pp. 1146–1157.

McEnally, M., and De Chernatony, L. (1999). The evolving nature of branding: consumer and managerial considerations. Academy of Marketing Science Review no. 02; http://www.vancouver.wsu.edu/amsrev/theory/hupfer03-2002.html.

Papadopoulos, N., and Heslop L. (2002). Country equity and country branding: problems and prospects. Journal of Brand Management, 9: 294–314.

Paswan, A.K., Kulkarni, S., and Ganesh, G. (2003). Loyalty towards the country, the state and the service brands. Journal of Brand Management, 10: 233–251.

Plummer, J.T. (1985). How personality makes a difference. Journal of Advertising Research, 24: 27–31.

Prentice, D.A. (1987). Psychological correspondence of possessions, attitudes and values. Journal of Personality and Social Psychology, 53: 993–1003.

Reynolds, T.J., and Gutman, J. (1984). Advertising as image management. Journal of Advertising Research, 24: 27–38.

Schiffman, L.G., and Kanuk, L. (1996). Consumer Behavior. Upper Saddle River, NJ: Prentice Hall.

Southgate, P. (1994). Total branding by design. London: Kogan.

Srikatanyoo, N., and Gnoth, J. (2003). Country image and international tertiary education. Journal of Brand Management, 10: 139–146.

Weitz, B., and Wensley, R. (2002). Handbook of Marketing. London: Sage.

Wells, W.D. (1990). A proposal for an international brand equity study. New York: DDB Needham Worldwide. Unpublished document.

Chapter 24

Emotion in Concepts

Part I: Emotion Elements in Product Concepts

Introduction

The issue of *emotionality* in concepts continues to arise. As product developers, ingredient suppliers, and marketers recognize the importance of emotions as a driver of food choice, they look for the communication of these emotions in concepts. In the academic literature the work with concepts deals with the nature of how we process information. Advertising research recognizes the difference between information and emotion in concepts (Golden and Johnson 1982; Lautman and Percy 1983). However, there is really a dearth of practical information in the scientific and business literatures on how to convey emotions in food concepts, what constitutes emotion in a concept, and even whether emotions are relevant. Sometimes the advertising agency may claim that the emotions are those hard-to-capture communications that are best presented by pictures. The data from studies with graphics show, however, that at least at the level of concepts it is hard to see how these graphics drive interest. They may drive emotional responses, but those emotional responses do not transfer to high-utility values for purchase intent. For example, the large-scale coffee study with 38 visuals shows no greater performance of visuals than text elements as drivers of acceptance (see Chapter 8).

This chapter looks at the issue of emotion from a more pragmatic, data-analytic perspective. The chapter presents the performance of elements that we refer to as *emotional*, comparing their performances to those of the remaining elements. This analysis shows the power of emotion as an acceptance driver. In this respect the It! studies come into play, because they provide a wealth of data on elements that we would call emotional. The information is available from the databases because they were deliberately inserted into the set of concept elements to study the emotional aspect of concepts.

Emotion Is Important

When respondents are asked to select the factors that drive the craveability of food, many select mood, whereas others select celebration, etc. These are emotion-laden situations. In the Crave It! study, for example, the respondents were instructed to select up to three aspects of the product and the situations that they felt were important as drivers of their craving. From their self-profiling it appears that selection or at least craveability of a variety of foods is driven by emotion. Table 24.1 lists the foods that are most often and least often associated with a variety of emotions, such as mood itself, celebration, associations, and family and friends. It is clear from this list that emotions play a role in driving interest, at least at the level of one's self-profiling. Depending on the particular food, however, the emotion involved will change, at least when respondents are asked to think about the role of emotion

We see from data in Table 24.1, and from some extrapolations, the following covariation of mood, or mood-related states and food:

1. Mood: chocolate, sugar, fat
2. Celebration: meat, heavy fat (cheese-cake)
3. Association: finger foods
4. Family and friends: meats
5. Escape: sweet, fats, coffee
6. Hormones: sweet, fat

Defining an Emotion Element in a Concept

The objective of information in a concept is to portray the product verbally and perhaps drive interest in purchase. Can emotion even play a role in food concepts or are researchers limited to the more objective descriptions of the features, the heritage, the usage, and the like? Researchers now talk a great deal about emotion in concepts, and various companies have put together initiatives to identify "emotion in food." What would this emotion look like or read like?

When one describes a food or beverage, how can the concept convey a feeling? One way to convey that feeling states the emotion directly; that is, telling consumers what and how they will feel and perhaps how they will recognize that they are actually feeling the emotion. Another approach portrays or evokes the emotion in more indirect terms. In both cases the goal is to stir up in the readers some emotional response without, however, talking about the specifics of the product or the consuming situation. Advertising agencies use emotion in their advertisements. Quite often the emotion part of the advertisement will be communicated by a set of images that convey and stir up emotion in the audience. Thus, an agency might use a picture of an older couple, walking hand in hand, to convey some emotionally tinged aspects of wellness.

The It! studies, concept driven rather than execution driven, attempt to deal with emotion by choosing terms that one might encounter in

a concept or an advertisement, without, however, becoming too "executional." The It! studies use concept elements with an emotional component. The research objective underlying these specific elements is to determine whether emotion elements perform as strongly, or perhaps even more strongly, than do statements about product features.

For this analysis we will look at the Drink It! study, which deals with beverages (see Chapter 22). Table 24.2 presents an example of some of these emotion elements from the beverage study. These elements were repeated in each of the studies, in virtually the same form, with the exception of a few that were very slightly modified or in some cases not used. With the ability to compare the utility values for emotion elements across the 29 beverages it becomes clear that in almost every case the emotion element that performs best among the other emotion elements is still only a modest scorer when compared with the other elements for the same beverage. Indeed, the winning elements for the product-level features score higher than do emotion elements. Most of the utility values for emotion elements lie within the range of $+5$ and -5, which means that they do not sway consumer interest. Thus, as a first approximation, text-based emotion elements do not do well with beverage concepts.

The emotion elements typically occupy only a small proportion of the total range of utilities for beverages. We see this small range graphically in Figure 24.1 and listed in more detail in Table 24.3. The only exception is the unusual case of flavored beer, where the emotion elements occupy about 90% of the range. Coffee appears to be the beverage where emotion elements are least important. From these results we conclude that emotion statements do poorly as drivers of acceptance because the utility range is small, hovering around 0. The data cannot imply that, in general, emotion is unimportant, but rather that based on the way the concepts were designed, the elements that one might call emotion un-

Table 24.1. Foods chosen because mood involved

Food	%	Food	%
Food that is directly craved because of mood			
Chocolate candy	47	Peanut butter	19
Ice cream	36	Chicken	19
Coffee	34	Cheese	17
Cola	34	Pizza	17
Pretzels	33	BBQ ribs	12
Nuts	31	Steak	12
Tortilla chips	29	Gravy	11
Food for celebration			
Steak	38	Cheese	5
BBQ ribs	25	Cola	4
Cheese cake	17	French fries	4
Olives	15	Chocolate candy	3
Chicken	3	Coffee	2
Pretzels	2	Potato chips	2
Cinnamon buns	1	Peanut butter	0
Food chosen because of associations			
Cheese	19	French fries	8
Cola	18	Tacos	8
Olives	18	Peanut butter	8
Nuts	16	Tortilla chips	8
Cheesecake	15	Pretzels	8
Chocolate candy	15	Coffee	6
Food chosen because of family and friends			
Tacos	28	Cola	9
BBQ ribs	27	Cheese	9
Steak	20	Ice cream	9
Pizza	19	Cinnamon buns	4
Tortilla chips	17	Chocolate candy	2
Hamburger	15	Peanut butter	1
Pretzels	15	Nuts	15
Gravy	15		
Food as an escape			
Cheesecake	20	Olives	5
Chocolate	20	French fries	4
Coffee	15	Peanut butter	4
Ice cream	14	Gravy	3
BBQ ribs	13	Tortilla chips	3
Chicken	13		
Food craved due to hormones			
Chocolate candy	28	BBQ ribs	4
Cinnamon rolls	13	Chicken	4
Ice Cream	10	Gravy	4
Cheesecake	10	Pizza	4
Pretzels	3	Steak	3
Cheese	1		

Data courtesy of It! Ventures, LLC.

derperform. It could be that either emotion is truly unimportant or that those elements describing emotion are the wrong ones.

How Emotion Elements Perform: Deconstructing Commercial Advertising for Fast-food Restaurants

A possibly fairer way to look at the impact of emotion elements in concepts deconstructs advertising copy, identifies which elements can be classified as *emotional*, and then performs the same type of analysis as done with

the Drink It! database. The key difference between the Drink It! study and deconstruction of advertising copy is that presumably the advertising copy already has been selected so that the emotion elements do well. Advertising copy is designed to convince and to sell, whereas the Drink It! study was to understand and create a database. We saw in Chapter 17 some of the conservatism when deconstructing other types of in-market communications.

A study on fast-food advertisements that is run on the Internet can help shed further

Table 24.2. Best-performing emotion elements for each beverage

Best element	Utility	Best element	Utility
A great way to celebrate special occasions		So refreshing . . . you have to drink some more	
Water	8	Water	3
White wine	5	Shakes	3
Quick and fun . . . ready to drink		Pure satisfaction	
Hot chocolate	4	Tequila	5
Flavor cider	4	Spritzer	4
Flavored alcohol	4	Energy drink	4
		Lemon lime	4
Simply the best			
Smoothie	8	You can imagine the taste even before you drink it	
Shakes	7	Spritzer	5
Tequila	6	Water	5
Soup	5	Hot tea	4
Flavor cider	4	With the safety, care and quality that make you trust it all the more	
Iced tea	4	White wine	4
Relaxes you after a busy day		It quenches THE THIRST	
Yogurt beverage	5	Hot chocolate	8
Tequila	4	Flavored coffee	4
Red wine	4	White wine	4
A wonderful experience . . . shared with family and friends		Tequila	4
White wine	7	When you think about it, you have to have it . . . and after you have it, you can't stop drinking it	
Iced tea	7	Fiber beverage	4
Tequila	6	So refreshing you want to savor how it makes you feel	
Hot tea	5	Energy drink	3
Energy drink	4		
Red wine	4		
Spritzer	4		
Premium quality			
Flavor cider	5		
Cola	4		

Data courtesy of It! Ventures, LLC.

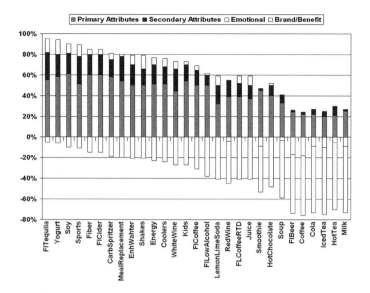

Figure 24.1. The relative range of the utility values for emotion elements compared with the other elements. The emotion elements are shown in *black*. Data are from the 2002 Drink It! study. Graph courtesy of It! Ventures, LLC.

light on this issue of emotion elements (Moskowitz et al. 2002). Originally run with 1942 respondents on the Internet, the study uses concept elements created by deconstructing corporate communications appearing on corporate Internet sites. Presumably, these elements comprise a better mix of emotional and factual elements, because one of the objectives of the Internet site is to promote the particular fast-food restaurant and its offerings.

After deconstruction, the text elements were classified as either emotion elements or nonemotion elements. The elements were then mixed and matched and presented to consumers on the Internet, who rated interest. IdeaMap implemented on the Internet enabled an estimation of the full utility model for each respondent. The classification questionnaire then enabled the creation of relatively large subgroups of respondents, based on usage patterns.

Table 24.4 lists the winning and losing utility values for the total panel and the source, as well as the same information about

the emotion elements. Table 24.5 shows how much of the total range of the elements is spanned by the different categories:

1. For the fast-food advertising, the full set of elements occupies a 9-point range. The emotion elements occupy a 6-point range. The ratio is 66%. This relative range occupied by the emotion attributes is higher than the relative range we saw in Drink It!

2. Emotion is important, but not as important as the food itself. The main course elements span 89% of the total range, whereas the emotion elements only span 66%.

3. The emotion elements are not always strong. Some do almost as well as the winning food elements (e.g., "Guaranteed to make your mouth water"). Others leave respondents indifferent (e.g., "Why celebrate when you can salsabrate?")

4. Thus, looking at data from the total panel, both for beverages and for actual fast food advertising, we find that emotion elements do not perform unusually well or unusually poorly compared with other elements. They occupy some or a lot of the

Table 24.3. Comparison of utility ranges of emotion elements compared with all 36 elements for the Drink It! database, and the proportion of the total range spanned by the emotion elements

	Utility range: emotion elements	Utility range: all elements	% Total range spanned by emotion elements
Flavored beer	17	19	90
White wine	8	15	53
Water	10	22	46
Smoothie	13	29	44
Soy beverage	8	18	42
Shakes	9	25	38
Hot chocolate	10	27	36
Tequila	7	20	36
Flavored cider	6	18	34
Red wine	7	23	31
Sports beverage	5	17	29
Yogurt beverage	5	19	28
Iced tea	8	31	27
Flavored alcohol drink	8	29	27
Meal replacement	6	21	27
Coffee	8	30	26
Lemon lime drink	6	25	24
Fiber beverage	4	20	22
Tea	7	33	20
Soup	9	42	20
Spritzer	6	32	20
Cooler	7	37	19
Energy drink	5	35	16
Cola	5	31	16
Flavored coffee	6	41	14
Milk	6	45	13
Juice	5	40	13
Iced coffee	4	35	12

range occupied by all of the elements. The importance of emotion in concepts must either lie in the execution of the concept, for example, in pictures, in music, in the way the message is conveyed, or in the segmentation. In the next section, we will see that segmentation provides part of the answer.

The Importance of Segmentation in the Performance of Emotion Elements

Segmentation has been stressed as a method to identify consumers having different *mindsets* (Chapter 7). Segmentation may also play a role for emotion elements as drivers of consumer liking. The best way to understand the effects of segmentation uses the fast-food data from deconstruction already discussed,

because, when one uses in-market communications to understand emotion, one moves closer to elements that have been created by advertising agencies. Table 24.4 shows the performance of the emotion-based concept elements. For the total panel the performance shows a narrow range, with the elements not performing particularly well.

A segmentation of the 1942 respondents suggests the existence of at least three clear segments, as shown in Table 24.6 (indifferent, social, good food, respectively). Two of the three segments do not respond to the elements described as conveying some aspect of emotion. The third segment (good-food seekers), constituting almost half of the respondents, shows strong responsiveness to the emotion elements. The winning emotional

Table 24.4. Winning and losing elements for fast foods, and the location of the emotion elements for fast food within that range

	Source	Utility
Winning and losing elements		
Grilled on an open flame for the tastiest flavor	Ranch1 (ranch1.com)	7
Our burgers are steam grilled on a bed of onions	Ranch1 (ranch1.com)	−2
Emotion elements		
Guaranteed to make your mouth water	Ranch1 (ranch1.com)	6
Fire up your taste buds like never before!	Carl's Jr. (carljr.com)	5
Satisfy your taste buds	Wendy's (Wendys.com)	4
Something special to satisfy your hunger	Wendy's (Wendys.com)	4
Delicious, and in hot demand	Carl's Jr. (carljr.com)	3
Your guests will love our delicious menu options	Church's Chicken (Churchschicken.com)	2
Food worth slowing down for	Boston-Market (Boston-Market.com)	1
Special meals for your kid's enjoyment	Wendy's (Wendys.com)	1
Why celebrate when you can salasbrate	Chi-chi's (chi-chis.com)	0

Table 24.5. How much of the element range for fast-food advertising is occupied by concept elements of different categories

Category	% Range
Main item	89
Emotion	66
Side order	67
Restaurant chain	56
Service	33
Options	22
Quality	22

element ("Guaranteed to make your mouth water") is their strongest-performing element (utility = +13). The poorest-performing emotion element is still substantially positive (utility = +4). Thus, the emotion elements do play a clear role, but only when these elements are polished and the respondents are segmented.

Following the insight provided by the fast-food data, we can now revisit the beverage data to determine whether a group of respondents in the dataset respond to emotion. Keep in mind that the elements in the Drink It! study were not created by advertising agencies as part of an integrated communications to consumers with the ultimate goal of promoting the restaurant and selling products. Rather, the elements in the Drink It! study were created for a scientific investiga-

tion, and thus the emotion elements are not necessarily polished.

The segmentation analysis for the Drink It! study identified three different subgroups. Only one segment of the three was responsive to the elements defined as emotion. We get a sense of the response to these emotion elements from the data listed in Table 24.7. The results from nine beverages (of the 29) suffice to give a sense of these segments. The elements are presented in descending order of utility value, based on the response of the segment that appeared most responsive to emotion. This segment has been called the *emotion segment* for purposes of this chapter. The table stops at the elements that cease performing well for the emotion-responsive segment.

From Table 24.7 we see three clear patterns:

1. *Emotions are not strongest performing for the total panel.* This can be seen by the low utility values for emotion elements compared with the utilities of other types of elements. This may again come from the Drink It! study dealing with elements in a relatively unromantic, clinical, text description.

2. *The base sizes vary for the emotion segment.* Emotions may play different roles in beverage concepts, depending on the specific beverage. In some cases, emotions are

Table 24.6. Performance of the fast-food elements by total panel and by segments, and the performance of the emotion concept elements

	Total	Segment		
		1	2	3
		Indifferent	Social	Good food
Base size	*1942*	*401*	*663*	*878*
Additive constant	*39*	*66*	*41*	*24*
Guaranteed to make your mouth water	6	−5	1	13
Fire up your taste buds like never before!	5	−7	1	13
Satisfy your taste buds	4	−4	−2	11
Something special to satisfy your hunger	4	−4	0	11
Delicious, and in hot demand	3	−5	−2	10
Your guests will love our delicious menu options	2	−3	−1	7
Food worth slowing down for	1	−3	−3	5
Special meals for your kid's enjoyment	1	−4	1	4
Why celebrate when you can salasbrate?	0	−6	−1	4

more important and the base size of the segment is larger. In other cases, emotions are less important and the base size is smaller. The data suggest for coolers that an emotion component hardly exists, because 27 (13%) of the 220 respondents fall into the emotion segment. In contrast, for iced tea there is probably a stronger emotion component, because 117 (48%) of 246 fall into the emotion segment.

3. *Emotion elements don't do as well as other elements.* The emotion elements are not the highest elements, even among the group defined as emotion. Emotions are only part of the picture, and often are less important than product features. This means that reactions to emotions exist, but these reactions are not the most critical.

Overview

The issue of emotions in concepts comes with a great deal of professional baggage. Most advertising is created by the agency, whose viewpoint is that emotion is important in concepts. This chapter does not, and cannot, determine whether emotion is critical. The chapter does reveal some patterns that bear on the issue of emotion.

1. Emotion elements elicit positive utilities in simple concepts.

2. For the total panel these utilities are not as high as the utilities from some (but not all) product features

3. Segmentation makes a difference. Some segments respond strongly to concept elements.

4. Emotion elements do better when created by experts in advertising agencies than when created by researchers who are interested in the performance of a variety of different concept messages. This makes sense. An advertising agency's business is to communicate to sell. A researcher's business is to use communications that help understand the way the consumer's mind works.

5. When the segmentation on communications is done by agencies, with well-polished emotion elements, one segment emerges that clearly responds very strongly to emotion. However, the response to emotion elements is not dramatically stronger in this segment than the response to other elements.

6. When the segmentation is done on concepts for a beverage across many beverages, the same type of emotion-responsive segment emerges, the base size of which varies across different beverage products as does the performance of the actual emotion elements.

Table 24.7. Performance of concept elements for nine beverages in the Drink It! study, by total panel and by that segment (out of three segments) identified as being most responsive to emotion elements

	Total panel	Emotion segment
Cooler: base size	220	27
Mixed with berry, citrus, fruit punch, peach, and tropical flavors . . . whatever you're looking for	21	38
*Looks great, smells great, tastes delicious	7	10
Quick and fun . . . ready to drink . . . no bartender required	3	9
*Pure satisfaction	1	9
*So refreshing you want to savor how it makes you feel	3	8
Tea: base size	234	74
Freshly brewed	6	11
From Lipton	4	10
*Drinking hot tea is so inviting	5	9
*Simply the best	3	6
Iced tea: base size	246	117
Iced sun brewed teas with a warm smooth flavor	7	10
From Lipton	8	10
*Drinking iced tea is so inviting	7	8
*So refreshing . . . you have to drink more	4	6
Yogurt beverage: base size	188	68
Mixed berry, strawberry, kiwi lime, lemon burst, apricot, peaches and cream . . . whatever you're looking for	14	13
*Simply the best	5	10
*Drinking yogurt is so inviting	3	8
*So refreshing you want to savor how it makes you feel	2	7
*Premium quality	3	7
*You can imagine the taste even before you drink it	2	7
Soy beverage: base size	194	90
Coffee, Mocha, Berry, Apple, Orange, Banana . . . whatever you're looking for	12	23
Frothy soy milk blended with fruit juices for that feeling of the exotic in a drink	11	20
*Soy milk . . . creamy, delicious, and highly nutritious	12	9
Sports drink: base size	232	68
From Gatorade	12	21
*It quenches THE THIRST	3	8
All natural	5	6
From Minute Maid	4	6
*So refreshing . . . you have to drink more	0	5
*You can imagine the taste even before you drink it	1	5
Water: base size	216	35
Refreshing flavors such as lemon, berry, orange, or tropical	10	35
Low-calorie alternative to sugar-ladened soft drinks	2	21
*Bubbly water in a premium glass container . . . for both the everyday and the more special occasion	−3	19
Resealable single serve container . . . to take with you on the go	5	17
Icy cold	5	12
Enhanced water that contains ingredients to energize you . . . specially formulated to keep you going	6	11
*With the safety, care and quality that make you trust it all the more	2	11
Made with mineral water . . . to deliver great taste	0	10

(continued)

Table 24.7. Performance of concept elements for nine beverages in the Drink It! study, by total panel and by that segment (out of three segments) identified as being most responsive to emotion elements *(cont.)*

	Total panel	Emotion segment
From Poland Spring	−3	10
Pure, fresh spring water . . . directly from the source	8	10
Multiserve containers . . . so you always have enough!	1	10
Spring water . . . contains the antioxidants your body needs	6	9
*So refreshing you want to savor how it makes you feel	3	8
Seltzer water . . . with just the right tang at the end	−12	8
*So refreshing . . . you have to drink another	2	8
*You can imagine the taste even before you drink it	0	7
*When you think about it, you have to have it . . . and after you have it, you can't stop drinking it	2	7
Juice: base size	235	20
Orange and white cranberry, apple, fruit punch, pear, raspberry, pineapple or tomato, carrot, or veggie blends . . . whatever you're looking for	6	38
Exotic blends naturally sweet with real pulp	2	19
All nectar juice with antioxidants	1	13
*Premium quality	3	11
With calcium, vitamins A and the energy-releasing B vitamins . . . or what ever you need	3	9
*So refreshing you want to savor how it makes you feel	1	8
Coffee: base size	226	40
*Cool and refreshing Iced Coffee	−19	25
Coffee and milk . . . blended just right for the perfect latte	−3	16
Dark French Roast coffee that is rich, bold, and roasted	−1	10
Fresh coffee made from 100% Colombian coffee beans	10	10
From Starbucks	−7	9
With all the milk, cream, and toppers you want . . . cinnamon, nutmeg, chocolate sprinkles, sugar cubes, whipped cream . . . whatever	−9	8
Fresh ground and brewed coffee	9	7
*Pure satisfaction	1	7

*Emotion elements.

Part II: Emotional Issues in Food Products

Introduction

Most of this book has dealt with concepts designed to understand how consumers respond to concepts about products or services. However, in his pioneering book on psychophysics, S.S. Stevens, the founder of modern psychophysics, chose the title *Psychophysics: An introduction to its perceptual, neural and social prospects* (Stevens 1975). Stevens foresaw the use of systematic, psychophysical methods as a tool to un-

cover the consumer mind with respect to social issues. Although Stevens did not live long enough to see his approach in action, the psychophysical way of thinking provides a useful approach to investigate these social issues. The systematic variation of concept elements, a heritage of psychophysical thinking, can be applied to databases about social issues and especially social issues with respect to food. This final section of the chapter deals with concept research with respect to two key, food-related issues: genetically engineered foods and obesity. Both of these topics are tangentially related to food concepts. The two topics demonstrate how to use the research principles to study *food consumerism*.

Traditional methods for dealing with social issues have been used for decades. For example, when dealing with social issues many researchers rely on secondary sources such as mentions in the media or on primary studies that use tracking methods (e.g., GfK Consumer Tracking, a survey of households and individual consumers that is repeated at regular intervals). Tracking studies constitute a marketing research tool that uses the same questionnaire over time to identify patterns of responses (Allan et al. 1998). Tracking studies would be akin to using the same classification questionnaire over time and in different situations. The responses to the classification questions show how the respondents profile themselves.

When it comes to social issues, researchers can use the conjoint analysis in the same way as one uses conjoint analysis with food product issues. Conjoint analysis estimates the utility values associated with the aspects of the social issues. It provides answers that are fundamentally different from those provided by tracking studies, just as it provides fundamentally different answers from those provided by self-profiling studies. It also shows how the specific elements in a descriptive vignette describe responses, only the responses are emotional ones from consumers about personal, food-related issues rather than about the food itself.

One of the interesting aspects of conjoint measurement is its ability to transcend some of the problems associated with the *biases* engendered by social desirability, otherwise known in common parlance as *politically correct answers* (Edwards 1957; Davison 1983; Phillips and Clancy 1972). When participating in a conjoint study about a sensitive issue, typical respondents do not know the underlying experimental design. Thus, after being confronted with a concept or vignette about a social issue, these respondents do not necessarily know how to answer the question to please the interviewer. The vignette is too complex. Consequently, the respondents have to answer somewhat more honestly, because there is no clear clue about what rating is expected and no way to please or fool the researcher.

Applying the Approach to Social Issues in the Food World

In early 2003, with the increasing popularity of the mega-studies, a group met to discuss whether a similar approach would be possible with social issues. Headed by Hollis Ashman of the Understanding and Insight Group, the participants decided that the It! approach could be used on an experimental basis in order to identify the elements of a social situation that generated anxiety. The study that emerged from the working meeting is known as *Deal With It!* (see Chapter 20 on the It! studies). The objective of this particular mega-study was to identify what particular features in an anxiety-provoking situation drove the consumer response of "Able to deal with this." The study of social policy by questionnaire is not new for researchers nor is it particularly new for conjoint measurement (Louviere 1988). Other researchers have already recognized that the conjoint approach has value in identifying the elements of social policy that drive respondent interest. These were the novel applications and issues addressed by the It! mega-study approach:

1. *Variety of studies.* The study would cover a variety of different anxiety-provoking situations, not just be limited to one topic. This objective was in keeping with the rationale for the mega-study. The study involved 15 issues, as shown by the *study wall* (Figure 24.2). Thus, the respondents were given the opportunity to participate in one of a variety of studies dealing with emotionally difficult topics.

2. *Multiple aspects and concrete statements exemplifying the issue and feelings.*

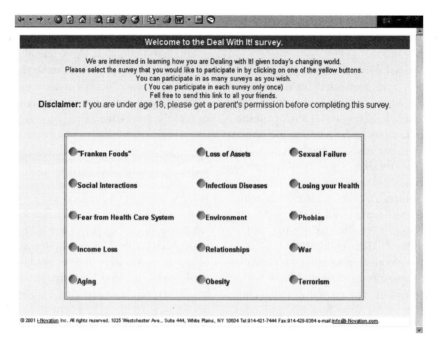

Figure 24.2. The wall for Deal With It!

The vignettes comprised more concrete elements rather than general questions. The focus on the specific statement of an issue, rather than a general statement, comes from the focus on the specifics and the concrete encouraged in conjoint analysis. This concreteness contrasts with a lot of research on social issues, which works with general questions rather than specific vignettes.

3. *Change of scale to make the study easier, but also change of scale direction to maintain the standard analytics.* The initial focus was to measure the degree to which a description provoked anxiety. The objective was to remain consistent with the It! studies, yet make the study easy for the respondents. In the It! research the top of the scale corresponds to the region of interest (e.g., highly interested or highly craveable). A pretest of the Deal With It! studies showed that many of the participants did not like to respond to concept elements with a question that asked them to rate "anxiety" or "difficulty in dealing with this situation." The team changed the question

to "How easy is it for you to deal with the situation?" but also reversed the scale, as well. The direction of the Deal With It! scale on the computer screen was changed as well to read: 9 = hard to deal with, versus 1 = easy to deal with. Respondents had no problems with the scale wording or the direction of the scale. In this way the analytic strategy used for the It! studies could be applied here.

Topics of Interest to Food Researchers

Two of the topics are of potential interest to those involved with food. These are genetically engineered foods, which occasionally in common parlance are called *frankenfoods* (see Clorfene-Casten 1999), from Mary Wollstonecraft Shelley's famous monster in her book *Frankenstein or the Modern Prometheus* (1818). The other is obesity, which is becoming an important topic for the food industry, especially with the increasing prevalence of childhood obesity (Sothern et al. 1999; Berg 2003).

The results from these two social-issues studies are presented in Tables 24.8 and 24.9, respectively. Both tables show the strongest or most anxiety-provoking elements for the total panel and for the two segments that were extracted from the respondents.

Genetically Modified Foods

The total panel shows a very low level of anxiety, as indexed by the additive constant of 24 (Table 24.8). This low constant implies that, without any elements in the concept, 24% of the respondents would say that they have a hard time dealing with the issue of these genetically modified foods. Basically, the issue of genetically modified foods is not at the top of the mind, at least overall, for the 121 respondents. What does worry the respondents is the active interference of the government in this issue. Reading between the lines, one might conclude that the issue here is lack of trust in the government regarding food and genetics or perhaps lack of trust in any authority whatsoever, including the food industry.

There are two clear segments, which will turn out to be the same types of respondents for both genetically modified foods and obesity, respectively:

1. *Fearful of the issue itself, but not really focused (segment 1).* These individuals constitute half the population. Their additive constant is extremely low: 19. This means that, without anything else involved but the idea of genetically modified foods, only 19% of the respondents rate the concepts about this topic as 7–9, or hard to deal with. They are responsive to a mélange of different ideas, with no key message driving their anxiety. They are as frightened of the use of the genetic modifications to drive drug production as they are anxious about being helpless.

2. *Fearful of intervention by any power group at all (segment 2).* These individuals also show a low additive constant: 29. Any mention of authority involved in genetically modified food frightens rather than reassures these individuals. They also comprise half the population and appear to be quite distrusting.

Table 24.8. Strongest anxiety-provoking elements for genetically modified foods: total panel and two-concept response segments

	Total	Segment	
		1	2
Base size	*121*	*59*	*62*
Additive constant	*24*	*19*	*29*
Total panel			
You trust the government will keep the earth and you safe	10	−1	21
You trust the local government will keep the earth and you safe	9	−1	18
Segment 1: Afraid of the issue itself			
You are scared . . . inside and out	5	11	−2
Genetically engineered pharmaceutical drug crops to deliver medicines by food . . .	6	10	1
You think about it when you are all alone . . . and you feel so helpless	3	9	−3
Affecting children . . .	4	9	−1
Segment 2: Afraid of intervention by authorities			
You trust the government will keep the earth and you safe	10	−1	21
You trust the local government will keep the earth and you safe	9	−1	18
You believe international cooperation will keep the earth and you safe	4	−6	14
You trust the Food and Drug Administration will keep the earth and you safe	4	−4	12
You believe the businesses impacted will work to keep the earth and you safe	5	−2	11

Table 24.9. Strongest anxiety-provoking elements for obesity: total panel and two-concept response segments

	Total	Segment 1	Segment 2
Base size	*123*	*58*	*65*
Additive constant	*37*	*42*	*32*
Total panel			
You believe that the food industry will work to help you find the right foods to eat	11	−2	23
You believe a plastic surgeon will help you get through this	9	−10	26
You just can't control the eating . . .	8	12	4
People you work with are affected by your size . . .	7	9	5
Segment 1: Afraid of the issue itself			
You just can't control the eating . . .	8	12	4
People around you are so judgmental . . .	4	10	0
You've added a lot of extra weight . . .	4	9	−1
People you work with are affected by your size . . .	7	9	5
You are uncomfortable because of your weight doing what everyone does naturally . . .	1	8	−5
People look at your body and judge you . . .	1	7	−3
People around you are embarrassed . . .	3	7	0
Segment 2: Afraid of intervention by authorities			
You believe a plastic surgeon will help you get through this	9	−10	26
You believe that the food industry will work to help you find the right foods to eat	11	−2	23
You believe work will help you get through this	−1	−10	7

Obesity

Obesity is probably more familiar to people than genetically modified foods and more anxiety provoking (Table 24.9). The additive constant is 37, meaning that more than one in three people are highly anxious about obesity, even without any elements present. This additive constant is far higher than the constant for genetically modified foods, suggesting the prevalence of anxiety with regard to obesity. Two segments emerged, both of which are more anxious at a base level about obesity than their corresponding segments were about genetically modified foods:

1. *Total panel.* The respondents fear losing control and losing social status and acceptance.

2. *Fearful of the issue itself, but not really focused (segment 1).* Segment 1, again about half the population, fears the issue itself, just as we saw for genetically modified foods. The key difference is that the respondents in this segment, when dealing with obesity, are afraid of what they themselves are doing. They are absorbed in what other people will think of them and appear to be focused on the external.

3. *Fearful of intervention by any power group at all (segment 2).* Segment 2 is anxious when they read about others in control who can help them, whether this be a plastic surgeon or the food industry.

Conclusions

These data suggest that social issues and public policy surrounding foods can be treated as concepts, just as foods and beverage products and services are treated as concepts. The Deal With It! mega-study shows that, when the respondents evaluate the concepts, they do so with one of two minds: either afraid of the topic itself or afraid of intervention by someone powerful. An analysis of the other 13 studies for Deal With It! confirms that this segmentation holds for the other issues as well and may represent a key organizing principle for issues.

References

Allan, J., Carbonell J., Doddington, G, Yamron, J., and Yang, Y. (1998). Topic detection and tracking pilot study: Final report. In: Proceedings of the DARPA Broadcast News Transcription and Understanding Workshop, 1998.

Berg, F. (2003). Underage and Overweight: America's Childhood Obesity Epidemic—What Every Parent Should Know. Long Island City, NY: Hatherleigh.

Clorfene-Casten, L. (1999). Frankenfoods. Conscious Choice, Dragonfly Chicago LLC; www.conscious-choice.com.

Davison, W.Ph. (1983). The third-person effect in communication. Public Opinion Quarterly, 47: 1–15.

Edwards, A. (1957). The Social Desirability Variable in Personal Assessment and Research. New York: Dryden.

Golden, L.L., and Johnson, K.A. (1982). The impact of sensory preference and thinking versus feeling appeals on advertising effectiveness. In: Bagozzi, R., and Tybout, A., editors. Advances in Consumer Research. Proceedings of the Association for Consumer Research, 10: 203–208.

Lautman, M.R., and Percy, L. (1983). Cognitive and affective responses in attribute-based versus end-benefit oriented advertising. In: Advances in Consumer Research. Proceedings of the Association for Consumer Research, 11: 11–17.

Louviere, J. (1988). Analyzing Decision Making: Metric Conjoint Analysis. London: Sage.

Moskowitz, H.R., Itty, B., Machaiah, M., and Ma, Z. (2002). Learning from the competition: II. A case history dissecting in-market Quick-Serve-Restaurants communications through conjoint analysis. Food Service Technology, 2: 19–34.

Phillips, D., and Clancy, K.J. (1972). Some effects of "social desirability" in survey studies. American Journal of Sociology, 77: 921–940.

Sothern, M. S., Hunter, S., Suskind, R., Brown, R., Udall, J.N., and Blecker, U. (1999). Motivating the obese child to move: the role of structured exercise in pediatric weight management. Southern Medical Journal, 92: 579–584.

Stevens, S.S. (1975). Psychophysics: Introduction to Its Perceptual, Neural, and Social Prospects. New York: John Wiley and Sons.

Part VI

The Grand Overview

Chapter 25

Concept Development and the Consumer-insights Business

Introduction

Creating concepts has been the domain of the advertising agency, marketers, and occasionally marketing research and development (R&D). To create concepts, one has to have insights. The *insights* business is a relevant penultimate chapter to this book, because today, more than ever before, the insights business is fractionated into warring factions. Understanding what businesspeople accept from consumer insights is therefore important, because it points to the best people to participate in the concept development exercise.

A great deal of today's press on the role of consumer research is presented from the *normative point of view* or "what should be the case." Much of the information in business journals deals with the general information needs, such as monitoring trends and identifying specific wants. Very little business literature deals with specifics that might be provided by asking the businessperson who uses insights to achieve competitive advantage to identify the type of information needed. The absence of structured demands in business leads to a less than optimal corporate structure for high-throughput, productive development of concepts. The food industry in particular suffers through some of these problems, even though it was perhaps the first industry to openly welcome market research in general and concept testing in particular.

The chapter provides a structured analysis about what business professionals expect from *insights providers*, who often constitute the first step in concept development. Business tends to rely on its providers, such as consumer researchers, brand planners, and graphics designers. What types of insights are these three groups expected to provide? The insights will govern the nature of the concepts that the providers generate and perhaps even the believability of those concepts to the insights buyer.

Today's Competitive, Inquisitive, and Unforgiving Environment

Consumer researchers are continually being asked to specify and occasionally to substantiate their contributions to the business process. Over the past 50 years the research industry has grown and evolved to the point where it provides eagerly awaited information about the structure of the competitive framework and customer wants/needs. The evolution of research has not been without its hazards, however. Today business has come to rely on knowledge. When the information is incorrect there is often hell to pay, whether the error lies in substance or even just in superficial appearance.

Unfortunately, as the value of consumer knowledge to business increased, so have the competitive pressures on knowledge providers:

1. Knowledge is becoming a commodity.
2. Knowledge buyers don't differentiate between knowledge workers (consumer researchers), creatives who use knowledge for marketing services (advertising agencies and graphics design firms), and those who integrate knowledge for higher-level strategic work (consultants).

3. Knowledge is dated, and methods to provide it are faddish. Today's methods are tomorrow's fodder for stories about the metaphor that failed.

4. Concept research has a limited life. What is important as a topic for research one day may become irrelevant tomorrow. Yet, at the same time, concept research is increasingly accepted as a key aspect of competitive advantage for business.

This chapter identifies the specific information needs about consumers/customers that are voiced by business managers. The chapter further considers the type of customer information/insight that the business manager expects from different classes of sensory analysts, consumer researchers, advertising-agency brand planners, and graphics designers. All of these professionals are at one or another time associated with that most vague of labels, *consumer insight*. All use concept research in one form or another, or contribute to concept development.

Consumer Insights: A Continuing, Elusive Notion That Has Created Its Own Cadre of Jobs

Where did the widely used term *consumer insights* come from? On what is it based? Valdés (2000) puts it quite succinctly: "Any experienced marketer will agree that in order to identify categories that may offer opportunity for growth, you must first identify which categories are more important to your target consumer. Marketers who recognize important lifestyle nuances can tailor their marketing programs to reflect these important variables." Most researchers are aware of this general definition.

According to Cagan and Vogel (2001), insights can be traced to the confluence of business needs and consumer demands, which pulls ever more on the insights of the businesses to build excitement, create needs,

and then offer their products and services as solutions:

People use products to improve their experience while doing tasks. They relate these experiences to their fantasies and dreams. Successful products fulfill a higher emotional value state, whether it is the excitement and security of driving in an SUV, the comfort and effectiveness of cooking in the kitchen, the relaxation and escapism of sipping coffee in a coffeehouse, or the independence and adventure of using a two-way communication device. The mantra that form follows function is no longer relevant; we now find ourselves in a period where *form and function must fulfill fantasy*.

Craig and Vogel continue with this line of thinking:

The demand by consumers for better products has continued to increase during the last three decades. During the 1980s and early 1990s quality development programs, reengineering and concurrent design were the initiatives that drove American and international companies to constantly improve their products. At the beginning of a new century, the emphasis has shifted from the back end to the front of the product development process.

Following this line of reasoning, Cagan and Vogel suggest that the knowledge worker recognize four key factors for success in this new world of excitement, need creation, and solution, and, for the purposes of this book, product concept development. Those four key factors would create these so-called consumer insights:

1. *The ability to identify product opportunities.* As cultures continue to change, opportunities emerge for new products. These products do not merely solve existing prob-

lems; they also create possibilities for new experiences.

2. *Heightened understanding of customer needs translated into actionable insights leading to product attributes of form and features.* For products to be successful, they must have features and forms that consumers quickly recognize as *useful*, *usable*, and *desirable*.

3. *A true integration of engineering, industrial design, and marketing.* Merely putting teams together in a multidisciplinary context does not suffice. The team members and the team itself must be supported and managed effectively in an atmosphere in which each discipline respects and appreciates the perspective of the other.

4. *Vision.* The identification of product opportunities should be the core force that drives companies that manufacture products, supply services, and process information.

Consumer Insights as a Competitive Advantage: New Growth for an Historical Discipline

Over the past decade, businesses have come to value the contribution of knowledge about consumers. One need only consult a search engine such as Google. Putting in the words "consumer insight" generates 6450 hits. Putting in the words "customer insight" generates a staggering 672,000 hits. One need only look a little further at conferences and short courses to understand even further the commercial value of such insight.

To a great degree, this appreciation has come from the efforts of consumer researchers who developed methods to better understand the consumer. Whereas years ago in a more halcyon, forgiving, slow-moving environment much of the effort focused on the process of executing tests correctly and according to a structure and suitably disciplined but relaxed program, today researchers and their in-house corporate clients

have stepped back from the lockstep discipline and begun to look at their true contribution to the corporation. Even at the R&D level, sensory analysts, once the purveyors of tests using the expert panel as the "golden tongue" and "nose," now offer their expertise as in-house purveyors of consumer insights.

Some of this newly emerging recognition has found itself reflected in new types of contributions by researchers. Furthermore, many others have joined the fray. Other professionals, traditionally trained as researchers, find themselves employed in the insights business. Indeed, the appeal of the notion of insights has been so strong that many companies have rebranded their market-research departments as consumer (or customer) insights departments. At the same time there is a growing interest by business-level consultants in the information business. Marketing consultants now actively promote their use of research and, of course, insight as a core benefit that they provide to the corporation.

The insights business has a long, venerable, and now rapidly revitalizing history. Advertising agencies were among the first to recognize the value of information. During the golden era of the 1960s to the middle 1980s, many agencies boasted of their large research departments, whose services they eagerly offered to their clients and whose services they billed back to their clients. A number of today's now-legendary market-research leaders came from these agencies, and got their start in the golden era, some 25–40 years ago. Many of these leaders graduated to *suppliers* in the market-research industry and are now on the verge of retiring after distinguished careers where they created the insights business. Few of the original legendary researchers remain in advertising agencies, which during the maelstrom of the 1990s disbanded their research groups. The sensory researchers are just beginning to enter that history of insights, having concentrated for years on small-scale projects of lesser strategic import to insights, such as discrimination tests and profiling. All of the insights providers have, in one way or another, been involved in creating concepts at the early stage of development.

There are two key types of insights provided by these agency researchers, and now their successors, the brand planners (Czerniawski and Maloney 1998):

1. *Brand-positioning statement.* Brand positioning can be defined as the way the customers have to think about the brand relative to competitors. Positioning comprises six essential elements: target groups, need, competitive framework, benefit, reason why, and brand character. These types of statements are the ones dealt with in this book.

2. *Ad strategy.* The ad strategy provides guidance and direction for the development of the brand's advertising campaign. It consists of the who, what, and why in addressing a specific brand marketing issue or objective.

Package designers did not actually discover the value of consumer insights until recently and did so when consumer insight began to become a key competitive advantage as it had been previously in the advertising agency business. For the most part, package (or, in general, graphics) designers were content to go on their intuition about what was relevant to consumers. Consumer research was often used only as a "disaster check." More recently, however, Young (2003), himself a graphics designer, has reiterated the importance of up-front research work, the need for insights, and the fact that the insights make the work better. Young recounts the story of the design for Listerine and how insights helped:

> For example, in a recent Listerine study, brand users consistently depicted the "barbell" shape and the stacked black brandmark accurately, yet were often unable to identify other elements of the brand's visual identity. This insight guided branding and de-

sign strategy, as the "barbell" became the focus of a new global packaging structure and the brandmark served as the foundation for extensions into new product categories.

Rating data constitute a large proportion of the consumer researcher's stock and trade. The consumer researcher's job is to create insight, or at least it has become so in the last several years. The early history of market research occasionally belies this as a new requirement. Often, and to a great degree even now, market researchers are expected to track consumers, identify emerging issues, measure responses to products, segment the marketplace, and perform a host of many other knowledge tasks. The notion of *insights* for a market researcher is itself not particularly clear. Market researchers are typically reinforced for their ability to execute studies on a timely, cost-effective basis, producing data to be acted on by other individuals.

Sensory analysts have entered into the fray of insights, but only recently. As already noted, their efforts typically focused on small-scale, internal R&D-driven projects. They were not part of the insights business, but rather concentrated on supporting product development. During the past decade, however, they have been invited to participate in strategic research, with an eye toward consumer insights about product-based issues. They are now becoming a force to contend with by the others in the insights business.

The Ambiguity of Data-based Insights versus Flawless Test Execution

Both the consumer (market) researcher and the sensory analyst base their contributions on data. Their work product is reports. In contrast, the advertising agency's work product is an advertisement, and the graphics designer's work product is a package or label.

The notion of insights as a part of the data is ambiguous. What is the raison d'être of consumer researchers and sensory analysts? Is the flawless execution of the study the important factor, with insights left to the researcher's clients who commissioned the study? Or should researchers themselves be rewarded for the insights? Which factor is the driving factor behind the research: good data or good insights? And, furthermore, what is the role of the researcher? Is insight merely an epiphenomenon of the research?

Consumer researchers and sensory analysts themselves cannot easily answer this question. The debate about the role of the researcher in the insights business continues and indeed forms the kernel of this chapter. Recently, researchers have come out strongly for a formal position in the insights business by stressing their role in the so-called *fuzzy front end*. For example, Urban and von Hippel (1988) have talked about a formalized process by working with certain kinds of individuals called *early adopters*. They suggest that insights can be formalized by working with certain kinds of individuals who are eager to try new products and experiment with them. These early adopters provide significant insight, communicate their likes and dislikes with others, and adopt new products quickly. Feedback from early adopters helps marketers address business opportunities and fine-tune strategies. Importantly, Urban and von Hippel do not talk about the *who* in the corporation, who gives these insights by working with the Early Adopters. The *who* as a defined job description is less important than the task as a defined set of activities.

There are other research approaches to formalize insights into the consumer mind. Some individuals promote ideation and brainstorming, others promote trend analysis, and still others promote ethnography. These research-based approaches all strive to formalize the task of learning about the mind of the consumer without talking about the *who* in the corporation.

It is worthwhile contrasting consumer researchers or sensory analysts involved in the insights business with agency planners and graphics designers:

1. Consumer researchers have as their main purpose the insights they bring to a business. In contrast, agency planners and designers use insights in their jobs. Insights are not their jobs, but rather are by-products of their jobs or inputs to their jobs.

2. As a part of their job, a nonresearcher can do research activity. For example, agency planners can do research themselves, as can package designers. No one will criticize either professional for going outside of his or her expertise to gather this data, or at least no one will do so publicly. Thus, the jobs (planning and design) can include an insights function without anyone complaining.

3. The consumer researcher is often actively or occasionally covertly discouraged from providing more than mere data. Some companies have legislated the limits to the researcher's activities. If the researcher provides insights, he or she may not necessarily be appreciated for taking over the planning or design function. The insights provided by researchers may be put into a "box," sanitized, reduced in terms of direct actionability, and then only later used by decision makers in the agency or manufacturer. Researchers have to be satisfied with offering their insights to someone who will use them, rather than immediately running with the insights to an ad campaign, product promotion, or package/product design.

Insights, Actionability, Concept Development, and a Reserved Seat for the Researcher at Management's Table

Management today has accepted the value of consumer insights as a strategic business advantage. No longer does the maverick CEO feel confident about making decisions by gut feel, at least not publicly. Anyone providing key, systematic, actionable information is invited by management to share that information. Figuratively speaking, this recognition is called *sitting at the management table*. The phrase alludes to that growing acceptance by management of the importance of valid information in the decision-making process.

One of the key issues for management is the *actionability* of the data. These data, but really insights driven by data, must lead to some action that moves the business ahead. Researchers and others in the corporation are awash in data from all sources. Data itself no longer constitute the choke point in the market, for almost anyone can obtain data by studies or by direct purchase. Information that cannot be easily obtained by subscribing to information services can often be obtained by commissioning a custom study to generate primary data. In one way or another, the marketer, the product developer, and the researcher can answer most of the questions about new and current products, advertisements, and consumer attitudes. The issue now becomes what type of data, what type of information, what type of insights are relevant, not just the mere ability to provide data, information, and insights. In this world of easy information, a researcher has to stand out as being vital. Merely rebranding the traditional researcher as a provider of insights may work in the short term, but might not work in the long term. Researchers have to provide data that can be acted on, even though it is not a researcher's job to create the product, the advertising, or the package design.

If there is so much data, then what specifically should be the job of the knowledge worker specializing in insights? If anyone today can get data, from commissioning a study to running one's own study on the Internet at low cost virtually overnight, then what do clients want from professionals who are in the insights business? Researchers often profess to offer both data collection and insights. We know also that the traditional role of research in agencies was to identify what consumers want so that better advertising could be produced. It used to be that design firms prided themselves on their artistic talents and that their offerings were miniworks of art that could not, for reasons never really explained, be tested. We now hear, however, in an increasingly loud voice from graphics designers, that they use consumer insights to guide them in their creative task and that the design merges, quite seamlessly, consumer needs/requirements with art and with business savvy. What is the truth here?

The question comes down to a simple one. What specifically do clients want or expect today in the way of insights about consumers from the insights-oriented researchers at the *research supplier* company, from the brand planners at the agencies, or from the graphics designers at the design house? These people have different jobs, different constituencies, and different talents. They all deal at some level with information and thus represent different points on the spectrum of work product: what they produce, insights from what they learn, and accountability for corporate success from the direction they provide (see Table 25.1). They all have access to so many alternative sources of information, and they all can bring insights along with their professional activities. What part of the insights business should each bring, if any at all? What do clients specifically expect from them in terms of the nature of the tools for insights and in the nature of the insights themselves? Is it insight at all that clients want as an intrinsic part of the service package and work product? Or is it just good data, good advertising copy, and good design, respectively, with insights simply some added evidence beforehand that the information, copy, or design will work in the marketplace? In the end, the issue is what do each of the professionals bring in the way of information to create better product concepts?

Table 25.1. Profile of three professions studied here

	Insight-oriented market research professional	Ad agency brand planner	Graphics/package designer
Work product	Report	Recommendation	Physical product/design
Accountability	For correct process	For right information leading to ad campaign	For actual graphics or package
Involvement in process	Very high, major involvement	Contracts it out	Low, may attend some focus groups
Seen as	Knowledge worker	Provider of insight and direction	Creator, artist in a business setting
Professional awards in the field	Few	None	Yes, constitute artistic and professional recognition
Type of work environment	Research supplier outside or corporate employee	Works within an adagency	Independent designer or corporate employee
Role of knowledge in job	Considered a knowledge worker	Uses knowledge and insight, but assumed to be an integrator of knowledge provided	Assumed to be an artist, who may use knowledge in some cases

The Insights Client as Consumer: Getting Inside the Client Mind

If clients are consumers of information, then what do they really want? What do these professionals want from those who sell them information, understanding, and insights? We know that there is a call for more integrated understanding of the consumers or customers rather than just piecemeal research. This call for information should signal that there are customer needs to be met. Research applied to understanding the insights user becomes valuable. Rather than focusing on consumers in the outside marketplace, research with the insights buyer focuses on business consumers. Furthermore, as information users at the client side become increasingly sophisticated, it may be that their demands change with respect to data and to insights. Having a mechanism to understand user demands in synch with today's increasingly sophisticated, demanding environment provides a contribution to the consumer knowledge industry.

An in-going assumption to guide the research was this mantra: "Consumers don't necessarily know what they want, but they will know it when they see it." The same as-

sumption and mantra might just as well be applied to the world of professionals; that is, knowledge users may not necessarily know what they want from researchers, agencies, and graphics designers, but they may well recognize it when they see it. Thus, we apply here conjoint measurement to understand the insights buyer rather than the consumer. The application is not new. A host of papers have been published about using conjoint analysis among professionals as respondents, such as travel agents (Renaghan and Kay 1987; Hu and Hiemstra 1996). Thus, the research reported here falls into a new application of an old research paradigm. Finally, one of the beauties of conjoint measurement is that it is relatively impervious to politically correct answers. With the elements varying in mix/match combinations, it is hard to identify the appropriate answers. Consequently, most of the answers have to be intuitive rather than carefully reasoned.

The first study of knowledge workers in business looks at the general issue regarding what the knowledge worker provides and how the worker provides the knowledge. The study looks at three different types of knowledge or insights providers: market researchers, agency brand planners, and graphics designers. Each of these groups uses consumer insights directly in its professional activities. Each group occupies a unique niche. The brand planner is an advisor, the graphics designer an artist, the market researcher is a knowledge specialist. The question motivating the research is what type of insight is expected from and valued from each, and how do these expectations vary with the information user in the company (general management vs R&D vs marketing professional, and so on)

Research Approach

The study used two high-level research techniques, first to identify the features of information desired by business users and then second to measure the importance of those features. Both have been presented in this book in previous chapters:

1. *Internet-enabled ideation.* The objective here is to identify, from the business information user, what types of information are desired. Rather than using focus groups limited to 6–10 businesspeople in a single venue, the research used Internet-based ideation (brandDelphi), followed by a brainstorming session to work on the inputs. Businesspeople were invited to participate in an Internet-based ideation. This approach has been previously used for consumer products to identify new opportunities (see Chapter 3 and Flores et al. 2003).

2. *Internet-enabled concept development and optimization.* The objective here is to identify which particular ideas or elements regarding business-relevant information are most important for the three information providers (marketing research, agency brand planners, and graphics design). The research used IdeaMap.Net to determine the impact of each element as a driver of interest for information provided by one of the three informational providers (see Chapter 5 and Moskowitz et al. 2001). The conjoint measurement system incorporates within it seg-

mentation based on the utility values, enabling the research to identify new, emergent groups of businesspeople with defined information needs and with interests in specific types of information.

Part 1: Identifying Information Needs Through Internet-enabled Ideation

In Part 1, a total of 320 respondents were recruited to participate in an Internet-facilitated ideation. The study was set up with the same three questions: What kind of information about consumers/customers do you want from your <Professional>? The place marker <Professional> was replaced by one of three professional-level marketing-service professionals: graphics design company, market researcher, or advertising agency brand plannery. Each of three respondent groups (100–105 per group) rated the three questions in a unique order (e.g., marketer researcher for question 1, advertising agency brand planner for question 2, and graphics design company for question 3). This set of three orders was done to reduce potential order bias.

Table 25.2 lists in unedited form some of the elements that emerged from the ideation session. Approximately 80 unique elements emerged from the ideation, with most partially relevant to insights. The elements were subsequently edited for the conjoint portion of the study.

Part 2: Identifying the Important Elements by Conjoint Analysis

Ideation produces new ideas, but does not indicate how strongly these ideas will perform in the body of a concept. The input to the conjoint study was a set of 36 elements created from the inputs of the aforementioned ideation, along with the results of a brainstorming session. The elements were divided into four logically distinct categories: partnership and philosophy, insights, tools, and deliverables and fees.

Table 25.2. Elements emerging from the brandDelphi for marketing research supplier

Delphi question: Beyond what you normally get from your MARKETING RESEARCH SUPPLIER, what type of additional information on consumer/customer insight would you like? Instructions: Read through the ideas/opinions. If you would like to delete any idea/opinion, click the TRASH CAN corresponding to the idea/opinion. NOTE: If you have accidentally deleted an idea/opinion, click "Restore" to undelete.
Ideas/opinions What are the times that they prefer to shop online? Have a clear quality policy or statement. I'd like the surveys to be a little bit shorter and more direct; that way more people will want to do them. Rate of success and failure. I also like to know each company's extent of social responsibility Why is the money not being distributed to the employees where it belongs? Company employees should be consulted more on what they need to do their job more efficiently. I would rather not participate in this question. Thank you. Background knowledge in our field. I always want the best goods or service for the least money. I like to deal with a company that cares. I'm really looking for more get into their head-statistics on likes, dislikes, hobbies, memberships, etc. I'd also like to know relative levels—where does this person or group fall when compared to their entire zip code or sector or other key label?

The respondents were invited by an e-mail list provider (Open Venue, Toronto), based on respondent qualifications (which had to be business related). The respondents were invited by e-mail and instructed to go to a site, at which they were again invited to participate. The e-mail provided three different links: one for each study. The respondents could choose whichever link they wished, knowing the profession that they would rate on insights. The reward for the interview was an entry into a sweepstakes. The Internet-based interview lasted 15–20 minutes and followed this order:

1. Invitation

2. Orientation

3. Evaluation of 60 randomized concepts, set up in an experimental design, for that respondent

4. Completion of a classification questionnaire

5. Return to the e-mail list supplier

Results: The Total Panel for the Three Groups

Table 25.3 presents the results from the total panel of respondents. A respondent was free to participate in one, two, or all three of the studies, but only once. The results for each respondent were analyzed to determine whether the data were consistent. This consistency check is possible because one has the utility values and can estimate the degree to which the respondent's data track the presence/absence of elements (see Chapter 9 and Moskowitz 2002). Only the respondents who showed no evidence of guessing were included in the analysis. The data from eight respondents were discarded, leaving 184 respondents.

The additive constant shows the basic level of interest by a respondent in the insights provided by the specific professional (marketing researcher, brand planner, or graphics designer). The additive constant shows the conditional probability that a respondent—a businessperson involved in products or services—would be interested in the insights from the professional if no other elements were present to qualify what the insights would be, how they would be presented, and what would be their impact. Respondents were most interested at a basic level in insights from the advertising agency and the graphics design house, and least interested in insights (without qualifications)

Table 25.3. Utility values for the additive constant and the 36 elements

	Agency planner	Marketing research	Graphics/package designer
Base size	*64*	*57*	*63*
Additive constant	*59*	*44*	*53*
Maximum utility value	4	14	8
Minimum utility value	−10	−11	−6
Standard deviation of the 36 utility values	3.6	5.4	3.2
E01 A partner that can help you identify unmet needs and wants among consumers	2	3	3
E02 A partner that makes you feel that you truly know the customers . . . everything about them . . . hobbies, passions, fears, dreams . . . and more	0	4	5
E03 A partner that understands how the macro trends shaping the world today are impacting your customer	−6	0	0
E04 A partner that does more than just describe your customer . . . they help you to understand the implications to your business	2	9	4
E05 A partner that can help you understand how technology is changing the way your customer makes decisions	0	3	5
E06 Combining information from a variety of data sources to tell a complete story about your customers	−4	4	5
E07 Our planners go undercover and live with your consumers to truly understand them in their real world	−6	1	−2
E08 Big creative ideas from our think tank to jump-start your business	−2	−1	3
E09 Consumer insight is a blend of art and science . . . we strive to provide the best of both	4	−1	1
E10 Learn what products your customers use . . . the brands they prefer . . . and why	3	14	5
E11 What motivates your customers to make a choice? Cost, convenience, habit, or something else?	4	6	4
E12 Helps you to understand what one single change your customers would make to the products and services they use to improve their day	1	3	1
E13 What prompts your customers to act?	2	11	2
E14 Because understanding consumers requires understanding EVERTHING that impacts them, we have PhDs on staff in many different disciplines	−5	3	−1
E15 Insight into their online behavior . . . how do they use the computer for recreation . . . to research? To compare prices? To buy?	0	3	−2
E16 After all, actionable consumer insight isn't about what consumers think TODAY . . . it's about what they will think TOMORROW	−2	5	8
E17 Since most consumer thought is below the surface of consciousness, we use art and design to uncover their thoughts and emotions	−3	2	−1
E18 Anyone can provide "qualitative insights." We deliver hard data to test your hypotheses and drive actionable decisions	−6	5	5
E19 Geodemography helps you to know where they live so you can find others like them	−7	−7	1
E20 Our Total Human Analysis connects attitudes, behaviors and societal influences to better understand why people do what they do	0	3	−1

Table 25.3. Utility values for the additive constant and the 36 elements

		Agency planner	Marketing research	Graphics/package designer
E21	Our genetic modeling identifies the relationship between variables so that you can focus on those that matter most	−4	−4	−2
E22	Our statisticians can perform advanced analyses on data and find the story that you did not even know was there	1	8	3
E23	We focus just on secondary research . . . there is already a world of information out there . . . you just have to know where to look	−6	−10	−5
E24	Through our custom proprietary research tools, we help you gain competitive advantage	−3	3	2
E25	Our behavioral analysis focuses only on activities and programs which impact sales	1	2	0
E26	Syndicated and secondary reports provide valuable information at a much lower cost	−5	0	2
E27	The most sophisticated data mining tools available . . . running on Cray Supercomputers . . . to tease out new insights from all your data	−3	−7	−6
E28	We provide only hard, quantifiable facts so that you can make better business decisions	−4	1	4
E29	Vivid scenario planning can help you see your consumers in the context of alternative future states	−10	4	2
E30	Our deliverables are more than strategic recommendations . . . we design and execute the program tactics	−4	0	2
E31	We provide consumer insights that lead to innovative solutions	−3	3	2
E32	Specializing in creating names and logos for your company, products and services	−2	−11	4
E33	We collect consumer insights and develop a wide range of tactical elements like packaging, merchandising, and promotions	−2	−5	0
E34	One flat monthly fee provides you with access to our full staff when you need them	3	−4	6
E35	All our fees are quoted upfront . . . never a surprise	4	8	6
E36	We are confident in providing you with powerful actionable results . . . so our fees are based on your success in the marketplace	2	2	6

from marketing research. Indeed, the additive constant for the advertising agency (constant = 59) is substantially higher than the constant for marketing researcher (constant = 44), meaning that, at the outset, one is more likely to feel one gains insights from the agency than from the market-research supplier. Although the marketing researcher may provide insights equal to those provided by the brand planner, the specifics delivered by market researcher will make the case for believing the researcher. In contrast, the advertising brand planner by virtue of his or her

job is believed, so the specific deliverables can be weaker and yet the level of belief in the two professionals could end up being the same.

This value, from the additive constant, is only part of the story, however. The elements show the greatest variability from high to low for market research, meaning that, although the market researcher starts out with a lower additive constant, this research can occasionally reach greater interest by incorporating the proper elements. It also means that, to reach a high utility value, with a great deal of interest by businesspeople, the specifics or elements promised by the market research must do a lot of the work. In contrast, the elements for the brand planner or the graphics designer need to do less work because they start out higher in their additive constant.

A sense of the extreme elements can be obtained from Table 25.4, which lists the winning and losing elements for the three professions. These results suggest that brand planners in advertising agencies show the smallest range of positive and negative elements, whereas market-research suppliers show the widest range. Furthermore, it also suggests that, with the proper selection of elements, there is the possibility of creating a strong concept for market research, but one has to be sure not to put in poor performers. The risks of misstating the types of insights and the reasons underlying those insights are greatest for the market researchers.

The Respondent's Job Determines What Insights Are Important

The respondents were instructed to check off their job title. Of these, there were six specific job titles with ten or more respondents. The remaining respondents fell into categories with fewer than ten respondents or checked the box "other." The results show that the respondent's job drives the utility values (i.e., what they want) more than does the nature of the insights provider. Some key trends appear in Table 25.5, which shows the additive constant and the top three elements for each type of respondent, independent of the particular insights provider. The 184 individual models from all three studies were pooled together before creating Table 25.5:

1. *The additive constants differ dramatically by respondent type.* Keep in mind that the additive constant reflects the likelihood of the respondent to accept insights from a third-party supplier. The constant is highest for administrators (65) and then managers, and lowest for information technology (IT)

and R&D (36 and 33, respectively). The least technical groups are most open to these insights, whereas the most technical is least open. For concept development this means one may not get the necessary help from R&D that one needs, simply because of their attitude toward knowledge providers. Perhaps R&D comprises individuals who are more "concrete" in their viewpoints and cannot go beyond themselves to create new ideas.

2. *Administrators*, although most open, are really responsive to general ideas, such as learning what motivates the customers.

3. *Managers*, slightly less open, respond to statements about process with known costs.

4. *Salespeople* show inconsistent patterns. They respond to known fees, to actionable insights, and to high-tech solutions. They show no single driving factor.

5. *Marketing personnel* want warmth, a sense of comradeship, and a soft sell. They respond strongly to statements about a blend of art and science, a phrase that turns off a number of the other groups.

6. *IT* responds to specifics, such as price, specific actions to drive responses, and hard data.

7. *R&D* wants to know the *why*, to control the present and predict the future.

Concept-response Segmentation: Identifying Groups of Individuals with Like-minded Responses

Three clear segments emerged, using k-means segmentation appropriate for IdeaMap.Net studies (see Chapter 7). Table 25.6 lists the winning elements by segment:

1. *Segment 1 comprises individuals interested in relationships with the insights supplier.* These people show a very high constant (59), with the elements adding a moderate amount of draw to the supplier. These respondents are predisposed to insights providers.

Table 25.4. Winning and losing elements for the three insight-providing professionals

	Utility
Additive constant	*59*
Advertising agency brand planners: Winners	
What motivates your customers to make a choice? Cost, convenience, habit or something else?	4
Consumer insight is a blend of art and science . . . we strive to provide the best of both	4
All our fees are quoted upfront . . . never a surprise	4
Learn what products your customers use . . . the brands they prefer . . . and why	4
Advertising agency brand planners: Losers	
Anyone can provide "qualitative insights." We deliver hard data to test your hypotheses and drive actionable decisions	−6
Our planners go undercover and live with your consumers to truly understand them in their real world	−6
We focus just on secondary research . . . there is already a world of information out there . . . you just have to know where to look	−7
Geodemography helps you to know where they live so you can find others like them	−7
Vivid scenario planning can help you see your consumers in the context of alternative future states	−10
Graphics design: Winners	
After all, actionable consumer insight isn't about what consumers think TODAY . . . it's about what they will think TOMORROW	9
One flat monthly fee provides you with access to our full staff when you need them	7
All our fees are quoted upfront . . . never a surprise	7
We are confident in providing you with powerful actionable results . . . so our fees are based on your success in the market place	7
Anyone can provide "qualitative insights." We deliver hard data to test your hypotheses and drive actionable decisions	6
A partner that can help you understand how technology is changing the way your customer makes decisions	6
Graphics design: Losers	
We focus just on secondary research . . . there is already a world of information out there . . . you just have to know where to look	−6
The most sophisticated data-mining tools available . . . running on Cray Supercomputers . . . to tease out new insights from all your data	−7
Market research: Winners	
Learn what products your customers use . . . the brands they prefer . . . and why	14
What prompts your customers to act?	11
A partner that does more than just describe your customer . . . they help you to understand the implications to your business	9
All our fees are quoted upfront . . . never a surprise	8
Our statisticians can perform advanced analyses on data and find the story that you did not even know was there	8
What motivates your customers to make a choice? Cost, convenience, habit or something else?	6
Market research: Losers	
Geo-demography helps you to know where they live so you can find others like them	−7
The most sophisticated data mining tools available . . . running on Cray Supercomputers . . . to tease out new insights from all your data	−8
We focus just on secondary research . . . there is already a world of information out there . . . you just have to know where to look	−10
Specializing in creating names and logos for your company, products and services	−11

Table 25.5. The three winning elements for respondents in each job description

	Utility
Administrator	
What motivates your customers to make a choice? Cost, convenience, habit or something else?	8
Learn what products your customers use . . . the brands they prefer . . . and why	7
One flat monthly fee provides you with access to our full staff when you need them	6
Management	
One flat monthly fee provides you with access to our full staff when you need them	13
All our fees are quoted upfront . . . never a surprise	12
We are confident in providing you with powerful actionable results . . . so our fees are based on your success in the market place	10
Sales	
All our fees are quoted upfront . . . never a surprise	14
Helps you to understand what one single change your customers would make to the products and services they use to improve their day	13
Our statisticians can perform advanced analyses on data and find the story that you did not even know was there	10
Marketing	
A partner that does more than just describe your customer . . . they help you to understand the implications to your business	11
Learn what products your customers use . . . the brands they prefer . . . and why	10
Consumer insight is a blend of art and science . . . we strive to provide the best of both	10
Information technology	
All our fees are quoted upfront . . . never a surprise	17
Helps you to understand what one single change your customers would make to the products and services they use to improve their day	17
Anyone can provide "qualitative insights." We deliver hard data to test your hypotheses and drive actionable decisions	15
Research and development (R&D)	
After all, actionable consumer insight isn't about what consumers think TODAY . . . it's about what they will think TOMORROW	19
A partner that can help you understand how technology is changing the way your customer makes decisions	16
Learn what products your customers use . . . the brands they prefer . . . and why	15

2. *Segment 2 comprises the price/performance or smart shopper segment.* These people are process oriented, with an emphasis on the bottom line. The elements have to work much harder because they begin with an additive constant of 50.

3. *Segment 3 comprises the technological empowerment group.* These individuals are interested in the latest processes to produce insights. The elements really have to work hard because their additive constant is 43. Fortunately, they have many elements that do quite well in convincing the insights buyer.

4. The concept-response segments transcend the conventional demographics and job descriptions (Table 25.7).

What Should the Market (Consumer) Researcher Do in Light of the Low Initial Response to the Role as Insights Provider?

It is somewhat disappointing that market researchers and, by extension, sensory analysts get less respect for insights than, say, graphics designers. Market researchers today pride

Table 25.6. Winning elements for the three key concept response segments

	Utility
Segment 1: Relationship (additive constant = 59)	
Learn what products your customers use . . . the brands they prefer . . . and why	10
A partner that does more than just describe your customer . . . they help you to understand the implications to your business	9
Consumer insight is a blend of art and science . . . we strive to provide the best of both	8
A partner that can help you understand how technology is changing the way your customer makes decisions	8
Segment 2: Process, price, and performance (additive constant = 50)	
All our fees are quoted upfront . . . never a surprise	17
We are confident in providing you with powerful actionable results . . . so our fees are based on your success in the market place	13
One flat monthly fee provides you with access to our full staff when you need them	10
Segment 3: Technology, state-of-the-art empowerment (additive constant = 43)	
Our statisticians can perform advanced analyses on data and find the story that you did not even know was there	20
Our Total Human Analysis connects attitudes, behaviors and societal influences to better understand why people do what they do	13
Our behavioral analysis focuses only on activities and programs which impact sales	11
The most sophisticated data mining tools available . . . running on Cray Supercomputers . . . to tease out new insights from all your data	10
A partner that can help you identify unmet needs and wants among consumers	10
Through our custom proprietary research tools, we help you gain competitive advantage	9
A partner that makes you feel that you truly know the customers . . . everything about them . . . hobbies, passions, fears, dreams . . . and more	9

themselves on their role of knowledge provider and their ability to serve up insights. What might be happening, however, is a confluence of roles. On the one hand, market researchers have always prided themselves on executing their job flawlessly, on finding the right suppliers at the right price. They, like the sensory analysts in R&D, have systematized much of what they do, have converted it to process, and then have presented the process to management, with the process couched in terms of value for the money. They work within a budget. They are accustomed to doing tracking studies, hiring out the execution and analysis. Their reward comes from bringing the project in under cost and within the time frame. In many cases the market researcher has become a broker of services, not particularly adept at the insights business, per se, but very able to discover bargains and ensure a seamless delivery of high-quality information. Sensory analysts are inexorably following that path, although, as we will see below, they have not gone very far.

This process capability, whose value cannot be dismissed, is a work product in which insights do not thrive. A researcher's output differs from the work product offered by a brand planner at an agency and by a graphics designer at a design house. The brand planner helps the agency to craft better advertising and is judged, for better or worse, on the performance of the advertising in driving sales. Ultimately, for the brand planner, it is not a matter of seamless delivery of information, but rather a matter of sales and market success. Seamless delivery of insights is the brand planner's method, but not the raison d'être. For the graphics designer, in turn, credibility in insights may come from the designer having to produce something tangible whose specific performance can be tested. Ultimately, for the graphics designer, therefore,

Table 25.7. Proportion of respondents falling into the segments, by key classification question

	Segment		
	1	2	3
			Technological
	Relationship	Price	empowerment
Total panel	*26*	*38*	*26*
Study in which they participated			
Market-research study	37	32	32
Ad agency study	27	36	38
Design firm study	16	44	40
Gender			
Female	26	39	35
Male	26	35	39
Age			
<30	29	38	33
<40	42	27	31
<50	21	39	39
<60	17	45	38
<75	14	57	29
Position in the company			
Junior	75	25	0
Middle management	37	29	34
Senior management	31	34	34
President	15	45	40
Owner	19	41	40
Buying capacity			
Supplier of insights	46	15	38
Use but do not buy from supplier	40	30	30
Decider	24	38	38
Do not use	16	48	36
Loyal to one supplier	29	38	33
Switch among suppliers	20	36	44
Market			
Midwest	20	37	43
Northeast	35	37	28
South	24	36	40
West	27	41	32
Job title			
Marketing	41	25	33
Research and development (R&D)	40	27	33
Sales	40	20	40
Administration	12	56	32
Information technology (IT)	9	45	45
Management	7	43	50

it is again not a matter of seamless delivery of information, but a matter of testable market success.

Neither brand planners nor graphics designers have chosen process as their main job. Market researchers, in contrast, are nei-

ther rewarded on creativity nor rewarded on advertising success. They are rewarded for process and not necessarily for the results of that process. Despite protestations to the contrary, most market researchers feel comfortable with process and rather opt for the accountability of process than the more merciless accountability to be found in the marketplace. Sensory analysts are in the same boat, but earlier in the development cycle.

Can Researchers Span the Range of Consultant and Data Provider?

One of the ongoing plaints to be heard at conventions of researchers is the palpable concern that researchers are losing impact, as consulting companies take over the job of providing insights. Not to be outdone, sensory analysts complain that market researchers are taking over their jobs. To some degree this fear is justified, again because both research groups are perceived to be the owner of process rather than the owner of insights.

The process brings with it security, however, at least to some. The tumultuous decades have seen the research community split into at least two camps: those who specialize in ad hoc research designed to solve problems, and those who specialize in continuous research of a contractual nature. Despite protestations to the contrary, it is the natural human inclination, the *homo economicus* residing in all people, to minimize expenditure and maximize profit. This inclination means that, despite protestations to the contrary, most companies will take the contractual, continuous research and assign to the business the personnel with whom they can make a profit. The other individuals, perhaps the more experienced, will be assigned where possible to soliciting and servicing the ad hoc business, where one's talent is always put to the test and where the daily competition for projects is brutal.

It should come as no surprise, therefore, that the market-research community values the ongoing tracking study and other forms of continuous research. The sensory community, in turn, values the ongoing contract to do profiling, discrimination, and acceptance tests. Business stability is generally more valued than is business instability. At the same time, however, the valued stability is purchased at a price: the reduction of expensive brainpower in favor of cheaper labor to service the contract. Consultants, living as they do on brainpower, even with big contracts, find this situation an easy opportunity as the research community moves toward its own business stability. That stability rewards tracking research and sensory support services, and simultaneously punishes problem-solving research. The longed-for stability opens the way for consultants of all types to become the problem solvers and the providers of costly consumer insights, product concepts included. Researchers, obeying the laws of economics, have ceded that high ground, despite their statements to the contrary.

The Role of Process in a World of Insights

If consumer researchers are going to compete in the knowledge economy, the big question turns to the proper role. In today's world, abundant with knowledge and perfused with technologies that make research almost a commodity available at the touch of one's keyboard [e.g., self-authoring systems (see Chapter 16)], what is the role of process-oriented researchers? Indeed, is there a role? Consultants daily promote themselves as the purveyors of insights and creators of better concepts. They attack the corporation at a higher level than do market researchers, who through their own culture feel that they must deal with work within the hierarchy dominated by middle management. Consultants do not typically obey a hierarchy, opting for the highest point of entry they can find and then selling down, often to a resistant market researcher or hostile sensory analyst. Researchers in turn, whether in the company or outside, are process bound and tend to stay within the guidelines set. Market-research suppliers report to market-researcher clients in-house because that is simply the *process* to be followed and the implicit guideline for doing business. Sensory suppliers report to their in-house sensory clients or, more often, to in-house R&D.

Given this behavioral obeisance to authority, to lines of demarcation, and to the apotheosis of process over result, what then is the researcher to become insofar as the insights business is up for grabs? Will the researcher, whether market researcher or sensory analyst, remain stuck in the ways of a process facilitator whose primary job is to produce neat-looking tables on demand at the lowest price? Will the researcher be able to break out of this mold and grow into a true consultant as the researcher once was in the fairy-tale days of yesteryear (that's only 25 years ago, but still yesteryear). Or will the ongoing saga of the researcher be played out again and again, on the small stage of individual corporations and in the large stage of research conferences, with the story never quite ending. Will the researcher remain in a limbo, aspiring to insights, but being content to have job security based on process? We are in for interesting times and should stay tuned.

Beyond the Insights to the Sensory Professional

The sensory professional provides a very good subject with which to finish this chapter. Much of this book has dealt with concepts, which to a great degree have been in the purview of the market researcher. Sensory analysts are beginning to make their voices heard in consumer research. Once confined to simple profiling using expert panels, and to mundane tests of difference and acceptance, sensory analysts are branching out into the world of consumer insights. It is not unusual now to see sensory professionals becoming involved in concept development.

Given this change in the role of the sensory professionals, it is interesting to see how they are perceived and how they perceive themselves. The sensory professional represents the next generation of professional to be involved in concept creation, optimization,

and evaluation. Do professionals in the field, including the sensory analysts themselves, perceive the role of the sensory analyst to involve a deep understanding of the consumer? Do sensory analysts see themselves graduating to concept developers? Or do they see themselves as process facilitators?

In the summer of 2001, Hollis Ashman and Jacqueline Beckley of the Understanding and Insight Group, as well as the senior author (H.R.M.), decided to conduct an experiment by studying how sensory analysts and their colleagues perceived them. The objective was to use the conjoint analysis procedure to profile the sensory analyst rather than the food concept. In a sense the approach was to treat the sensory analyst like a product and identify the features of this new product. Hollis, who had experience in a variety of companies both in food and nonfood areas, was the driving force to understanding what makes a professional. An engineer by education and intensely curious about professionalism, she took the lead in the experiment, created the concept elements, and put the study together in a 2-week period. Originally designed to understand the professional in a food company, this study was run among the sensory professionals belonging to the *Sensory e-group*. This e-group, a Yahoo!-based list server, comprised more than 500 individuals with an interest in problems related to the subjective evaluation of foods and beverages.

The e-group would prove to be a perfect place to run the study of *sensory analyst as product*. For one, the e-group comprised Internet-enabled participants, so an Internet-based conjoint study would be simply a link away. Most people who wanted could participate. Second, the e-group comprised individuals who were heterogeneous with respect to factors such as interests, education, and job, but who were passionate about the field of product evaluation. Third, the study comprised issues that could be helpful to educators in the field, as well as to corporate practitioners.

The study itself comprised 24 elements combined into 40 combinations. The respondents evaluated the combinations as fit to a sensory professional and then completed a classification question to indicate gender, age, profession, education, and so on. Although the invitation was by Internet to the sensory e-group, we requested where possible that the respondent should pass the link to others in the organization.

The results were quite clear. We begin with the ratings for the total panel and then move immediately to the segments. Table 25.8 shows the utility values for the interest model:

Total panel. The total-panel data ($n = 137$ respondents) suggest that the sensory professional is perceived to be more of a consultant rather than a risk taker. The sensory professional teaches, but does not actively change the business by looking for new method. Certainly publications, etc., do not matter.

Segment 1: Academic orientation (26%). This segment perceives the sensory professional to be a technical expert, much as a scientist is a technical expert. This segment sees the sensory professional as publishing, keeping up with the field, etc., while also being a staff service person.

Segment 2: Helpful staff (44%). These individuals are pure service people who show the way with knowledge and approaches, but do not actively pursue any risks in the field by trying new technologies. They are true corporation people, selfless, trying to bring the truth to the group.

Segment 3: Business growers (30%). These are business-oriented people who use their professional abilities to move the business forward. They have more of a marketing and take-charge mentality, look to be the most proactive of the group, and have the most forward-looking mind-set.

Expected response of segments to concept development and research. Given these three mind-sets, it now becomes more evident

what types of pressures and ferments go on in the sensory analysis field. The academically oriented tend to be the most welcoming for new ideas and should be the ones to become quickly involved in concept research. The helpful staff, almost half the group, tend not to be interested in too much beyond doing a good job and thus are anchors maintaining the field at the status quo. They should not be expected to stretch beyond their current work to concept work until several years have passed. The business builders, like the academics, tend to be more future and opportunity oriented.

How the Characteristics of Individuals Drive Their Perception of a Sensory Professional

At the end of the concepts or vignette evaluations, each respondent profiled himself or herself in terms of education, job, etc. We saw that the position of an individual in the corporation drives his or her view of the insights professional. We find the same thing for the sensory profession. The individual himself or herself brings in criteria to what a sensory professional should be. Often the features that appear most appropriate for a sensory specialist might be aspirations of the respondent. Key examples, which are presented in Table 25.9, can be summarized as follows:

1. *Education level.* The BA/BS respondent sees the sensory specialist as a team player, an expert, and a resource for points of view. The MA/MS and the PhD respondents see the specialist as more of an innovator. Both higher-degreed respondents thus see the sensory professional as more of a leader for new ideas rather than simply as an expert.

2. *Job title.* The marketer sees the specialist as someone with a reservoir of knowledge to be called on. The product developer sees the sensory specialist as providing direction and expertise. The market researcher sees the sensory specialist as an individual

Table 25.8. Results from the sensory professionalism study

		Segment		
		1	2	3
		Academic	Helpful	Business
	Total	oriented	staff	grower
Number of respondents	*137*	*36*	*60*	*41*
Additive constant	*65*	*62*	*69*	*63*
Category 1: Orientation to new ideas				
E01 Provides a role model for individuals new to the field	2	1	2	4
E02 Remains committed . . . with a drive to succeed	1	−7	1	8
E03 Applies creativity and critical thinking to move the business forward	−1	−8	−4	9
E04 Oriented towards new possibilities . . . open to change and new learning	0	4	−1	−3
E05 Remains authentic to his/her personal values while considering the values of others or the values of the organizational culture (politics)	−1	−5	−3	5
E06 Provides an opinion and guidance in critical situations	3	−1	3	9
Category 2: Responsibility and personal growth				
E07 Continues to seek out new internal and external ways to do business	0	−3	5	−5
E08 Personally tries out new and innovative approaches	−2	5	−1	−8
E09 Shows others how to integrate product, consumer, and market knowledge in the project	5	8	7	1
E10 Adept at applying knowledge and follow-through on the tasks required to complete the job	4	−1	9	2
E11 Actively promotes new and innovative approaches through the organization	5	9	5	1
E12 Takes action when discovering that something was done wrong or inappropriately by a functional group	0	−8	4	1
Category 3: Interaction with others				
E13 Passionate about listening to the needs and ideas of others	0	0	1	−1
E14 Actively provides point of view in professional discussions	5	7	3	7
E15 Shows humility in presenting his/her ideas while accepting constructive criticism and contrary opinions without being defensive	−7	−3	2	−24
E16 A team player	4	9	8	−5
E17 Often accepts a leadership role	−3	−15	−1	6
E18 Provides vision and resourcefulness	2	3	0	5
Category 4: Technical competence				
E19 Maintains thorough knowledge of technical literature	0	12	−12	8
E20 Makes difficult decisions under pressure	−2	−11	1	1
E21 Recognized as an expert in his/her field	5	15	−3	7
E22 Uses coaching and negotiation to motivate coordinated action to achieve goals	−5	−6	−6	−3
E23 Maintains close liaison with other practitioners in the field	3	11	−2	3
E24 Publishes articles in various journals and books	−19	8	−41	−10

Table 25.9. Utility values for the strongest-performing elements for selected subgroups of respondents

	Utility
Degree: BA/BS ($n = 51$)	
A team player	10
Recognized as an expert in his/her field	7
Provides vision and resourcefulness	6
Actively provides point of view in professional discussions	6
Maintains close liaison with other practitioners in the field	6
Degree: MA/MS ($n = 59$)	
Actively promotes new and innovative approaches through the organization	7
Shows others how to integrate product, consumer, and market knowledge in the project	5
Remains committed . . . with a drive to succeed	5
Adept at applying knowledge and follow-through on the tasks required to complete the job	5
Recognized as an expert in his/her field	5
Degree: PhD ($n = 27$)	
Shows others how to integrate product, consumer, and market knowledge in the project	8
Adept at applying knowledge and follow-through on the tasks required to complete the job	7
Actively provides point of view in professional discussions	7
Job: marketing ($n = 5$)	
Recognized as an expert in his/her field	20
Maintains thorough knowledge of technical literature	14
Makes difficult decisions under pressure	14
Job: product development ($n = 28$)	
Shows others how to integrate product, consumer, and market knowledge in the project	12
A team player	6
Actively provides point of view in professional discussions	6
Job: market research ($n = 11$)	
Maintains close liaison with other practitioners in the field	10
Provides a role model for individuals new to the field	9
Actively promotes new and innovative approaches through the organization	8
Job: sensory specialist ($n = 79$)	
Adept at applying knowledge and follow-through on the tasks required to complete the job	7
Actively promotes new and innovative approaches through the organization	7
Shows others how to integrate product, consumer, and market knowledge in the project	7
Experience $<$ 10 years in field ($n = 11$)	
Shows others how to integrate product, consumer, and market knowledge in the project	16
Actively provides point of view in professional discussions	7
Personally tries out new and innovative approaches	6
Experience 10–15 years in field ($n = 17$)	
Provides vision and resourcefulness	16
Recognized as an expert in his/her field	14
Actively promotes new and innovative approaches through the organization	12
Actively provides point of view in professional discussions	12
Experience 15–20 years in field ($n = 30$)	
A team player	8
Provides an opinion and guidance in critical situations	7
Adept at applying knowledge and follow-through on the tasks required to complete the job	7
Experience 20+ years in field ($n = 78$)	
Applies creativity and critical thinking to move the business forward	4
Adept at applying knowledge and follow-through on the tasks required to complete the job	2
Remains committed . . . with a drive to succeed	2

who carries out a process for providing knowledge. The sensory specialist sees himself or herself as providing innovative approaches and spreading knowledge through the corporation. Thus, the two groups that have to accomplish something—the marketer and the product developer—see the sensory specialist to be someone to call on. The market researcher and the sensory specialist see the sensory specialist as a teacher and a doer of tasks.

3. *Years of experience.* Relative novices in the field (0–10 years) see the sensory specialist as bringing new ideas to the company. Those with the middle range of experience (10–20 years) see the sensory specialist as an expert and a resource. Those longest in the field (20 or more years) see the sensory specialist as moving the business ahead and look inside the specialist for an ongoing level of commitment. A person who has been in the field 20 or more years is no longer responsive to the elements. The utility values are very small. Little impresses them. The long-timer is not likely to get involved in concept research and unlikely to move into the newly evolving areas requiring ideation.

References

Cagan, J., and Vogel, C. (2001). Creating Breakthrough Products: Innovation from Product Planning to Program Approval. Upper Saddle River, NJ: Prentice Hall.

Czerniawski, R.D., and Maloney M.W. (1998). How to be a better client advertiser. www.adcollege.com/dispatches/dispatch_1.pdf.

Flores, L., Moskowitz, H.R., and Maier, A. (2003). From weak signals to successful product development: using advanced research technology for consumer driven innovation. In: Proceedings of the Second ESOMAR Conferences, Technovate, Cannes. Amsterdam: ESOMAR [European Society of Marketing Research].

Hu, C., and Hiemstra, S.J. (1996). Hybrid conjoint analysis as a research technique to measure meeting planners' preferences in hotel selection. Journal of Travel Research, 35: 62–69.

Moskowitz, H.R. (2002). Establishing data validity in conjoint: Experiences with Internet-based 'megastudies.' Journal Of Online Research, 1; pdf file: www.IMRO.org.

Moskowitz, H.R., Gofman, A., Katz, R., Itty, B., Manchaiah, M., and Ma, Z. (2001). Rapid, inexpensive, actionable concept generation & optimization: the use and promise of self-authoring conjoint analysis for the foodservice industry. Foodservice Technology, 1: 149–168.

Renaghan, L.M., and Kay, M.Z. (1987). What meeting planners want: the conjoint analysis approach. Cornell H.R.A. Quarterly, 28: 67–76.

Urban, G., and von Hippel, E. (1988). Lead user analyses for the development of new industrial products. Management Science, 34: 569–82.

Valdés, I. (2000). Why you should know these important consumers. Consumer Insight Magazine, December; cited on the Internet.

Young, S. (2003). A designer's guide to consumer research. www.prsresearch.com/ins_art_designers_guide.html.

Chapter 26

Scientific and Business Realpolitik: Insights from Selling New Ideas for Concept Research

Introduction: The Ways of the World

Despite protestations to the contrary, the world does not wait for, nor relish, new ideas. This truism, distressing as it may seem, pervades the research community as much as anywhere else in the business world, and perhaps even a bit more. Most researchers are risk averse, and although they publicly announce their support of new ideas and new methods, in practice they are the first to critique anything new.

Why such a negative-sounding exhortation in a book about concepts? Concept research seems to be the most established of methods in the business world, for, as was stated in the first chapter of this book, the concept provides the blueprint for the product or service. It stands to reason that the research community should thus welcome innovative thinking that leads to new ways for analyzing responses to concepts. However, *Realpolitik* intrudes, and the shattered dreams of innovative researchers are often scattered about in its wake. Guised as stringent tests of validity and reliability, the stubborn efforts to maintain the status quo inevitably defeat all but the most determined.

Despite this Cassandra-like dirge, new ideas do eventually find their way into the research community, although not without a fight and not without the cadre of those who would rather see these new ideas somehow magically disappear. What is it about new ideas that capture the imagination, that so entrance the young novice that he or she is ready to throw down a professional gauntlet?

The study of how these new ideas are embraced by the intellectually brave is a book in itself, which talks about the confrontation of the new and the established, the young and the old, and in classical terms the ancients and the moderns. This book has not the room to deal with the issue in the way that one might wish, but we can at least explore some of the senior author's experiences with IdeaMap, a research approach that at the time of its introduction in 1992–1993 was novel and iconoclastic. The history of how IdeaMap was introduced, the nature of the responses, and the meaning of those responses provide fascinating insight into another aspect of the scientific method: convincing others of the validity and usefulness of one's own ideas.

From Chaos to Procedure, from Process to Pontification

A cursory history of consumer research as a profession reveals the growth of the field from a cottage industry with few accepted standards to a professionalized, organized discipline combining both practitioners and academics. From a field starting in the 1920s and 1930s, when researchers literally had to invent the tools in order to conduct the study, consumer research has evolved into a discipline where standards, or at least accepted practices, govern everything from the recruitment of respondents, to the types of questions one asks, to the types of analyses one does, to the format of data tables, and even to the level of recommendations that one should make. These procedures are not carved in stone, but rather have evolved from chaos in order to make it technically and financially feasible to acquire and report consumer-based data.

As the research field matured and continues to mature, and as procedures become sacrosanct, either because they have been established beyond question as being correct or, more often, because they have been used so frequently, those very same procedures often ossify. All too often the *process of research* takes over, until the actual implementation of the study becomes as important as the reasons why the study was commissioned in the first place. The process becomes standardized. The freewheeling spirit of inquiry dies under the weight of professionalization and *processization*. Processes pervade all aspects of the research business, as researchers evolve from problem solvers to managers of those outsourced vendors who actually provide the consumer data. Process becomes the sine qua non for many research buyers, who limit the types of research that they do to that for which they have norms or to that for which they have preferred suppliers, who conduct the same study, year in and year out, in a standardized, routinized, occasionally rote fashion, albeit executed superbly.

Standardization and routinization of the research process should not come as a surprise in light of both the pressures to reduce the cost of the data collection process and the demand on the decreasing research staff to acquire and manage ever-increasing amounts of information. When the stress on in-house researchers is high and the demands so strong, it is probably the nature of people in general and researchers in particular to jettison new ideas and stick with proven processes.

When research began to gain wide acceptance, it was touted as a strategic tool for marketers and developers because it provided insight into the consumer's mind. Perhaps in those early days the true appreciation of research may have been less than we think, but

one cannot help feeling that, over the years, professionalizing consumer research has reduced acceptance rather than increased it. One need only read books published 30–50 years ago to feel that somehow we researchers have lost some of our "soul" as the field has developed. Academic journals devoted to consumer research are filled with increasingly sophisticated mathematical models, but in the candid opinion of many practitioners these sophisticated models are appreciated neither by the research suppliers nor by the clients. As a consequence, research is becoming less relevant, often falling below the corporate radar screen and being consigned to the purchasing department, in the worst case, or to a vastly reduced staff (in size and quality), in the best case.

The Actual Process of Selling One's Ideas: What to Expect and How to Decode the Covert Messages

A good idea alone will not make it through the thickets of rejection. Just having an idea that can solve many existing problems does not suffice. The idea must be sold, and sold actively in a way that gives it sufficient traction to take hold and prosper. The consumer-research and sensory-analysis communities are replete with good ideas that never made it because they were developed and commercialized, but never sold well enough. In the end these good, or occasionally not so good, ideas simply faded away, atrophied, and were relegated to the dustbin of business history.

Selling does not mean simply arguing that the approach works and proffering some proof that it does. Everyone argues that one's method provides the answer needed and capably marshals one, two, three, or more points to support one's argument. Buyers of research methods do not wait patiently for a research supplier or vendor to produce a new approach, listen attentively to why it works, and then passively agree. Most buyers are accustomed to litanies of reasons why a new idea is better. Therefore, the selling must be different. It must break through indifference and actually get the buyer to purchase the service. In concept research, therefore, this type of selling must motivate the buyer to try the method when the buyer is already awash in alternatives offered by ferociously competing suppliers.

How then does the supplier sell the idea to a jaded buyer? What should the seller do? What types of strategies work and what strategies seem good at the outset but are filled with peril because they lead down to disastrous outcomes? Here are 21 points for sellers and innovators to consider. The points do not answer the question about what to do as much as indicate some of the perils that can be avoided and some of the bumps in the road that will inevitably be encountered and cannot be avoided. Knowing they exist helps, however.

The Research World and Its Zeitgeist

The Irresistible Impact of Zeitgeist

The *zeitgeist* is the intangible spirit of the times. In research, zeitgeist refers to the trends, the way people think, their readiness to adopt ideas, and the framework in which they place these ideas. Researchers are not immune from the zeitgeist. Sometimes ideas that are rejected in one period as being absolutely silly and impossible are readily accepted in another. It is not necessarily that the ideas have changed, but rather the people have or, more accurately, the zeitgeist has changed. It is hard to put a finger on the precise meaning of zeitgeist in concept research, other than to say that it encompasses trends. Today's zeitgeist comprises an acceptance of experimental design and modeling, along with the recognition that other tools such as *ethnography* (i.e., in-context) observation are important. Zeitgeist is best seen retrospectively.

The Impact of Professionalization

Over the past 40 years the research industry has become increasingly *professionalized*, which means that there are norms and procedures to follow and expectations about how studies will be designed, conducted, and reported. The profession places consultants higher than marketer researchers, and the researchers higher than the field service. Professionalization has exerted an effect on those trying to sell new research ideas to corporate buyers. With professionalism, a buyer no longer stands alone waiting to solve a problem with whatever means are available. Professionalism connotes norms of behavior, prescribed standard ways to do things. Professionalism enables buyers a number of short cuts or standard routines of action. Professionalism, however, hampers new ideas. Not only must buyers understand these new ideas, but they also must also judge whether other people in the same league would accept the solutions. The buyers stand with mind half-focused on their own need and half-focused on whether this solution is appropriate in the eyes of others. Even the research approach that works may be rejected because it might be seen to be inappropriate by other people whom the buyer feels are professionals.

The Tyranny of Best Practices

The business world adores best practices, because they enable one to point to a specific set of activities that illustrate how things should be done. Best practices are convenient if the research doesn't know how to solve a particular problem. The best practices are recommendations from a group. The problem is that best practices, like so many other prescriptive systems, stop innovation dead in its tracks. When the best practices come out, they are publicized. At first, they are talked about as a recommended approach, recognizing that there may be other approaches. Eventually, however, best practices take on a life of their

own. As those who have been intimately involved in best practices retire, the practices
stay on as recommendation, but lose their immediacy and become staid practices, never
quite the state of the art, but always safe because they have been blessed by enjoying a
history of prior use. In truth these recommended best practices have a place, but are
adopted without means of revision and without recognizing they represent practices at
one point of time. They ultimately become
corporate shackles. Good examples of this include the monadic testing of concepts, used
because it provides the purest reaction to the
concept.

Statistics and Statistical Thinking

Statistical thinking has always been important
in research. Two forms of statistics pervade
consumer research in general and concept research in particular. The most common form is
inferential statistics, whose goal is to determine whether two or more stimuli (e.g., concepts or concept elements) differ from each
other. The other type of statistics is modeling,
whose goal is to develop relations between
variables, generally based on equations (e.g.,
curve fitting), where the experimentally varied
concept elements are related to the consumer
rating. Statistics are very important, for they
keep researchers honest. They provide best estimates as to whether two concepts differ from
each other and with what probability. Statistics are also occasionally used to destroy new
ideas. Many suppliers have encountered
clients who demand the latest in inferential
statistics and modeling, even if the clients
don't fully understand what is required to use
the statistics and what will be obtained. A little
knowledge of the statistical vocabulary in the
hands of an aggressive or insecure client can
cause one to go through endless machinations.
This is especially the case when a client who
knows little about modeling requires endless
reruns of a model. Such situations occur more
often than one would like to admit.

The Tyranny of Databased Norms

Norms are exactly what they say they are,
viz., expected external performance on other
measures for previously run studies, with
given scores on ratings. One of the key issues
in research is what the data really mean.
Many suppliers cannot easily describe what a
buyer should expect, given the performance
of the concept on specific rating scales. In
these cases it helps a vendor to provide
norms, because then the researcher knows
more or less what to expect. This happy result is very useful, especially when the researcher uses the concept-evaluation test to
make a go/no-go decision. It is always helpful to know what to expect for a given concept score, especially when one is ready to
launch. The problem with norms comes from
their use in the exploratory situations, where
there are just promises or concept elements
rather than fixed, polished, final-stage concepts. Many insecure research buyers insist
that, even at this very early stage, the researcher provide norms so that the buyer can
know what to expect from early ideas. The
idea of the tyranny of norms comes from the
fact that, although norms may be perfectly
okay and quite useful in late stages of research, they should not necessarily apply at
the very early stages. Research buyers who
insist on norms even at this early stage do not
clearly understand the difference between
development and late-stage research.

Validity as a Matter of Opinion

As mentioned in Chapter 9 on validity and reliability, one of the most resounding phrases
that the senior author ever heard was told to
him by S.S. Stevens, his PhD professor,
around 1968. It was Stevens's contention that,
in the end, there was no such thing as real validity, at least not in research about human behavior. Validity, sad to say, was merely a matter of opinion. That stark statement means
that, in selling new ideas to a buyer, one can-

not really point to the fact that it works. Validity can be dismissed instantly by a suspicious buyer simply refusing to accept the criteria. What is important for this chapter is that it is impossible to convince buyers who do not want to purchase. In their mind, what they believe in is valid and nothing else. They don't need to be convinced, because they cannot be convinced. The laws of human behavior and the validity of a test method are not like the validity of a physical law, which can be demonstrated in front of someone by an *experimentum crucis*. One can demonstrate the principles of gravity such that it is impossible for a critic to deny the demonstration. In contrast, when it comes to behavior and to concept research, there are no such immediate demonstrations that simply wash away all resistance. No matter how strong the data, no demonstration that the approach works is 100% convincing. There are always counterarguments to everything. In such situations it is probably better not even to continue. One cannot win. The only consolation is that the buyer who has just rejected the concept has got to do something. Buyers have to believe in something as having validity or ultimately they lose their job. One cannot stonewall everything forever, even though to the seller it may seem that way.

The Buyer's World

Cost as the Key Driver to Adopt Only Certain Types of Methods

Many companies have come under pressure from costs dictated in part by obeisance to Wall Street's criteria of financial performance. Such emphasis on money leads to a mandate from management to look for ways to reduce operating costs. Many companies, therefore, look to research managers to reduce their operating costs for obtaining information. The covert statement, never really made, is that it doesn't really matter how good the data are. It is the cost of the data that can be measured, and it is this cost that is public. Many research departments, therefore, focus a lot of their attention on methods to reduce the research cost. It used to be that one would call a few research companies their *strategic partners*. That euphemism was another way to reduce costs by first becoming a major factor in a research company's volume, giving the research buyer leverage to hammer down the prices. The cost issue had another, more pernicious effect. It focused interest of research buyers on cheaper methods to accomplish the same objectives. The true goal was not to do things better, but rather to do the same things cheaper. Much as a researcher would try to present new ideas, the easier sell was to gather the knowledge more cheaply. Whether this meant cutting prices, using new vehicles for research, or inventing methods that were based on less expensive focus group, the concentration on cost reduced the chance of any new ideas taking hold.

The Absence of Direct Accountability for Corporate Performance and the Creation of a Fearful Yet Complacent Culture

Over the years, the senior author has observed an interesting phenomenon involving accountability. The pattern seems to be that, when a person is accountable for producing something that can be measured, there is a greater likelihood of that person adopting something new that differs from what has been done before. The *new something* may be somewhat risky, but that also may generate more money for the individual. In contrast, those who are not accountable for profits or sales, but who are held to task for process, generally are risk averse. They tend to focus on the procedures, holding to the orthodox methods for which they avoid punishment. As individuals, they reward adherence to process and make every appearance of welcoming new ideas, but generally judge everything by criteria that can be best described as risk aversion. Their criteria for accepting new ideas are not necessarily those that they state for public consumption. In public they often proclaim interest in new ideas, new methods, and increased corporate profits. With the rather limited accountability, however, and with limited rewards for corporate profits, they tend to err on the side of doing less, but doing that less in a more demonstrable manner that proclaims their interest in advanced methods. Many researchers fall into this camp, not necessarily because they are constituted psychologically in that way, but rather because the contingencies of reinforcement are such that these researchers are more easily punished by failure to adhere than by success from accomplishment. Knowledge workers in the information-acquisition business (i.e., consumer researchers and sensory analysts) fall into this category. Individuals in external supplier companies, doing the exact same job, tend to be held more accountable for corporate profits and thus pursue new ideas far more vigorously. One of the most fascinating about-faces in behavior occurs when a hitherto corporate-based researcher joins a supplier company. The accountability that occurs effects the dramatic change from focus on process to focus on accomplishment. Indeed, many companies now recognize this change and improvement in behavior, leading them to spin off nonessential support jobs to externally paid consultancies and contingent workers. The result is a rapid increase in one's accountability (even to oneself) and the associated increasing interest in new ideas.

Fear of the New

The plethora of business books talking about new ideas, new processes, and the need to "break it even if not broken" provide a wonderful array of inspirational messages to the person who is presenting new ideas. These business books exhort the developer of new approaches to persevere and, at the same time, tell those in corporations that the vitality of their company depends on breaking the old and embracing the new. There is only one problem. People don't like *new*, no matter what they say. People like what is familiar to them. They buy when they feel comfortable. New is frightening. New is not frightening to the supplier trying to sell the approach to the corporation, because the approach comes from the person who will benefit most from the discomfort. One strategy is to present the new approach as being *not so new*. Quite often one can convince a potential buyer that the ideas are not really new, but rather extensions of what the buyer is already doing. In that case the seller of the new idea can overcome the buyer's fear of the new. Other than that, the seller may need to last several years, until the approach is no longer perceived as really being new. It helps the buyer enormously when presenting a new idea to the corporation to state unequivocally that the approach is not really new to the industry, but is probably merely new to the corporation. That positioning gives the buyer and his group a feeling that they are not guinea pigs who try new ideas, but rather prudent businesspeople who have waited until now to try this now-established method. A good rule of thumb is that it takes about 3–4 years from the time an approach first emerges into public awareness until the seller can convince the buyer that the approach has been around for years and is only new to the corporation.

Migration to the Comfortable in an Era of Increasing Pressure

One of today's issues in consumer research is that a plethora of alternative methods are available to buyers to address a problem and yet, at the same time, there is less time to evaluate these alternative research methods fairly and adequately. In the late 1970s to the mid-1980s, research buyers had sufficient time to evaluate various techniques. Many research buyers prided themselves on staying au courant. The pace at that time was demonstrably slower. Often a buyer would invite the research supplier to a meeting to present the new idea. Buyers were constrained to work with their chosen suppliers, but many buyers prided themselves on being adventurous. There was little punishment for trying new ideas. Corporate profits were growing, there was no pressure from Wall Street to increase profits by streamlining everything in sight, and for the most part the research profession viewed itself as a true provider of corporate knowledge. People took risks. The risks of being wrong were not particularly great. The pace was slow. The fax had not been invented, e-mail was not even thought of, and most researchers did not have personal computers on which they could run extensive analyses. Over time, however, the pressure on corporations increased. Research buyers became younger, had less experience, and became increasingly nervous about making mistakes. That nervousness translated itself into reduced interest in new techniques as a key feature of one's job. As the younger researchers saw older corporate colleagues "dehired" only to morph into lower-paid consultants without the corporate perks and benefits, these younger researchers turned increasingly conservative and risk averse. The consequence was inevitably the diminished, and even the loss of, interest in new techniques and advances. Migration had begun in earnest to what was comfortable and safe.

The Seller's World

Why Argumentation Rarely Suffices

The excited novice often tries to provide the buyer with evidence that an approach works. All too often, however, the buyer operates under constraints, rules, and wishes quite unknown to the novice presenting the new in-

formation. The buyer may be required to work with a limited number of approved suppliers, with the approval coming from considerations having nothing to do with the specific power of the technique, but rather legislated according to some preexisting financial considerations. Today (2005), many companies have opted to limit their supplier list to a few companies over which they can exert some type of control, usually financial. The approved supplier may be chosen beforehand based on a basic level of adequate performance, but also with the agreement that the supplier will rebate some percentage of the billing. Argumentation simply does not work in this case because the decision to purchase is entirely out of the buyer's control. In some very unusual cases the argument for using a new method is so compelling that the buyer goes out of his or her way to set up a presentation of the method to other members of the staff. All too often, however, that presentation goes nowhere because everyone to whom the new approach is presented is, at the same time, constrained by the very same rules.

Buying Styles: Contrasting the Practitioner and the Academic

How should sellers of new ideas present themselves? When it comes to concept research, two opposing behavioral styles and worldviews continue to intersect, known by the names of *practitioner* and *academic*, respectively. These names do not pertain to the position of the individual in the work environment. Rather, they pertain to the way a person perceives the job, with particular respect to the execution of concept research. They apply both to buyers of research and to sellers or vendors of research. The practitioners are often content with running simple concept studies, reporting the results, and applying fairly straightforward statistical tests such as t tests or analysis of variance. Occasionally, practitioners apply more sophisticated analyses to the data, usually because this analysis has either been requested by a client or is currently in favor and faddish. The academics often find the more technical aspects attractive. As a result, they typically opt for more complex approaches, harder-to-understand methods, and harder-to-interpret results. An example of this is conventional conjoint analysis using full-profile methods versus discrete choice analysis. Depending on the nature of the buyer—practitioner or academic—one may have an easy or a hard time selling a new idea. It is not particularly easy to sell to either. Practitioner buyers are not particularly interested in new ideas, but rather want to get the job done. What appeals to practitioner buyers is speed and cost. Accuracy is generally assumed, and the accuracy issue to practitioners is generally only one of cost of entry. In contrast, academic buyers are interested in the nuances of the approach, the differences between the approach and methods currently used. Typically, academic buyers want to know what algorithm lies behind a method, how this algorithm compares with other algorithms currently used, what the benefits are, and so on. In some cases academic buyers actually use the interaction with research suppliers to get a free education. Academics may also not be particularly interested in the research nuances, but rather simply believes that their job is to learn about the methods, no matter what the cost and time implications are to the suppliers who are unfortunate enough to encounter them. Such academic buyers can be recognized by the daily, hour-long conversations by phone, the interminable revisions, and the interweaving of procedural issues, hypothetical cases, and hard-to-solve problems brought into the discussion (viz., as a test of the limits of the new method). One happy thing often occurs. The academics, unable to really make a decision in sufficient time and always "going to school" on the suppliers, usually move out of the corporation, especially when the corporation looks

carefully at costs of employees and the benefits that they bring. The academics then move into a supplier role and quite often rapidly become practitioner sellers because the situation demands that they do so.

Recognizing the Total Rejector in Research

Most people who have spent a long time in the research business are aware of individuals who make it their policy to investigate all new methods, but have virtually no record of having adopted anything new. These individuals set themselves up to be arbiters of the field within their corporation. Such individuals may not be the official gatekeepers of the corporation, but they do very well in screening potential suppliers. They generally reject everything that comes their way, preferring to stay with the tried and true, but do their rejecting only after an ostentatious exploration. Their motives do not necessarily include power, because in corporations one does not obtain power by rejecting but rather by achieving. Their motives are control and perhaps latent hostility toward others, but most of all a desire to enjoy the respect of others. They are similar to those reviewers for scientific journals who pride themselves on a high rejection rate or on forcing students to write and rewrite their master's or doctoral thesis until the manuscript is "just right." These individuals can be recognized by statements such as "We have to wait for the right time to introduce this idea . . . no one is ready yet." In truth, no one will ever be ready. They can also be recognized by their interminable requirements for additional information at every step of the purchase process. When the supplier feels that the person will reject everything, no matter what is offered, there is a strong probability that the feeling is grounded in reality and that one is confronting an *eternal rejector*. The entire department, not only the individual, might be rejectors, and the rejection behavior might well be a corporate characteristic rather than an annoying individual one.

The Wyatt Earp Syndrome

Wyatt Earp was a famous gunman of the Old West who was reputed to be a crackerjack shot. Whenever Wyatt Earp came to town, a number of the most competitive townspeople would look at the occasion as one to go up against Earp and kill him in a dual. The Wyatt Earp syndrome refers to the challenge behavior that a person encounters when introducing a new idea that somehow makes others jealous or uncomfortable. If the individual doing the introducing is a graduate student, a junior professional, or a very established senior, then no one bothers to challenge. If they do challenge, however, then it is done on a one-to-one personal basis. In the case of a senior professional who is well known, quite often the meeting is more public and a number of different corporate members are invited. These members typically come from different disciplines, and during the presentation they attempt to find holes in the research approach. This type of massive "star chamber" inquiry is not done when a junior professional presents. The reason is simple. The junior professional can be easily defeated by the assembled professionals. There is no challenge in defeating someone so easily in public. The truly senior presenters, however, can take on all of the challengers, making it particularly attractive for a younger corporate professional to score points and make a good impression by somehow attacking and defeating the presenter. Occasionally, the Wyatt Earp situation ends up in the defeat of Earp. More often, however, the presenter—Wyatt Earp—wins. This win doesn't mean acceptance. It just means that Wyatt Earp has lived to present another day, but perhaps never, however, to the individuals whose challenges he just survived and who would have been far happier to see him succumb to their attacks.

Watching Out for Hidden Agendas

One key lesson that everyone in business and science eventually learns is that there are always hidden agendas. When someone is asked by another person to present new ideas, there is generally a reason that remains unstated, at least in the beginning. Sometimes the hidden agenda is nothing more than the need to learn about the new techniques. Other times the hidden agenda may be somewhat more pernicious. The requester may have been asked to hire a researcher to solve a problem and may have already selected a vendor, but is now being asked to check with other vendors. The hidden agenda here is that one is summoned under somewhat dubious pretenses to show up so that the requester can say that the request has been fulfilled. Even more pernicious, and unfortunately increasingly common, is the request for a presentation so that the requestor can claim conversance with the new methods. In today's intensely competitive environment, and in light of shrinking workloads, researchers in companies quite often make a big fuss over interviewing suppliers for projects. The interview process itself becomes formalized, enabling all those concerned to say that they are keeping up with the newest techniques. Sometimes there are even full days set aside, so-called supplier fairs, where suppliers compete to present. Although these supplier fairs are good ideas because they expose the client/buyer to new ideas, their ultimate validity can be established only by looking at how many new ideas are tried. If all are presented but none are tried, then one can be sure that the hidden agenda is not to learn new ideas, but rather to be seen as being involved in screening new approaches.

What Should One Emphasize or at Least Present?

In a world awash with competitors, what should one emphasize in the presentation or at least mention? Do the traditional truths of selling really work? Indeed, can one even sell to the insights community, such as researchers? Or do researchers simply buy, and buy what they want, choosing from among a variety of those who offer their techniques?

Selling the Sizzle: Does It Really Work in Consumer Research?

Salespeople are often advised to sell the benefits of a product rather than its specific features. When applied to selling research this strategy occasionally works, but not often. Selling research to a corporation means providing an answer. Usually the person who buys the research feels that his or her reputation is involved with the research service. This implicit involvement of one's reputation and ultimate security means that buyers will always be cautious when buying. If the research fails, then buyers feel that they have compromised themselves. This means that the best strategy is not to sell the sizzle: buyers don't want sizzle, but rather safety first and performance second. Selling sizzle means that the person introducing the new research idea may be perceived to be inflating the deliverables. It is probably a better idea to sell the approach, as potentially boring as it might seem, and after selling it suggest that some of the more attractive deliverables flow naturally from that approach. The buyer will be less suspicious and more likely to purchase.

Selling versus Collaborative Problem Solving

Many successful researchers position themselves as consultative salespeople at the start and collaborators through the research process. Indeed, in consumer research the nature of buyers is such that they are most approachable with messages about how the seller of the new research idea can collaborate with the buyer to solve a problem. Research buyers are more likely to buy from a colleague who can work with them rather than buying from a salesperson. Indeed, one of the great epithets in the research business is the term *salesperson*. To researchers, the word *sales* connotes taking advantage, whereas the word *collaborative* speaks of solving problems together. Researchers have a strong affiliation need, which resonates to the word collaborate.

The Value of Publications

In the research community the publication of methods and results occupies a curious location in the researcher's psychological space. On the one hand, most research buyers are employed by industry. The key evaluative criterion by one's superiors is whether the purchase moved the business ahead or, in crassest terms, whether the purchase advanced the superior's career. The notion of publishing an article seems irrelevant to most superiors of the corporate researcher. There are, however, others in the corporation, including the corporate researcher but also the researcher's internal clients, who react positively to a research seller having been published. Those positively disposed individuals may be brand managers and especially those recently graduated from business schools where the value of publications is stressed, top managers who are impressed by credentials, publications being one of them, and scientists in the corporation who value publications as part of their scientific training. The author has found that publications generate three distinct benefits:

1. They legitimize a person by showing that the ideas of the research provider have been accepted by some parts of the professional community.

2. Publication forces research providers to clarify their thoughts, and this clarity of thinking comes through in meetings.

3. Publishing can be fun, and it is always delightful to share with doubting clients and

research buyers one's observations about research topics. Setting these points down on paper gives the research seller a document to give to the buyer, which subtly, but ever so forcefully, substantiates the seller's reputation. Finally, publishing moves one's name out from a constricted environment to a far broader one. The pen is mightier than the sword, and nothing is so delightfully rewarding as a potential purchaser chirping about having read the research provider's work. The experience breaks the ice and enables the buyer and the seller to engage in a more earnest dialogue.

The Impact of a Demonstration

One of the most powerful ways to sell is to demonstrate one's approach. This is best done when the approach is computerized. Demonstrations for the senior author began in earnest when he developed a version of IdeaMap that automatically presented the stimuli, acquired data from a respondent, immediately estimated the untested data, and then presented the results. The author began with a simple demonstration comprising some 12 elements, which had been previously prepared and located on a set of semantic scales, as per the standard IdeaMap setup (see Chapter 5). The program was prepared to run on a personal computer. The respondents would evaluate 20 concepts, each comprising 2–4 elements, including a picture. The computer recorded the response time for each respondent. The evaluations for the 20 combinations typically required 4–5 minutes. At the end of the evaluations the program estimated the utilities of the 12 elements and of the 28 elements that were not tested but whose elements were estimated by the IdeaMap algorithm. The optimizer enables respondents to maximize or minimize acceptance. This demonstration eliminates a lot of the resistance to a new idea. It may not convince a buyer to purchase, but it at the very minimum defuses any latent skepticism.

The buyer may resort to other stonewalling techniques, but the validity of the approach has been demonstrated.

Recognizing the Moment: The Value of Years and Cumulative GRPs (Gross Rating Points)

Arlene Gandler, managing director of Moskowitz Jacobs Inc., has suggested that one does not necessarily convince critics by the force of a single argument. Rather, it is the passage of time that does a lot of the work. That is, as the critics continue to be exposed to the research approach in the literature over months or better over the years, the approach begins to take on a reality as if it were always available, always present, always used. The resistance by critics to new ideas that so clearly manifests itself wanes over time because the idea is no longer new. The combination of longevity and the evolving track record creates a sense of validity for the method. It is hard for critics to fight methods that have been around for 5–10 years. Critics can then change their tune, arguing that the method is no longer new and exciting. This argument is just fine. Buyers may like to hear about new, but they buy what is safe and makes them comfortable.

As buyers are exposed to the new technology for concept research or indeed any research method, the rule of *cumulative GRPs* begins to operate. One never knows which particular exposure drove an individual to buy the product. One only knows that the more often one's message gains exposure, the higher the sales are, all other factors held constant. Along with Arlene's observation of GRPs as a driver of acceptance is the equally wise observation by Professor E.G. Boring, the author's intellectual grandfather. E.G. (Garry) Boring was the well-known author of books on the *History of Sensation and Perception in Experimental Psychology* (1929) and the *History of Experimental Psychology* (1950). Boring was also the doctoral professor of the aforementioned S.S. Stevens, Boring's student and the senior author's doctoral professor. All were products of the *tough thinking* so dominant at that time at Harvard University. It was Boring's contention that one never convinced one's enemies, but rather one outlived them and, in some happy circumstances, also outpublished them. Boring maintained that the battle of ideas was long term and Darwinian. Stevens reiterated this Darwinian notion by encouraging publication and by averring that a person's real work and value would not emerge until 25–30 years later. Both Boring and Stevens recognized that it was time, GRPs, and publications, not the immediately powerful and elegant presentation of one's ideas, that would in the end win the game decisively.

A Decade of Experiences in Selling in IdeaMap: Observations on the Process

A good way to understand the sales, adoption, and modification of new technologies in concept research comes from the in-depth examination of a specific research approach. Since much of this book deals with conjoint analysis and concept development, we can continue the theme by looking at how buyers respond to new methods of conjoint analysis. Over the past decade and a half the author has been involved in the computerization of conjoint measurement and the expansion of the approach through the IdeaMap technique. The application of IdeaMap and its associated technologies is woven throughout this book. A recounting of the experiences in developing the procedure, rather than the theory and applications, provides readers with a somewhat different, more personal perspective on approaches in concept research. The key issues addressed are the nature of the presentation of the idea to companies, their reactions, and analysis of these reactions in light of what feelings of a personal rather

than a scientific nature might be motivating the reactions.

Introducing the IdeaMap Approach for Concept Research to This World

Traditional methods for concept testing work with gestalt or complete ideas. The concept test becomes in its own way a beauty contest. The nice reports, the massive tables, and the language selected in a politically sensitive way to couch the results all really end up talking about which concept won and which concept lost. The introduction of conjoint measurement in the 1960s, and really more basically the notion of experimental design, provided a way for researchers to better understand what particular features of a concept really drives interest. In this way it would be possible both to isolate the key drivers as well as to create newer and better concepts. The different uses of conjoint measurement in this book bear witness to that potentiality.

The IdeaMap approach was developed in the early 1990s with the goal to make concept research even more powerful and simpler. It is impossible for a normal person to sit through hundreds of concepts, evaluating the different ones, in order for the researcher to understand that individual's mind. Perhaps for the braver souls willing to donate their time, life, and sanity to marketing science, such an effort would be possible, noble, and rewarded. It would not be so for more normal, time-pressed people, and especially not so for children. Yet, the senior author deemed it vital to be able to understand how any single individual would respond to hundreds of concept elements without having to endure an inordinately long, boring, and ultimately painful test session. How could one therefore apply the principles of conjoint measurement, not to a dozen or a few dozen concept elements, but, say, to a mass of 300, 400, or even 500 elements? Furthermore, it was vital that the data be usable even with a base size of perhaps one person. A prudent researcher would never, of course, rely on the opinions of one person, although we all do in our daily lives. Yet, by having the individual utility function for that person available for say *200–400 concept elements*, one would have developed an extraordinarily powerful tool. The replication of the study with two, three, five, ten, a dozen, a hundred, or several hundred respondents would be simply aggregating the same powerful individual model, this time across many people. Sampling many respondents would not be done to make the study feasible; that had been done at the individual level. Rather, sampling many respondents would be done for the conventional statistical reasons; namely, the estimate of the mean or central tendency would be that much tighter.

Often those involved in science feel that the process of presenting one's ideas to colleagues engenders a generally fair, occasionally biased, review of the ideas, followed by adoption if the idea is reasonable and if sufficient data can be mustered. This noble, naïve idea of science should be contrasted with the senior author's observation of what actually happens in the business and academic worlds. The IdeaMap technology for concept research, with its base in conjoint analysis but with its departure from the strict rules, provides a good case history for studying the response and nature of adoption. The remaining paragraphs in this chapter present some impressions of responses to IdeaMap in its early history.

Creating New Ideas

The actual genesis of new ideas is not particularly difficult, especially if one has done it before. New ideas can be likened to the combination of old elements in new mixtures. Relatively few components of conjoint measurement can be said to be radically new. As IdeaMap has been presented in this chapter and in Chapter 5, it can be seen to combine well-accepted principles. Perhaps

the key new aspect is the estimation of untested elements by a numerical analysis procedure. Even the procedure itself is not particularly new, but rather based on a computation scheme in physics called the *relaxation method* (Szabo and Babushka 1991), which is known to mathematicians in numerical analysis. The author happened to come upon the approach in his first year of graduate school when reading about mathematical models and numerical analysis for a mathematical psychology course. Thus, nothing really in the IdeaMap method can be said to be novel. Only the combination of elements is really novel.

General Reactions to New Ideas

Armed now with a sense of the IdeaMap approach, a bit of its novelty, and the desire to sell it to the business community, let us move forward to the history of its introduction. We begin with the general and move to the specific.

The popular and now even the academic business literature is replete with methods proffered to the businessperson that promise breakthrough thinking. One need only look at books on creativity and innovation to discover that the average businessperson is inundated with suggestions. Thus, it is interesting to see the types of reactions to presentations of the IdeaMap method. After the author had presented the approach at a variety of meetings, three distinct patterns of responses continued to emerge:

1. *Politeness masking relative disinterest.* Quite a number of individuals felt that the approach was interesting, but not particularly appropriate for them. No one in this group was particularly negative to the general idea, but rather it appeared that many of these individuals were simply overexposed to research vendors hawking new ideas. It may be that, with the many voices out there in the business environment, new paradigms will simply have to take a lot longer to be accepted,

because many individuals have begun mentally to screen out all new ideas. They are overwhelmed, not negative.

2. *Interest in parts of the approach, especially technical ones.* With the increasing awareness of new methods, researchers are being exposed to methods that are totally new to them. The mathematical aspects are one of these areas. Often the conversations about the IdeaMap method degenerated into technical discussions between the author and statisticians, leaving a dazed consumer researcher or marketer wondering what IdeaMap was really all about. Not all of these types of technical discussions constituted a highjacking of the meeting; some were honest inquiries into method. However, the preponderance of the technical discussions appears to have led nowhere in the years that the method was being introduced.

3. *Interest in the process because of a desire to be* au courant. We discovered in presenting these methods that some individuals were interested in newer and better ways to develop concepts and then products. These individuals were interested in the potential use of the IdeaMap paradigm as a way to solve problems hitherto considered to be too expensive or too difficult to solve and were interested in the larger picture of product development rather than in learning specific methods to master them. They wanted to know what was out there to solve their problems.

To recap for a moment, the IdeaMap method is a conjoint analysis-based procedure that comprises several steps. These steps are both conventional and, in some cases, radically different from what has been typically done (see Table 26.1). The departures from conventional research bestows more power to IdeaMap because it can handle far more concept elements (up to several hundred) and because it creates an individual utility model for each respondent, even though each respondent only evaluates a part of the element set.

When the senior author began to sell the IdeaMap approach in 1991, he faced a set of situations, responses, and new needs that are important to record here as examples of what one will face and perhaps what they mean. When reduced to its basics, IdeaMap is really very simple; namely, a method to computerize the presentation of concepts in multiple-media format, obtain the data from an individual, and then instantly analyze the data. Today, in 2005, this approach seems rather straightforward. However, in 1991–1996, the approach was radical, threatening, and in some ways unnerving to researchers accustomed to conventional methods.

In the initial stages of development the author recognized this approach was new and had to be demonstrated. The demonstrations were at first rather technical in nature, but eventually became increasingly simpler. The reason behind the initial emphasis on technical detail had a great deal to do with the types of audiences willing to listen to a presentation about IdeaMap. These audiences were for the most part technical people rather than marketers or general managers. There were two reasons for this interest by technical people:

1. *Corporate guardians.* Technical people view themselves as the corporate guardians of new ideas. They are not necessarily given either that role or that title. Rather, as part of their self-perception technical people tend to be the first to say that they are interested in a new approach. This is both their public persona and, as it turned out to be case, their actual behavior. So it should come as no surprise that, when the IdeaMap approach was announced by advertising and by mail, the first to evidence interest were the technical people.

2. *Curiosity about the inner mechanics.* The technical people want to know how an approach works. Technical individuals feel that they don't do their job unless they dissect a method. Other people in the corporation want to know what the approach and

Table 26.1. IdeaMap compared with other conjoint analysis methods

	IdeaMap	Conjoint (full profile)	Conjoint (adaptive)
Nomenclature	Basic group = category Category comprises a related set of elements	Basic group = attribute Attribute comprises a related set of levels	Basic group = attribute Attribute comprises a related set of levels
Limits	Only on number of elements (>400)	Each attribute typically has the same number of levels	Each attribute typically has the same number of levels
Number of combinations (concepts) tested	No limit Each respondent sees different set of combinations	Usually fewer than 100 Each respondent sees a specific set from a pre-determined set of combinations	By individual—relatively few for each individual
Mode of testing	Computer	Computer, paper, and pencil	Computer
Psychological atmosphere of test	Stimulus–response; encourage rapid, gut-feel responses	Studied, rational	Studied, rational
Number of combinations seen by a respondent	60–150	40 (approximately)	20 combinations (approximately) + evaluation of single elements
Kid's version	Yes	No	No
Number of total combinations in the study	100 different for each respondent 100 respondents see a total of 10,000 combinations	~200	Unknown
Eliminate irrational combinations?	Yes—through a list of pairwise constraints (combinations that can't appear): up to 999 constraints	Yes—but must be built in at the start of design. Easy to do because everyone sees same combinations	No
Respondent task	Sees limited number of combinations, limited number of elements, easier interview. The combinations are different for each respondent	Sees limited number of combinations—usually the same from respondent to respondent	Sees all elements, selects ones that would be relevant Program discards remaining (unchosen) elements Respondent then goes through a shorter conjoint task
Number of elements	400 (with data imputation)	Maximum of 30–40, although more can be tested, but only 30–40 elements available per respondent	Maximum of 30–40, although more can be tested, but only 30–40 elements available per respondent
Individual utility model for each individual for all elements in the study	Yes—so that IdeaMap provides full set of utilities for each person	No	No

(continued)

Table 26.1. IdeaMap compared with other conjoint analysis methods *(cont.)*

	IdeaMap	Conjoint (full profile)	Conjoint (adaptive)
Segmentation of respondents	Yes—through relating each person's utility values (full set) to underlying semantic scales. System independent of scale size (viz., free of scaling bias, number usage)	Yes—based on clustering respondents by the pattern of their utilities. Problems—with scale usage (artifacts can affect utilities)	No
Purely graphic version available	Yes—called StyleMap, so that package designs can be developed using the same designed approach	No	No
Optimization	Optimization is done by integer optimization, using well-defined algorithms (e.g., branch and bound). The algorithm can take into account pairwise constraints (viz., that certain elements cannot appear together)	Optimization often done by choosing simply the best elements	None, since each respondent sees different elements that are relevant to the respondent. No way to optimize for a group of individuals, although one can optimize for a single individual
Modeling	Dummy variable modeling	Effects modeling	Simple selection, followed by critical pairs
Availability in language	Available most language	Available in most languages	Available in most language
Do-it-yourself versions	IdeaMap Wizard allows anyone with access to set up program to type in elements, put in visuals, write in questions, set up the study, and run (after registration)	No—programming needs a professional	No—programming needs a professional
Flexibility in design	The experimental designs are set up for dummy-variable modeling, which means that the concepts do not all have the same size and also that in some categories one or more categories are missing	Designs are set up for effects modeling—every category must appear in each concept, albeit with a different	Designs are set up for effects modeling—every category must appear in each concept, albeit with a different element

(continued)

Table 26.1. IdeaMap compared with other conjoint analysis methods *(cont.)*

	IdeaMap	Conjoint (full profile)	Conjoint (adaptive)
Flexibility in design	The designs can encompass anywhere from 12 independent variables in 20 combinations to 60 independent variables in 96 combinations	Designs are specifically developed for each study	Designs are specifically developed for each study
Types of stimuli	Emotional as well as rational; text, graphics (pictures, video), auditory	Typically rational, typical text	Typically rational, typically text
Level of modeling	Individual model, with all elements	Group model	Individual model (but idiosyncratic, limited to elements in individual's consideration set, no others)
Types of uses	Product design, communication Package design	Product design	Product design
Interpretation of results	Very intuitive—easily understood and explains the meaning of utilities (as pecentage of respondents who go from disinterested to interested when element is present)	Harder to understand and explain utilities because of effects modeling	Hard to understand and explain utilities
Web available	Yes, in full format	Yes	No (as of 1 October 1999)
Ongoing utility of data	High, because there are hundreds of elements, so that users can return to the database again and again	Moderate—because of limited number of elements in study the users may go back a few times—but does not usually serve as a long-lived database	Low—usually used to solve a single problem

technology accomplish and only secondarily want to know the innards. As much as one tries to focus attention on the benefits, when a technical person invites the presenter, the methodological validity is judged by individuals who fancy themselves statistical experts or at least who judge methods by the rules of statistical validity. It is no surprise, therefore, that almost all of the initial questions were statistical and technical. Even when the audience to whom the method was being presented did not really know statistics, many insisted on technical questions and technical (if irrelevant) criticisms. The biggest stumbling block appeared to be the explanation about how the utilities of untested elements were imputed. The approach, dimensionalization, and subsequent estimation methods were not easy to accept. To many in the audience in the early 1990s the data imputation seemed too great a stretch of imagination, although a decade later the fusion of data from multiple sources is now considered a standard, well-accepted technique.

Once the method began to gain interest, there was the continuing issue of price. Two issues emerged most frequently—cost per interview and base size:

1. *Cost per interview.* Many researchers tried to compute cost per interview and compare it with the standard costs. This led the author to observe that the use of cost per interview or other common metrics was not really an evaluation, per se, because the cost bases of conventional and IdeaMap are so very different. Rather, it appeared that focusing on the cost per interview was a strategy to cope with the new idea and to fit this new approach into one's frame of reference. Those not familiar with conjoint measurement in the early 1990s used the cost metric as their initial foray into understanding the approach.

2. *Base size.* It appeared very important for consumer researchers to work with many respondents in a study, perhaps because the large base size represented validity. Thus, the author had initially suggested that the IdeaMap approach would work with as few as 30 people. This positioning was later replaced by the statement that, for a given price, the researcher would be able to test 150 respondents. The evaluation by 150 respondents for slightly more money was attractive because it was a much lower cost per interview (the key initial metric) and because researchers were accustomed to data from large base sizes that they could analyze many different ways.

The IdeaMap approach was introduced by a relatively large-scale advertising effort in professional magazines and newspapers, such as *Advertising Age*. The belief was that the advertising would create awareness and interest. The campaign was based on the simple question "What is the big idea?" Preceded by the aforementioned sales efforts, the IdeaMap approach was formally introduced at the 1992 Advertising Research Foundation (ARF) annual expo in New York City in March. There was a great deal of interest in this new concept-research technique, perhaps spurred on by the automatic, instant feedback and by the IdeaMap booth being one of a handful of booths to feature a computer.

The four major lessons learned at the ARF expo can be summarized as follows:

1. *Advertising.* It helps to advertise, but the advertising creates awareness. Advertising did not create a need for IdeaMap. People did ask about the approach at the convention, often mentioning that they had seen the advertising.

2. *Entertainment value.* People will line up to see something interesting.

3. *Interest does not mean sales.* Immediate interest and "ah-ha" experiences do not translate to subsequent sales

4. *Senior clients critical to convince.* Sales presentations to senior buyers are far more impactful than are sales presentations to juniors who do not buy, but it is the juniors who are always the most impressed.

Despite the strong introduction of IdeaMap to the research community, the full adoption of the approach would take several years. Three issues continued to emerge that probably reflect more on the way research is commissioned, conducted, evaluated, and used than on the IdeaMap method itself:

1. *The need to relate the data to standard concept scores.* There was a continuing attempt to relate the approach to conventional concept tests, and specifically the performance of each individual concept. It became increasingly clear that most researchers had been trained to understand the notion of how well or poorly a concept performed. One consequence of this training was that it was

hard for many researchers to understand the idea that concepts were vehicles in which to embed elements, and that it was the element, not the concept, that merited attention. This basic notion underlying conjoint measurement became clearer to most clients with explanation, but the difference between fixed concepts as gestalts and concepts as vehicles for their components continues to demand explanation even today. The growing popularity of conjoint analysis in particular and experimental design in general has not eliminated the need to explain differences between prototypes underlying the design gestalt concepts.

2. *Discomfort with modeling.* Most researchers have been trained to deal with data, but not with regression modeling and certainly not with dummy-variable regression. The discomfort disappeared quickly after it was demonstrated to buyers that the approach was a simple extension of standard regression.

3. *Understanding parts of the utility model, especially the additive constant.* One of the biggest stumbling blocks was the realization that most of the researchers did not understand the principles of the regression approach. Even those familiar with regression modeling needed to anchor their statistical understanding with something concrete.

The Human Comedy: How Different Members of the Insight Business Reacted to IdeaMap

IdeaMap is a good model to study the reactions to innovation. It combines at once new ideas about modeling and information with solid, historical foundations. It is sufficiently new to be challenging to old methods, yet sufficiently grounded in the old methods not to be easily dismissed. The reactions to IdeaMap tell as much about the in-going assumptions and mind-sets of the audience as they do about the method. The audience to whom IdeaMap has been presented is gener-

ally sophisticated and always looking for newer and better ways to solve business problems. At the same time, buyers are caught on the horns of a dilemma: they are torn between the old loyalties that have sustained them over the years and these new computerized methods.

Coming as it did in the early and middle 1990s and having lasted more than a decade, IdeaMap generated a number of instructive reactions. Some have been rational, whereas others have been quite emotional. The reactions have often been quite positive from those who have to make decisions; for instance, marketing, management, and R&D. Reactions have ranged from positive through neutral to strongly negative from those who are in the insights business and whose work product is a smoothly running process. This latter group comprises marketing researchers, agency brand planners, and sensory analysts. The same pattern appears to hold both in the United States and internationally, although the reaction has tended to come earlier in the United States. Europe, Latin America, Asia, and Australia/New Zealand have been slower to react. We can summarize the reactions of individuals in the different corporate functions as follows:

1. *General managers.* These individuals have to grow the business. They reacted to IdeaMap by asking how it could be applied and, without prodding, identified the different applications. Many of the general managers actually understood the logic of IdeaMap quite well—a very welcome surprise. General managers are bound neither by politics, nor by convention, nor by previous commitments or relationships with other suppliers. To general managers, success in problem solving is paramount. If the approach answers the problem, then general managers usually bless the procedure and encourages their subordinates to use it when appropriate. General managers do not get lost in the details of the method or attempt to reverse engineer it.

2. *Marketers.* These individuals have to sell and are interested in anything that helps them sell. Most marketers had some experience with regression analysis and had no problems with the method. In fact, most of the marketers were fairly quantitative. A number confessed that they were only marginally quantitative, but that at the same time did not want to get lost in the details. They simply wanted to know what the method would deliver, how it was executed, and what its strengths and weaknesses were. They wanted to know applications and felt comfortable dealing with specific opportunities. They, like the general managers, were generally positive.

3. *Marketing researchers.* These individuals are responsible for insights. Many of the researchers felt uncomfortable with conjoint measurement in general and with IdeaMap in particular. Quite a number also volunteered quite categorically that they felt the IdeaMap output to be too quantitative for their internal marketing clients. Marketing researchers talked on some occasions about applications, but often spent time comparing IdeaMap with conventional methods. Marketing researchers tended to remain between positive and neutral in attitude. They were interested in the cost benefits and needed be reassured that they were not guinea pigs for this new method. Many researchers stated that they were under pressure to produce better data and were being pushed by marketing, either expressly or covertly, to improve their arsenal of techniques.

4. *R&D product developers.* These individuals are responsible for creating the actual products. Most of them were inexperienced in concept research, but were willing to listen and to take direction from their marketing researchers or sensory analysts. Many were fascinated with the method, especially because they stated that they had always wanted to have clear direction for development. R&D product developers seemed not to have any covert agendas, perhaps because the they are judged on their ability to provide

a physical product, rather than on the information-gathering process.

5. *Sensory analysts.* These individuals are responsible for guidance in product features. They support the product developers. Traditionally, sensory analysts tend to be conservative, suspicious of new ideas, and demanding of proof and demonstration. Many of them were modestly interested in the approach, but very few could see where it fit in, at least in the early and mid-1990s. Later responses in the early 2000s would be positive, but only when the approach was simplified and made very inexpensive. Most of their questions dealt with the statistical validity of the approach and the precise way that the data were transformed from ratings to utility values. A number of them felt that concept work was not in their area of expertise and deferred to the consumer researchers and the marketers. Among all the groups presented to, the sensory analysts were the least enthusiastic and often felt that IdeaMap was totally irrelevant to their current needs. Unlike consumer researchers, sensory analysts generally did not say that they were being pushed to explore new methods. Those who mentioned pressure generally did so with respect to turnaround time and budget. This insular attitude of sensory analysts began to change, however, in the later 1990s, when they became members of cross-functional teams and had responsibility for product concepts. In almost all meetings the sensory analysts made a point of saying that if they liked a technique they wanted to understand it fully in order to bring it in to the corporation. The sensory analysts were generally interested in *build*, not *buy*, approaches to research.

Insights from IdeaMap: The Eight Stages of Adopting a New Research Approach

Based on the introductory years of IdeaMap, there appears to be a sequence of eight steps between the point that the companies attend an initial IdeaMap presentation to the point that they adopt it as a standard method. The eight steps are presented next in short paragraphs, comprising both the observation of what happens at the stage and comments about what that observation might mean. The stages assume the following states of knowledge:

1. *What the buyer knows.* The old ways of gathering information about the concept do not work particularly well. Better ways are needed. However, if a buyer adopts the new method, then there is the possibility that the buyer will be asked to defend the method or at least explain it in public. There is even greater likelihood that the method and the buyer will be exposed publicly if the method works, primarily because everyone in the corporation is interested in success. The buyer also knows or at least fears that there may be implicit punishment if the method fails, although for the most part the buyer often does not know any particular individual who was fired for trying a method. The buyer suffers a nameless dread and feels trapped. The buyer has to risk potential safety and success versus failure on a technology only half-understood. Success means potential exposure to others.

2. *What the seller (vendor) knows.* The research vendor knows how the method works and can muster a case history or two. The research vendor does not know the internal climate of the company unless he or she has previously worked there.

Stage 1: Superficial comparison with other methods. This is the attempt to put the approach into a frame of reference with which the buyer is comfortable. As already noted, most buyers are accustomed to testing a limited set of concepts and selecting winners. They know of conjoint measurement, but there is far more talk about the method than actual use by clients.

Comment. The audience may be trying to focus on the opportunity, but they are not necessarily clear about what is really being offered. This act of comparison with other methods may be a mechanism to cope with new ideas, similar to the Linnaean approach of classification. The earliest approach to understanding is to put something into a class about which one understands several features. By lumping IdeaMap with known procedures the buyer feels more comfortable.

Stage 2: Interest in passively experiencing the approach. At this stage the buyer sees some merit in the system. However, the system is technical, and to many this technical foundation is frightening. Buyers sense, however, that the method may have some redeeming qualities. They want to be exposed to the research approach, but in a safe way. They do not want to be committed to purchase, because generally they don't have signature authority for new methods. By viewing an interview in action, from a distance they feel that they can cope with the method and feel less vulnerable. At this stage they are less interested in analysis and more interested in the actual field work.

Comment. This second stage is the start of incorporating the research approach into the buyer's repertoire. The stated desire to look at an actual interview is again a coping mechanism. By seeing the interview buyers can begin to feel comfortable with the mechanics of the system. Understanding the mechanics of the system gives them a sense of control or at least a sense that they can explain what happens in an interview.

Stage 3: Larger audience buy in by means of a presentation to other corporate members. At this stage the interested party invites other corporate members to see the technology. Often there are structured occasions, such as a "lunch and learn" or "supplier fair," at which the approach can be discussed. The presentation usually comprises audience members with status above and below that of the inviter. The reason for this range of status is quite simple. The inviter wants approval from perceived superiors and is sufficiently interested in the approach to want to win over lower-level professionals in the company.

Comment. The larger audience provides a buffer in time and in space. The second presentation, this time in a public setting, alleviates any responsibility for incorrect decision making. There is a clear attempt to obtain either consensus about the decision or approval about experimenting with the method. The second presentation provides an additional opportunity to learn the technology better without any risks.

Stage 4: Presentation of a brief, or a request for proposal. At this stage the potential buyer is internalizing the method as a way to solve a problem. Providing a brief constitutes the first steps in a complex dance between the research provider, the potential buyer, and the corporation. The brief is usually accompanied by a request for a written proposal and often a suggestion that the seller present the proposal to the appropriate staff involved in the project. This fourth stage can take place at any time after the original idea is proposed as a method to the buyer. Sometimes things happen quickly, in weeks. More often it takes months and quite frequently a year or two. The new approach does not die in the meantime. Rather, it gestates. There appears to be some necessary time between initial introduction and presentation of a brief, perhaps to allow for the ideas to sink in.

Comment. The brief usually represents the first stage in letting down one's guard against the new idea. However, the brief is a double-edged sword. Some briefs represent genuine interest in a method. Other briefs may constitute old, intractable problems that the researcher and the corporation have never been able to solve. The new research method might be a way to solve it. Nothing can be lost by allowing a new research supplier to try his or her hand at the difficult problem.

Stage 5: The first project. The first project is characterized by excessive design, analysis, and hand holding. The project may either be a conventional one (now posed as one to be solved by IdeaMap) or a unique one that has been difficult to solve, but which appears solvable by IdeaMap. In either case the results typically are presented to the purchasing group in a large-scale meeting at which the attendees are encouraged to comment.

Comment. Even when done correctly the first project may lead to a dead end. Half of the success is in the correct design and execution of the study. Half is in the mind of the buyer, independent of the results. The notion that validity is a matter of opinion never applied so well as it does here. At this stage many agendas are floating around, with some individuals trying to solve problems, some trying to show their intelligence, some trying to exert political influence.

Stage 6: Adoption of IdeaMap as part of the research repertoire. This occurs after successful completion of multiple projects. Users begin to think in IdeaMap terms; that is, in concepts that comprise elements systematically. New individuals come into the arena as users. There is a continuing need to resell the approach and reexplain it, but the audience is not hostile.

Comment. There is clearly a warmer interaction with users as compared with the first set of interactions. There is clearly greater willingness to modify problems to fit the IdeaMap paradigm rather than present the research vendor with hitherto intractable problems. This stage is best identified by the emergence of clusters of projects, either at one time or over time.

Stage 7: Demand for expansion of the consumer interface at the data-acquisition stage. At this stage, the user begins to think of these structured studies for other uses. The focus now turns to other types of stimulus presentation such as graphics, video, sound, and exploration of other methods for acquiring responses, and other types of rating questions.

Comment. At this stage, IdeaMap becomes an approach and metaphor rather than a specific, limited method. Stage 7 usually comes about 1–2 years after the client has become more comfortable. It represents an expansion of IdeaMap as a research approach rather than a single technique

Stage 8: Demand for increased power and flexibility in data analysis and data reporting. This stage is characterized by demand for additional types of statistical analysis and different types of modeling. The quantitative requirements at this stage generally differ from those at the start of the relationship with the buyer. IdeaMap results now start being used for more decisions. The analytic approach now must expand to incorporate other ways to look at the data.

Comment. This type of expansion is not usually done by those who originally were interested. Rather, the demand usually comes from the statistical/quantitative group who want a new tool. Often there are offers from

clients to collaborate. This stage represents final step in internalizing the approach.

Overview

The eight stages resemble those of a religious conversion. These stages are not fixed in stone, but they present the typical sequence. The final two stages, expansion of stimulus presentation and data analysis, provide the catalysts for true development. One might use Schopenhauer's three stages of a new idea as a summary:

1. Disregard, disinterest, consignment to irrelevance

2. Open antagonism

3. Acceptance as a self-evident truth

References

Boring, E.G. (1929). Sensation and Perception in the History of Experimental Psychology. New York: Appleton-Century-Crofts.

Boring, E.G. (1950). History of Experimental Psychology. New York: Appleton-Century-Crofts.

Szabo, B., and Babushka, I. (1991). Finite Element Analysis. New York: John Wiley and Sons.

Chapter 27

Two Views of the Future: Structured Informatics and Research Unbound

Part I: Structured Informatics

Information

During the past 50 years consumer research has grown from a simple cottage industry to a sophisticated industry that provides valuable information and insight. Fifty years ago a lot of interest focused on questions that we would now call tactical. These questions dealt with product or concept acceptance. Today, however, much of market research is strategic. Strategic questions deal with issues as diverse as the value of a brand, the opportunities in the market for a new product, and the existence of hitherto unexpected segments. In the course of their professionalization and that of the field, consumer researchers have evolved from information pioneers to knowledge workers, much valued in some corporations but dispensed with entirely in others.

Along with the professionalization of consumer research has come an extensive amount of data that now reside in corporation file cabinets. These data deal with reactions of consumers to concepts and products, trends as consumers see them, and so forth. In a sense, corporations sit on enormous intellectual capital relevant to themselves. The decades of concept research represent a great deal of this intellectual capital. Quite often the companies themselves are not even aware of the vast amount of untapped knowledge they possess. Indeed, with the great turnover of personnel, with mergers, and with the tendency of corporations to reinvent the wheel, it is no wonder that corporations remain unaware of this capital, which they could use to a particularly great extent beyond the initial applications for which the data were developed. What is required is a new way to think about data, a new way to systematize, and a new way to use the data. These new methods must incorporate two aspects of data analysis—mechanical and intuitive—in a way that makes the data an asset that can be valued by accounting methods.

It is worth noting here that several companies have begun to work with their amassed data in a way that allows them to use the information on an ongoing basis. For instance, some companies may request that the supplier provide all of the information in an electronic format rather than in the traditional paper format. Electronic formats can be scanned for key words. Other companies go further, requiring that the reports be issued in a specific format or contain an accompanying form detailing certain standard aspects of the report contents. That form provides the necessary information for the knowledge workers to access the information.

This final chapter presents an approach to the five evolutionary steps of intellectual capital as exemplified by concept studies. The premise is that an exceptional amount of information capital is resident in the concept research. This represents gold that has already been dug but not necessarily processed. These are the five levels of capital:

1. Knowledge residing in the minds of the few, who have gained it through direct experience, but inaccessible to anyone else

2. Lots of data available for access, and some knowledge, but no realization about how to obtain the data

3. The one-off, large-scale, cross-sectional landmark study that everyone refers to, but which stays in splendid isolation

4. Systematized landmark studies with regular updates that represent a major effort toward concept informatics

5. Added value analytics accompanying the systematized landmark study, along with the ability to do short add-on studies to widen and deepen some knowledge areas that are of momentary importance

Level 1: Knowledge Resides in the Minds of a Few Who Have Gained It Through Direct Experience

All companies employ experienced individuals who have been through situations, problems, dilemmas, or disasters, whether in basic research, marketing, promotion, etc. These individuals possess a repository of information locked in their minds, their attitudes, and in the way that they approach data. Such experienced, battle-tested professionals often can answer questions rather easily by using the fortuitous, happy combination of their own experience and accepted scientific principles. One need only go through a company and ask about who is the repository of information. Typically, every company will have one or more of these individuals, who evolve into repositories because of personality characteristics. They store information, they can access the information pretty quickly when asked a question and, because of fortuitously outlasting everyone else, they evolve into a walking knowledge base.

The problem with these individuals is the reality of the human condition. Since employees are mortal, at some foreseeable point these experts will no longer be around to answer the questions. Death or, more typically, termination occurs at some time during an employee's life. All employees are subject to the vagaries of company politics; they may be terminated for reasons beyond their con-

trol. Furthermore, employees may be terminated because the company merges with another company or, more often, is taken over by another company. In the quest for profitability of operations these walking repositories of corporate information lose their jobs. Finally, no matter how cooperative these individuals may be, or how long they stay at the company, the information in their minds is rarely well organized, generally not available to others in its depersonalized form, and therefore cannot be easily acted on.

In the research business we see examples of this every day on the client side. Many of the individuals with whom we work in corporations are veritable repositories of what the company did years ago. They can be called on to recount what appeared to work and what did not. These experienced individuals are very helpful to research suppliers because they actually have personally suffered through many of the problems that research addresses today, although undoubtedly with older products and services no longer extant.

On the *supplier side* in market research we also see this type of knowledge resident in highly experienced researchers, often with 15+ years. Experienced project directors or higher-level suppliers leave companies for other companies and lose track of what they worked on. People may retire from the company or leave for other reasons. They often just simply lose interest in research. The bottom line is that their research knowledge about how to do things, their hands-on experience, and their wisdom cannot be accessed easily by other researchers who need that wisdom.

A variety of situations can emerge wherein the knowledge base lies in the hands of a very few. Examples include small corporate research suppliers. Small supplier companies rarely have enough time for consensus decision making. Survival depends on the rapid assessment of a situation and a rapid decision. A company might survive because a single talented individual spots opportunities and grabs the opportunity that the situation presents. The shared knowledge base is often minimal in those companies that were founded by a stroke-of-genius type of entrepreneur. In such companies, often there is no room for major professional development by anyone else beyond the owner(s), so other talented individuals stay for a while before leaving for better growth opportunities. Finally, all too often companies that remain in flux for a very long time don't have the breathing space and luxury to create a large-scale, publicly accessible database of information and knowledge. The company keeps changing, evolving. Many of the old-timers simply leave the troubled waters. As a result the information base shrinks to those individuals who, by some accident, remain in the corporation as if they were the only ones not spun out by the corporate centrifuge.

Potentially severe consequences can result when the intellectual capital of a company is wrapped up in the mind of the people that work in the company and when the company lacks a formal knowledge management system. There are three key issues here:

1. *The extent of the knowledge is hard to determine.* The knowledge base cannot be delineated. No one knows exactly what knowledge resides in the mind of any particular individual. Of course, we know what type of knowledge a person has when that person answers a question. If there are no questions, then that knowledge remains untapped and unstructured.

2. *Changes in the knowledge base are hard to track.* The increase or decrease in knowledge over time, either generally or in particular subject areas, can't be measured. In a changing environment it is difficult to know whether the company is maintaining its knowledge base or whether the knowledge is slowly eroding, thus weakening the company.

3. *Lost or jumbled knowledge is hard to reconstruct.* The knowledge residing in the mind of an individual who leaves the com-

pany, dies, or becomes incapacitated can't be recaptured. How can that knowledge be saved or re-created? In the market-research business this is akin to the loss of knowledge about what works versus what doesn't, and what to do versus what not to do. Senior individuals who leave take with them many of the company experiences and much of the wisdom. That knowledge may comprise content, skill, or both.

Level 2: Lots of Data, Little Knowledge, Even Less Wisdom

With the ongoing professionalization of consumer research, many companies have grown accustomed to commissioning studies that address specific momentary objectives. These companies lock away a great deal of information in their corporate coffers. That rich store of information comprises disparate reports about projects. The nature of commissioned research is to answer questions so that, more often than not, the ad hoc studies deal with specific, time-limited, scope-limited issues. At the same time, corporations commission tracking studies and other forms of continuous or syndicated research. Finally, a corporation may buy syndicated research, either on a one-off basis or, more often, on a continuing basis. The combined mass of studies generates piles of reports and, over time, ossifies into bookcases of bound volumes. The amount of information extant in these corporations can often be amazing, and it does gratify those who have to work with old data that they often find answers to new problems in this mass of old information.

To a great degree the information as currently arranged does not generate corporate wisdom despite the impressive amount of data and large number of reports. Companies occasionally now do ask for their reports in searchable electronic formats, along with key words to aid searches. What is lacking, however, is a way to bring the information together into a coherent whole. Certainly, having material available to the company in an electronically accessible format is far better than having critical knowledge reside in the minds of a few individuals, yet, for the most part, the information is scattered about. Sorting through lots of studies to absorb their results and then abstracting a pattern to answer one's needs take a great deal of time. All too often these studies have been done for different purposes. The real wisdom and the ultimate intellectual capital to emerge from these disparate studies lie in the patterns inferred by researchers rather than in direct, solid information available from the studies. Many researchers have discovered this unhappy reality, much to their chagrin, when their visits to the corporate archives of market research turn up report after report, but with little connection among these reports. The picture is much like a fossil bed of consumer research projects. There are very few connections between the reports and, although sufficient information exists in the data, drawing useful conclusions for today's work is very difficult. At least, however, the data still exist, even if they are not the most useful information, nor in the most useful form.

Level-2 knowledge represents a great deal of what corporations believe to be their knowledge base. By dint of commissioning a great deal of research over the years, corporations amass a lot of relevant information that is neither adequately organized nor well presented in a format to optimize the intellectual capital. It is difficult for anyone inexperienced with the actual subject matter to obtain a great deal of coherent information, although the specific study information is completely available on a project-by-project basis.

All things considered, level-2 knowledge represents the beginning of a database that can evolve into intellectual capital for the corporation. One tip-off about the primitive state of level-2 information is the realization that this mass of data is not consulted particularly frequently. Often, in response to a question, one hears the blanket assertion that "the corporation knows the answers—they are somewhere in the files." On following up this question, however, one quickly discovers that no one knows how or where to find these data or indeed where to begin to look. Few researchers regularly delve into the archives to answer research reports are usually well filed and reasonably indexed. The overarching problem is that there is no general way to summarize the information contained therein and no systematic method to search the database productively. The studies are simply not connected. There is no organizing principle to guide researchers. There is a need for a new area of technology here, perhaps best called *market-research informatics*.

Level-2 information often generates in its wake at least six *counterproductive* actions by the corporation:

1. *Reflex cessation of ongoing research.* As soon as someone with budget power in the corporation recognizes the massive amount of available information, all too often they reflexively cut the research budget allocated for additional studies. The rationale is that, when confronted with the information about an abundance of corporate information, the person responsible for budget decrees the current information *must be digested* before additional work is undertaken. This reflex action occurs more often in corporations than one would like to admit and generally in those corporations where the knowledge base is not centralized. Lack of centralization means that no one really knows what knowledge is necessary, what knowledge is duplicated, and what knowledge is missing. Most practitioners inside and outside the corporation recognize that this cessation of research activities is dictated far more by budget considerations than by a pragmatic consideration about digesting knowledge. If the truth were known, the knowledge available in the previous reports often has little current value and motivates one only slightly to attempt to digest it.

2. *Replacement of knowledge by rote systems.* No one likes anxiety. Nothing in corporations succeeds in reducing anxiety as much as one's ability to point to a system that appears to answer the problem. If the corporation maintains files on previous projects, then the collection of information provides a convenient excuse. Since the data are presumed to already exist simply because there are previous data, there is no perceived need to re-create such through a new research project. The inevitable result is the subtle, continuing loss of current information because some probably obsolete data in the corporate archives block the way to getting new data.

3. *Work on what has already been paid for rather than on what is newly relevant; or busy work efforts to codify old data as a way to justify one's previous efforts.* A lot of effort can be placed on codifying old data. This is very much like building a very extensive and expensive addition or modification to an inexpensive, small house. At the end of the day, one is still left with a small house, albeit one that has been significantly modified. It is not clear whether one will ever get back the investment made in upgrading the house. A similar story holds with data. One can codify old data, put into place data-retrieval systems, and make the system searchable by computer programs (e.g., knowledge management systems). In the end, however, one is still left with a well-documented, but not particularly useful, mass of relatively unconnected information.

4. *Data not updated because there are no simple templates to follow.* If the data structures result from the accumulation of disparate studies, then there is neither a simple structure nor a template to guide updates. Consequently, there is little provision left to upgrade and modernize the data, unless some of the data come from tracking studies that by their very nature lend themselves to updating. Ad hoc research studies commissioned to answer specific problems stand by themselves, adding somewhat to the general store of knowledge but not really fitting in with much of the other information, except at a very superficial level. How does one properly update this database? What does one do with the data if the world has changed? How can studies be reconciled with one another? What are general trends that can be discerned? These are questions that can be easily asked but not easily answered, given the type of data that many companies have.

5. *No clear "big picture."* A collection of ad hoc information is just that—a collection of information. There is no "big picture" in these studies. Perhaps the big picture emerges through the talented insights of someone who can abstract the data to generate a coherent overview. However, the big picture clearly depends on the talents of an analyst who is capable intellectually and has been correctly briefed to search for and then to apprehend the bigger picture underlying the data. What happens, however, when the data need to be accessed by other people beyond the talented individual? Not everyone can be briefed to be prepared. Indeed, over a period of years the cast of characters changes, and the personal memory of why any particular study was commissioned fades into obscurity.

6. *No clear operational next steps.* One of the key issues in research is the need to make it actionable. Each ad hoc study possesses its own logic of actionability. The studies, disparate and often incommensurate, are commissioned in order to answer specific problems, often of a local momentary nature. As a result, the studies themselves typically enjoy actionability at a local rather than a global level. However, what is the actionability of the data once the masses of data are combined in an easy-to-understand, easy-to-access format? What would be the next step to be taken even if all of the ad hoc studies and all of the focus groups were to be combined into the database? The actionability of the larger combined set is just not clear. There is no clear data structure to suggest next steps. Perhaps there might be some next

steps, but the data are not arrayed for anything except momentary insight.

Level 3: One-off, Larger-scale, Cross-sectional Landmark or Foundation Study

Relatively large-scope landmark studies may constitute one true possible foundation for intellectual capital based on consumer research. A *landmark study* is defined here as a study of a wide variety of in-market stimuli (e.g., concepts, products, and packages) organized in such a way that the data can be readily accessed and used again and again. The study is a *cross-sectional* analysis of what exists in the marketplace, following the deconstruction approach. Broad-scale studies need not be large in terms of the number of respondents they interview, but should be large in terms of the number of stimuli they encompass. Simply stated, the more stimuli that are tested, the better. Furthermore, if the stimuli comprise both in-market and private stimuli developed by the company, then the landmark study may be even more valuable. Examples of landmark studies might be large-scale deconstruction studies (Chapter 17) and the It! foundational studies (Chapter 20).

What makes a landmark study so valuable that it can create corporate intellectual capital? Why is the landmark study more valuable than a series of disparate studies—even a series of studies that goes back a decade or more? Here are three reasons why such a landmark study could constitute the beginnings of significant intellectual capital:

1. *The data enable a structured analysis of the competitive frame, which is always relevant to developers and marketers.* Any study that encompasses a relatively large number of relevant stimuli provides critical information that could not be obtained from smaller-scale studies limited to a few stimuli. The marketer and product developer can compare stimuli against each other to measure how each stimulus performs. If the stimuli are brand communications, then the marketer can determine how well the different brands communicate to the same set of consumers. If the stimuli are products rather than concepts, then the marketer can rank the different in-market products on acceptability.

2. *Results generate norms that can be calibrated against market performance to construct a predictive system.* With multiple stimuli, one can always compare stimuli against one another and to market performance in order to develop norms. Norms show not only how different products or communication elements perform, but also help decision making. A landmark study becomes far more valuable when it contributes to the norms. Furthermore, if a landmark study is private, then this dataset can provide information that creates a competitive advantage in the marketplace. To the degree that the norms correspond to actual product performance (e.g., shares, shipment for products, and recall for advertisements), the database becomes even more valuable.

3. *Multiple stimuli reveal patterns that could not be uncovered with smaller-scale studies.* By assessing communications or products from different sources, marketers or developers can uncover patterns leading to good versus poor performance. Most companies do not bother to evaluate communications or products. Two exceptions are their own offerings to identify likely winners or weed out losers, and market leaders to determine how well the new entry will perform against the product considered to be the key competitor. A cross-sectional database and subsequent analyses provide the beginnings of significant insight into why market leaders lead and why market laggards lag. This information is referred to again and again by current and new users of the database who are interested either in a broad overview of the

category or who require specific direction to create new, winning entries.

Considering the attractiveness of a cross-sectional database on consumer reactions to the competitive frame of products or communications, it is surprising that most companies do not implement these studies regularly to create a systematized knowledge base. Corporate cultures do not value continuing streams of cross-sectional data obtained from ad hoc studies of consumers. Corporate cultures do, however, value market-level data when such data are purchased from the market and represent *objective* measures of performance (e.g., A.C. Nielsen). Such market-level data provide information about purchase patterns, but not about attitudes and other subjective reactions. The notion of creating a parallel set of data using cross-sectional stimuli from the consumer's point of view (i.e., attitudes toward actual stimuli or concepts about the stimuli) seems not to have been internalized into the marketing, research and development (R&D), or consumer research functions.

A wide gap appears to exist between data that one purchases about in-market products as they perform in the marketplace and data that one creates. Furthermore, for companies that create such a normative database the probability is low that the company will invest in updating this cross-sectional database of consumer reactions to in-market products or communications. Here are two hypotheses about why companies are so reluctant to invest in this wider view and therefore do not invest in ongoing landmark studies:

1. *Knowledge fullness and data satiety.* The richness of the database often works in an unexpected way, preventing the re-creation or the updating of the database anew each year. All too often the single cross-sectional database is mined for years to answer different problems. The database often continues to provide many new insights each time and provides clear direction to marketers and developers. A living, "ever-green" database, or at least one that is continually used, provides a false sense of security; that is, the database appears to be au courant simply because it continues to be used and no one believes that consumer attitudes change. By the time someone realizes that the category has changed, several years have passed and, more often than not, a whole new cast of characters arrives on the scene. These individuals need to be convinced yet again about the value of such a database. Occasionally, such convincing works. In contrast, people realize that external markets change rapidly and often unexpectedly. Therefore, it is accepted that market data need to be updated. Consequently, market data do not become stale.

2. *That which is not budgeted for an ongoing basis as a line item is not real in a general sense, and takes on reality and validity when used to answer specific problems that are localized in time and space.* In the research world, ongoing tracking studies are typically contracted on a yearly basis, whether needed or not. As a result, management becomes accustomed to receiving updated information periodically. The goal of researchers involved in tracking studies or market data is to *maintain* consistency in the data and to *explain away*, if necessary, any differences of the data from the last reporting period. The key is that the tracking study is *budgeted for* and embeds itself deeply in the corporate culture. In contrast, a cross-sectional landmark study of attitudes or in-market products is usually commissioned by a single forward-looking individual—whether marketer, developer, or researcher, respectively—who is keenly aware of the value of such information. Unfortunately, the landmark study does not have the opportunity to embed itself in the corporate culture, where its value can be demonstrated by the combination of cross-sectional information and periodic updating. The authorization leading to the study is typically a custom project to answer a specific question. The

corporation, accustomed to ad hoc studies, classifies the cross-sectional landmark study as just one of the usual run of custom studies, ignoring entirely the potential for databasing, even though, later on and behaviorally, the study will be referred to again and again. The cross-sectional landmark study never really enters the corporate culture as an information-building activity to be performed routinely to systematize corporate knowledge.

Level 4: Systematized Landmark Studies with Regular Updating

Intellectual capital from research really begins to build up when the cross-sectional landmark study is repeated at regular intervals. As the corporate players begin to understand that they can measure the entire competitive frame on a systematic basis and look at the changes of the information over time, the database becomes increasingly valuable. Regular updating generates the following three key benefits:

1. *Trends.* Updating reveals the relative strengths and weaknesses of different competitors over time.

2. *Relation to external market behaviors.* Updating enhances the opportunity for more rigorous analysis of the data. Cross-sectional information can then either be used for itself as a measure of brand or communication strength or else tied back on a product-by-product basis to other activities in the market. The change manifests itself as an evolution from searching for a single number to searching for causal relations underlying the data. Such causal relations are difficult, but not impossible, to uncover when the study is done in one time period. Causal relations are far easier to uncover when the cross-sectional study is repeated systematically.

3. *Discipline.* One might think researchers shape the data and the thinking. If, in fact, the cross-sectional database is updated over time, then the sheer repetition transforms the mind-set of the corporation from haphazard answering of some questions to a disciplined analysis and search for patterns.

Level 5: Added-value Analytic Systems That Accompany the Landmark Cross-sectional Database

The highest level of intellectual property is a system that combines the properties of large-scale cross-sectional data, regular updating, and associated software in order to detect patterns in the data and trends that can be used. The associated software comes in at least three varieties:

1. *A system that identifies newly emergent segments in the population.* These segments may not be previously discovered. A system that can automatically discover segments based on the response patterns adds substantial value to the cross-sectional database. The author has presented a version of such a segmentation approach using current in-market analysis of communications (Moskowitz 2000; Moskowitz et al. 2002) with relevant, interpretable segments that lead to immediate action. Thus, an overarching segmentation system (Wells 1975; Mitchell 1983) would probably not be relevant. Segments must emerge from the data themselves rather than imposed on the data from external theory (Moskowitz 1996a, 1996b; Moskowitz and Bernstein 2000).

2. *A system that predicts membership in the segment from external data.* One of the key uses of segmentation is to take information from the database and create new products or communications based on this information, which are in turn directed to the particular segment that will be most receptive. If the segment is truly homogeneous in terms of the messages to which it responds, then the segmentation system improves market performance. Many times, though, the segments that emerge transcend the conventional, easy-to-find subgroups in the population that are divided by geodemographics.

There needs to be a method that relates the member in a segment to the classification information about the individual. In this way individuals in the population not in the database can be assigned to the segments (Moskowitz and Gofman 2003). There are many available data-mining tools that can be used for this specific purpose (Rudd 2000).

3. *A system that detects the emergence and trajectory of segments over time.* One of the key knowledge issues in a company is to identify the direction, strength, and nature of trends. The fashion industry struggles with this problem continually. However, almost every company needs a system by which to understand what is a fad and what is a trend, as well as a system by which to identify corporate offerings that best take advantage of the trend. For instance, if the trend is toward healthier foods, then where is this trend heading? Over time do more individuals fit into the segment called "oriented to healthier foods"? What specific features of in-market products and communications lead to new product ideas or new messages? The analysis system that can handle this problem can provide significantly enhanced intellectual capital.

Perspectives

This section of the chapter presents the basis for developing a new form of intellectual capital based on market research. The approach, using cross-sectional analyses of in-market communications or products, coupled with higher-order analytic techniques holds the promise of simultaneously measuring a large-scale competitive frame, identifying new segments, and tracking over time.

Part II: Research Unbound

Introduction: From Chaos to Procedure, from Procedure to Process

A cursory history of both sensory analysis and market research as professions reveals the growth of the field from a cottage indus-try with few accepted standards to professionalized, organized disciplines encompassing both practitioners and academics. From two fields starting about the same time in the 1920s and 1930s, when researchers had to literally invent the tools in order to conduct the study, sensory analysis and market research have each separately, yet in parallel, evolved into disciplines where standards, or at least accepted practices, govern everything. Researchers working with consumers have best practices dealing with the recruitment of respondents, the types of questions to ask, and onto the types of analyses, the format of data tables, and even the level of recommendations that one should make. These procedures are not carved in stone; rather, they have evolved from chaos to make it technically and financially feasible to acquire and report consumer-based data.

As the field of concept research matures, and as procedures become sacrosanct, either because they have been established beyond question as being correct or, more often, because they have been used so frequently, those very same procedures often ossify. All too often the *process of research* takes over, until the actual implementation of the study becomes as important as the reasons why the study was commissioned in the first place. The process now becomes standardized. The freewheeling spirit of inquiry dies under the weight of professionalization and *processization*.

Processes pervade all aspects of the concept-research business, as researchers evolve from problem solvers to managers of those outsourced vendors who actually provide the consumer data. Process becomes the sine qua non for many research buyers, who limit the types of research that they do to that for which they have norms, or to studies for which they have preferred suppliers who conduct the same study, year in, year out. Standardization and routinization of the research process should not come as a surprise in light of both the pressures to reduce the cost of the data collection process and the demand on the decreasing research staff to acquire and manage ever-increasing amounts of information. When the stress on in-house researchers is high and the demands so strong, the nature of people in general and researchers in particular is probably to jettison new ideas and take refuge in proven process.

Evolving Toward a Diminished Relevance of the Formalized Research Profession

When research began to gain wide acceptance, it was touted as a strategic tool for developers and marketers because it provided insight into the consumer's mind. Perhaps in those early days the true appreciation of research may have been less than we think, but one cannot help feeling that, over the years, professionalizing sensory analysis and market research has reduced acceptance rather than increased it. One need only read books published 30–50 years ago to feel that somehow we researchers have lost some of our soul as the field has developed. Academic journals devoted to sensory analysis and market research dealing specifically with concepts and communications are filled with increasingly sophisticated mathematical models, but in the candid opinion of many practitioners these sophisticated models are appreciated neither by the research suppliers nor by the clients. Few clients can understand the techniques that they buy beyond simple concept tests and few can explain the approach and the analytic strategy even at a discursive level, with mathematical modeling stripped away. As a consequence, formalized, professional research for concept and communication studies is becoming less relevant, often falling below the corporate radar screen and being consigned to the purchasing department in the worst case, or to a vastly reduced staff (in size and quality) in the best case.

Evolving the Field by Transferring Control of Knowledge Acquisition to Research Users

One way to escape the dilemma of increasing need of research data, yet increasing irrelevance and expense of the formal researchers, is to change the nature of research from a process, guided by a select few "objective" professionals, to a transaction open to all who wish to do the transaction, especially those who have a close and nonobjective connection with the problem. Instead of a formalized, ritualistic process, research may evolve into a momentary act of gathering and interpreting data, with anyone needing the knowledge empowered to perform that act. This notion is similar to the evolution in banking for dispensing cash and updating one's records. Banks hired tellers to report information to customers and to service customer needs (e.g., cash withdrawal or check deposit). Tellers were part of a costly, time-intensive process. Customers had to wait in line at the bank to be serviced. Today this process has changed considerably. In most countries, customers can go to a local ATM (automatic teller machine), insert any one of a number of different cards, and do much of the banking that used to take an hour or more. The old process has given way to a 3- to 5-minute transaction that hardly occupies any of the customer's attention. In research this particular paradigm is known as *self-authoring* because research users author the test stimulus, do their own study, and interpret their own data. Chapter 16 treats the topic in depth. We look at self-authoring in this chapter from the aspect of how it affects professional research providers.

What can we learn from the change in the banking paradigm to apply to concept research? Must a professional researcher be involved in a concept study, or can tools be developed to empower the untrained, but involved, educated individual to do research?

These involved individuals may not understand how to conduct research, but if the tools exist to help them do research automatically and correctly, then they will be most grateful. Indeed, it is they in the first place who are most closely involved with the problem leading to the research and to whom the correct answer will mean so much. If this paradigm shift can be created, then perhaps research itself may enjoy a renaissance of value to involved users, such as product developers and marketers. Those professions may also enjoy their own personal renaissance along the way.

Designing the Transaction Paradigm for Concept Research: Consumers, Computers, Creatives, Concepts

What should be the features of a new research paradigm for concepts look like when it is based on the notion of momentary transaction rather than process? We must keep in mind that, in the spirit of this book, our focus remains on concepts to understand the consumer and to communicate. We can distinguish at least five such features:

1. *Democratic availability.* The research design and setup may be done by virtually anyone who is properly equipped and oriented. Democracy comes from the widespread availability of intelligent computer technology, at a reasonable price, for average people.

2. *Expert system resides in the machine rather than in the research professional.* Anyone who needs to conduct the research should be able to set up a study following prompts from the computer system. These prompts should be both practical (to accomplish the task) and informative (to educate the users about the implications of each act done in setting up the study). They should be context sensitive so the users need not wade through mountains of research advice.

3. *Computers should present the test stimuli and acquire the data in a batch mode, on demand, without researchers needing to be present.* The same system that sets up a study should present the stimuli and acquire the data. The data acquisition should be transparent to the consumers. For example, automatic skip patterns, automatic experimental designs, and the like, should be built into the system, even if the users are unaware that these are necessary.

4. *Computers should analyze the data, using both simple data summarization as well as complex statistical methods.* As the research transaction migrates from the professionals to the end users (research nonprofessionals) the analysis will have to become increasingly fast and automatic, as well as more intelligent and user friendly. Already many computer programs designed for data analysis feature default options. These programs assume that users probably do not profoundly understand the underpinnings of the analysis nor probably care about many of the available analysis options. Users simply want the results to answer a question. Unfortunately, however, researchers pride themselves on knowing exactly how to analyze the data and being able to add value through their insights through complex transformations of the data. The transaction mode of research with nonresearch professionals probably will not value such handcrafted analysis because of additional research costs and additional delays due to the involvement of another professional in the process. The additional research insights may not be worth waiting for in the fast-paced, hypercompetitive business world of today.

5. *Computers should generate standard reports, with appropriate text in an easy-to-use format, such as PowerPoint.* The transaction paradigm may well find its greatest use in the generation of reports, which will have a limited amount of text to link key tables. The automatically produced results will look

more like an integrated set of *factoids* rather than a well-composed document where care has gone into crafting the report. To a craftsman researcher such computerization will seem like the proverbial cookie-cutter system, but the computerization and automation will make the findings easier for research nonprofessionals to understand.

Chapter 16 on self-authoring conjoint analysis for concepts considers just such an approach. With the increasing sophistication of users, the demystification of the research process, and the ever-present push to increase profitability by decreasing the cost factors in a study (e.g., salaries), we can be sure that the foregoing five steps would be at the forefront of new methods for concept research.

Beyond the Self-authoring System to Mind Genomics and to Structured, Query-based, Concept-response Databases

By itself, self-authoring systems provide only the single building blocks for the new age of concept research. When one combines a series of self-authoring studies into one coherent study, the result is a new system to create an unusually powerful database the likes of which have not been previously seen by researchers. Most databases comprise either publicly available information restructured into a user-friendly format or proprietary databases that are expensive to develop and to maintain.

The self-authoring system by itself provides only the initial part of the concept-response database. The Internet-facilitated interview, coupled with the *wall* approach to concept studies, provides the other part. Chapter 20 discusses the structure of the It! study, comprising a set of studies linked by common elements and by a common classification questionnaire. Chapter 18 on *first principles* presents the same basic idea of a linked set of studies, this time using the exact same elements for a product category, with each study in this linked set differing by end use or

by emotion state. The It! studies, which use the self-authored system for conjoint analysis, are run efficiently on the Internet with thousands of respondents participating in a single evening and provide an enormous amount of high-value, structured information. These foundational studies accomplish on a grand scale what a single study does on a smaller scale. Internet-enabled interviews make the acquisition of data very straightforward, inexpensive, and quick. Utilities to create the wall enable researchers to expand the self-authoring system from a single study to literally an overnight database.

Running a large number of linked studies has been likened to the *genomics* approach (Moskowitz et al. 2003). Rather than working laboriously with a single gene, a researcher can now work with hundreds or thousands of genes while using equipment that enables one to map out aspects of the many different genes on a chromosome. Certainly, the genomics approach is not merely a single experiment, run dozens of times, over years. Rather, the approach looks for specific patterns across thousands of genes, performing this search in a disciplined, systematic, and cost-effective way. The foundational studies, It! and First Principles, using self-authored conjoint analysis, are modeled on this genomics approach and, in some metaphoric way, might be said to be identifying the genomics of the mind. With increasing numbers of studies linked with larger, transnational access to respondents, and with increasingly powerful analysis and reporting systems available today online and automated, the time is close when the transaction approach to creating a valuable database of the consumer's mind will become the norm rather than the exception. This large-scale approach should become popular by the end of this decade, if not sooner. Three powerful and synergistic benefits will drive this popularity:

1. Research power from conjoint analysis
2. Research scope from the simultaneous, parallel array of topics

3. Executional discipline from a standard approach to knowledge gathering, implemented easily and widely on the Internet

Given the power of the database, the next logical approach might be concept insights and concepts *on demand*. Today's information technologies enable researchers to interrogate all sorts of databases, such as Lexis/Nexis. There is no reason why a query-based system cannot be built for the It! and First Principles databases in order to identify insights and create concepts for segments on demand. At that point the cycle will be complete. The traditional research will have evolved in part into a transaction, both for creating the knowledge via self-authoring systems and for accessing it by pay-as-you-go *query systems*. The consequence will be that the system provides data, insights, and concepts on demand at the push of a button.

Class Warfare: The Forthcoming Fight Between Tradition-based Research Craftspeople and Knowledge Transacters

Changes in the structure of concept research will undoubtedly lead to internecine warfare between traditionalists vested in the process itself and research users interested in substantive problems, with only superficial appreciation of research. Some of this warfare will be conducted at the level of the supplier or *vendor*, pitting older and established professionals who have based their business on relationships against possibly younger, more technology-savvy individuals looking for opportunities to make their mark. When marketing managers discover that they can cut costs and time by doing some of the research themselves, other battles may break out at the level of the research user. The fight will be especially fierce in the large companies that have an entrenched community of research professionals opposing an equally entrenched and no less motivated marketing group. These research professionals may be

the traditional consumer researchers that we know today or the newly emerging sensory researchers waving the flag of consumer insights as their raison d'être.

The winners in this new paradigm will be those willing to embrace the change, who feel that they no longer need to control the information process, and who are willing to share with research users the process of both acquiring and interpreting data. Perhaps, over time, craftsman researchers will disappear to be replaced by data consultants brought in because of an expertise in assessing the business implications of data.

The war, when it erupts out of the new paradigm, will not be a particularly pleasant one, because it will result in the loss of jobs for some researchers and perhaps eventually the disappearance of the research function as we know it. Rather than hire a specialist to manage the research process, corporate management may empower other, non-research-oriented employees to perform research or knowledge-development activities that today we would subsume under research. If

1. data acquisition and information use evolve to a momentary transaction,

2. the transaction is embodied in off-the-shelf technology,

3. data can be accessed by simple queries, and

4. the transaction successfully and rapidly delivers actionable guidance in minutes or hours rather than in weeks,

then there is no reason to assume that the traditional research function can go on the way it has for these many years. Arguments based on the sheer economy of "doing one's own research," with consequent reductions in cost and time, will force researchers continually to justify themselves, evolve, and perhaps in some cases vanish when the justification is not sufficiently strong.

We already see this trend in the replacement of time-consuming ad hoc quantitative research by focus groups, with the sessions quite often conducted by the brand manager rather than by a professional moderator. The focus group represents one type of short-lived research transaction between the consumer and the marketer. The speed with which a focus group can be recruited, run, and debriefed is so attractive that all too often marketers merely dispense with the more laborious quantitative study in favor of a longer series of groups. It is not unusual to have marketers proudly announce that they have substituted a series of 24 focus groups for a large quantitative study, because they feel that groups are easier, more productive, and probe more deeply into the mind of the consumer. The next step in this evolution is quantitative research.

New Knowledge Workers in Concept Research: The Technical Groups in a Corporation

Although in many companies market research has evolved to the management of outside data vendors run through the marketing department, at the very same time the R&D and the engineering staffs cry for consumer feedback. These groups are neither close emotionally nor intellectually to consumer researchers. Technical employees must create products and services, not simply manage data acquisition. Sensory analysts often fill this role, but these sensory analysts specialize in the sensory characteristics of the product, not in the concept work. In the main, sensory analysts position themselves as the low-cost research suppliers and consequently are given low budgets for consumer testing. For the most part, in-house sensory specialists who work with concepts do so with relatively primitive, inexpensive forms of research.

As research evolves from process to transaction and databasing concepts becomes increasingly easy, technical employees will find it easier, cheaper, and ever more seductive to do the research themselves. The product de-

velopers and sensory analysts will become the researchers. Self-reliance has always appealed to the more technical groups in companies and fits with the way that engineers and R&D perceive themselves. Consequently, we may see the proliferation of do-it-yourself, transaction-oriented research at the R&D level. This proliferation will be spurred on by the happy confluence of cost, speed, and power in the automated research, since affordable, rapid feedback for guidance is precisely what developers need. The same type of growth may occur in service-based businesses, as the research process gives way to short spurts of concept generation and rapid consumer testing and feedback, or even database creation, data mining, and synthesis of new ideas.

Homework: The Hidden Benefit of the Revolution

As research becomes more of a transaction it will motivate users to do their homework. Today's pressures on researchers often force them and their marketing associates to forgo disciplined research in favor of either "shoot from the hip" or more esoteric, yet fun-filled, ideation sessions, which are billed as freeing one's creative abilities from the shackles on "in the box" thinking. Both of these latter strategies, shoot from the hip and fun-filled ideation sessions, replace quantitative research when the researchers are enjoined to think out of the box. The sessions generate novel ideas, but few ideas really stand the test of time and even fewer are worth putting into a database.

The speed and simplicity of transaction-oriented research should make systematic exploration and development of concepts easier, including those concepts that are touted as lying outside of the box. For such novel ideas, the inputs to the research need merely comprise a collection of different product categories whose features are to be merged into a new product rather than comprising "close in" components from the same product category. In the end, therefore, the transaction paradigm and its ensuing rapid feedback should promote more disciplined research. Since in transaction-oriented research no one waits long for an answer, the research user can change direction in the afternoon. Research will then evolve into a series of short excursions comprising development-testing-refinement-retesting. This short set of steps will replace long, potentially risky and costly single research studies. Again, this paradigm is quite familiar to scientists who conduct many small pilot studies rather than one large study.

The New Role of Universities

In the early days before the proliferation of business curricula the concept-research business attracted individuals with diverse backgrounds, often in the liberal arts. Today the profession is filled with business majors. There are not many sensory researchers in concept research today, perhaps because the field has not sparked academic research.

Today's consumer research students will be exposed to the idea of transaction-oriented research at the university level because the notion of transaction fits the short-term and low-cost nature of university courses in research. Since universities reach students in their most intellectually formative years, they will consequently become a fertile field for educating students to accept and ultimately demand this new type of research. It may well turn out to be the students who will change the research paradigm as they demand the same rapid gratification of knowledge and insight in their business projects that they enjoyed in their university research classes.

The Coming Golden Age

Despite the problems of turf, budgets, technological prowess, and lack of a substantive literature for scientists, concept research is headed toward a golden age. As the power of consumer information becomes increasingly widespread, the golden period will emerge, fanned by the flames of hypercompetition and fueled by easy-to-do research and easy-to-create proprietary databases. When companies can make more money by knowing at any time what their customers want, management will encourage advancements in concept research, starting with test design and proceeding to field execution, data analysis, modeling, and synthesis of newer and better ideas. Those involved in concept research will be afforded tools to understand consumers in more profound ways.

The best way to understand this potential is to draw a parallel with personal computers versus mainframes. Those operating on mainframes could accomplish a lot, but primarily in data analysis. When it came to computer-aided design, only the richest corporations could afford to dedicate mainframes to such use. The remaining mainframe users were stuck doing relatively low-level design. The capabilities of users have multiplied with personal computers, the analogue of transaction-oriented research. Personal computers have evolved to the point where any individual willing to lay out a few thousand dollars for a fast computer and programs can acquire and analyze data with the most powerful statistical systems, design interactive multimedia, and set up a mall on the Internet. The growth of personal, affordable, off-the-shelf technology is astounding. The personal computer has leveled the playing field so that the smallest user can be as sophisticated as the largest company when it comes to applications on the computer. The result has been nothing short of a massive demand for computer programmers facile in the different aspects of the personal computer. The same can be said of Internet research. The low cost, rapid turnaround, and ease of use makes Internet research a very fast-growing field. The same type of growth may be in store for concept research, albeit with those individuals ready to change paradigms and evolve from research as process to research as transaction.

References

Mitchell, A. (1983). The Nine American Lifestyles, New York: Macmillan.

Moskowitz, H.R. (1996a). Segmenting consumers on the basis of their response to concept elements: an approach derived from product research. Canadian Journal of Market Research, 15: 38–54.

Moskowitz, H.R. (1996b). Segmenting consumers world-wide: an application of multiple media conjoint methods. In: Proceedings of the 49th ESOMAR Congress, Istanbul. Amsterdam: ESOMAR [European Society of Marketing Research], pp. 535–552.

Moskowitz, H.R. (2000). The mind of the utility customer: high level customer research to understand & optimize. Paper presented at Customer Management for the Energy Industry, New Orleans.

Moskowitz, H.R., and Bernstein, R. (2000). Variability in hedonics: indications of worldwide sensory & cognitive preference segmentation. Journal of Sensory Studies, 15: 263–284.

Moskowitz, H.R., German, J.B., and Saguy, I.S. (2003). Unveiling health attitudes and creating good for you foods via informatics and innovative Web-based technologies. CRC Critical Reviews in Food Science and Nutrition (in press).

Moskowitz, H.R., and Gofman, A. (2003). System and method for content optimization. US Patent 6,662,215 B1, 16PP.

Moskowitz, H.R., Itty, B., Shand, A., and Krieger, B. (2002). Understanding the consumer mind through a concept category appraisal: toothpaste. Canadian Journal of Market Research, 20: 3–15.

Rudd, O.P. (2000). Data Mining Cookbook. New York: John Wiley and Sons.

Wells, W.D. (1975). Psychographics: A critical review. Journal of Consumer Research, 12: 196–213.

Index